Adaptive Food Webs

Stability and Transitions of Real and Model Ecosystems

Presenting new approaches to studying food webs, this book uses practical management and policy examples to demonstrate the theory behind ecosystem management decisions and the broader issue of sustainability. All the information readers need to use food-web analyses as a tool for understanding and quantifying transition processes is provided. Advancing the idea of food webs as complex adaptive systems, readers are challenged to rethink how changes in environmental conditions affect these systems. Beginning with the current state of thinking about community organization, complexity, and stability, the book moves on to focus on the traits of organisms, the adaptive nature of communities, and their impacts on ecosystem function. The final section of the book addresses the applications to management and sustainability. By helping to understand the complexities of multispecies networks, this book provides insights into the evolution of organisms and the fate of ecosystems in a changing world.

John C. Moore is a Professor; founding Head of the Department of Ecosystem Science and Sustainability; and Director of the Natural Resource Ecology Laboratory at Colorado State University. His research on food web structure, function, and dynamics is positioned at the interfaces of community ecology, ecosystem ecology, and evolution. He pioneered the concepts that detritus is an important source of energy and stability in ecosystems.

Peter C. de Ruiter is Director of the Institute for Biodiversity and Ecosystem Dynamics at the University of Amsterdam and occupies the Chair of Theoretical Ecology at the Wageningen University, The Netherlands. His research focuses on the structure and stability of food webs, and the relationship between biological diversity, ecosystem functioning, and environmental quality, in particular in soils.

Kevin S. McCann is Canadian Research Chair in Biodiversity at the University of Guelph, Canada. A leading theoretical ecologist with an expertise in food webs and biodiversity, his research examines the biological structure underlying diversity and the critical relationship between biostructure, ecosystem function, and stability. In 2013, McCann was only the second Canadian to be elected as a lifetime fellow to the Ecological Society of America for research achievements.

Volkmar Wolters is a Professor and Head of the Department of Animal Ecology and Systematic Zoology at Justus Liebig University, Germany. His research focuses on soil food webs, pollination webs, agricultural biodiversity, and community ecology at the landscape scale. He is president of the Ecological Society of the German speaking countries (GfÖ), and is therefore engaged with policy-makers and key stakeholders in biodiversity and ecology issues.

Adaptive Food Webs

Stability and Transitions of Real and Model Ecosystems

EDITED BY

JOHN C. MOORE
Colorado State University, CO, USA

PETER C. DE RUITER
Wageningen Universiteit, The Netherlands

KEVIN S. McCANN
University of Guelph, ON, Canada

VOLKMAR WOLTERS
Justus-Liebig-Universität Giessen, Germany

CAMBRIDGE
UNIVERSITY PRESS

University Printing House, Cambridge CB2 8BS, United Kingdom

One Liberty Plaza, 20th Floor, New York, NY 10006, USA

477 Williamstown Road, Port Melbourne, VIC 3207, Australia

314–321, 3rd Floor, Plot 3, Splendor Forum, Jasola District Centre, New Delhi – 110025, India

79 Anson Road, #06–04/06, Singapore 079906

Cambridge University Press is part of the University of Cambridge.

It furthers the University's mission by disseminating knowledge in the pursuit of education, learning, and research at the highest international levels of excellence.

www.cambridge.org
Information on this title: www.cambridge.org/9781107182110
DOI: 10.1017/9781316871867

© Cambridge University Press 2018

First published 2018

Printed in the United Kingdom by TJ International Ltd. Padstow Cornwall

A catalog record for this publication is available from the British Library.

Library of Congress Cataloging-in-Publication Data
Names: Moore, John C. (John Christopher), editor.
Title: Food webs : stability, state transitions, and the adaptive capacity of ecosystems / edited by John Moore, Colorado State University, CO, USA, Peter de Ruiter, Wageningen Universiteit, The Netherlands, Kevin McCann, University of Guelph, ON, USA, Volkmar Woltors, Justus-Liebig-Universitat Giessen, Germany.
Description: New York, NY : Cambridge University Press, 2017.
Identifiers: LCCN 2017023592 | ISBN 9781107182110
Subjects: LCSH: Food chains (Ecology)
Classification: LCC QH541.15.F66 F664 2017 | DDC 577/.16–dc23
LC record available at https://lccn.loc.gov/2017023592

ISBN 978-1-107-18211-0 Hardback

Contents

See color plate section between pages 304 and 305.

Contributors

Beatrix E. Beisner
Department of Biological Sciences, University of Quebec at Montreal, QC, Canada

Sofia Berg
Ecological Modelling Group, School of Bioscience, University of Skövde, Sweden, and Department of Physics, Chemistry and Biology, Division of Theoretical Biology, Linköping University, Sweden

Louis-Félix Bersier
Department of Biology Ecology and Evolution, Université de Fribourg, Switzerland

Amrei Binzer
Department Evolutionary Ecology, Max Planck Institute for Evolutionary Biology, Germany

Klaus Birkhofer
Department of Ecology, Brandenburg University of Technology, Germany

Nico Blüthgen
Department of Biology, Technical University Darmstadt, Germany

Antonio Bodini
Department of Chemistry, Life Sciences, and Environmental Sustainability, University of Parma, Italy

David Andrew Bohan
Pôle ECOLDUR, INRA, France

Cristina Bondavalli
Department of Chemistry, Life Sciences, and Environmental Sustainability, University of Parma, Italy

Amanda L. Caskenette
Freshwater Institute, Fisheries, and Oceans Canada, Manitoba, Canada

Andrew J. Davis
Biochemistry Department, Institute for Chemical Ecology, Max Planck Society, Germany

Peter C. de Ruiter
Biometris, Plant Sciences Group, Wageningen University, The Netherlands, and Institute for Biodiversity and Ecosystem Dynamics, University of Amsterdam, The Netherlands

Eva Diehl
Department of Animal Ecology, Justus Liebig University, Germany

Carsten F. Dormann
Biometry and Environmental System Analysis, University of Freiburg, Germany

Amy Downing
Department of Zoology, Ohio Wesleyan University, OH, USA

Bo Ebenman
Department of Physics, Chemistry and Biology, Division of Theoretical Biology, Linköping University, Sweden.

Anna Eklöf
Department of Physics, Chemistry and Biology, Division of Theoretical Biology, Linköping University, Sweden

Colin Fontaine
Centre d'Ecologie et des Sciences de la Conservation, Muséum National d'Histoire Naturelle, France

Ursula Gaedke
Department Ecology/Ecosystem Modelling, Potsdam University, Germany

Núria Galiana
Ecological Networks and Global Change Group, Theoretical and Experimental Ecology Station, CNRS and Paul Sabatier University, France

Christian Guill
Department Ecology/Ecosystem Modelling, Potsdam University, Germany

Martin Hartvig
Systemic Conservation Biology Group, J.F. Blumenbach Institute of Zoology and Anthropology, Georg-August University of Göttingen, Germany, and Center for

Macroecology, Evolution and Climate, Natural History Museum of Denmark, University of Copenhagen, Denmark

Céline Hauzy
Department of Physics, Chemistry and Biology, Division of Theoretical Biology, Linköping University, Sweden, and UPMC, Ecologie et Evolution, France INRA, USC 2031 Ecologie des Populations et Communautés, France

Hans Heesterbeek
Faculty of Veterinary Medicine, University of Utrecht, The Netherlands,

Amber Heijboer
Biometris, Plant Sciences Group, Wageningen University, The Netherlands, and Ecology and Biodiversity, Institute of Environmental Sciences, Utrecht University, The Netherlands

Frank Jauker
Department of Animal Ecology, Justus-Liebig-University, Germany

Tomas Jonsson
Department of Ecology, Swedish University of Agricultural Sciences, Sweden, and Ecological Modelling Group, School of Bioscience, University of Skövde, Sweden, and Department of Physics, Chemistry and Biology, Division of Theoretical Biology, Linköping University, Sweden.

Ferenc Jordán
Danube Research Institute, MTA Centre for Ecological Research, Hungary, and The Microsoft Research – University of Trento Centre for Computational and Systems Biology, Italy

Alexandre Jousset
Ecology and Biodiversity, Institute of Environmental Sciences, Utrecht University, The Netherlands

Sonia Kéfi
Institut des Sciences de l'Evolution de Montpellier, CNRS, Université Montpellier, France

Toni Klauschies
Department Ecology/Ecosystem Modelling, Potsdam University, Germany

Michio Kondoh
Faculty of Science and Technology, Ryukoku University, Japan

Jennifer Adams Krumins
Department of Biology, Montclair State University, NJ, USA

Jan J. Kuiper
Department of Aquatic Ecology, Netherlands Institute of Ecology, The Netherlands, and Aquatic Ecology and Water Quality Management Group, Department of Environmental Sciences, Wageningen University, The Netherlands

Yuanheng Li
Systemic Conservation Biology Group, J.F. Blumenbach Institute of Zoology and Anthropology, Georg-August University of Göttingen, Germany, and German Centre for Integrative Biodiversity Research (iDiv), Germany, and Faculty of Biology and Pharmacy, Institute of Ecology, Friedrich Schiller University Jena, Germany

Miguel Lurgi
Ecological Networks and Global Change Group, Theoretical and Experimental Ecology Station, CNRS and Paul Sabatier University, France

Christian Mazza
Department of Mathematics, University of Fribourg, Switzerland

Kevin S. McCann
Department of Integrative Biology, Summerlee Science Complex, University of Guelph, ON, Canada

José M. Montoya
Ecological Networks and Global Change Group, Theoretical and Experimental Ecology Station, CNRS and Paul Sabatier University, France

Wolf M. Mooij
Department of Aquatic Ecology, Netherlands Institute of Ecology, The Netherlands, and Aquatic Ecology and Water Quality Management Group, Department of Environmental Sciences, Wageningen University, The Netherlands

John C. Moore
Natural Resource Ecology Laboratory, Department of Ecosystem Science and Sustainability, Colorado State University, CO, USA

Akihiko Mougi
Faculty of Life and Environmental Science, Shimane University, Japan

Christian Mulder
National Institute for Public Health and the Environment, The Netherlands

Russell E. Naisbit
Department of Biology – Ecology and Evolution, University of Fribourg, Switzerland

Anje-Margriet Neutel
British Antarctic Survey, UK

Karin A. Nilsson
Department of Ecology and Environmental Science, Umeå University, Sweden

Rudolf P. Rohr
Department of Biology, Ecology and Evolution, University of Fribourg, Switzerland

Axel G. Rossberg
School of Biological and Chemical Sciences, Queen Mary University of London, UK

Giampaolo Rossetti
Department of Chemistry, Life Sciences, and Environmental Sustainability, University of Parma, Italy

Liliane Ruess
Institute of Biology, Humboldt-Universität zu Berlin, Germany

Torbjörn Säterberg
Department of Physics, Chemistry and Biology, Division of Theoretical Biology, Linköping University, Sweden

Alix M. C. Sauve
Institute of Ecology and Environmental Sciences of Paris, Sorbonne Universités, France, and Centre d'Ecologie et des Sciences de la Conservation, Muséum National d'Histoire Naturelle, France

Marco Scotti
GEOMAR Helmholtz Centre for Ocean Research Kiel, Germany

Valentina Sechi
Wageningen University, Department of Soil Quality, The Netherlands

Sanja Selaković
Faculty of Veterinary Medicine, University of Utrecht, The Netherlands

Stefan Sellman
Department of Physics, Chemistry and Biology, Division of Theoretical Biology Linköping University, Sweden

Henrik G. Smith
Department of Biology, Lund University, Sweden, and Centre for Environmental and Climate Research, Lund University, Sweden

Floor H. Soudijn
Institute for Biodiversity and Ecosystem Dynamics, University of Amsterdam, PO Box 94248 1090 GE Amsterdam, The Netherlands

Elisa Thébault
Institute of Ecology and Environmental Sciences of Paris, Sorbonne Universités, France

Ross M. Thompson
Institute of Applied Ecology, University of Canberra, Australia

Michael A. S. Thorne
British Antarctic Survey, UK

Michael Traugott
Mountain Agriculture Research Unit, Institute of Ecology, University of Innsbruck, Austria

Richard Williams
Microsoft Research, 7JJ Thomson Avenue, Cambridge, UK

Kirk O. Winemiller
Section of Ecology and Evolutionary Biology, Department of Wildlife and Fisheries Sciences, Texas A&M University, TX, USA

Volkmar Wolters
Department of Animal Ecology, Justus Liebig University, Germany

Guy Woodward
Department of Life Sciences, Imperial College London, UK

Catherine M. Yule
School of Science, Monash University, Malaysia

Preface

Adaptive Food Webs is a synthesis of talks from the fourth decadal conference on food webs, after the publishing of the seminal book by Robert May entitled *Stability and Complexity in Model Ecosystems*, published by Princeton University Press in 1973. That book provided a theoretical grounding for the study of food webs or communities as a whole based on the mathematics of simple and complex dynamic systems. The relationship between a defined and static structure or architecture of a food web and its dynamic state to stability – the tendency of a system to return to equilibrium following a minor disturbance – was central. A significant conclusion was that ". . . The real world is no general system. Nature represents a small and special part of parameter space." (May 1973, page 76).

Significant shifts in thinking and advances in computational technology have occurred over the past four decades. The first decadal conference in 1982 – "Current Trends in Food Web Theory" – and second decadal conference in 1993 – "Food Webs: Integration of Patterns and Dynamics" – largely focused on what emerged from May's work, by studying the relationships between the food web structure and food web dynamic. The third decadal conference in 2003 – "Dynamic Food Webs: Multispecies Assemblages, Ecosystem Development and Environmental Change" – presented a concept of the food web with a dynamic architecture. It is clear that the boundaries of communities are not well defined, the nature of the interactions among species are complex and often non-linear, the architectures of communities are dynamic, and the species that comprise communities adapt to and coevolve within their communities. The field recognizes the complex adaptive system nature of food webs in principle, if not by name, and is now a fusion of approaches using linear to non-linear equation-based and agent-based constructs, with static architectures to dynamic architectures to adaptive evolving species and architectures.

The fourth decadal symposium – "Adaptive Food Webs: Stability and Transitions of Real and Model Ecosystems" – embraces the notion of food webs as being complex adaptive systems by exploring dynamic structures and processes, through both changes in external drivers and changes in the adaption of species within the food web. This book is organized into three parts: Part I discusses the current state of thinking about community organization, complexity, and stability; Part II focuses on traits of organisms, the adaptive nature of communities, and their impacts on ecosystem function; and Part III addresses how our current understanding of food webs can influence ecosystem and resource management, and the broader issue of sustainability.

May aptly recognized that the real world does not represent a general system. Moving forward, we envision a better understanding of the real world through how multispecies networks develop. How the network itself feeds back to influence organisms as a selective force, providing insights into the evolution of organisms, in ways that can explain how these improbable networks develop and persist.

Introduction

John C. Moore, Peter C. de Ruiter, Kevin S. McCann, and Volkmar Wolters

Many systems being studied today are dynamic, large and complex: traffic at an airport with 100 planes, slum areas with 10^4 persons or the human brain with 10^{10} neuron(e)s. In such systems, stability is of central importance, for instability usually appears as a self-generating catastrophe. Unfortunately, present theoretical knowledge of stability in large systems is meager: the work described here was intended to add to it.

Gardner and Ashby (1970)

A variety of ecologically interesting interpretations can be, and have been, attached to the term "stability."

May (1973)

Climate change, eutrophication, land-use practices, deforestation, intensification of agriculture, and harvesting from fisheries are changing ecosystems across the globe. The study of food webs provides a framework to address these environmental challenges. Food webs are descriptions of the trophic interactions among consumers and resources. The information contained within these descriptions includes aspects of ecosystem structure (i.e., species richness, network architecture), ecosystem function (i.e., primary production, decomposition, biogeochemical cycles), and dynamics (i.e., population and process rates and change, stability and persistence) that all ecosystems share. Understanding how food webs respond to natural and anthropogenic disturbances, be they gradual or abrupt, is important to our basic knowledge of systems, to the formulation of environmental policies, and the implementation of management practices.

Ecologists have long understood the observed patterns in distribution of species and their biomass resulted in part from offsetting processes of births and death, immigration and emigration, competition for resources, production and predation, and basic energetic properties, and that they were in some way related to their stability or ability to persist (Elton, 1927; Lotka, 1956; Hutchison, 1959; Hairston *et al.*, 1960; Paine, 1966). In the past 50 years, there have been several attempts to tie structural, functional, and dynamic aspects of ecosystems together in a unifying way. Two contemporary works: one from the community ecology perspective provided by MacArthur and Wilson (1967) – *The Theory of Island Biogeography* – and one from the ecosystem ecology perspective provided by Odum (1969) – "The strategy of ecosystem development" – summarized the thinking at that time. *The Theory of Island Biogeography* offered a fresh perspective that blended nearly a century of empirical data on the distribution of species across the globe collected by naturalists with mathematical representations of the processes of colonization and extinction to explain these observations. The theory proved remarkably robust to experimentation (see Simberloff and Wilson, 1969, 1970) and remains a recognized triumph for ecology. "The strategy of ecosystem development" juxtaposed key aspects of structure, function, and dynamics for early and late successional communities based on decades of observations, but did not offer an underlying explanation as to why these patterns emerged.

We present here a synthesis of talks from a conference that continues a tradition that started a decade after the publication of the seminal book entitled *Stability and Complexity in Model Ecosystems* by Robert May (1973). The book proved pivotal as it revitalized the applications of mathematical models used in ecology connecting the population models used to study the dynamics of individual species and the model used to investigate pairwise interactions (e.g., consumer–resource; competition and mutualism), with the descriptive and experimental studies of whole communities and food webs. The book provided a theoretical grounding for the study of food webs or communities as a whole that was based on the mathematics of simple and complex dynamic systems. The relationship between the structure or architecture of food webs and their dynamic state to stability – the tendency of a system to return to equilibrium following a minor disturbance – was central.

Prior to its publication, developments in the mathematics of dynamic systems explored the possibilities of complex dynamic states emerging from simple dynamic equations, and the dependency of these structures and ensuing dynamics on changes in internal and external factors. Two such developments included advances in our understanding and importance of non-linear dynamics (Lorenz, 1963; see May, 1977 and Cushing *et al.*, 2003), and the study of complexity and stability of dynamic systems (Gardner and Ashby, 1970). Rosenzweig (1971) introduced the concept of the paradox of enrichment, wherein the stability of simple non-linear systems was governed by not only available energy, but under certain conditions the dynamic states transitioned from a stable equilibrium to inherently unstable states with increased rates of energy inputs.

This chapter begins with the opening paragraph of the seminal paper by Gardner and Ashby (1970), wherein they challenged a long-standing tenet that diverse complex systems were more stable than less diverse simple systems. Their work was motivated more by the study of systems in general than ecology, but the results and implications to natural systems were clear. Two elements of the structure of the simple linear system – the size of the system defined by the number of interacting nodes and the number of interactions among the nodes relative to the number of nodes (connectedness or connectance) – were inextricably tied to the stability of the system.

May (1972) connected these ideas to ecological food webs by noting that, indeed, with all else being equal, large complex systems tended to be less stable than less diverse and complex systems. However, May ended the paper with a caveat that provided a path for ecologists to reconcile the dissonance that the results revealed relative to long-held beliefs about diversity and stability. In natural systems as opposed to mathematical abstractions, all else is not equal, as matter and energy need to be conserved, and that nodes are species that are the outcome of evolutionary processes. These natural physical and biological laws shape the connectedness of communities as systems arranged into "blocks" of species with a high degree of interactions among the species within the blocks, and fewer interactions among the blocks.

Subsequent to the publication of *Stability and Complexity of Model Ecosystems* food-web ecology developed as a field of study in its own right, moving beyond a descriptive science. The relationship between structure and stability, the pattern of interactions, and the determinants of diversity had with it and others (MacArthur and Wilson, 1967) a framework and body of theory to work within. The decade that followed produced important theses on the relationships between the ecosystem structure and the stability of simple model systems (Pimm and Lawton, 1977, 1978, 1980) and more complex descriptions of real food (Cohen, 1978; Pimm, 1982). May and colleagues (May, 1976; May and Oester, 1976) expanded on the idea that simple systems under certain circumstance generated complex dynamics.

The first decadal conference – "Current Trends in Food Web Theory" hosted by DeAngelis, Post, and Sugihara in 1982 at Fontana Village Inn, in Gatlinburg, NC – focused on the key elements of the approaches that emerged from May's work. While the conference focused on the approaches of the times, with its emphases on networks, topology, connectance, and stability, it also confronted head-on the often competing and yet complementary views of community and ecosystem ecology (DeAngelis *et al.*, 1983). May (1973) recognized that closed ecosystem models that included decomposers, biomass, and energy flow were important – but not the focus of his tome. DeAngelis (1975, 1980) offered early treatments of this, melding the approach of May with that of those modeling energy and material flows commonly used in ecosystem ecology. The conference addressed a clear need to incorporate ecosystem concepts and biogeochemistry into the dialogue by addressing the interactions between organisms and their environment as an integrated system. The decade that followed generated several advances in this direction with the infusion of empirically based food-web descriptions to study complexity and stability, the concepts of the trophic cascade (Carpenter *et al.*, 1985, 1987), the compartmentalized nature of communities (Moore and Hunt, 1988), and the relationships among food-web structure, dynamics, and ecosystem processes (read function; DeAngelis, 1992).

The second decadal conference entitled "Food Webs: Integration of Patterns and Dynamics" hosted by Polis, Winemiller, and Moore in 1993 at Pingree Park Campus of Colorado State University, west of Fort Collins, Colorado – explored the relationships between dynamic states and food-web structure (Polis and Winemiller, 1996). The inclusion of biomasses and life-history characteristics of organisms and ecosystem processes into descriptions had become routine. Trophic cascades, top–down and bottom–up controls, temporal and spatial variation in structure and dynamics, and patterns of interactions within communities that were important to stability were discussed. What separated this conference from its predecessors was a healthy discussion of the role of detritus, biogeochemical cycles, and the concept of resource subsidies in shaping food webs and their dynamics. Food-web theory was largely based on primary production as the resource, yet large quantities of primary production goes unconsumed while living and enters the detritus pool as a donor-controlled resource. This changed after the Pingree Park conference.

The third decadal conference entitled "Dynamic Food Webs: Multispecies Assemblages, Ecosystem Development and Environmental Change" hosted by de Ruiter, Wolters, and Moore in 2003 in Giessen, Germany, broached the ideas of a food web with a dynamic architecture. In their prologue to the conference book, de Ruiter *et al.* (2005) noted that the conference presentations focused on the structure, function, and dynamics of food webs, but with greater detail on the biological properties of individuals, populations, and compartments within communities. The strengths of interactions, particularly the importance of weak links among compartments within food webs, were emphasized. In hindsight, the emphasis on the heterogeneity of structures and dynamic states over space and through time represented an important departure from past conferences. These ideas reshaped our thinking and notions of stability and persistence of communities, recognizing that food-web architectures are not static, but rather change over space and through time (McCann, 2012; Moore and de Ruiter, 2012; Rossberg, 2013).

Contributions of the fourth decadal symposium entitled "Adaptive Food Webs: Stability and Transitions of Real and Model Ecosystems" hosted by Moore, de Ruiter, McCann, and Wolters are presented here. The conference vetted the notion of dynamic structures and processes through both changes in external drivers and changes in the adaption of species within the food web. The focus on adaption adds an element of dynamic that individuals bring to systems that had been largely ignored. As food-web ecologists we often ask to what extent systems are governed by the internal constraints of their static or dynamic architecture and governed external perturbations. If the relationships between structure, function, and dynamics are indeed inextricably interrelated, then studying how the traits of species change in response to a changing environment holds much promise. Traditional equation-based modeling approaches that capture the average behaviors of populations of species or functional groups were prominent at the conference, but agent-based modeling approaches that focus on the characteristics of individuals were present as well. The second quote at the start of this chapter is from May (1973), wherein he recognizes and acknowledges the pliable nature of the term stability. Stability is still an important theme, but clearly has expanded beyond the restrictive definition of neighborhood or local stability, as concepts of persistence and feasibility are broached as well.

With the overarching theme of *Adaptive Food Webs*, the chapters in this book focus on topics that are germane to ecosystem sustainability given the current state of the planet: biodiversity loss and protection, natural resource use and environmental quality, climate change, eutrophication, intensification of agriculture and harvesting from fisheries, and spread of pests and diseases. In many respects, the book is one realization of food-web science. May (1973) envisioned that the abstract general strategic models would be used alongside the more specific and targeted tactical models, as we see them both being "sympathetically handled, [. . .] each providing new insights for the other."

The book is organized into three parts to capture the recent advances in food-web theory. Part I delves into the current state of thinking about community organization, complexity, and stability. Part II focuses on traits of organisms, the adaptive nature of communities, and their impacts on ecosystem function. Part III addresses the importance of understanding food webs to management and the broader issue of sustainability. A review of the themes presented here and those of the past decadal conferences trace the development of our thinking of food webs over the past four decades. The field is now a fusion of tactical to strategic models using linear to non-linear equation-based and agent-based constructs, based on static to dynamic architectures to adaptive evolving architectures. What might we expect in the next decade? Big data and the rapid development and proliferation of genomics and metabolomics are obvious places to look. The information gleaned from these approaches may help us understand how the development of multispecies networks, read food webs, feeds back to influence organisms as a selective force, providing insights into the evolution of organisms beyond that which studies of single species and pairwise interactions can provide, and into the fate of ecosystems in a changing world.

References

Carpenter, S. R., Kitchell, J. F., and Hodgson, J. R. (1985). Cascading trophic interactions and lake productivity. *Bioscience*, **35**, 634–639.

Carpenter, S. R., Kitchell, J. F., Hodgson, J. R., *et al.* (1987). Regulation of lake primary productivity by food web structure. *Ecology*, **68**, 1863–1876.

Cohen, J. E. (1978). *Food Webs in Niche Space*. Princeton, NJ: Princeton University Press.

Cushing, J. M., Costantino, F. F., Dennis, B., Desharnais, R. A., and Henson, S. M. (2003). *Chaos in Ecology: Experimental Nonlinear Dynamics*. San Diego, CA: Academic Press.

DeAngelis, D. L. (1975). Stability and connectance in food web models. *Ecology*, **56**, 238–243.

DeAngelis, D. L. (1980). Energy flow, nutrient cycling, and ecosystem resilience. *Ecology*, **61**, 764–771.

DeAngelis, D. L. (1992). *Dynamics of Nutrient Cycling and Food Webs*. London, UK: Chapman and Hall.

DeAngelis, D. L., Post, W., and Sugihara, G. (1983). *Current Trends in Food Web Theory*. Oak Ridge, TN: Oak Ridge National Laboratory.

de Ruiter, P. C., Wolters, V., and Moore, J. C. (2005). *Dynamic Food Webs: Multispecies Assemblages, Ecosystem Development and Environmental Change*. San Diego, CA: Academic Press.

Elton, C. (1927). *Animal Ecology*. New York: Macmillan.

Gardner, M. R. and Ashby, W. R. (1970). Connectance of large dynamic (cybernetic) systems: critical values for stability. *Nature*, **228**, 784.

Hairston, N. G., Smith, F. E., and Slobodkin, L. B. (1960). Community structure, population control, and competition. *American Naturalist*, **94**, 421–425.

Hutchinson, G. E. (1959). Homage to Santa Rosalia, or Why are there so many kinds of animals? *American Naturalist*, **93**, 145–159.

Lorenz, E. (1963). Deterministic, non-periodic flow. *Journal of Atmospheric Sciences*, **20**, 448–464.

Lotka, A. J. (1956). *Elements of Mathematical Biology*. New York, NY: Dover Publications.

MacArthur, R. H. and Wilson, E. O. (1967). *The Theory of Island Biogeography*. Princeton, NJ: Princeton University Press.

May, R. M. (1972). Will a large complex system be stable? *Nature*, **238**, 413–414.

May, R. M. (1973). *Stability and Complexity in Model Ecosystems*. Princeton, NJ: Princeton University Press.

May, R. M. (1976). Simple mathematical models with very complicated dynamics. *Nature*, **261**, 459–467.

May, R. M. (1977). Thresholds and breakpoints in ecosystems with a multiplicity of stable states. *Nature*, **269**, 471–477.

May, R. M. and Oster, G. F. (1976). Bifurcations and dynamic complexity in simple ecological models. *American Naturalist*, **110**, 573–599.

McCann, K. S. (2012). *Food Webs*. Princeton, NJ: Princeton University Press.

Moore, J. C. and Hunt, H. W. (1988). The compartmentation of real and model ecosystems. *Nature*, **333**, 261–263.

Moore, J. C. and de Ruiter, P. C. (2012). *Energetic Food Webs: An Analysis of Real and Model Ecosystems*. Oxford, UK: Oxford University Press.

Odum, E. P. (1969). The strategy of ecosystem development. *Science*, **164**, 262–279.

Paine, R. T. (1966). Food web complexity and species diversity. *American Naturalist*, **100**, 65–75.

Pimm, S. L (1982). *Food Webs*. London, UK: Chapman Hall.

Pimm, S. L. and Lawton, J. H. (1977). The number of trophic levels in ecological communities. *Nature*, **268**, 329–331.

Pimm, S. L. and Lawton, J. H. (1978). On feeding on more than one trophic level. *Nature*, **73**, 542–544.

Pimm, S. L. and Lawton, J. H. (1980). Are food webs divided into compartments? *Journal of Animal Ecology*, **49**, 879–898.

Polis, G. A. and Winemiller, K. O. (eds.) (1996). *Food Webs: Integration of Patterns and Dynamics*. New York, NY: Chapman and Hall.

Rosenzweig, M. L. (1971). Paradox of enrichment: destabilization of exploitative ecosystems in ecological time. *Science*, **171**, 385–387.

Rossberg, A. G. (2013). *Food Webs and Biodiversity*. Oxford, UK: Wiley Blackwell.

Simberloff, D. S. and Wilson, E. O. (1969). Experimental zoogeography of islands: the colonization of empty islands. *Ecology*, **50**, 278–296.

Simberloff, D. S. and Wilson, E. O. (1970). Experimental zoogeography of islands: a two-year record of colonization. *Ecology*, **51**, 934–937.

Part I

Food Webs: Complexity and Stability

1 Food Webs versus Interaction Networks: Principles, Pitfalls, and Perspectives

Carsten F. Dormann and Nico Blüthgen

1.1 Introduction

Food webs have aroused interest and scientific analysis for many decades (Cohen, 1978). Especially Odum's school of ecosystem thinking sought to quantify fluxes in ecosystems, based on feeding guilds (Odum, 1953). Later, the theoretical analysis of interactions among species took a prominent role, arguing that information fluxes are as important as energy fluxes (Pimm, 1982). These interaction-network ideas still dominate models and experiments today (Rossberg, 2013). In contrast, interactions between two (trophic) levels ("bipartite" or "two-mode" networks) are a more recent ecological mainstream activity. Pollination networks featured verbally in early scientific works (dating back to comments in the third chapter of Darwin, 1859), but it was only in the 1980s that data describing such interaction networks specifically received analytical attention (starting with the work of Jordano, 1987). Today, food-web ecologists and network ecologists are still two largely separate scientific communities, with different data, methods, aims, and interpretations. Attempts to bridge this gap are relatively few (Ings *et al.*, 2009).

Any food-web workshop will typically bring together people from both sides and those already straddling the fields. Still, studies show a considerable separation, despite substantial intellectual overlap. In this chapter, we present an (necessarily incomplete) overview of current differences between food-web and network ecology with the aim of highlighting the underlying similarities. We believe that both fields can profit from the expertise and experience present in the other, and we suggest specific steps toward incorporating so far neglected issues tackled in the other field. Specifically, we organize this chapter into four main dimensions (scientific focus; data; nodes and links; and methods) after a brief section defining the terms we use.

1.2 Definitions

Food webs describe who-eats-whom-relationships in an $n \times n$ adjacency matrix. Since every food-web entity may interact with any other, this matrix has the dimensions of the number of entities (species, guilds) and is called *one-mode* or *unipartite*. If interactions are restricted to those between, and not among, two trophic levels, the resulting

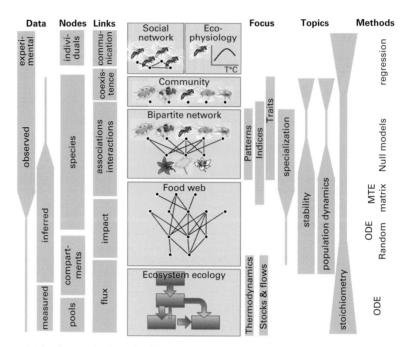

Figure 1.1 Main changes in the scientific context when moving from individual-based ecophysiological studies to energy and nutrient flux-based ecosystem-level studies of food webs and networks. ODE and MTE refer to ordinary differential equations and metabolic theory of ecology, respectively.

$k \times l$ matrix describes *bipartite* networks. Food webs are typically interpreted along trophic relationships (and often contain only data on trophic interactions, as revealed, e.g., in the 113 webs in http://ipmnet.org/loop/foodweb.aspx). Interaction networks, in contrast, include a large diversity of relationships between species, e.g., mutualism, facilitation, or commensalism (Figure 1.1).

Usually, a network focuses on a specific function for a better interpretation and does not attempt to mix pollinators and predators in a single matrix, although some attempts have now been doing so (Pocock *et al.*, 2012). In line with most current publications we shall henceforth use "food web" for one-mode trophic relationships, and "interaction network" for other relationships, which are most often bipartite (even those of plant facilitation networks: Verdú and Valiente-Banuet, 2011). Note that other definitions have been proposed (e.g., food webs being binary adjacency matrices, while networks have weighted links: Allesina, 2009).

We refer to any kind of food web or network as *binary* if the data in the adjacency matrix are 0/1 (for absent and existing links, respectively) and as *weighted* if data are quantitative (e.g., predation rates or number of observed interactions). Each cell with a value different from 0 is called a *link*, while the actual events underlying a link are called *interactions*.

For reasons of presentational clarity we make statements about the prevalent patterns, and the reader may want to mentally add qualifying phrases such as "mainly," "largely,"

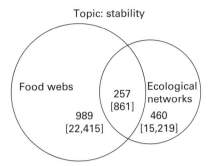

Figure 1.2 Venn diagram showing the number of publications on food webs or ecological networks that include the topic "stability." Web of Science search in September 2014 using the search terms "ecolog* network*," "food-web," and "stability." Slight modifications of the terms yield similar results. Total numbers of papers without "stability" are given in square brackets. The term "stability" was found in 5.4% of the food-web articles and 4.5% of the network articles.

or "generally" to most of them. At the same time, we cite and discuss studies success-fully reaching beyond the dominant research modes. Figure 1.2 summarizes the main trends we discuss in this chapter.

1.3   Scientific Focus and Applications

Species interactions have been studied for very different reasons. Food webs originally were models of ecosystems, representing the main pathways of mass flow (typically carbon or total biomass: Odum, 1953). Scientific questions were related to population sizes of particular species, for example those commercially or culturally important to humans (fish harvests, top-predator abundances). Soon the stability of such ecosystems was being analyzed (i.e., resistance to disturbance, resilience to overexploitation; Figure 1.2), and the role complexity plays for species coexistence and hence food-web stability (May, 1973). The first such analyses were largely theoretical, using oversim-plified "Tinkertoy models" (Pimm, 1982; Montoya *et al.*, 2006), but models of real food webs and applications to human effects soon followed. The interest in such strategic theoretical models has not diminished and still dominates the literature on food webs today (see Fussmann and Heber, 2002; Melian and Bascompte, 2002; Murdoch *et al.*, 2002; Solé *et al.*, 2002; Ives and Cardinale, 2004; Allesina and Tang, 2012; Kéfi *et al.*, 2012; Thompson *et al.*, 2012 for some of many examples).

The literature on *bipartite* networks, in particular mutualist and host-parasitoid net-works, has a very different focus. Here specialization of its members is of central interest, particularly coevolution in mutualistic networks (such as plant-pollinator or seed-disperser networks: Schleuning *et al.*, 2012; Morris *et al.*, 2014). Related to specialization, many studies investigated how trait matching between interacting species contributes to the distribution of links or their relative weight (Vázquez *et al.*, 2009). Another topic largely confined to bipartite networks is the asymmetry of interaction strengths. One may argue that indeed all network metrics – from connectance over betweenness to nestedness – are quantifying the effect of specialization, from different angles or at different levels. The description of these patterns is still the main preoccupation of network ecologists, although only few studies experimentally or empirically demonstrate causes and consequences of such patterns, e.g., in predicting land-use effects on species' population declines (Winfree *et al.*, 2007; Weiner *et al.*, 2014).

Two themes common to both food-web and network research are the relation between diversity (and hence complexity) and coexistence or stability, and the community organization from subgroups (modularity). The mathematical aspects differ between the approaches, however, as we will see below.

1.4   Data

Possibly because mathematical properties of food-web models are complex, model structure received more attention than data (and still does). The typical empirical data are simply lists of species or taxa, at least in traditional food webs. Links are then inferred from co-occurrences of species, reports of interactions in the literature or cafeteria experiments in the lab ("refectory experiments"), rather than based on field observations or *in situ* food-choice experiments ("picnic experiments"). Reasons and challenges for inferring links and even interaction strengths are multifold (Morales-Castilla *et al.*, 2015). Whenever actual fluxes have been measured in food webs, these studies tend to take a more ecosystem-level standpoint (Neutel and Thorne, 2014). This is in stark contrast to bipartite interaction networks, where interactions are observed in the field and are thus much more certain to be real. However, gut content analyses – morphological or molecular identification of the organisms consumed by an individual predator – recently contributed to more empirical, quantitative data on interactions. Stable-isotope analyses or fatty acids play an important role in resolving trophic relationships, but particularly molecular gut content analyses allow for a better taxonomic resolution of prey items consumed by predators and are increasingly employed in food-web studies (Traugott *et al.*, 2013). The methodological progress of food webs based on gut contents now increasingly approaches data properties of interaction networks – and inherits their advantages as well as disadvantages.

Empirical data on interactions are not without problems, however. The effect of variation in sampling intensity and the resulting number of observations per species (ranging from singletons to species with many hundreds of observations), have been investigated in several food-web and network studies (Goldwasser and Roughgarden, 1997; Martinez *et al.*, 1999; Banašek-Richter *et al.*, 2004; Nielsen and Bascompte, 2007; Dormann *et al.*, 2009). Interaction networks are notoriously incomplete and many links are missed in any given sample (Sørensen *et al.*, 2011). The influence of such variation in number of observations on network patterns is now routinely accounted for by using null models (see below). In food webs, data uncertainty, statistical artifacts incurred through externally driven species abundance, and sampling intensity are non-issues and (implicitly) assumed irrelevant relative to model structural uncertainty (Martinez *et al.*, 1999).

This assumption is likely to be wrong. Many links possible *in principle* or elsewhere may be absent at the study site. Binary data may suggest a generalist behavior of a predator despite huge but unquantified differences between prey preferences, effectively making the predator seem to behave in a highly specialized way. Thus quantitative links give a more realistic impression of the importance of a link. For example, easily a third of all links in a network are singleton observations, adding (more or less) random noise with large effects on qualitative network structure but small effects on weighted metrics (Blüthgen, 2010).

Binary adjacency matrices are generally insufficient to gauge the importance of a link for food-web and network structure. The *quality* of link information is hugely important for food webs and networks alike. Over the last decade or so food webs became more quantitative, but still lag network analyses when it comes to assessing the effects of sampling intensity on food-web structure.

1.5   Nodes and Links: What Actually *is* Your Network?

Any ecosystem, even those in extreme environments, easily comprises hundreds, thousands, or even millions of "species" (although it remains a point of contention how to delineate microbial species). Soils as well as benthic sediments can be immensely species rich, even above the bacterial realm (think of fungi, nematodes, algae, crustaceans). It is thus common practice (and technical necessity) in food-web science to lump species into guilds (e.g., "decomposers") or taxonomic units (e.g., "diatoms"). Larger species are thus often represented as a single compartment (e.g., "sea otter"), while lower trophic levels are pooled (e.g., "kelp"). The nodes of a food web are thus heterogeneous: sometimes a species, sometimes a guild. Taxonomic resolution is indeed a long-standing debate in food-web ecology, since it heavily affects food-web structure and hence stability analyses and population dynamics (Williams and Martinez, 2008; Boit *et al.*, 2012).

Bipartite networks are much better resolved, and even if some species remain unidentified: nodes are species. An interaction in a network actually refers to an observed event, rather than a potential connection (see previous section). This is, in itself, not sufficient for most of the questions that network ecologists try to address. One line of interest is in rates (pollen transfer, parasitism, etc.), rather than observed events. One visit by a pollinator need not be sufficient to transfer the required pollen for fertilization, and a single parasite may itself become hyper-parasitized. The higher resolution of bipartite networks thus needs to be backed up by additional measurements of specific interaction efficiency (as provided by Vázquez *et al.*, 2005), rather than generic conversion coefficients (for herbivores or predators). Another line of interest is in specialization and partitioning of interactions (niche theory). Both niche-based and frequency- or rate-related questions require appropriate treatment of the observed interaction data or careful consideration of sampling limitations (Blüthgen, 2010).

Food webs commonly aggregate some of their members into manageable ecological units, while bipartite networks remain resolved to species level. In combination with quantitative information on the strength of link (see last section) this information determines interaction probabilities and hence the flow of energy and control in both food webs and networks. Lumping may be inevitable, but its consequences may be severe and are not well understood. Also common to both types is the lumping in space (i.e., from different locations) or time (i.e., over hours, seasons, years). Again, spatio-temporal pooling of interactions obscures the potentially fine-balanced nature of species interactions (Fründ *et al.*, 2011). Food webs or networks from adjacent ponds or valleys are likely to have overlapping species, but their internal structure may differ substantially due to some differences (e.g., predatory fish in ephemeral ponds, shifts in flower abundance due to grazing). In seasonal environments, species often have temporal niches and phenologies, and food-web structure and dynamics will change constantly (Olesen *et al.*, 2008; Boit *et al.*, 2012). Both food-web and network ecology still have to develop strategies to sample and represent spatio-temporal dynamics (Fortuna *et al.*, 2013; Wells *et al.*, 2014; Cazelles *et al.*, 2016).

1.6   Methods

Food-web models describe the temporal (and occasionally spatial) population dynamics of their members (be they species or guilds). Ordinary differential equations (ODEs; in the case of spatial models: partial differential equations) and difference equations are most commonly employed and in the more simple cases their behavior at equilibrium can be analyzed using algebraic stability analysis (Case, 2000). Achieving coexistence of the whole food web is already a major success, since these models are very sensitive to initial conditions and parameter settings (Rossberg, 2013). Comparisons with data are mostly informal or qualitative. Quantitative matches between food-web models and observed data are rare (Reuman *et al.*, 2008; Boit *et al.*, 2012).

Analyses of bipartite networks are static, with a few notable exceptions (Bastolla *et al.*, 2009; Benadi *et al.*, 2012; James *et al.*, 2012; Suweis *et al.*, 2013). This seems surprising, as the system under investigation is much simpler. However, this simplicity also reduces the possibilities for coexistence (Benadi *et al.*, 2012), and in some cases this is only achieved by allowing resource space to increase with species numbers (Bastolla *et al.*, 2009). While it is possible to actually fit network models to data (Wells and O'Hara, 2014), the ecological realism of such models currently remains low. Instead, the vast number of publications on bipartite networks contents itself with describing patterns, rather than understanding the underlying dynamic processes. Network indices are plentiful, directly or indirectly related to specialization and typically over-interpreted. If abundances of species are not modeled as part of the network structure, observed patterns cannot be distinguished from random interactions of differently abundant species, as seen when using null models that maintain species abundance (Vázquez and Aizen, 2003, 2006; Blüthgen *et al.*, 2008; Dormann *et al.*, 2009; Joppa *et al.*, 2009). Such simple, parsimonious explanations are surprisingly often ignored. For example, James *et al.* (2012: p. 229) emphasized that "the question of why real ecological networks often have a highly nested architecture remains unanswered."

Network analyses should profit greatly from embracing the dynamic-modeling approach common in food-web ecology. Capturing the way species are interacting more realistically will, in the long run, certainly prove superior to statistical descriptions of static snapshots. Using null models for food webs may be useful, but is seen by theoretical ecologists as an inferior, transient option.

1.7   Conclusions

Different areas of ecology come with a different research tradition. As food-web ecology and interaction-network ecology start to overlap, each sub-discipline can benefit from the teething experiences of the other. In particular, the more statistical viewpoint of interaction networks seems to acknowledge the importance of sampling errors and biases introduced by binary data, at least more so than the arena of food-web ecology currently does. In exchange, food-web ecologists have a long-standing tradition, and the tools, to embrace static data as snapshots of a dynamic system, a position that would also serve network ecologists well. Once the main technical and semantic trenches have been bridged, both directions will have to address unresolved issues, of which we find two particularly challenging but important. First, the common aggregation of data across time (e.g., over months and years) and space (different locations), distorting the actual interactions at any point in time and space into a non-existent average. Second, the structural model error emerging from sampling problems (spurious interactions and non-detected real ones). The real test for both fields is predictive accuracy beyond the data – and a tight integration with experimental studies.

References

Allesina, S. (2009). Cycling and cycling indices. In *Ecosystem Ecology*, ed. S. E. Jorgensen, Amsterdam, Elsevier, pp. 50–57.

Allesina, S. and Tang, S. (2012). Stability criteria for complex ecosystems. *Nature*, **483**, 205–208.

Banašek-Richter, C., Cattin, M.-F., and Bersier, L.-F. (2004). Sampling effects and the robustness of quantitative and qualitative food-web descriptors. *Journal of Theoretical Biology*, **226**, 23–32.

Bastolla, U., Fortuna, M. A., Pascual-García, A., *et al.* (2009). The architecture of mutualistic networks minimizes competition and increases biodiversity. *Nature*, **458**, 1018–1021.

Benadi, G., Blüthgen, N., Hovestadt, T., and Poethke, H. J. (2012). Population dynamics of plant and pollinator communities: Stability reconsidered. *The American Naturalist*, **179**, 157–268.

Blüthgen, N. (2010). Why network analysis is often disconnected from community ecology: A critique and an ecologist's guide. *Basic and Applied Ecology*, **11**, 185–195.

Blüthgen, N., Fründ, J., Vázquez, D. P., and Menzel, F. (2008). What do interaction network metrics tell us about specialization and biological traits? *Ecology*, **89**, 3387–3399.

Boit, A., Martinez, N. D., Williams, R. J., and Gaedke, U. (2012). Mechanistic theory and modelling of complex food-web dynamics in Lake Constance. *Ecology Letters*, **15**, 594–602.

Case, T. J. (2000). *An Illustrated Guide to Theoretical Ecology*. Oxford, UK: Oxford University Press.

Cazelles, K., Araújo, M. B., Mouquet, N., and Gravel, D. (2016). A theory for species co-occurrence in interaction networks. *Theoretical Ecology*, **9**, 39–48.

Cohen, J. E. (1978). *Food Webs and Niche Space*. Princeton, NJ: Princeton University Press.

Darwin, C. R. (1859). *The Origin of Species by Means of Natural Selection or the Preservation of Favoured Races in the Struggle for Life*. London: J. Murray.

Dormann, C. F., Blüthgen, N., Fründ, J., and Gruber, B. (2009). Indices, graphs and null models: Analyzing bipartite ecological networks. *Open Ecology Journal*, **2**, 7–24.

Fortuna, M. A., Krishna, A., and Bascompte, J. (2013). Habitat loss and the disassembly of mutalistic networks. *Oikos*, **122**(6), 938–942.

Fründ, J., Dormann, C. F., and Tscharntke, T. (2011). Linné's floral clock is slow without pollinators: Flower closure and plant–pollinator interaction webs. *Ecology Letters*, **14**, 896–904.

Fussmann, G. F. and Heber, G. (2002). Food web complexity and chaotic population dynamics. *Ecology Letters*, **5**, 394–401.

Goldwasser, L. and Roughgarden, J. (1997). Sampling effects and the estimation of food-web properties. *Ecology*, **78**, 41–54.

Ings, T. C., Montoya, J. M., Bascompte, J., *et al.* (2009). Ecological networks: Beyond food webs. *Journal of Animal Ecology*, **78**, 253–269.

Ives, A. R. and Cardinale, B. J. (2004). Food-web interactions govern the resistance of communities after non-random extinctions. *Nature*, **429**, 174–177.

James, A., Pitchford, J. W., and Plank, M. J. (2012). Disentangling nestedness from models of ecological complexity. *Nature*, **487**, 227–230.

Joppa, L. N., Bascompte, J., Montoya, J. M., *et al.* (2009). Reciprocal specialization in ecological networks. *Ecology Letters*, **12**, 961–969.

Jordano, P. (1987). Patterns of mutualistic interactions in pollination and seed dispersal: Connectance, dependence asymmetries, and coevolution. *American Naturalist*, **129**, 657–677.

Kéfi, S., Berlow, E. L., Wieters, E., *et al.* (2012). More than a meal... Integrating non-feeding interactions into food webs. *Ecology Letters*, **15**, 291–300.

Martinez, N. D., Hawkins, B., Dawah, H. A., and Feifarek, B. P. (1999). Effects of sampling effort on characterization of food-web structure. *Ecology*, **80**, 1044–1055.

May, R. M. (1973). *Stability and Complexity in Model Ecosystems*. Princeton, NJ: Princeton University Press.

Melian, C. J. and Bascompte, J. (2002). Food web structure and habitat loss. *Ecology Letters*, **5**, 37–46.

Montoya, J. M., Pimm, S. L., and Solé, R. V. (2006). Ecological networks and their fragility. *Nature*, **442**, 259–264.

Morales-Castilla, I., Matias, M. G., Gravel, D., and Araújo, M. B. (2015). Inferring biotic interactions from proxies. *Trends in Ecology and Evolution*, **30**, 347–356.

Morris, R. J., Gripenberg, S., Lewis, O. T., and Roslin, T. (2014). Antagonistic interaction networks are structured independently of latitude and host guild. *Ecology Letters*, **17**, 340–349.

Murdoch, W. W., Kendall, B. E., Nisbet, R. M., *et al.* (2002). Single-species models for many-species food webs. *Nature*, **417**, 541–543.

Neutel, A.-M. and Thorne, M. A. S. (2014). Interaction strengths in balanced carbon cycles and the absence of a relation between ecosystem complexity and stability. *Ecology Letters*, **17**, 651–661.

Nielsen, A. and Bascompte, J. (2007). Ecological networks, nestedness and sampling effort. *Journal of Ecology*, **95**, 1134–1141.

Odum, E. P. (1953). *Fundamentals of Ecology*. Philadelphia, PA: Saunders.

Olesen, J. M., Bascompte, J., Elberling, H., and Jordano, P. (2008). Temporal dynamics in a pollination network. *Ecology*, **89**, 1573–1582.

Pimm, S. L. (1982). *Food Webs*. Chicago: Chicago University Press.

Pocock, M. J. O., Evans, D. M., and Memmott, J. (2012). The robustness and restoration of a network of ecological networks. *Science*, **335**, 973–977.

Reuman, D. C., Mulder, C., Raffaelli, D., and Cohen, J. E. (2008). Three allometric relations of population density to body mass: Theoretical integration and empirical tests in 149 food webs. *Ecology Letters*, **11**, 1216–1228.

Rossberg, A. G. (2013). *Food Webs and Biodiversity: Foundations, Models, Data*. Oxford, UK: Wiley.

Schleuning, M., Fründ, J., Klein, A.-M., *et al.* (2012). Specialization of mutualistic interaction networks decreases toward tropical latitudes. *Current Biology*, **22**, 1–7.

Solé, R. V., Alonso, D., and McKane, A. (2002). Self-organized instability in complex ecosystems. *Philosophical Transactions of the Royal Society B: Biological Sciences*, **357**, 667–681.

Sørensen, P. B., Damgaard, C. F., Strandberg, B., *et al.* (2011). A method for under-sampled ecological network data analysis: Plant-pollination as case study. *Journal of Pollination Ecology*, **6**, 129–139.

Suweis, S., Simini, F., Banavar, J. R., and Maritan, A. (2013). Emergence of structural and dynamical properties of ecological mutualistic networks. *Nature*, **500**, 449–452.

Thompson, R. M., Brose, U., Dunne, J. A., *et al.* (2012). Food webs: Reconciling the structure and function of biodiversity. *Trends in Ecology and Evolution*, **27**, 689–697.

Traugott, M., Kamenova, S., and Ruess, L. (2013). Empirically characterising trophic networks: What emerging DNA-based methods, stable isotope and fatty acid analyses can offer. *Advances in Ecological Research*, **49**, 177–224.

Vázquez, D. P. and Aizen, M. A. (2003). Null model analyses of specialization in plant–pollinator interactions. *Ecology*, **84**, 2493–2501.

Vázquez, D. P. and Aizen, M. A. (2006). Community-wide patterns of specialization in plant–pollinator interactions revealed by null models. In *Plant–Pollinator Interactions: From Specialization to Generalization*, ed. N. M. Waser and J. Ollerton. Chicago, IL: University of Chicago Press, pp. 200–219.

Vázquez, D. P., Morris, W. F., and Jordano, P. (2005). Interaction frequency as a surrogate for the total effect of animal mutualists on plants. *Ecology Letters*, **8**, 1088–1094.

Vázquez, D. P., Chacoff, N., and Cagnolo, L. (2009). Evaluating multiple determinants of the structure of plant–animal mutualistic networks. *Ecology*, **90**, 2039–2046.

Verdú, M. and Valiente-Banuet, A. (2011). The relative contribution of abundance and phylogeny to the structure of plant facilitation networks. *Oikos*, **120**, 1351–1356.

Weiner, C. N., Werner, M., Linsenmair, K., and Blüthgen, N. (2014). Land use impacts on mutualistic networks: Disproportional declines in specialized pollinators via changes in flower composition. *Ecology*, **95**, 466–474.

Wells, K. and O'Hara, R. B. (2014). Species interactions: estimating per-individual interaction strength and covariates before simplifying data into per-species ecological networks. *Methods in Ecology and Evolution*, **4**, 1–8.

Wells, K., Feldhaar, H., and O'Hara, R. B. (2014). Population fluctuations affect inference in ecological networks of multi-species interactions. *Oikos*, **123**, 589–598.

Williams, R. J. and Martinez, N. D. (2008). Success and its limits among structural models of complex food webs. *Journal of Animal Ecology*, **77**, 512–519.

Winfree, R., Williams, N. M., Dushoff, J., and Kremen, C. (2007). Native bees provide insurance against ongoing honey bee losses. *Ecology Letters*, **10**, 1105–1113.

2   What Kind of Interaction-Type Diversity Matters for Community Stability?

Michio Kondoh and Akihiko Mougi

2.1   Introduction

How do so many species coexist in nature? This long-lasting question in ecology (Hutchinson, 1959) was given a new direction by the theoretical prediction, derived using a mathematical model of random communities, that a more complex community network, characterized by higher species richness or more interspecific interactions, is less likely to be stable (May, 1972). The major issue that arose was why a complex community, which should be unstable according to ecological theory, can persist in real nature. An approach that has commonly been taken to tackle the issue was to change the assumption of random community (Lawlor, 1978). This approach seems to have been successful to some extent. Indeed, recent studies that looked into the real community network have identified structural non-randomness of community networks that can help the persistence of species in a community (Neutel *et al.*, 2002; Emmerson and Raffaelli, 2004; Kondoh, 2008; Thébault and Fontaine, 2010; Stouffer and Bascompte, 2011).

A number of studies were carried out to describe the detailed structure of real community networks, such as food web (de Ruiter *et al.*, 2005) and mutualistic web (Bascompte and Jordano, 2013), and to find their structural patterns that may contribute to community stability. It is no doubt that analysis of those specific networks greatly contributed to our understanding of how ecological communities are organized and how their structure is related to community dynamics. However, at the same time, little is understood regarding the structural and dynamics properties of community networks with multiple interaction types. Actually, it is only recently that different interaction types have been put together in a single picture describing how diverse species and variety of interactions are built up into a complex network of ecological community (Ohgushi, 2005; Mélian *et al.*, 2009; Thébault and Fontaine, 2010; Kéfi *et al.*, 2012; Pocock *et al.*, 2012; Toju *et al.*, 2014).

While a number of studies have been carried out to understand how species diversity contributes to population, community dynamics, and ecosystem processes (e.g., May, 1972; Hooper *et al.*, 2005), it is only recently that the ecological consequence of interaction diversity has been theoretically explored (Allesina and Tang, 2012; Mougi and Kondoh, 2012, 2014a, b, c; Georgelin and Loeuille, 2014; Sanders *et al.*, 2014; Rúa and Umbanhowar, 2015). A central question is what, if any, is the contribution of interaction diversity to community stability. Mougi and Kondoh (2012, 2014b) gave a potential answer to this question by presenting the model-derived hypothesis that the coexistence of different interaction types itself can stabilize an otherwise unstable, complex community (interaction-type diversity hypothesis). They analyzed a population-dynamics model of multispecies community, where population growth is influenced by up to three kinds of interspecific interactions: antagonism (prey–predator, host–parasite, host–parasitoid, cheating), mutualism, and competition. They looked at how community stability changes with changing the relative proportion of different interaction types or network complexity and found: (1) that there is an optimal way of mixing different interactions that maximizes community stability; and (2) that the mixing of different interaction types can reverse the classically negative complexity–stability effect into a positive one. Those theoretical findings imply two corresponding mechanisms through which interaction-type diversity enhances community stability (Mougi and Kondoh, 2012, 2014b). First, coexistence of different interactions itself can stabilize community dynamics. Second, diversity of interaction should increase connectance, which contributes to community stability through the positive complexity–stability effect.

If one is to interpret the ecological implications of Mougi and Kondoh (2012, 2014b), it should be noted that there are at least two alternative ways to classify interaction types, one based on the sign combination of interspecific effect (i.e., sign-based definition of interaction type), while the other on fitness components that the interaction influences (fitness component-based definition). The former way is more traditional and often used in earlier community ecological literatures (Burkholder, 1952). For example, antagonism, mutualism, and competition have been respectively defined as interactions where two species have (positive, negative), (positive, positive), and (negative, negative) effects to the counterpart's fitness. Note that with this interaction-type definition it does not matter what is the actual cause of fitness change – say, only the sign combination matters. Thus both Batesian mimicry and parasitism can be categorized into the same type, although the processes behind those two interactions are totally different. The latter definition is rather based on fitness components and thus focuses on the cause of interspecific effect. For a plant species, for example, both pollinator and seed disperser are mutualistic partners, as those interactions increase fitness of both sides. Yet those mutualistic partners are "different" in their roles in plant life history. Therefore having more interactions with one group, pollinator or seed disperser, is unlikely to compensate for shortage of the other interaction. Thus they can be considered "different" in the latter definition. In other words, with the fitness component-based definition, there should be upper limits in fitness gain a species can get from a single interaction type.

Here a new question emerges regarding the interaction-type diversity hypothesis – that is, what "kind" of interaction diversity is necessary for community stability? This question has been unanswered as the earlier studies overlooked those alternative ways of defining "different" interactions (Mougi and Kondoh, 2012, 2014b; Suweis *et al.*, 2014). Interaction-type diversity hypothesis was originally based on the model that assumes both a sign-based and fitness component-based definition. Indeed, Mougi and Kondoh (2012, 2014b) assumed not only that different interaction types had different sign combinations of interaction coefficient ([+, −], [+, +], [−, −] for antagonism, mutualism, and competition, respectively), but also that the total fitness gain a species can get from an interaction type (antagonism or mutualism) had an upper limit. This was realized by assuming that the fitness gain that a species can get from a single interaction of antagonism (or mutualism) is inversely proportional to the number of antagonistic (mutualistic) partners. In contrast, Suweis *et al.* (2014), who assumed that antagonism and mutualism are different only in their sign combinations and share the same fitness component, reported that mixing mutualism and antagonism does not stabilize a community, but can give rise to a positive complexity–stability effect. The disagreement between the models seems to stem from which definition was chosen for interaction diversity, yet a model analysis with more systematic setting as to interaction-type definition is necessary to answer which kind of interaction-type diversity stabilizes complex communities, especially when no study has been carried out as to what happens if interaction type is defined based solely on the fitness component.

In this chapter we present four community models that assume two "different kinds" of interaction types, but in different definitions. Two interactions are (Model I) antagonism and mutualism that contribute to different fitness components (both sign- and fitness component-based definition), (Model II) antagonism and mutualism that contribute to the same fitness components (sign-based definition); (Model III) two different kinds of antagonism contributing to different fitness components (fitness component-based definition), or (Model IV) two different kinds of mutualism contributing to different fitness components (fitness component-based definition). By comparing the effects of interaction-type diversity to community stability in those models, we reveal that the two different interaction diversities, sign-based and fitness component-based, contribute to community stability and its relation to network complexity in different manners.

2.2 Model

Consider an S-species community, where a randomly chosen two species interact with each other with probability C (connectance). An interaction is of Type X (antagonistic or mutualistic) with probability, P_X, while it is of Type Y (antagonistic or mutualistic) with probability, P_Y (= $1 − P_X$). The population dynamics of species i is described as:

$$\frac{dN_i}{dt} = N_i \left(r_i - s_i N_i + \sum_{j=1, j \neq i}^{S} a_{ij} N_j \right), \qquad (2.1)$$

where N_i is the population level of species i, r_i is the intrinsic growth or mortality rate of species i, s_i is density-dependent self-regulation, and a_{ij} is the interaction coefficient between species i and species j. The interaction coefficient takes a positive value either when species i utilizes species j in antagonistic interaction or when those two species are mutualistic partners, while it is a negative value when species j utilizes species i in antagonistic interaction. We used a cascade model for antagonistic interaction, where for each pair of species $i, j = 1, \ldots, n$ with $i < j$, species i never consumes species j, while species j may consume species i.

Two interspecific interaction types, X and Y, are antagonism or mutualism and may or may not share the same fitness component. When the two interaction types share the same fitness component, there should be a maximum limit in the fitness gain that a species can get from interactions of both types. Therefore the fitness gain that a species gets from interaction Type X (or Y) should, on average, decrease with the fitness gain that the species gets from interaction Type Y (X). The mutual dependence of fitness gain from interaction Types X and Y was realized in our model by assuming "fitness gain allocation" between interactions, where the amount of fitness gain that a species gets from an interaction type is proportional to its "potential preference." When the two interaction types do not share the same fitness component, in contrast, the fitness gain that a species can get from interaction Type X (or Y) would be independent of the fitness gain through interaction Type Y (X). We used the following four models, in which interaction Type X and Y are "different" according to different definitions: both sign- and fitness component-based definition used in Model I; sign-based definition used in Model II; fitness component-based definition used in Model III (antagonism), and Model IV (mutualism). In those four models the values of a_{ij} are determined in different manners (Figure 2.1).

> **Model I**: Antagonism (X; if $a_{ij} > 0$, then $a_{ji} < 0$) and mutualism (Y; if $a_{ij} > 0$, then $a_{ji} > 0$) coexist. Fitness gain allocation was made only within the same interaction type. Specifically, the antagonistic interaction strength of consumer i to its resource species j decreases with increasing number of resource species that species i utilizes in antagonistic interactions; the mutualistic interaction strength of mutualist i to its partner species j decreases with increasing number of mutualistic partners of species i. The fitness gain allocation was made based on the potential preference of species i to species j, A_{ij}, assigned to all the antagonistic and mutualistic interactions. More specifically, the interaction strength was determined as $a_{ij} = e_{ij} A_{ij} / \sum_{k \in \text{resource in antagonism of sp. } i, k \neq i} A_{ik}$ and $a_{ij} = e_{ij} A_{ij} / \sum_{k \in \text{resource in mutualism of sp. } i, k \neq i} A_{ik}$ for antagonism and mutualism, respectively, where e_{ij} is the conversion efficiency when species i utilizes species j. Please note that the negative effect that resource species j receives from consumer i was determined as $a_{ji} = -a_{ij}/e_{ij}$ for antagonistic interactions.
>
> **Model II**: Antagonism (X; if $a_{ij} > 0$, then $a_{ji} < 0$) and mutualism (Y; if $a_{ij} > 0$, then $a_{ji} > 0$) coexist. Fitness gain is allocated among all the interactions. The antagonistic interaction strength of consumer i to its food-resource species j decreases with increasing total number of its resource species and mutualistic partners; similarly, the mutualistic interaction strength of mutualist i to its partner species j decreases with increasing total number of its food-resource

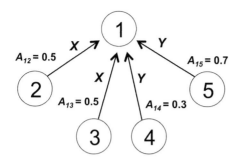

	Model I	Model II	Model III	Model IV
X	ant.	ant.	ant.	mut.
Y	mut.	mut.	ant.	mut.
a_{12}	$0.5\ e_{12}$	$0.25\ e_{12}$	$0.5\ e_{12}$	$0.5\ e_{12}$
a_{13}	$0.5\ e_{13}$	$0.25\ e_{13}$	$0.5\ e_{13}$	$0.5\ e_{13}$
a_{14}	$0.3\ e_{14}$	$0.15\ e_{14}$	$0.3\ e_{14}$	$0.3\ e_{14}$
a_{15}	$0.7\ e_{15}$	$0.35\ e_{15}$	$0.7\ e_{15}$	$0.7\ e_{15}$

Figure 2.1 Interaction coefficients, a_{1k}, in the four models (Models I to IV). Species 1 interacts with four species, Species 2 to 5. Interactions with Species 2 and 3 are of Type X (antagonistic or mutalistic), while interactions with Species 4 and 5 are of Type Y (antagonistic or mutualistic). Potential preference to species k (A_{1k}) is set to 0.25, 0.3, 0.5, and 0.7 for $k = 2$, 3, 4, and 5, respectively.

species and mutualistic partners. The fitness gain allocation was made based on the potential preference of species i to species j, A_{ij}, as

$$a_{ij} = e_{ij}\,A_{ij} \Big/ \sum_{k\,\in\,\text{resource in antagonism or mutualism of sp. } i,\ k \ne i} A_{ik}$$

for antagonism and mutualism. The negative effect that resource species j receives from consumer i was determined as $a_{ji} = -a_{ij}/e_{ij}$ for antagonistic interactions.

Model III: Two different kinds of antagonism (X and Y; if $a_{ij} > 0$, then $a_{ji} < 0$) may coexist. Fitness gain allocation was made only within the same antagonism type, X or Y. The interaction strength, a_{ij}, of Type X (or Y) antagonism decreases with increasing number of resource of species i in Type X (or Y) antagonism. The fitness gain allocation was made based on the potential preference of species i to species j, A_{ij}, as

$$a_{ij} = e_{ij}\,A_{ij} \Big/ \sum_{k\,\in\,\text{resource in Type X antagonism of sp. } i,\ k \ne i} A_{ik} \quad \text{and}$$

$$a_{ij} = e_{ij}\,A_{ij} \Big/ \sum_{k\,\in\,\text{resource in Type Y antagonism of sp. } i,\ k \ne i} A_{ik}$$

for Type X antagonism and Type Y antagonism, respectively. The negative effect that resource species

j receives from consumer i was determined as $a_{ji} = -a_{ij}/e_{ij}$ for antagonistic interactions.

Model IV: Two different kinds of mutualism (X and Y; if $a_{ij} > 0$, then $a_{ji} > 0$) may coexist. Fitness gain allocation was made only within the same mutualism type, X or Y. The interaction strength, a_{ij}, of Type X (or Y) mutualism decreases with increasing number of mutualistic partner of species i in Type X (or Y) mutualism. The fitness gain allocation was made based on the potential preference of species i to species j, A_{ij}, as $a_{ij} = e_{ij} A_{ij} / \sum_{k \in \text{resource in Type X mutualism of sp. } i, \, k \neq i} A_{ik}$ and

$a_{ij} = e_{ij} A_{ij} / \sum_{k \in \text{resource in Type Y mutualism of sp. } i, \, k \neq i} A_{ik}$ for Type X mutualism and Type

Y mutualism, respectively.

Parameters, s_i, e_{ij}, and A_{ij} are randomly chosen from a uniform distribution between 0 and 1. The intrinsic rate of change, r_i, is determined to hold $dX_i/dt = 0$ after imposing an equilibrium density of each species, X_i^*, from a uniform distribution between 0 and 1. Stability analysis was based on a Jacobian community matrix (May, 1972). Our measure, community stability, represents the probability of local equilibrium stability, which was estimated as the frequency of locally stable systems across 1000 sample communities (Chen and Cohen, 2001).

2.3   Result

The effect of interaction-type mixing on community stability and its dependence on community complexity (S, C) varied between the models.

When interaction-type diversity was based on both sign and fitness component (Model I; sign- and fitness component-based definition of interaction type), as assumed also in Mougi and Kondoh (2012, 2014b), community stability tended to have its maximum at the intermediate degree of mixing (Figure 2.2a, b).

Pure antagonistic communities were very stable, but tended to be strongly destabilized with a small contamination of mutualism. With an increasing proportion of mutualism, the community stability showed a unimodal pattern and had the minimum at the pure mutualistic community. The complexity–stability relationship varies with varying degree of interaction-type mixture; a positive complexity effect was observed at the intermediate mixing level.

When interaction-type diversity was only sign based (Model II; sign-based definition), as assumed in Suweis *et al.* (2014), there is no stabilizing effect of mixing interaction types (Figure 2.2c, d). More specifically, the community stability was maximized at the pure antagonistic community and monotonously decreases with increasing proportion of mutualistic interactions. A positive complexity–stability relationship was, however, observed under a wide parameter range. Increasing species richness tended to have a stabilizing effect on population dynamics to the community

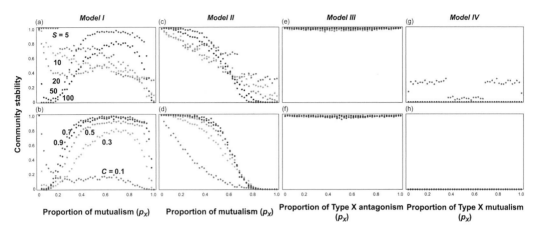

Figure 2.2 Effect of interaction-type mixing on community stability with varying complexity for Model I (a, b), Model II (c, d), Model III (e, f), and Model IV (g, h). For upper panels (a, c, e, g), red ($S = 5$), orange ($S = 10$), green ($S = 20$), blue ($S = 50$) and purple ($S = 100$); $C = 0.3$. For lower panels (b, d, f, h), red ($C = 0.1$), orange ($C = 0.3$), green ($C = 0.5$), blue ($C = 0.7$), and purple ($C = 0.9$); $S = 50$. Community stability was measured as the proportion of locally stable community models. (A black and white version of this figure will appear in some formats. For the color version, please refer to the plate section.)

with connectance ($C = 0.3$) and a smaller mixing level ($p_X < 0.3$), while increasing connectance always stabilized the community with given species richness ($N = 50$).

When there are two different kinds of antagonism (Model III; fitness component-based definition), the community tended to be very stable and changing the mixing level of two interactions only had a minor effect on community stability. Increasing species richness slightly stabilized the community, while increasing connectance slightly lowered community stability. These complexity effects were, however, very weak compared with the complexity effect that was observed in Models I and II. Therefore, to get a more clear view, we analyzed a model with weaker self-regulation intensity ($s_i = 0.0–0.5$), which decreases the relative stability of the community (Figure 2.3) and thus allows us to look at how changing the level of interaction-type mixing or complexity potentially affects community stability. The model analysis showed that interaction-type mixing destabilizes the community; increasing species richness stabilized the community, while increasing connectance stabilized the community.

When two different kinds of mutualism (Model IV; only fitness component-based definition) were considered as components of interaction-type diversity, community tended to be unstable and mixing level had virtually no effects on community stability, except that, for the minimum complexity level ($S = 5$, $C = 0.3$), community stability was lowest at the intermediate mixing level. Virtually no complexity effect was observed. However, with stronger self-regulation intensity ($s_i = 0.0–10.0$), the picture is rather different (Figure 2.3). Increasing species richness always lowered community stability, while increasing connectance can slightly stabilize the community at the intermediate mixing levels.

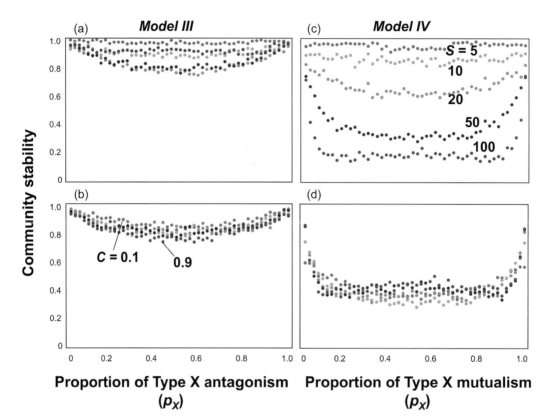

Figure 2.3 Effect of interaction-type mixing on community stability with varying complexity for Model III (a, b; antagonism) with weaker self-regulation and IV (c, d; mutualism) with stronger self-regulation. For upper panels (a, c), red ($S = 5$), orange ($S = 10$), green ($S = 20$), blue ($S = 50$), and purple ($S = 100$); $C = 0.3$; $s_i = 0.0$–0.5. For lower panels (b, d), red ($C = 0.1$), orange ($C = 0.3$), green ($C = 0.5$), blue ($C = 0.7$), and purple ($C = 0.9$); $S = 50$; $s_i = 0.0$–10.0. Community stability was measured as the proportion of locally stable community models. (A black and white version of this figure will appear in some formats. For the color version, please refer to the plate section.)

2.4   Discussion

Community stability, its dependence on the level of interaction-type mixing and community-network complexity varied between the models. Taken together with the former studies (Mougi and Kondoh, 2012, 2014b; Suweis *et al.*, 2014), this clearly suggests that how we define "different" interaction types or how we measure the interaction-type diversity is essential in the study of community dynamics and its dependence on community complexity.

Interaction-type diversity, achieved by mixing two interaction types, enhances community stability only when the diversity is both sign- and fitness component-based. In the absence of fitness component-based diversity (Model II), in agreement with the earlier study (Suweis *et al.*, 2014), the community diversity monotonously decreased with increasing proportion of mutualism (Figure 2.2c, d). Based on the same pattern,

Suweis *et al.* (2014) argued that sign diversity does not stabilize community, but the stabilizing effect of mixing interaction types in Mougi and Kondoh (2012) must be attributable solely to the fitness component-based assumption of interaction diversity. Our present analysis, however, clearly demonstrates that this is not the case. In the absence of sign-based diversity (Model III or IV), the community stability was maximized when there was no diversity in interaction type (Figure 2.2e–h, Figure 2.3), similar to the model without fitness component-based diversity (Model II). This contradicts the interpretation by Suweis *et al.* (2014) and suggests that sign-based diversity is necessary for interaction-type mixing to stabilize a community. The interaction-type diversity that stabilizes a community should be of both fitness components and sign combination.

The interaction-type diversity that reverses the otherwise negative relationship between complexity and stability is of sign combination. Present model analysis suggests that a positive complexity–stability relationship emerges as long as sign-based interaction-type diversity is considered (Models I and II). In the models not assuming sign-based interaction-type diversity (Models III and IV), in contrast, community stability was lowered, or only slightly affected, by increasing community complexity. Mutualistic and antagonistic interactions often contribute to the same fitness component in nature. For example, an animal seed disperser, which helps plants dispersing their offspring, receives nutrition or energy as a reward, while gains nutrition or energy by eating other plant materials (antagonism) as well. Thus the fitness component is identical for mutualistic and antagonistic interactions. Present study suggests that this type of interaction-type diversity can give rise to a positive complexity–stability effect. In contrast, coexistence of different kinds of mutualism (or antagonism) is unlikely to make a complexity–stability relationship positive.

Some earlier community models assumed the coexistence of multiple interaction types (sign-based), yet still predicted a negative complexity–stability effect. For example, the original model proposed by May (1972) assigned random values to elements of community matrix and thus in effect assumed a variety of sign-based interaction types, including (+, +), (–, –), (+, –), (+, 0), and (–, 0). More recently, Allesina and Tang (2012), in addition to May's setting, analyzed model communities that incorporate multiple interaction types, such as competition, antagonism, and mutualism, and found that the complexity–stability relationship is always negative, although the threshold for instability varies between the models. The difference between their and the present studies (Model I and II) is of assumption about how interaction strength depends on link numbers per species. In the earlier models (May, 1972; Allesina and Tang, 2012), the average effect magnitude of interspecific interaction was assumed to be independent of either species richness or connectance. We would argue that those models assumed an unrealistic situation, for example, that a species can get an infinitely large fitness gain in an infinitely large mutualistic community and therefore it may not be appropriate to apply the model prediction directly to large, complex ecological communities.

The present study may help future community ecological studies to characterize diverse interspecific interactions. The recent research trend of incorporating different "kinds" of interactions together in a single-community network provides a new picture of natural, complex communities (Mélian *et al.*, 2009; Kéfi *et al.*, 2012; Pocock *et al.*, 2012; Toju *et al.*, 2014) and may allow us to find a new structural property of the interaction network or its relation to population to community dynamics. Yet, as this line of research is advanced, a new question arises as to what is interspecific interaction and how those interactions can be classified. As community ecology traditionally characterized interspecific interactions according to the sign combination of fitness effect, pollination and seed dispersal are considered to be of the same category, mutualism. However, this sign-based definition is not the single, right way of defining interaction types. Indeed, the present study demonstrates that a fitness component-based definition can also be essential to relate community-network structure to community dynamics.

It should be noted that interspecific interactions in real ecosystems are extremely diverse and would not be able to be captured fully by the sign- or fitness component-based classification we introduce in this study. Interaction strength may vary over time and space (Kondoh, 2003; Ramos-Jiliberto *et al.*, 2012; Mougi and Kondoh, 2014a); the sign combination and fitness components that an interaction contributes may depend on environmental conditions (Johnson *et al.*, 1997; Thompson, 2005); some interactions may act to modify another interspecific interaction (Werner and Peacor, 2003; Ohgushi, 2005; Ohgushi *et al.*, 2007, 2013). Interactions that have the same sign combination and contribute to the same fitness component can differ in another aspect that we have overlooked. Those diverse types of interactions, which could not be captured by the present simple framework, should exist in nature, may have a specific contribution to community dynamics and require another framework to be captured appropriately. If we try to understand community dynamics as a collection of population dynamics driven by interspecific interactions (Gause, 1932, as a classical example), it would become necessary to develop an appropriate framework that allows us to incorporate those diverse types of interspecific interactions into a community-dynamics model. This will become a big challenge for the future study of community networks.

References

Allesina, S. and Tang, S. (2012). Stability criteria for complex ecosystems. *Nature*, **483**, 205–208.

Bascompte, J. and Jordano, P. (2013). *Mutualistic Networks*. Princeton, NJ: Princeton University Press.

Burkholder, P. R. (1952). Cooperation and conflict among primitive organisms. *American Scientist*, **40**, 601–631.

Chen, X. and Cohen, J. E. (2001). Transient dynamics and food-web complexity in the Lotka–Volterra cascade model. *Proceedings of the Royal Society B: Biological Sciences*, **268**, 869–877.

de Ruiter, P. C., Wolters, V., and Moore, J. C. (2005). *Dynamic Food Webs: Multispecies Assemblages, Ecosystem Development, and Environmental Change*. San Diego, CA: Academic Press.

Emmerson, M. and Raffaelli, D. (2004). Body size, patterns of interaction strength and the stability of a real food web. *Journal of Animal Ecology*, **73**, 399–409.

Gause, G. F. (1932). Experimental studies on the struggle for existence. *Journal of Experimental Biology*, **9**, 389–402.

Georgelin, E. and Loeuille, N. (2014). Dynamics of coupled mutualistic and antagonistic interactions, and their implications for ecosystem management. *Journal of Theoretical Biology*, **346**, 67–74.

Hooper, D. U., Chapin, III, F. S., Ewel, J. J., *et al*. (2005). Effects of biodiversity on ecosystem functioning: A consensus of current knowledge. *Ecological Monograph*, **75**, 3–35.

Hutchinson, G. E. (1959). Homage to Santa Rosalia, or Why are there so many kinds of animals? *American Naturalist*, **93**, 145–159.

Johnson, N. C., Graham, J. H., and Smith, F. A. (1997). Functioning of mycorrhizal associations along the mutualism–parasitism continuum. *New Phytologist*, **135**, 575–585.

Kéfi, S., Berlow, E., Wieters, E., Navarrete, S., *et al*. (2012). More than a meal. . . Integrating non-feeding interactions into food webs. *Ecology Letters*, **15**, 291–300.

Kondoh, M. (2003). Foraging adaptation and the relationship between food-web complexity and stability. *Science*, **299**, 1388–1391.

Kondoh, M. (2008). Building trophic modules into a persistent food web. *Proceedings of the National Academy of Sciences of the United States of America*, **105**, 16631–16635.

Lawlor, L. R. (1978). A comment on randomly constructed model ecosystems. *American Naturalist*, **112**, 445–447.

May, R. M. (1972). Will a large complex system be stable? *Nature*, **238**, 413–414.

Mélian, C. J., Bascompte, J., Jordano, P., and Křivan, V. (2009). Diversity in a complex ecological network with two interaction types. *Oikos*, **118**, 122–130.

Mougi, A. and Kondoh, M. (2012). Diversity of interaction types and ecological community stability. *Science*, **337**, 349–351.

Mougi, A. and Kondoh, M. (2014a). Adaptation in a hybrid world with multiple interaction types: A new mechanism for species coexistence. *Ecological Research*, **29**, 113–119.

Mougi, A. and Kondoh, M. (2014b). Stability of competition–antagonism–mutualism hybrid community and the role of community network structure. *Journal of Theoretical Ecology*, **360**, 54–58.

Mougi, A. and Kondoh, M. (2014c). Instability of a hybrid module of antagonistic and mutualistic interactions. *Population Ecology*, **56**, 257–263.

Neutel, A.-M., Heesterbeek, J. A. P., and de Ruiter, P. C. (2002). Stability in real food webs: Weak links in long loops. *Science*, **296**, 1120–1123.

Ohgushi, T. (2005). Indirect interaction webs: Herbivore-induced effects through trait change in plants. *Annual Review of Ecology, Evolution, and Systematics*, **36**, 81–105.

Ohgushi, T., Craig, T. P., and Price, P. W. (2007). *Ecological Communities: Plant Mediation in Indirect Interaction Webs*. Cambridge, UK: Cambridge University Press.

Ohgushi, T., Schmitz, O., and Holt, R. D. (2013). *Trait-Mediated Indirect Interactions: Ecological and Evolutionary Perspectives*. Cambridge, UK: Cambridge University Press.

Pocock, M. J. O., Evans, D. M., and Memmott, J. (2012). The robustness and restoration of a network of ecological networks. *Science*, **335**, 973–977.

Ramos-Jiliberto, R., Valdovinos, F. S., Moisset de Espanés, P., and Flores, J. D. (2012). Topological plasticity increases robustness of mutualistic networks. *Journal of Animal Ecology*, **81**, 896–904.

Rúa, M. A. and Umbanhowar, J. (2015). Resource availability determines stability for mutualist–pathogen–host interactions. *Theoretical Ecology*, **8**, 133–148.

Sanders, D., Jones, C. G., Thébault, E., *et al.* (2014). Integrating ecosystem engineering and food webs. *Oikos*, **123**, 513–524.

Stouffer, D. B. and Bascompte, J. (2011). Compartmentalization increases food-web persistence. *Proceedings of the National Academy of Sciences of the United States of America*, **108**, 3648–3652.

Suweis, S., Grilli, J., and Martian, A. (2014). Disentangling the effect of hybrid interactions and of the constant effort hypothesis on ecological community stability. *Oikos*, **123**, 525–532.

Thébault, E. and Fontaine, C. (2010). Stability of ecological communities and the architecture of mutualistic and trophic networks. *Science*, **329**, 853–856.

Thompson, J. N. (2005). *The Geographic Mosaic of Coevolution*. Chicago, IL: University of Chicago Press.

Toju, H., Guimarães, P. R., Olesen, J. M., and Thompson, J. N. (2014). Assembly of complex plant–fungus networks. *Nature Communications*, **5**, 5273.

Werner, E. E. and Peacor, S. D. (2003). A review of trait-mediated indirect interactions in ecological communities. *Ecology*, **84**, 1083–1100.

3 Symmetry, Asymmetry, and Beyond: The Crucial Role of Interaction Strength in the Complexity–Stability Debate

Anje-Margriet Neutel and Michael A. S. Thorne

3.1 Introduction

What is it that makes a biological community stable? What gives it the ability to resist disturbances, or the ability to adapt and change gradually and not fall apart at the smallest perturbation? Do species interact in such a way that these "webs" of interactions make stable organizations? And are the stabilizing properties in the interactions among species related to general properties of community or ecosystem structure?

One of the classic ideas in ecology is that complex communities or ecosystems are more stable, in an unspecified sense, than simple ones (Elton, 1927; McArthur, 1955; Odum, 1971). This is often illustrated by the example of a well-developed tropical rainforest and a monoculture in an arable field. The rainforest, a highly diverse system, with small fluctuations in population abundances and steady nutrient cycles, is regarded as a stable system, while the relatively simple agricultural system does not have that many feedback mechanisms (Levins, 1974) and is very susceptible to, for example, pest outbreaks or adverse weather conditions.

In contrast is the idea that complexity in communities leads to instability (Levins, 1968; Garner and Ashby, 1970; May, 1972). This idea originates from the studies of mathematical models that represent dynamic equilibria of randomly interacting species. The larger the number of species or the higher the proportion of realized interactions (connectance) in a system, the less likely the system is to return to the equilibrium state after a small perturbation from equilibrium.

More than anything, these examples make clear that when talking about community stability we may think of many aspects of systems, types of disturbance, and stability (Grimm and Wissel, 1997). Stability may be associated with such concepts as constancy, the ability to resist disturbances, the ability to return to an equilibrium after disturbance, or the speed with which the system returns to equilibrium. In this chapter we will use the traditional mathematical framework, that of Jacobian "community" matrices, linearizations (around the equilibrium) of classic population dynamical models (May, 1973). Stability here is defined as the ability of the community, when perturbed from an equilibrium state, to return to the same equilibrium state. If all the eigenvalues of a matrix have negative real parts, the system is stable. Equilibrium is a state where for each species the population size is constant, i.e., growth rates equal loss rates. This approach to stability allows us to evaluate large, complex systems, and by using classic population-dynamical models we can obtain the parameter values of the matrix elements from field observations. Our observations show how the values of the matrix elements, the so-called interaction strengths (the effects of species on each other near equilibrium, see Berlow *et al.* [2004] for a discussion of this definition), shape the feedbacks in the communities that govern system stability. Looking at the strengths of feedbacks gives a new perspective on such questions as: Are more complex systems more vulnerable? What are the key processes in system stability? What is the relation between community structure and stability?

3.2   Are More Complex Communities Less Likely to be Stable?

In May's models, species interact randomly, with random interaction strength. The communities are modeled as community matrices, Jacobian matrices that are linear approximations of systems of differential equations, evaluated at equilibrium (Levins, 1968; May, 1973). The entries of a community matrix are the interaction strengths, partial derivatives of the differential equations, representing the effects of the species on each other's dynamics near equilibrium. The models show that more complex communities are less likely to be stable, given a certain level of self-regulation (self-damping) of the species: $a\sqrt{(nC)} < 1$, where n is the number of species; C is connectance, the actual, divided by the possible number of interspecific interactions; a is the average strength of the interactions; and 1 is the fixed level of self-damping assumed for the species (May, 1972). An important ecological question that resulted from May's work (and that of his contemporaries, for example, Levins, 1968; Gardner and Ashby, 1970) is: Given that natural systems are complex, how can these systems exist, and what makes them stable, despite their complexity? Natural systems are not random, of course. Then, what is this non-randomness, that it allows complexity and stability to agree so well (McCann, 2000; Montoya *et al.*, 2006)?

One of the answers to the above question is that in biological communities, species feed on each other. These feeding relations form food webs, interconnected food chains in which consumers (predators) have a negative effect on (decrease the growth rates of) their food sources (prey), and food sources (prey) have a positive effect on (increase the growth rates of) their predators. And there is a hierarchy within these chains from top species to bottom species; it is very rare that cycles occur of the form: X eats Y, Y eats Z, and Z eats X again. This hierarchical predator–prey structure limits the connectivity of the network and generates many short negative feedbacks (see below) between predators and their prey, which have been shown to contribute to system stability (Levins, 1974; DeAngelis, 1975; Pimm, 1982).

But although food webs may be relatively stable compared to other types of ecological interactions, this still leaves the question: Are more complex food webs less likely to be stable than simpler ones? In the decades after May's generalized result, food-web studies certainly point in that direction. Pimm and Lawton (1978) show that food webs with more omnivorous relations (feeding on different levels in a food chain), with a more complex structure, tend to be less stable than simpler ones.

3.3   Interaction Strength and Stability

Traditionally, relatively little attention has been given to the strengths of the interactions in theoretical evaluations of food-web structure, not because they have been considered unimportant, but because they are so hard to determine, much harder than the presence or absence of interactions (Paine, 1988; Berlow *et al.*, 2004). In material terms, the classic theoretical definition of the (negative) effect of a predator on its prey is the loss rate of the prey population due to predation by this predator, relative to the predator population size (measured in biomass, for example). The (positive) effect of the prey on the predator is then the growth rate of the predator population due to predation on this prey, relative to the prey population size. Ecologists have long recognized the fact that there are strong and weak links in a community and that this will have important consequences for system stability. It has been argued that the more prey species a predator feeds on, the less strong the effect on each prey will be (thus while an increasing number of interactions would be destabilizing, the lower average strength of the interactions would compensate for this [May, 1972; Hastings, 1982]). It has also been argued that predator populations have a much stronger effect on the dynamics of their prey than vice versa (Pimm and Lawton, 1977). Empirical studies, in which the strengths of interactions were determined by manipulation experiments, have made clear that in natural communities most interactions are very weak (Paine, 1988; Wootton, 1994; Menge, 1995; Polis and Strong, 1996). Both empirical and theoretical studies have indicated that the patterning of strong and weak interactions may be important for system stability (Yodzis, 1981; de Ruiter *et al.*, 1995) and have shown how the many weak interactions enhance system stability (Paine, 1988; Polis and Strong, 1996; McCann *et al.*, 1998; Neutel *et al.*, 2002).

In the past decade and a half, food-web ecology has provided ample evidence of the fundamental importance of interaction strengths in the stability of complex food webs (Bersier *et al.*, 2002; Kondoh, 2003; Banašek-Richter *et al.*, 2004; Drossel *et al.*, 2004; Emmerson and Raffaelli, 2004; Wooton and Emmerson, 2005; Brose *et al.*, 2006; Rooney *et al.*, 2006; Banašek-Richter *et al.*, 2009; Gross *et al.*, 2009; Montoya *et al.*, 2009; Novak *et al.*, 2011; Ulanowicz *et al.*, 2014). It is only recently that we have had enough well-documented food webs to examine whether real systems (that is, model systems with interaction strengths directly parameterized from real, observed systems) actually decrease in stability with increasing complexity. And it seems that this is not the case (Neutel *et al.*, 2007; Neutel and Thorne, 2014; Jacquet *et al.*, 2016). If we parameterize the interaction strengths of observed food-web structures randomly, we find that more complex ecosystems tend to be less stable, even if we use the simplest complexity measure, the number of species (trophic groups) (Figure 3.1a). But if we

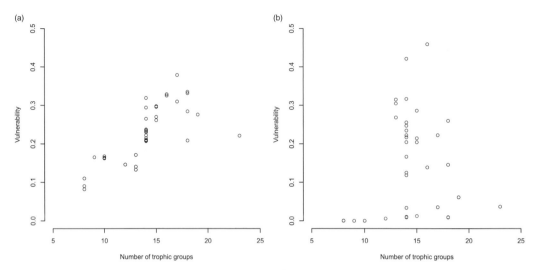

Figure 3.1 Correlation between the number of of trophic groups in 39 observed food webs (Neutel and Thorne, 2014) and food-web stability, measured by the largest real part of the eigenvalues of the corresponding Jacobian community matrix λ_d, where λ_d was measured with the self-damping (diagonal) effects set at zero. We called the resulting positive λ_d the vulnerability of the network (the higher the value, the less stable is the web, or, in other words, the more self-damping is needed to stabilize the web). (a) Interaction strengths are parameterized randomly from the same positive and negative intervals (symmetric). (b) Interaction strengths are empirically based.

obtain the interaction strengths from observation, the relation between complexity and stability disappears. Diversity does not enhance instability (Figure 3.1b).

3.4   Feedbacks

So, why do realistic interaction strengths have such an impact and what then governs the stability of food webs, if it is not their complexity? To answer these questions we have to look more closely at the feedbacks in the system and quantify them (Maruyama, 1963; Levins, 1974; Dambacher *et al.*, 2003). Feedbacks are circuits of interactions, creating ongoing cycles (loops) of effects, which all together determine the stability of the system. The strength of a feedback loop is the product of these effects. A negative feedback loop tends to counteract a disturbance; a positive feedback loop tends to amplify it. A predator–prey relation creates a negative feedback loop, through the positive effect of the prey on the predator population (increasing its growth) and the negative effect of the predator on the prey population (increasing its mortality). There are also longer feedback loops, both negative and positive ones. For example, if a top predator (X) feeds not only on an intermediate predator (Y) but also on a prey (Z) at the bottom of the same food chain, then this omnivorous feeding relation creates a positive feedback loop X→Y→Z→X (the product of two negative effects and one positive).

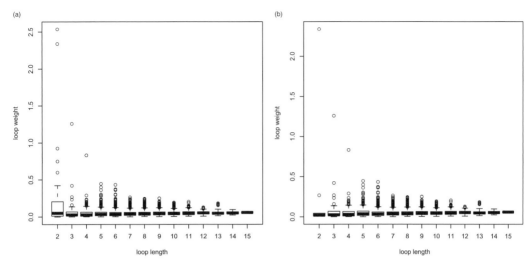

Figure 3.2 Feedback spectra. (a) Complete feedback spectrum of an observed ecosystem (Antarctic dry tundra; Neutel and Thorne, 2014), with length of feedback loops versus their weights (Neutel *et al.*, 2002). (b) The spectrum of the feedback loops with links to or from detritus. Both example spectra are representative of other observed food webs (Michell and Neutel, 2012).

3.5 Recycling of Organic Matter Does Not Affect Community Stability

An important feature of observed ecosystems is the recycling of organic matter. Energy does not just flow from the primary producers through the web of interactions to the top predators, but is also returned to the base of the web in the form of feces, exuviae, and carcasses. All this dead organic matter – detritus – is food for the microbes and other decomposers and thus feeds back into the ecosystem. Ecosystems contain many feedback loops emerging from interactions with detritus. In fact, the majority of all the feedback loops in such food webs (in a web of twenty species you will find thousands of feedback loops) (Figure 3.2a) contain detritus (Figure 3.2b) (Mitchell and Neutel, 2012).

Theoretical studies of food webs have largely ignored the process of recycling (but see DeAngelis, 1992). Detritus is generally thought to have a stabilizing effect on food-web dynamics, which is supported by simple food-web models (Moore *et al.*, 1993; Neutel *et al.*, 1994; Moore *et al.*, 2004). But in most of our best documented food webs (and especially in the more mature ecosystems) detritus feedbacks do not have an influence on the overall stability of the system – when detritus interactions are removed from those systems, system stability remains the same (Neutel and Thorne, 2014).

3.6 Random Parameterization of Predator–Prey Interaction Strengths Leads to Artificial Feedback Signatures

If we leave the detritus interactions aside, we are left with the feedbacks formed by predator–prey interactions and it is in these feedbacks that we have to look for an explanation of the lost complexity–instability relationship. When interactions are parameterized randomly from the same negative (–1, 0) and positive (0, 1) intervals (May, 1973), this symmetry in positive and negative effects generates a very regular, homogeneous feedback spectrum (Figure 3.3a, d) – all the positive and negative feedbacks will have roughly the same average size, or weight (Neutel *et al.*, 2002) (the longer the feedback, the closer its size approaches the mean value of the sampling interval).

When we incorporate in our random parameterizations an asymmetry between negative and positive effects (sampling from intervals [–10, 0] and [0, 0.1]), following Pimm and Lawton (1977), assuming that a predator will generally have a stronger effect on its prey population than vice versa, the spectrum is very different (Figure 3.3b, e). Feedback loops with relatively many negative effects will be relatively strong, creating asymmetries between the strengths of positive and negative feedback loops (Neutel and Thorne, 2014). But although we find an asymmetry between positive and negative effects in our observed systems, the spectrum of a system with empirical interaction strengths looks completely different from both the symmetric and asymmetric random-type ones (Figure 3.3c). The negative two-link loops are by far the strongest, with longer loops tending to be relatively weak (Neutel *et al.*, 2002; Mitchell and Neutel, 2012). The three different feedback signatures underlie the stability of the systems.

3.7   Two- and Three-Link Feedbacks Key in Food-Web Stability

To capture the relation between feedback structure and stability, we need only look at the two-link and three-link feedbacks (Neutel and Thorne, 2014). The stronger the total three-link feedback relative to the total two-link feedback, the more positive the largest eigenvalue of the community matrix (assuming no self-damping of the species), and hence the less stable (we call this "more vulnerable") will be the ecosystem (Figure 3.4).

In fact, in the symmetrical parameterizations of the observed food-web structures we don't even have to look at the three-link loops, just the two-link loops are enough. System stability is effectively determined by the number of links (L) relative to the number of species (n), which is exactly what May found ($a\sqrt{(nC)} < 1$, where $C = 2L/((n-1)n)$) for his randomly constructed matrices (May, 1972). Negative two-link predator–prey feedback loops in themselves are stabilizing, but in complex networks with symmetric parameterization of the interaction strengths, more links per species (complexity) leads to less stability.

3.7.1   Symmetry: Excessive Negative Feedback

Levins (1974) explains elegantly what happens when a predator–prey system (or more generally, a system where for each effect of X on Y there is a reciprocal effect of Y on X with opposite sign) with random parameterization from symmetric intervals increases in size or connectedness. There are two necessary conditions for stability: (1) the overall

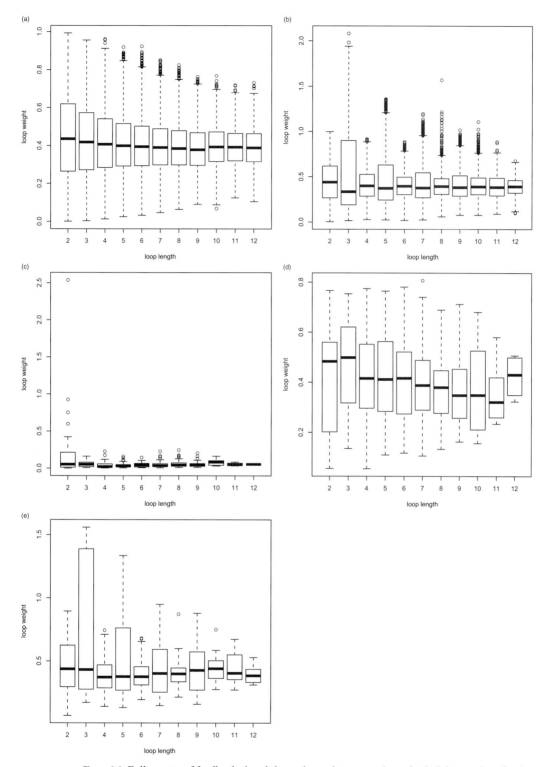

Figure 3.3 Full spectra of feedbacks involving only predator–prey (non-detrital) interactions for the different parameterizations of an observed ecosystem (Antarctic dry tundra, Neutel and Thorne, 2014). The symmetric (a) and asymmetric (b) spectra contain the two- and three-link feedbacks of 100 samplings of those parameterizations and show a signature already apparent in each symmetric (d) and asymmetric (e) samplings. The synthetic signatures are very different from the empirical feedback spectrum (c).

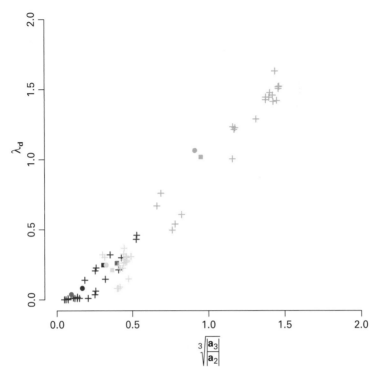

Figure 3.4 Correlation between key feedbacks and stability. Three-link and two-link predator–prey feedbacks $\sqrt[3]{\frac{|a_3|}{|a_2|}}$, where a_3 and a_2 are coefficients of the characteristic polynomial of the community matrices representing the sum of the feedbacks of the 3rd and 2nd levels, respectively (Neutel and Thorne, 2014), and system vulnerability λ_d for empirical parameterizations of the interaction strengths of the Antarctic dry and wet tundra systems and 21 soil food webs (where detritus does not affect stability) with plus signs for the 21 soil webs, compared with symmetric (yellow symbols) and asymmetric (green symbols) parameterizations (N = 71, R^2 = 0.97, P < 10^{-15}). Redrawn with permission. © 2014 Neutel and Thorne. *Ecology Letters* published by John Wiley & Sons Ltd and CNRS. (A black and white version of this figure will appear in some formats. For the color version, please refer to the plate section.)

feedback strength on each level (measured as summations of feedbacks and combinations of feedbacks of a given length) must be negative; and (2) higher level feedback must not exceed lower level feedback (Levins, 1974). Suppose that condition (i) is satisfied, which means that at each level the net feedback is negative. Then the negative feedback of three-link loops will always exceed that of lower level feedback, because the former involves combinations of the order n^3 while the latter only involves combinations of the order n^2.

3.7.2   Asymmetry: We Live in a Positive-Feedback World

What Pimm and Lawton did was take an essential step from systems limited by excessive negative feedback to systems limited by positive feedback. An asymmetry in predator–prey effects (Pimm and Lawton, 1977) creates an asymmetry at the feedback level. It causes loops with relatively many negative effects to be relatively strong. For three-link predator–prey feedbacks, this means that positive feedback will dominate (since positive three-link feedback loops have two negative effects and their negative counterparts have two positive effects). Thus asymmetric parameterizations lead to relatively unstable webs (that is, requiring much self-regulation for stability) compared to the symmetric ones (Figure 3.4), with effectively more omnivory (more three-link omnivorous loops relative to two-link loops) leading to less stability – the result of a completely different feedback structure. But because more connected webs are often also more omnivorous, incorporating an asymmetry in the random parameterization of interaction strengths does not change the existing paradigm that complexity begets instability.

3.7.3   Beyond Symmetry and Asymmetry

In empirical parameterizations, on the other hand, even though positive feedback also dominates over negative feedback in the three-link loops, this dominance is much less. We live in a positive-feedback world, but only just. Also, there is a large variation between interactions and between systems. Stability is not determined by the number of two- or three-link feedback loops, but by their total strength. Therefore it is not so surprising that empirical parameterizations show that the omnivorous structure with the strongest three-link relative to two-link feedback (Neutel and Thorne, 2014), or even just the strongest three-link positive predator–prey feedback loop (Neutel *et al.*, 2007; Mitchell and Neutel, 2012) is key to system vulnerability.

Food webs with interaction strengths sampled from distributions based on empirical flow and biomass rates are orders of magnitude more stable than their randomized counterparts (preserving the sign structure, see de Ruiter *et al.*, 1995; Neutel *et al.*, 2002; James *et al.*, 2015) and connectance-based metrics (May, 1972; Allesina and Tang, 2012), even if the asymmetry within the pairs of interactions is taken into account (Tang *et al.*, 2014), show no correlation with the empirically derived stability values (Neutel *et al.*, 2007; Neutel and Thorne, 2014). This is because such metrics do not capture the complex organization of strong and weak links in a community.

3.8   Nature Lies in a Corner of the Phase Space (or, the Tyranny of the Synthetic)

The above makes clear that empirically based interaction strengths do not just lead to a fine tuning of the system-dynamic properties, they completely change them; stability-complexity paradigms are no longer valid. As a consequence, we might then ask what it is that is defined (or suggested) by the synthetic constructions commonly applied in the field of theoretical ecology, and how representative they are of real systems that we observe in nature. Or, put another way, what are the constraints in nature that allow ecological systems to function?

Consider the full parameter space, or phase space, of all possible states resulting from an $N \times N$ matrix. Topology aside, what are the constraints set by randomly parameterized models within such a vast combinatorial space? The regularity of the distributions of parameter values drawn on in synthetic models results in an artificial simplicity (the symmetric feedback spectrum) at the system level, which means that a very specific region of the phase space is being occupied, necessarily different from the region that would be occupied by empirically derived matrices. It is then natural to ask what the true constraints are in nature that would define how much of the vast combinatorial space would be taken up by real, natural, systems. And, in so doing, we can begin to make clear the differences that mark real systems from those represented by synthetic constructions.

What are some of the possible constraints on natural systems? There are biological and energetic constraints, feeding is not 100% efficient, and energy gets lost along the food chain (Lindeman, 1942). Pimm and Lawton (1977) recognized that this constraint brings with it an asymmetry in the interaction between predator and prey. But they did not take into account how energy is conserved in the whole food web: when predators feed on more than one prey their total effect is spread out over the various prey types. Also, energy and nutrient input is limited – the environment cannot supply unlimited resources, and there are constraints on the metabolic rates of organisms. Studying the interplay between realistic architecture and interaction strengths, and defining the underlying biological constraints, has been a main focus of food-web ecology for the past decades. This has resulted, for example, in ideas on the relation between stability and energy-flow distribution, over trophic levels (Neutel *et al.*, 2002, 2007) and between parallel food-web compartments (Rooney *et al.*, 2006); between stability and body-size patterns, using metabolic scaling theory and macro-ecological correlations of body size and abundance (Emmerson and Raffaelli, 2004; Brose *et al.*, 2006); and between stability and adaptive behavior, in the form of prey switching (Kondoh, 2003) or emerging from evolutionary rules (Drossel *et al.*, 2004).

3.9   Feedbacks and a Comparative Ecosystem Ecology

How then do complex communities obtain and preserve their stability? In terms of the feedback properties, the weak and strong interactions are distributed within food webs in such a way that the weight of the shortest positive (destabilizing) feedback is kept relatively low, compared to that of the negative two-link feedback. This means that only a little self-regulation (self-damping) or intraspecific interaction is needed for system stability – together, the self-damping and two-link feedbacks are a stabilizing force, counteracting the net positive three-link feedback (Neutel and Thorne, 2014). The importance of the organization of interaction strengths in food webs for the stability of community matrices has long been known (Yodzis, 1981). So has the relevance of feedback for matrix stability (Levins, 1974, 1977). But by quantifying the feedbacks in observed systems we can now study the biological, ecosystem-functional properties important for stability and compare ecosystems. We have previously identified (Neutel and Thorne, 2014) critical feedback substructures governing ecosystem vulnerability, which, translated into energetic properties, are

$$\left(\frac{1}{(1 - e_j)} \left(\frac{Q_{jk}}{Q_{ij}Q_{ik}} M_i + \frac{Q_{ik}}{Q_{jk}Q_{ij}} M_j + \frac{e_j}{e_k} \frac{Q_{ij}}{Q_{jk}Q_{ik}} M_k \right) \right)^{-\frac{1}{3}} \qquad (3.1)$$

where i is the bottom prey, j is the intermediate predator, and k is the omnivore. Q_{ij} and M_j are rates of consumption and non-predatory mortality respectively, and e_j is a biomass-conversion efficiency. The stronger the predation losses in a community relative to non-predatory mortality, the stronger the critical feedback and the stronger the self-damping forces (which form part of the non-predatory mortality) will have to be to keep the system stable. The relation between ecosystem functioning, feedback structure, and stability can be seen, for example, in Antarctic tundra ecosystems (Neutel and Thorne, 2014). Wet, nutrient-rich tundra communities with a rapid turnover of nutrients and strong predation pressure generate more positive feedback than dry moss communities with lower predation pressure and slower nutrient turnover. As measured by expression (Eq. 3.1), more self-damping, and hence more competition within the groups, would be needed to provide a stable balance in the wet tundra systems (Neutel and Thorne, 2014).

If life-history strategies form patterns at the ecosystem level that shape the functioning of ecosystems (Grime and Pierce, 2012), these patterns will have an effect on the internal dynamics of the systems, through the feedback signatures they generate. And, to go a step further, one may wonder whether the dynamic constraints on ecosystems themselves will shape the different strategies that species can adopt to function together in their environment. Competitive, opportunistic, and stress-tolerating population strategies (Grime, 1977), or, in a comparable triangular model-framework, equilibrium, opportunistic, and periodic population strategies (Winemiller and Rose, 1992) scale up to multi-trophic communities and shape ecosystems (Grime and Pierce, 2012). But could it be that some of these strategies simply result from the stability conditions? In undisturbed, highly productive environments where predation is strong, self-regulation (intraspecific or intragroup competition) will have to be strong, to counteract the strong positive feedbacks. But where destabilizing feedbacks in the community are weak, such as in stressful environments, where species protect themselves, and their non-predatory losses (natural death rates) are relatively large (organisms die of old age and not through predation), intraspecific competition need not be strong for communities to be stable.

This leaves us with an emphatic message for the theoretical network community: do not study the structure and dynamics of ecological systems without taking into account the (observed) functioning and traits of the organisms. Interaction strengths and the resulting feedbacks are not just another topic to be studied in relation to fundamental network-theoretical properties. They are the fundamentals of the networks themselves.

Acknowledgments

We thank John C. Moore and two anonymous reviewers for comments on the manuscript. This work is part of the British Antarctic Survey Polar Science for Planet Earth Programme. It was funded by the Natural Environment Research Council.

References

Allesina, S. and Tang, S. (2012). Stability criteria for complex ecosystems. *Nature*, **483**, 205–208.

Banašek-Richter, C., Cattin, M. F., and Bersier, L.-F. (2004). Sampling effects and the robustness of quantitative and qualitative food-web descriptors. *Journal of Theoretical Biology*, **226**, 23–32.

Banašek-Richter, C., Bersier, L.-F., Cattin, M.-F., *et al.* (2009). Complexity in quantitative food webs. *Ecology*, **90**, 1470–1477.

Berlow, E. L., Neutel, A. M., Cohen, J. E., *et al.* (2004). Interaction strengths in food webs: issues and opportunities. *Journal of Animal Ecology*, **73**, 585–598.

Bersier, L. F., Banašek-Richter, C., and Cattin, M. F. (2002). Quantitative descriptors of food-web matrices. *Ecology*, **83**, 2394–2407.

Brose, U., Williams, R. J., and Martinez, N. D. (2006). Allometric scaling enhances stability in complex food webs. *Ecology Letters*, **9**, 1228–1236.

Dambacher, J. M., Luh, H. K., Li, H. W., and Rossignol, P. A. (2003). Qualitative stability and ambiguity in model ecosystems. *American Naturalist*, **161**, 876–888.

DeAngelis, D. L. (1975). Stability and connectance in food web models. *Ecology*, **56**, 238–243.

DeAngelis, D. L. (1992). *Dynamics of Nutrient Cycling and Food Webs*. New York: Chapman and Hall.

de Ruiter, P. C., Neutel, A. M., and Moore, J. C. (1995). Energetics, patterns of interaction strengths, and stability in real ecosystems. *Science*, **269**, 1257–1260.

Drossel, B., McKane, A. J., and Quince, C. (2004). The impact of nonlinear functional responses on the long-term evolution of food web structure. *Journal of Theoretical Biology*, **229**, 539–548.

Elton, C. (1927). *Animal Ecology*. New York: Macmillan.

Emmerson, M. C. and Raffaelli, D. (2004). Predator prey body size, interaction strength and the stability of a real food web. *Journal of Animal Ecology*, **73**, 399–409.

Gardner, M. R. and Ashby, W. R. (1970). Connectance of large dynamic (cybernetic) systems: critical values for stability. *Nature*, **228**, 784.

Grime, J. P. (1977). Evidence for the existence of three primary strategies in plants and its relevance to ecological and evolutionary theory. *American Naturalist*, **111**, 1169–1194.

Grime, J. P. and Pierce, S. (2012). *The Evolutionary Strategies That Shape Ecosystems*. Oxford, UK: Wiley-Blackwell.

Grimm, V. and Wissel, C. (1997). Babel, or the ecological stability discussions: an inventory and analysis of terminology and a guide for avoiding confusion. *Oecologia*, **109**, 323–334.

Gross, T., Rudolf, L., Levin, S. A., and Dieckmann, U. (2009). Generalized models reveal stabilizing factors in food webs. *Science*, **325**, 747–750.

Hastings, H. M. (1982). The May–Wigner stability theorem. *Journal of Theoretical Biology*, **97**, 155–166.

Jacquet, C., Moritz, C., Morissette, L., *et al.* (2013). No complexity-stability relationship in natural communities. *q-bio*, arXiv:1307.5364.

James, A., Plank, M. J., Rossberg, A., *et al.* (2015). Constructing random matrices to represent real ecosystems. *American Naturalist*, **185**, 680–692.

Kondoh, M. (2003). Foraging adaptation and the relationship between food-web complexity and stability. *Science*, **299**, 1388–1391.

Levins, R. (1968). *Evolution in Changing Environments*. Princeton, NJ: Princeton University Press.

Levins, R. (1974). The qualitative analysis of partially specified systems. *Annals of the New York Academy of Science*, **231**, 123–138.

Levins, R. (1977). Quantitative analysis of complex systems. In *Mathematics and the Life Sciences*, ed. D. E. Matthews, Vol. 18 of Lecture Notes in Biomathematics. New York: SpringerVerlag, pp. 152–199.

Lindeman, R. L. (1942). The trophic-dynamic aspect of ecology. *Ecology*, **23**, 399–417.

MacArthur, R. (1955). Fluctuations of animal populations, and a measure of community stability. *Ecology*, **36**, 533–536.

Maruyama, M. (1963). The second cybernetics: deviation-amplifying mutual causal processes. *American Scientist*, **51**, 164–179.

May, R. M. (1972). Will a large complex system be stable? *Nature*, **238**, 413–414.

May, R. M. (1973). *Stability and Complexity in Model Ecosystems*. Princeton, NJ: Princeton University Press.

McCann, K. S. (2000). The diversity–stability debate. *Nature*, **405**, 228–233.

McCann, K., Hastings, A. G., and Huxel, R. (1998). Weak trophic interactions and the balance of nature. *Nature*, **395**, 794–798.

Menge, B. A. (1995). Indirect effects in marine rocky intertidal interaction webs: patterns and importance. *Ecological Monographs*, **65**, 21–74.

Mitchell, E. G. and Neutel, A. M. (2012). Feedback spectra of soil food webs across a complexity gradient, and the importance of three-species loops to stability. *Theoretical Ecology*, **5**, 153–159.

Montoya, J. M., Pimm, S. L., and Solé, R. V. (2006). Ecological networks and their fragility. *Nature*, **442**, 259–264.

Montoya, J. M., Woodward, G., Emmerson, M. C., and Solé, R. V. (2009). Press perturbations and indirect effects in real food webs. *Ecology*, **90**, 2426–2433.

Moore, J. C., de Ruiter, P. C., and Hunt, H. W. (1993). Influence of productivity on the stability of real and model ecosystems. *Science*, **261**, 906–908.

Moore, J. C., Berlow, E. L., Coleman, D. C., *et al.* (2004). Detritus, trophic dynamics and biodiversity. *Ecology Letters*, **7**, 584–600.

Neutel, A. M. and Thorne, M. A. S. (2014). Interaction strengths in balanced carbon cycles and the absence of a relation between ecosystem complexity and stability. *Ecology Letters*, **17**, 651–661.

Neutel, A. M., Roerdink, J. B. T. M., and de Ruiter, P. C. (1994). Global stability of two-level detritus-decomposer food chains. *Journal of Theoretical Biology*, **171**, 351–353.

Neutel, A. M., Heesterbeek, J. A. P., and de Ruiter, P. C. (2002). Stability in real food webs: weak links in long loops. *Science*, **296**, 1120–1123.

Neutel, A. M., Heesterbeek, J. A. P., van de Koppel, J., *et al.* (2007). Reconciling complexity with stability in naturally assembling food webs. *Nature*, **449**, 599–602.

Novak, M., Wootton, J. T., Doak, D. F., *et al.* (2011). Predicting community responses to perturbations in the face of imperfect knowledge and network complexity. *Ecology*, **92**, 836–846.

Odum, E. P. (1971). *Fundamentals of Ecology*, 3rd edn. Philadelphia, PA: Saunders.

Paine, R. T. (1988). Food webs: road maps of interactions or grist for theoretical development? *Ecology*, **69**, 1648–1654.

Pimm, S. L. (1982). *Food Webs*. London, UK: Chapman & Hall.

Pimm, S. L. and Lawton, J. H. (1977). Number of trophic levels in ecological communities. *Nature*, **268**, 329–331.

Pimm, S. L. and Lawton, J. H. (1978). On feeding on more than one trophic level. *Nature*, **275**, 542–544.

Polis, G. A. and Strong, D. R. (1996). Food web complexity and community dynamics. *American Naturalist*, **147**, 813–846.

Rooney, N., McCann, K., Gellner, G., and Moore, J. C. (2006). Structural asymmetry and the stability of diverse food webs. *Nature*, **442**, 265–269.

Tang, S., Pawar, S., and Allesina, S. (2014). Correlation between interaction strengths drives stability in large ecological networks. *Ecology Letters*, **17**, 1094–1100.

Ulanowicz, R. E., Holt, R. D., and Barfield, M. (2014). Limits on ecosystem trophic complexity: insights from ecological network analysis. *Ecology Letters*, **17**, 127–136.

Winemiller, K. O. and Rose, K. A. (1992). Patterns of life-history diversification in North American fishes: implications for population regulation. *Canadian Journal of Fisheries and Aquatic Sciences*, **49**, 2196–2218.

Wootton, J. T. (1994). Predicting direct and indirect effects: an integrated approach using experiments and path analysis. *Ecology*, **75**, 151–165.

Wootton, J. T. and Emmerson, M. (2005). Measurement of interaction strength in nature. *Annual Review of Ecology Evolution and Systematics*, **36**, 419–444.

Yodzis, P. (1981). The stability of real ecosystems. *Nature*, **289**, 674–676.

4 Ecologically Effective Population Sizes and Functional Extinction of Species in Ecosystems

Bo Ebenman, Torbjörn Säterberg, and Stefan Sellman

4.1 Introduction: Two Types of Extinctions

There are two types of species extinctions: true, or numerical extinctions, and functional extinctions. Numerical extinction – the traditional concept of extinction – occurs when the very last member of a species dies, while functional or ecological extinction occurs when a species becomes too rare to fulfill its ecological, interactive role in the ecosystem (Conner, 1988; Estes *et al.*, 1989; Novaro *et al.*, 2000; Jackson *et al.*, 2001; Redford and Feinsinger, 2001; Soulé *et al.*, 2003; Sekercioglu *et al.*, 2004; McConkey and Drake, 2006; Baum and Worm, 2009; Estes *et al.*, 2010; Anderson *et al.*, 2011; Cury *et al.*, 2011; Galetti *et al.*, 2013; Säterberg *et al.*, 2013; McConkey and O'Farrill, 2015; Sellman *et al.*, 2015). It has been estimated that the current rate of numerical species extinction is about 1000 times higher than the natural background rate of extinction, on par with that of the great mass extinctions (Pereira *et al.*, 2010; Barnosky *et al.*, 2011; Pimm *et al.*, 2014).

In contrast, the rate of functional extinctions is largely unknown. Critical abundance thresholds or ecologically effective population sizes of species, below which they cease to function in the system of which they are parts, have been established for only a few species (McConkey and Drake, 2006; Estes *et al.*, 2010; Cury *et al.*, 2011). However, trophic cascades and regime shifts observed in a variety of ecosystems following declining population size of a species clearly indicate the presence of such abundance thresholds of species in many ecosystems (Frank *et al.*, 2005; Casini *et al.*, 2009; Estes *et al.*, 2011; Smith *et al.*, 2011; Ripple *et al.*, 2014). Indeed, recent theoretical studies suggest that the frequency of functional extinctions might be disturbingly high and that even moderate declines in the densities of some species might lead to numerical extinctions of other dependent species (Säterberg *et al.*, 2013; Sellman *et al.*, 2015). In other words, a species can go functionally extinct well before the species becomes so rare that it loses its genetic and demographic viability and puts it in danger of a numerical extinction.

A striking example illustrating the difference between numerical and functional extinction is that of the Pinta Island tortoise (*Chelonoidis nigra abingdonii*) in the Galapagos. In 1971 only one individual, known as Lonesome George, of the Pinta tortoise remained alive. Although it took nearly forty years (2012) before Lonesome

George died of old age and the subspecies thus became numerically extinct, the Pinta Island tortoise had been functionally extinct since long back in time with important consequences for the vegetation and habitat structure of the island ecosystems (Froyd *et al.*, 2014).

Here we review the empirical evidence of functional extinctions and ecologically effective population sizes of species in ecosystems and present a recently developed theory on these topics. We also discuss the implications of this new theory for conservation biology and sustainable exploitation of natural populations. Finally, we identify gaps in our knowledge and point to possible avenues for future research in this field.

4.2   Declining Populations: An Important Component of Biodiversity Loss

Numerical extinction of species represents only one part of the actual loss of biodiversity (Butchart *et al.*, 2010; Dirzo *et al.*, 2014). Equally important are range contractions and declining local populations of many species (Loh *et al.*, 2005; Collen *et al.*, 2009). Indeed, for vertebrate species it is estimated that there has been a mean decline of nearly one-third in the number of individuals across species in the past four decades (Collen *et al.*, 2009; Butchart *et al.*, 2010) and for mammalian species one in two shows a negative population trend (Schipper *et al.*, 2008). There is also a trend of declining populations of many invertebrate species (Dirzo *et al.*, 2014). Many bird species show declining populations and it is estimated that the global number of individual birds have experienced a 20 to 25% reduction in the past 500 years (Gaston *et al.*, 2003). About 7% of the bird species in the world have become so rare that they no longer are considered to contribute significantly to ecosystem processes such as pollination and seed dispersal (Sekercioglu *et al.*, 2004; see also Anderson *et al.*, 2011). Likewise, 20 to 30% of all plant species, the primary producers, are estimated to be threatened, the main cause being high rates of habitat loss leading to range contractions and declining populations (Brummitt *et al.*, 2008; Joppa *et al.*, 2011).

Negative population trends are particularly pronounced in large-bodied herbivores and carnivores in terrestrial as well as in marine ecosystems (Jackson *et al.*, 2001; Baum *et al.*, 2003; Myers and Worm, 2003; Di Marco *et al.*, 2014; Ripple *et al.*, 2014). For instance, both coastal and oceanic populations of sharks have undergone large and rapid declines – some species, such as hammerhead and white sharks, by as much as 75% in 15 years (Baum *et al.*, 2003). Similarly, 22 of the 28 largest terrestrial mammalian carnivores are undergoing continuing population declines (Ripple *et al.*, 2014). This is very worrying, since large predators and herbivores are known to play important regulatory roles in many ecosystems (Baum *et al.*, 2003; Estes *et al.*, 2011; Ripple *et al.*, 2014).

4.3   Functional Extinctions and Ecologically Effective Population Sizes

It is well known that declining population size of a species increases the risk of extinction of the species itself (Lande, 1993). Estimating numerical extinction risk and minimum viable population sizes (MVP) of species is the business of traditional population viability analysis (Box 4.1). The concept of MVP is mainly based on genetic and demographic criteria and represents a single-species approach to conservation biology (Soulé *et al.*, 2003; Sabo, 2008). However, decreasing abundance of a species might affect other dependent species well before the declining species itself is threatened by extinction (Säterberg *et al.*, 2013). That is, a species might lose its ecological and interactive role in the community long before becoming so rare as to be endangered by processes such as demographic stochastic variation and inbreeding depression. Thus the minimum population size required for genetic and demographic viability of a species may not be sufficient for ecological viability.

Box 4.1 Definitions of Basic Concepts

Numerical extinction. This occurs when the very last member of a species dies (true extinction). This is illustrated below, where increased mortality rate of a given species leads to dwindling numbers and eventual extinction of the species itself (Figure 4.1a).

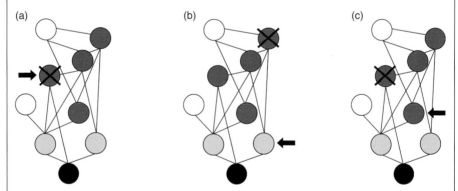

(a)　　　　　　　　　(b)　　　　　　　　　(c)

Figure 4.1 A schematic illustration of numerical and functional extinctions. Shown are the main players in the pelagic food web of the Baltic Sea; black (primary producer), pale gray (herbivores), dark gray (intermediate species), and white (top predators) (Berg *et al.*, 2011). The species marked with an arrow is the focal species being affected by an increased mortality rate and the crossed-out species is the species becoming extinct because of this mortality increase. (a) Increased mortality of a species leads to extinction of the focal species itself. (b) Increased mortality of a given species leads to extinction of another directly linked species. (c) Increased mortality of a given species leads to extinction of another indirectly linked species. Thus in (a) the focal species becomes numerically extinct, while in (b) and (c) the focal species becomes functionally extinct – a reduction in its population size leads to numerical extinction of another species.

Minimum viable population (MVP) size. The minimum population size needed for long-term persistence of a species, for example, the population size needed for a 95% probability of survival over 100 years (Shaffer, 1981). This approach does not usually account for interactions among species and, if it does, interactions are usually incorporated as a constant source of mortality rather than as a dynamical effect on the focal species' growth rate (Sabo, 2008).

Numerical extinction and MVP are the basic concepts in traditional population viability analysis (PVA).

Functional extinction. Broadly, a species can be considered functionally extinct when it has become so rare that it no longer fulfills its ecological role in the system (Soulé *et al.*, 2003). A possible, more precise, definition of functional extinction is that a focal species becomes functionally extinct when an increase in its mortality rate (or decrease in its birth rate) and a decrease in its abundance leads to extinction of another species in the ecological community rather than extinction of the focal species itself (Säterberg *et al.*, 2013) (Figure 4.1b, c).

Ecologically effective population (EEP) size. The population size of a species below which one or more species in the ecological network become extinct (Säterberg *et al.*, 2013). Or, more broadly, population sizes large enough to maintain critical interactions and prevent ecosystem degradation (Soulé *et al.*, 2003).

Functional extinction and EEP size are basic concepts in a more community-oriented approach in conservation biology.

When increased mortality rate (or decreased birth rate) and dwindling numbers of a focal species leads to extinction of another species in the community rather than extinction of the species itself, the focal species can be characterized as functionally or ecologically extinct even though it still exists (Säterberg *et al.*, 2013; Sellman *et al.*, 2015; Box 4.1). Even though one or more species become extinct following reduced abundance of the focal species, its functional role may not always be entirely lost. In such cases it might be more appropriate to consider the focal species as functionally quasi-extinct (i.e., partial loss of interactive, functional role). Much less is known about the risk and nature of functional extinctions and functional quasi-extinctions than about 'ordinary' numerical extinctions. And compared to estimates of MVP sizes there are very few studies that have tried to establish the EEP size of a species, below which the existence of other species in the community is jeopardized (Estes *et al.*, 2010; Cury *et al.*, 2011).

4.3.1   Functional Extinctions and Ecologically Effective Population Sizes in Real Ecosystems

Although clear-cut examples of functional extinction events are relatively scarce, some studies provide convincing evidence of their existence. A well-known example comes from the coastal waters of the North Pacific and southern Bering Sea where declining populations of sea otters (*Enhydra lutris*) led to dramatic increase in the abundance of their most important prey, sea urchins. In some places the sea urchins grew to such large numbers that their intense grazing on the underwater kelp forests

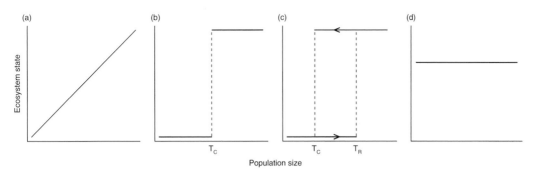

Figure 4.2 Schematic picture showing potential relationships between the abundance of a focal species and the state of the ecosystem of which it is a part. The state of the system could be the probability of persistence (i.e., no species extinctions caused by declining population size of the focal species). Thresholds are indicated by broken lines. At these thresholds small changes in abundances lead to large changes in the state of the ecosystem. (a) Absence of threshold – a linear positive relationship between the abundance of a focal species and the state of the system. That is, declining abundance of the focal species gradually decreases the probability of system persistence. (b) A marked threshold – an abrupt change in system state caused by a small change in the abundance of a focal species. Here, once the population falls below the threshold T_C (collapse threshold) the system collapses. (c) Hysteresis is present – collapse and recovery trajectories differ. Here, T_C and T_R denote the collapse and recovery threshold, respectively. (d) The state of the system is independent of the abundance of the focal species.

led to system collapse – a state shift to a barren seafloor (Estes and Palmisano, 1974). Since the kelp forests provide habitat for numerous species of fish and invertebrates, their collapse had large effects on the biodiversity in these waters. This collapse illustrates an extinction cascade unleashed in the absence of top–down regulation (i.e., predators limiting herbivores and thereby preventing them from overgrazing primary producers). For the Aleutian Archipelago, Estes *et al.* (2010) were able to estimate the number of sea otters required per kilometer of coastline to prevent such a state shift from occurring. This threshold can be seen as the EEP size of sea otters, below which the risk of a state shift is high.

Once such a state shift has occurred it might be difficult for the system to recover and shift back to its original state (May, 1977; Lundberg *et al.*, 2000; Scheffer *et al.*, 2001; Beisner *et al.*, 2003). Specifically, just as there is a threshold abundance required to prevent a state shift from occurring, there might be an abundance threshold level that must be reached, which is higher than the collapse threshold, in order for the system to recover and return to the original state. That is, the collapse and recovery trajectories might differ from each other – so-called hysteresis (Scheffer *et al.*, 2001, Beisner *et al.*, 2003) (Figure 4.2).

Critical abundance thresholds (EEP sizes) of species have also been identified for small pelagic fishes, such as anchovy and sardines, and krill – the main prey types of many seabirds – in several coastal marine ecosystems (Cury *et al.*, 2011). Fishing-induced reduction of fish and krill abundances below this threshold, negatively affect the recruitment of seabirds, and the breeding success of some bird species might potentially be so severely reduced as to put the population in question at risk of extinction even though the exploited prey species may be out of danger.

The examples above deal with predator–prey interactions where the function lost is either a top–down regulation of herbivore abundances as in the example with the sea otters and kelp, or a bottom–up supply of resources as in the example with small fish species and seabirds. Critical abundance thresholds of species have also been observed in mutualistic systems, such as plants and seed-dispersing animals. For example, a critical abundance threshold has been documented for a species of flying fox (*Pteropus tonganus*) that functions as a seed disperser for large-seeded fruit tree species (McConkey and Drake, 2006). When fruiting trees become crowded with flying foxes, agonistic behaviors caused by this crowding lead some individuals to grab fruits from the tree and fly elsewhere to consume them, thus dispersing seeds further away from the tree. However, if flying fox abundance is low the seeds will be dropped close to the tree, impairing the tree's dispersal. As flying foxes are the most important seed dispersers for many large-seeded tree species in the tropical Pacific region and because many flying fox species are endangered (IUCN, 2014) there might be a risk that the long-term survival of such tree species will be compromised (McConkey and Drake, 2006).

Evidence for functional extinctions also comes from a recent mesocosm experiment including one plant species, three herbivorous aphid species, and their associated specialized parasitoid wasp species (Sanders *et al.*, 2013, 2015). Here experimental reduction of the density of one of the parasitoids had cascading effects throughout the mesocosm community sometimes leading to complete extinction of the other parasitoid species. The reduction in the abundance of one parasitoid resulted in increased abundance of its aphid host. This led to intensified competition among the aphid species in the system and, as a result, the other two aphid species decreased in abundance, which in turn caused their specific parasitoids to become extinct.

Together with other studies on trophic cascades and regime shifts, as discussed above, these detailed case studies illustrate the potential importance of functional extinctions in ecological communities and the existence of critical abundance thresholds or EEP sizes of species below which other dependent species in the communities are put in danger of extinction. In the next section we present results from recent theoretical studies also suggesting potentially high rates of functional extinctions in ecosystems.

4.3.2   Frequency of Functional Extinctions in Model Ecosystems

In a recent theoretical study we developed a theoretical framework to explore the frequency and nature of functional extinctions in ecological networks (Säterberg *et al.*, 2013). Specifically, EEP sizes of species were analytically derived by increasing their mortality rates until an extinction event occurred in the network (see Box 4.2 for a summary). This analytical method was used to investigate the frequency and nature

Box 4.2 Theoretical Derivation of Ecologically Effective Population Sizes

To describe how EEP sizes can be derived for the species in a community we here use a simple food chain with three species: (1) primary producer, (2) herbivore, and (3) predator. Each species (indicated by a horizontal arrow in Figure 4.3) is exposed to an increased mortality rate and the species becoming extinct as a result of this perturbation is recorded (indicated by a cross).

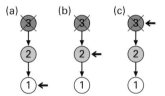

Figure 4.3 Schematic of a simple exercise to illustrate the theoretical derivation of ecologically effective population sizes using three-species food chains described by the equations below. The horizontal arrows indicate the species exposed to increased mortality and the crossed species indicates the species becoming extinct.

An analytical way of finding the species becoming extinct is as follows (Säterberg *et al.*, 2013): first, assume that the dynamics of the species in the food chain can be described by a generalized Lotka–Volterra model:

$$dN_1/dt = N_1(r_1 + \alpha_{11}N_1 + \alpha_{12}N_2)$$
$$dN_2/dt = N_2(r_2 + \alpha_{21}N_1 + \alpha_{23}N_3)$$
$$dN_3/dt = N_3(r_3 + \alpha_{32}N_2)$$

Here r_i is the intrinsic growth rate of species i and α_{ij} is the per capita effect of species j on the per capita growth rate of species i. The α_{ij}s constitute the elements in the interaction matrix **A** of the community (food chain).

The equilibrium density of each species i (i = 1, 2, 3) is given by:

$\hat{N}_i = -\sum_{j=1}^{3} r_j \gamma_{ij}$ where $\gamma_{ij} = -\partial \hat{N}_i / \partial r_j$. In words, $-\gamma_{ij}$, which is element ij in the inverse interaction matrix, $-\mathbf{A}^{-1}$, gives the effect of a small change in the intrinsic growth rate of species j on the equilibrium abundance of species i.

Now, suppose that the intrinsic growth rate of species j is decreased by an amount ε_j giving the new intrinsic growth rate $r'_j = r_j - \varepsilon_j$. The new equilibrium density of species i will then be: $\hat{N}'_i = \hat{N}_i + \varepsilon_j \gamma_{ij}$. It can be seen that $\hat{N}'_i = 0$ when $\varepsilon_j(i) = -\hat{N}_i / \gamma_{ij}$.

Now, for each species j we wish to find the minimum amount, $min_i\left(\varepsilon_j(i)\right)$, by which its intrinsic growth rate can be decreased before some species i in the network becomes extinct. We can then derive the EEP size of species j:

$$N_j(EEP) = \hat{N}_j + min_i\left(\varepsilon_j(i)\right)\gamma_{jj}$$

In the particular example illustrated above, a simple food chain with three trophic levels, increased mortality of species 1 (plant) leads to extinction of species 3 (predator); increased mortality of species 2 (herbivore) also leads to extinction of species 3; and increased mortality of species 3 leads to extinction of species 3 itself. Here the EEP size of species 3 is zero, since a decrease in its abundance, even its numerical extinction, will not cause extinction of any other species in the food chain.

of functional extinctions of species in eight natural food webs and in sequentially assembled model food webs. Sequentially assembled model webs, with 50 interacting species and a mean connectance of 0.12, were generated by relating parameter values to the average body mass of species and letting dynamical constraints act during the web build-up (for details see Lewis and Law, 2007; Säterberg et al., 2013). Models of natural food webs were parameterized using allometric relationships and, where available, data from gut analysis (Berg et al., 2011; Säterberg et al., 2013). These natural food webs were from freshwater, marine, and terrestrial habitats with species richness ranging from 9 to 67 species and connectance ranging from 0.06 to 0.32.

The frequency of functional extinctions was surprisingly high. Indeed, the probability that increased mortality rate and decreased abundance of a species led to extinction of another species rather than extinction of the species itself was on average as high as 0.49 for the natural food webs and 0.72 for the model webs. Furthermore, a relatively high proportion of all first extinctions were of species that were not directly linked to the species whose mortality rate was increased and population density decreased, that is, the extinct species was neither a consumer of nor consumed by the focal species. Here declining abundance of the focal species led to decreased or increased abundance of directly linked species, which in turn caused the extinction of another species not directly linked to the focal species. Results for three natural food webs – one terrestrial, one freshwater, and one marine web – are presented in Figure 4.4. It can be seen that functional extinctions occur at all trophic levels, demonstrating the importance of both top–down and bottom–up processes in these communities. Theoretical work investigating the effects of increased mortality rate of top predators in simple food web modules (3 to 9 species) found that it often led to extinction of another species occupying a high trophic position in the system rather than extinction of the top predator whose mortality rate was increased (Wollrab et al., 2012). This outcome is in line with the results from the experimental study on a simple six-species plant–aphid–parasitoid community described above (Sanders et al., 2013).

4.3.3   Ecologically Effective Population Sizes of Species in Model Communities

The analytical method used to derive EEP sizes of species is summarized in Box 4.2 (Säterberg et al., 2013; Sellman et al., 2015). Briefly, for each species in the community, the increase in its mortality rate or decrease in its birth rate needed to cause extinction of another species is calculated. Then, the new and lowered population abundance resulting

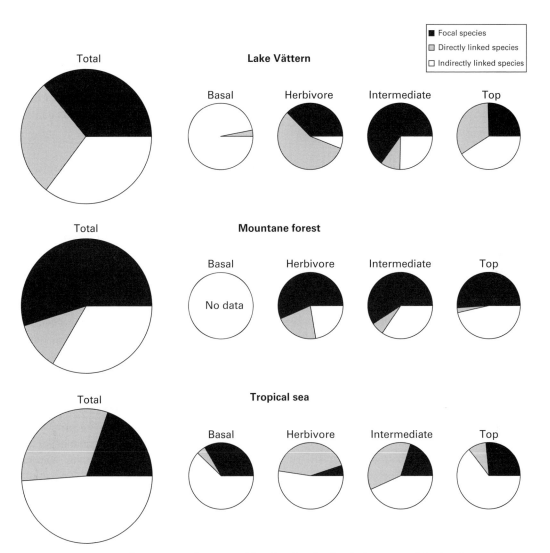

Figure 4.4 Frequency and pattern of functional extinction in three natural food webs: a pelagic freshwater food web (Lake Vättern, Sweden; Berg *et al.*, 2011), a terrestrial food web (Montane forest; Cohen *et al.*, 1990), and a marine food web (Tropical sea; Cohen *et al.*, 1990). Pie diagrams show the proportion of extinctions affecting the focal species (black), directly (gray), and indirectly (white) linked species. Focal species belong to one of the following trophic groups: basal species (primary producers), herbivores (species feeding only on primary producers), intermediate species (consumers that are neither top nor herbivore species), and top consumers (species that are not consumed). Increased mortality rate of a focal species often leads to extinction of another species, that is, the focal species often becomes functionally extinct.

from this increased mortality rate is solved for. This abundance can be seen as the EEP size of the species.

Now, for a given species, how much can its mortality rate increase (or its birth rate decrease) and its abundance decrease before it becomes functionally extinct? Results based on application of the method above (Box 4.2) suggest that even relatively small reductions in the population size of a species can lead to extinction of another species. Specifically, the reduction in population size of species needed to trigger the extinction of another species is inversely related to its body mass, implying that large-bodied species at the top of food chains can only be exposed to small increases in mortality rate and small decreases in abundance before becoming functionally extinct compared to small-bodied species lower in the food chains.

4.4   Implications

The results presented here demonstrate the potential importance of functional extinctions in ecological communities and have important implications when setting target sizes of populations in conservation biology, in that we derive EEP sizes of species below which other dependent species in the ecosystem are likely to become extinct. The EEP could be much higher than the currently looked at MVP. If this is the case, conservation strategy should focus on quantifying EEPs rather than MVPs in order to minimize extinctions in ecological networks. In the words of Soulé *et al.* (2003) "it is essential to consider the densities or population levels that maintain interaction effectiveness rather than mere persistence at minimal numbers." Indeed, strong arguments have been put forward for the need to adopt a more network-oriented approach in conservation biology because of the importance of interactions among species (Soulé *et al.*, 2003, 2005; Sabo, 2008). We believe that the theoretical framework presented here will aid a change in that direction. The potential risk of functional extinction also has implications for sustainable exploitation of natural populations. For some harvested species, target levels based on traditional maximum sustainable yield (MSY) reasoning might be too low for the species to uphold their ecological roles in the ecosystems (Smith *et al.*, 2011).

4.5   Outlook

The theoretical framework we have presented here constitutes a first step, although hopefully a useful step, toward a deeper understanding of functional extinctions in ecological communities. Here we point to some possible ways in which this theory can be developed and used to explore a number of open ecological questions.

Extinction cascades. Once a species falls below its EEP size and another species is lost from the community, the community might become dynamically unstable. In the worst case an extinction cascade can be triggered by the loss of the first species (Pimm, 1980; Borrvall *et al.*, 2000; Solé and Montoya, 2001; Dunne *et al.*, 2002; Allesina and Bodini, 2004; Ebenman *et al.*, 2004; Ebenman and Jonsson, 2005; Eklöf and Ebenman, 2006; Thebault *et al.*, 2007; Petchey *et al.*, 2008; Fowler, 2010; Curtsdotter *et al.*, 2011; Colwell *et al.*, 2012; Kaneryd *et al.*, 2012). A recent theoretical study indicates that the risk of community destabilization can be substantial (Säterberg *et al.*, 2013).

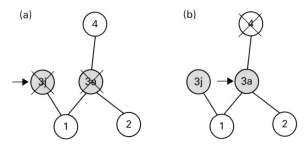

Figure 4.5 Functional extinction in a simple four-species food-web module. Here the juvenile and adult life stages of species 3 (gray) consume different prey species. (a) Increased mortality rate in juveniles (arrow) leads to extinction of the species itself (cross), that is, a numerical extinction. (b) Increased mortality rate in adults (arrow) leads to extinction of species 4 (cross), that is, focal species 3 becomes functionally extinct.

The role of network type and topology. Theoretical work suggests that the dynamics and stability of ecological networks depend on the pattern, type, and strength of species interactions (May, 1972; McCann, 2000; Nuetel *et al.*, 2002; Brose *et al.*, 2006; Gross *et al.*, 2009; Thebault and Fontaine, 2010; Stouffer and Bascompte, 2011; Allesina and Tang, 2012; Mougi and Kondoh, 2013). How will network topology, network type, and distribution of interaction strengths affect the risk of functional extinction? Is there any relationship between the degree of modularity (compartmentalization) of a network and the risk of functional extinction? Are antagonistic networks, such as food webs, more likely to be affected by functional extinctions than mutualistic ones? A recent theoretical study suggests that the frequency of functional extinctions might be higher in networks containing a mixture of antagonistic and mutualistic interactions than in networks with only one type of interaction (Sellman *et al.*, 2015).

Metacommunities. Another challenge is to disentangle how dispersal of individuals among local communities might affect the risk and impact of functional extinctions. In open communities dispersal can potentially rescue species from functional as well as numerical extinctions. The potential for rescue effects to operate will depend on the rate of dispersal relative to the rate of population decline and extinction. Furthermore, it will likely depend on the spatial scale of movement of species at different trophic levels. Predators may often range over larger areas than their prey (McCann *et al.*, 2005).

Stage-structured populations: ecologically effective abundances of life stages. In many species the different age classes or life stages have different ecological niches (e.g., Werner and Gilliam, 1984; Ebenman, 1987; Ebenman and Persson, 1988; Ebenman, 1992; Miller and Rudolf, 2011; Rudolf and Lafferty, 2011; de Roos and Persson, 2013). As a consequence individuals in the different life stages interact with different subsets of other species in the community. Is there a higher probability of functional extinction if the mortality rate of juveniles is increased than if that of adults is increased? Will increased mortality rate of one life stage lead to numerical extinction while increased mortality rate of another life stage lead to functional extinction (Figure 4.5)?

Increased mortality rate → Increased abundance *Hydra effect*	Increased mortality rate → Decreased abundance *Increased fishing intensity*
Decreased mortality rate → Increased abundance *Relaxation from disease agents*	Decreased mortality rate → Decreased abundance *Reverse hydra effect*

Figure 4.6 Potential effects of changed mortality rate on abundance – counterintuitive outcomes are shaded. Possible causes behind changed mortality rates are given for the expected outcomes (unshaded). In this review we focus on the scenario in the top-right quadrant.

Accounting for environmental variation. It would be valuable to develop the modeling framework further so it can be more readily applied to real communities exposed to stochastic environmental variation. Such a framework could be based on analysis of time series of species using multivariate autoregressive modeling (Ives *et al.*, 2003) to parameterize stochastic models of the communities. Using such a framework it would be possible to estimate EEPs for each species in a community. Here, the EEP could be defined as the minimum population size of a species needed for, say, a 95% probability of survival of all species in a community over a time period of 100 years.

Hydra effects and upper abundance thresholds of species. In addition to lower abundance thresholds, discussed here, there could also be upper abundance thresholds of species, above which other species in the community might become extinct. The theoretical framework presented here can also be applied to investigate such upper abundance thresholds of species. Because of indirect effects, increased mortality rate in a species caused by a stressor can sometimes lead to increased abundance of the affected species – the so-called hydra effect (e.g., Abrams, 2009; Wollrab *et al.*, 2012). Some stressors, such as eutrophication, might lead to decreased mortality rates and hence increased growth rates of species, which can be expected to result in increased abundance of the affected species. Due to indirect effects, a decrease in the mortality rate of a species could potentially also lead to decreased abundance of the species, though this would probably be a less likely outcome (see table within Figure 4.6 for the possible effects of stressors on mortality rates and abundances of affected species). Of course, when numerical extinction of a species is caused by an increased abundance of a focal species, the focal species cannot be considered as functionally extinct. Rather, the focal species has become "functionally invasive."

Stressors affecting multiple species simultaneously. Some stressors, such as hunting and lethal control of species (Colman *et al.*, 2014) and outbreak of disease agents (Wilmers *et al.*, 2006), might directly affect mainly one species, while other stressors, such as climate change and pollution, can be expected to strike more broadly. An interesting challenge is to develop the methods presented here such that the consequences of stressors affecting many species simultaneously can be analyzed.

Other criteria of functional extinction. Here we say that a species becomes functionally extinct when a decrease in its abundance caused by an increase in its mortality rate or a decrease in its birth rate lead to the extinction of another species in the community. This is one possible definition of functional or ecological extinction. Functional extinction can also be defined according to other criteria. For instance, we could say that a species becomes functionally extinct when its population has declined to a level at which the efficiency of some ecosystem services, such as pollination, have been reduced by 50%. And we can calculate EEP sizes with respect to this particular criterion.

Ecologically effective link strengths. Human-induced stressors on the natural environment, such as climate change, can also lead to changes in the strength of interactions between species (Gilbert *et al.*, 2014). We can use the analytical method described above to derive critical magnitudes of direct pairwise interactions. That is, how much can we increase or decrease the strength of an interaction link before some species in the community becomes extinct. Are communities particularly sensitive to changes in the strengths of certain types of links?

References

Abrams, P. (2009). When does greater mortality increase population size? The long history and diverse mechanisms underlying the hydra effect. *Ecology Letters*, **12**, 462–474.

Allesina, S. and Bodini, A. (2004). Who dominates whom in the ecosystem? Energy flow bottlenecks and cascading extinctions. *Journal of Theoretical Biology*, **230**, 351–358.

Allesina, S. and Tang, S. (2012). Stability criteria for complex ecosystems. *Nature*, **483**, 205–208.

Anderson, S., Kelly, D., Ladley, J., Molloy, S., and Terry, J. (2011). Cascading effects of bird functional extinction reduce pollination and plant density. *Science*, **331**, 1068–1071.

Barnosky, A., Matzke, N., Tomiya, S., *et al.* (2011). Has the Earth's sixth mass extinction already arrived? *Nature*, **471**, 51–57.

Baum, J. and Worm, B. (2009). Cascading top–down effects of changing oceanic predator abundances. *Journal of Animal Ecology*, **78**, 699–714.

Baum, J., Myers, R., Kehler, D., *et al.* (2003). Collapse and conservation of shark populations in the northwest Atlantic. *Science*, **299**, 389–392.

Beisner, B., Haydon, D., and Cuddington, K. (2003). Alternative stable states in ecology. *Frontiers in Ecology and the Environment*, **1**, 376–382.

Berg, S., Christianou, M., Jonsson, T., and Ebenman, B. (2011). Using sensitivity analysis to identify keystone species in size-based food webs. *Oikos*, **120**, 510–519.

Borrvall, C., Ebenman, B., and Jonsson, T. (2000). Biodiversity lessens the risk of cascading extinction in model food webs. *Ecology Letters*, **3**, 131–136.

Brose, U., Williams, R., and Martinez, N. (2006). Allometric scaling enhances stability in complex food webs. *Ecology Letters*, **9**, 1228–1236.

Brummit, N., Bachman, S., and Moat, J. (2008). Applications of the IUCN Red List: towards a global barometer for plant diversity. *Endangered Species Research*, **6**, 127–135.

Butchart, S., Walpole, M., Collen, B., *et al.* (2010). Global biodiversity: indicators of recent declines. *Science*, **328**, 1164–1168.

Casini, M., Hjelm, J., Molinero, J., *et al.* (2009). Trophic cascades promote threshold-like shifts in pelagic marine ecosystems. *Proceedings of the National Academy of Sciences of the United States of America*, **106**, 197–202.

Cohen, J., Briand, F., and Newman, C. (1990). *Community Food Webs: Data and Theory*. Berlin Heidelberg: Springer-Verlag.

Collen, B., Loh, J., Whitmee, S., *et al.* (2009). Monitoring change in vertebrate abundance: the Living Planet Index. *Conservation Biology*, **23**, 317–327.

Colman, N., Gordon, C., Crowther, M., and Letnic, M. (2014). Lethal control of an apex predator has unintended cascading effects on forest mammal assemblages. *Proceedings of the Royal Society B: Biological Sciences*, **281**, 20133094.

Colwell, R., Dunn, R., and Harris, N. (2012). Coextinction and persistence of dependent species in a changing world. *Annual Review of Ecology, Evolution, and Systematics*, **43**, 183–203.

Conner, R. (1988). Wildlife populations: minimally viable or ecologically functional? *Wildlife Society Bulletin*, **16**, 80–84.

Curtsdotter, A., Binzer, A., Brose, U., *et al.* (2011). Robustness to secondary extinctions: comparing trait-based sequential deletions in static and dynamic food webs. *Basic and Applied Ecology*, **12**, 571–580.

Cury, P., Boyd, I., Bonhommeau, S., *et al.* (2011). Global seabird response to forage fish depletion: one-third for the birds. *Science*, **334**, 1703–1706.

de Roos, A. and Persson, L. (2013). *Population and Community Ecology of Ontogenetic Development*. Princeton, NJ: Princeton University Press.

Di Marco, M., Boitani, L., Mallon, D., *et al.* (2014). A retrospective evaluation of the global decline of carnivores and ungulates. *Conservation Biology*, DOI: 10.1111/cobi.12249.

Dirzo, R., Young, H., Galetti, M., *et al.* (2014). Defaunation in the Anthropocene. *Science*, **345**, 401–406.

Dunne, J., Williams, R., and Martinez, N. (2002). Network structure and biodiversity loss in food webs: robustness increases with connectance. *Ecology Letters*, **5**, 558–567.

Ebenman, B. (1987). Niche differences between age classes and intraspecific competition in age-structured populations. *Journal of Theoretical Biology*, **124**, 25–33.

Ebenman, B. (1992). Evolution in organisms that change their niches during the life cycle. *American Naturalist*, **139**, 990–1021.

Ebenman, B. and Jonsson, T. (2005). Using community viability analysis to identify fragile systems and keystone species. *Trends in Ecology and Evolution*, **20**, 568–575.

Ebenman, B. and Persson, L. (1988). *Size-Structured Populations: Ecology and Evolution*. Berlin Heidelberg: Springer Verlag.

Ebenman, B., Law, R., and Borrvall, C. (2004). Community viability analysis: the response of ecological communities to species loss. *Ecology*, **85**, 2591–2600.

Eklöf, A. and Ebenman, B. (2006). Species loss and secondary extinctions in simple and complex model communities. *Journal of Animal Ecology*, **75**, 239–246.

Estes, J. A. and Palmisano, J. F. (1974) Sea otters: their role in structuring nearshore communities. *Science*, **185**, 1058–1060.

Estes, J., Duggins, D., and Rathbun, G. (1989). The ecology of extinctions in kelp forest communities. *Conservation Biology*, **3**, 252–264.

Estes, J., Tinker, T., and Bodkin, J. (2010). Using ecological function to develop recovery criteria for depleted species: sea otters and kelp forests in the Aleutian archipelago. *Conservation Biology*, **24**, 852–860.

Estes, J. A., Terborgh, J., Brashares, J. S., *et al.* (2011). Trophic downgrading of planet Earth. *Science*, **333**, 301–306.

Fowler, M. (2010). Extinction cascades and the distribution of species interactions. *Oikos*, **119**, 864–873.

Frank, K., Petrie, B., Choi, J., and Leggett, W. (2005). Trophic cascades in a formerly cod-dominated ecosystem. *Science*, **308**, 1621–1623.

Froyd, C., Coffey, E., van der Knapp, W., *et al.* (2014). The ecological consequences of megafaunal loss: giant tortoises and wetland biodiversity. *Ecology Letters*, **17**, 144–154.

Galetti, M., Guevara, R., Côrtes, M., *et al.* (2013). Functional extinction of birds drives rapid evolutionary changes in seed size. *Science*, **340**, 1086–1090.

Gaston, K., Blackburn, T., and Goldewijk, K. (2003). Habitat conversion and global avian biodiversity loss. *Proceedings of the Royal Society B: Biological Sciences*, **270**, 1293–1300.

Gilbert, B., Tunney, T., McCann, K., *et al.* (2014). A bioenergetics framework for the temperature dependence of trophic interactions. *Ecology Letters*, **17**, 902–914.

Gross, T., Rudolf, L., Levin, S., and Dieckmann, U. (2009). Generalized models reveal stabilizing factors in food webs. *Science*, **325**, 747–750.

IUCN (2014). IUCN Red List of Threatened Species. Version 2014.1.

Ives, A., Dennis, B., Cottingham, K., and Carpenter, S. (2003). Estimating community stability and ecological interactions from time-series data. *Ecological Monographs*, **73**, 301–330.

Jackson, J., Kirby, M., Berger, H., *et al.* (2001). Historical overfishing and the recent collapse of coastal ecosystems. *Science*, **293**, 629–638.

Joppa, L., Roberts, D., and Pimm, S. (2011). How many species of flowering plants are there? *Proceedings of the Royal Society B: Biological Sciences*, **278**, 554–559.

Kaneryd, L., Borrvall, C., Berg, S., *et al.* (2012). Species-rich ecosystems are vulnerable to cascading extinctions in an increasingly variable world. *Ecology and Evolution*, **2**, 858–874.

Lande, R. (1993). Risk of population extinction from demographic and environmental stochasticity and random catastrophes. *American Naturalist*, **142**, 911–927.

Lewis, H. and Law, R. (2007). Effects of dynamics on ecological networks. *Journal of Theoretical Biology*, **247**, 64–76.

Loh, J., Green, R., Rickets, T., *et al.* (2005). The Living Planet Index: using species population time series to track trends in biodiversity. *Philosophical Transaction of the Royal Society B: Biological Sciences*, **360**, 289–295.

Lundberg, P., Ranta, E., and Kaitala, V. (2000). Species loss leads to community closure. *Ecology Letters*, **3**, 465–468.

May, R. (1972). Will a large complex system be stable? *Nature*, **238**, 413–414.

May, R. (1977). Thresholds and breakpoints in ecosystems with a multiplicity of stable states. *Nature*, **269**, 471–477.

McCann, K. (2000). The diversity–stability debate. *Nature*, **405**, 228–233.

McCann, K., Rasmussen, J., Umbanhowar, J., and Chase, J. (2005). The dynamics of spatially coupled food webs. *Ecology Letters*, **8**, 513–523.

McConkey, K. and Drake, D. (2006). Flying foxes cease to function as seed dispersers long before they become rare. *Ecology*, **87**, 271–276.

McConkey, K. and O'Farrill, G. (2015). Cryptic function loss in animal populations. *Trends in Ecology and Evolution*, **30**, 182–189.

Miller, T. and Rudolf, V. (2011). Thinking inside the box: community-level consequences of stage-structured populations. *Trends in Ecology and Evolution*, **26**, 457–466.

Mougi, A. and Kondoh, M. (2013). Diversity of interaction types and ecological community stability. *Science*, **337**, 349–351.

Myers, R. and Worm, B. (2003). Rapid worldwide depletion of predatory fish communities. *Nature*, **423**, 280–283.

Neutel, A., Heesterbeek, J., and de Ruiter, P. (2002). Stability in real food webs: weak links in long loops. *Science*, **296**, 1120–1123.

Novaro, A., Funes, M., and Walker, S. (2000). Ecological extinction of native prey of a carnivore assemblage in Argentine Patagonia. *Biological Conservation*, **92**, 25–33.

Pereira, H., Leadley, P., Proenca, V., *et al.* (2010). Scenarios for global biodiversity in the 21st century. *Science*, **330**, 1496–1501.

Petchey, O., Eklöf, A., Borrvall, C., and Ebenman, B. (2008). Trophically unique species are vulnerable to cascading extinction. *American Naturalist*, **171**, 568–579.

Pimm, S. (1980). Food web design and the effects of species deletion. *Oikos*, **35**, 139–149.

Pimm, S., Jenkins, C., Abell, R., *et al.* (2014). The biodiversity of species and their rates of extinction, distribution, and protection. *Science*, **344**, 1246752.

Redford, K. and Feinsinger, P. (2001). The half-empty forest: sustainable use and the ecology of interactions. In *Conservation of Exploited Species*, ed. D. Reynolds, G. Mace, K. Redfor, and J. Robinson. Cambridge, UK: Cambridge University Press. pp. 370–399.

Ripple, W., Estes, J., Bechta, R., *et al.* (2014). Status and ecological effects of the world's largest carnivores. *Science*, **342**, 1241484.

Rudolf, V. and Lafferty, K. (2011). Stage structure alters how complexity affects stability of ecological networks. *Ecology Letters*, **14**, 75–79.

Sabo, J. (2008). Population viability and species interactions: life outside the single-species vacuum. *Biological Conservation*, **14**, 276–286.

Sanders, D., Sutter, L., and van Veen, F. (2013). The loss of indirect interactions leads to cascading extinctions of carnivores. *Ecology Letters*, **16**, 664–669.

Sanders, D., Kehoe, R., and van Veen, F. (2015). Experimental evidence for the population-dynamic mechanisms underlying extinction cascades of carnivores. *Current Biology*, **25**, 1–4.

Säterberg, T., Sellman, S., and Ebenman, B. (2013). High frequency of functional extinctions in ecological networks. *Nature*, **499**, 468–470.

Scheffer, M., Carpenter, S., Foley, J., Folke, C., and Walker, B. (2001). Catastrophic shifts in ecosystems. *Nature*, **413**, 591–596.

Schipper, J., Chanson, J., Chiozza, T., *et al.* (2008). The status of the world's land and marine mammals: diversity, threat, and knowledge. *Science*, **322**, 225–230.

Sekercioglu, C., Daily, G., and Ehrlich, P. (2004). Ecosystem consequences of bird declines. *Proceedings of the National Academy of Sciences of the United States of America*, **101**, 18042–18047.

Sellman, S., Säterberg, T., and Ebenman, B. (2015) Pattern of functional extinctions in ecological networks with a variety of interaction types. *Theoretical Ecology*, DOI 10.1007/s12080-015-0275-7.

Shaffer, M. L. (1981). Minimum population sizes for species conservation. *BioScience*, **31**, 131–134.

Smith, A., Brown, C., Bulman, C., *et al.* (2011). Impact of fishing low-trophic level species on marine ecosystems. *Science*, **333**, 1147–1150.

Solé, R. and Montoya, J. (2001). Complexity and fragility in ecological networks *Proceedings of the Royal Society B: Biological Sciences*, **268**, 2039–2045.

Soulé, M., Estes, J., Berger, J., and Martinez Del Rio, C. (2003). Ecological effectiveness: conservation goals for interactive species. *Conservation Biology*, **17**, 1238–1250.

Soulé, M., Estes, J., Miller, B., and Honnold, D. (2005). Strongly interacting species: conservation policy, management, and ethics. *BioScience*, **55**, 168–176.

Stouffer, D. and Bascompte, J. (2011). Compartmentalization increases food-web persistence. *Proceedings of the National Academy of Sciences of the United States of America*, **108**, 3648–3652.

Thebault, E. and Fontaine, C. (2010). Stability of ecological communities and the architecture of mutualistic and trophic networks. *Science*, **329**, 853–856.

Thebault, E., Huber, V., and Loreau, M. (2007). Cascading extinctions and ecosystem functioning: contrasting effects of diversity depending on food web structure. *Oikos*, **116**, 163–173.

Werner, E. and Gilliam, J. (1984). The ontogenetic niche and species interactions in size-structured populations. *Annual Review of Ecology and Systematics*, **15**, 393–425.

Wilmers, C., Post, E., Peterson, R., and Vucetich, J. (2006). Predator disease out-break modulates top–down, bottom–up and climatic effects on herbivore population dynamics. *Ecology Letters*, **9**, 383–389.

Wollrab, S., Diehl, S., and de Roos, A. (2012). Simple rules describe bottom–up and top–down control in food webs with alternative energy pathways. *Ecology Letters*, **15**, 935–946.

5   Merging Antagonistic and Mutualistic Bipartite Webs: A First Step to Integrate Interaction Diversity into Network Approaches

Elisa Thébault, Alix M. C. Sauve, and Colin Fontaine

5.1   Introduction

The study of ecological networks with different types of interactions is still in its infancy. Most descriptors of network structure have been developed for networks with a single type of interaction, and mainly for unipartite networks such as food webs (Bersier *et al.*, 2002). Similarly, our understanding of species coexistence and community stability derives from theoretical studies considering one interaction type, mostly competition or predation (Leibold, 1996; McCann *et al.*, 1998).

The study of networks with multiple interaction types faces many challenges: we need to define a typology of non-trophic interaction types (Olff *et al.*, 2009; Kéfi *et al.*, 2012), collect new empirical data sampling simultaneously various interaction types (Pocock *et al.*, 2012), develop network descriptors accounting for the diversity of interaction types (Fontaine *et al.*, 2011; Sauve *et al.*, 2016a), and build new theories to understand its implications on community and ecosystem functioning (Kéfi *et al.*, 2012; Mougi and Kondoh, 2012; Sanders *et al.*, 2014). The purpose of this chapter is to present one possible framework to study the structure and dynamics of networks with a diversity of interaction types, exploring links between network structure and community stability.

5.2   Interlinked Bipartite Webs: A Framework for Studying Networks with More Than One Interaction Type

During the last decade, the study of bipartite networks – webs describing interactions between two given guilds in an ecosystem – has emerged as an important field of research in network ecology. Bipartite networks have been described for many different types of interactions: pollination (Memmott *et al.*, 2004), seed dispersion (Jordano *et al.*, 2003), herbivory (Cagnolo *et al.*, 2011), predation (van Veen *et al.*, 2008), parasitism (Vazquez *et al.*, 2005), symbiosis (Montesinos-Navarro *et al.*, 2012), and facilitation (Verdú and Valiente-Banuet, 2008). There is now an increasing knowledge of the structure and dynamics of these networks. In particular, we know that their structure and stability depend on the interaction type considered (Thébault and Fontaine, 2010; Fontaine *et al.*, 2011). On one hand, nestedness (network structure where specialist species interact with subsets of the species that interact with more generalist ones, Figure 5.1a) has often been described in mutualistic networks (Bascompte *et al.*, 2003). On the other hand, antagonistic networks often appear more modular (i.e., they exhibit groups of species that interact more within groups than among groups, Figure 5.1b) than mutualistic webs (Fontaine *et al.*, 2011; Krasnov *et al.*, 2012).

Although bipartite networks mainly focus on one type of interaction at a time, the same guilds can be involved in various networks, each one displaying a different interaction type: for instance, plants are present in pollination, seed-dispersion, and herbivory webs; insect herbivores are involved in herbivory and host–parasitoid webs, etc. Networks sharing guilds are thus easy to assemble into super-networks with multiple types of interactions. The structure of these interlinked networks can then be described considering the structure of each sub-network as well as the way these sub-networks are combined (Figure 5.1). The latter involves focusing on the species that link the two sub-networks (termed hereafter "linking species") to describe how these species connect the two sub-networks ("interlinked patterns," Figure 5.1d–g). Such an approach allows addressing new questions about how the diversity of inter-action types shapes ecological and evolutionary processes in ecological communities (Fontaine *et al.*, 2011).

A few recent papers have started to study the stability of networks combining different types of interactions (Melián *et al.*, 2009; Mougi and Kondoh, 2012; Pocock *et al.*, 2012). However, this research is still in its infancy because most aspects of the topological properties remain unknown for networks with multiple interaction types. Below we focus on interlinked networks made of one mutualistic and one antagonistic sub-network connected by a common guild of species and investigate the relationship between their structure and stability (Figure 5.1c). We ask the follow-ing questions: (1) What are the consequences of sub-network structure on stability? Do we get the same results in interlinked networks and in isolated networks? (2) What is the importance of interlinked patterns on stability?

5.3 Consequences of the Structure of Antagonistic and Mutualistic Sub-Networks on Interlinked Network Stability

We first assess whether the structure of the antagonistic and the mutualistic parts of super-networks have the same implication for community stability as predicted when antagonistic and mutualistic networks are studied in isolation. To do so, we build a model describing the dynamics of species densities over time according to the guild (mutualists, antagonists, or linking species) they belong to (Figure 5.1c). The general model is defined by the following set of equations (Sauve *et al.*, 2014):

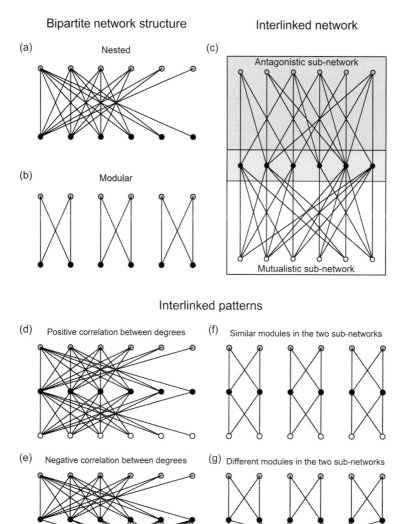

Figure 5.1 The structure of bipartite and interlinked networks. (a–b) Schematic representation of nested (a) and modular (b) bipartite networks. (c) Structure of an interlinked network composed of antagonistic and mutualistic bipartite webs. The species that link these two webs are represented in black, hereafter the "linking species." The structure of an interlinked network can be described by the structure of its bipartite sub-networks but it depends also on the way the two sub-networks are linked together (interlinked patterns, e.g., d–g). For example, linking species might be generalist in both sub-networks (positive correlation between their degrees) (d), or generalist in one sub-network and specialist in the other (negative correlation between degrees) (e). When the two sub-networks are modular, modules in one sub-network might overlap (g), or not (f), modules in the other sub-network.

$$\frac{dM_i}{dt} = r_{Mi}M_i - I_{Mi}M_i^2 + \sum_{j=1}^{S_L} \frac{c_{ji}^{(mut)}L_jM_i}{\alpha_{ji}^{(mut)-1} + \sum_{k,c_{ki}^{(mut)}>0}L_k} \tag{5.1}$$

$$\frac{dA_j}{dt} = r_{Aj}A_j - I_{Aj}A_j^2 + \sum_{i=1}^{S_L} \frac{c_{ij}^{(ant)}L_iA_j}{\alpha_{ij}^{(ant)-1} + \sum_{k,c_{kj}^{(ant)}>0}L_k} \tag{5.2}$$

$$\frac{dL_m}{dt} = r_{Lm}L_m - I_{Lm}L_m^2 + \sum_{j=1}^{S_M} \frac{c_{mj}^{(mut)}M_jL_m}{\alpha_{mj}^{(mut)-1} + \sum_{k,c_{mk}^{(mut)}>0}M_k}$$

$$-\sum_{j=1}^{S_A} \frac{c_{mj}^{(ant)}A_jL_m}{\alpha_{mj}^{(ant)-1} + \sum_{k,c_{kj}^{(ant)}>0}L_k} \tag{5.3}$$

M_i, A_j, and L_m are the densities of mutalist species i, antagonist species j, and linking species m; r_{Mi}, r_{Aj}, and r_{Lm} are their respective intrinsic growth rates; and I_{Mi}, I_{Aj}, and I_{Lm} their density dependent self-limitation. $c_{ij}^{(mut,ant)}$ is the maximum rate of interaction between mutalist or antagonist species j and linking species i, and $\alpha_{ij}^{(mut,ant)}$ is the half-saturation constant of the functional response for the corresponding interaction. This model is a development of the model proposed by Thébault and Fontaine (2010), which simulated the dynamics of separated antagonistic and mutualistic networks of varying structures and then compared structure–stability relationships between bipartite antagonistic and mutualistic networks. Following Sauve *et al.* (2014), we analyze the impacts of the respective structures of antagonistic and mutualistic sub-networks on the stability of the total community. We focus here on the effects of three aspects of the structure of antagonistic and mutualistic sub-networks – connectance, modularity, and nestedness – on community persistence, defined here as the proportion of persisting species once an equilibrium is reached. We compare results from the dynamics of antagonistic and mutualistic networks in isolation to the one of interlinked networks coupling antagonistic and mutualistic interactions.

When antagonistic and mutualistic networks are studied in isolation, connectance has a positive effect on species persistence in mutualistic networks whereas it has a negative effect in antagonistic networks (Figure 5.2; Thébault and Fontaine, 2010).

In interlinked networks, the connectance of the mutualistic sub-network has a positive effect on species persistence whereas the connectance of the antagonistic sub-network has a negative effect on persistence (Figure 5.2; Sauve *et al.*, 2014). The consequences of mutualistic and antagonistic connectance thus remain qualitatively similar whether networks exhibit different types of interactions or not. However, we note that the overall network connectance – ignoring interaction types – is not related with species persistence (Figure 5.3).

When antagonistic and mutualistic networks are studied in isolation, Thébault and Fontaine (2010) showed that the consequences of connectance on persistence are mainly mediated by changes of modularity and nestedness. Modularity has a negative effect on

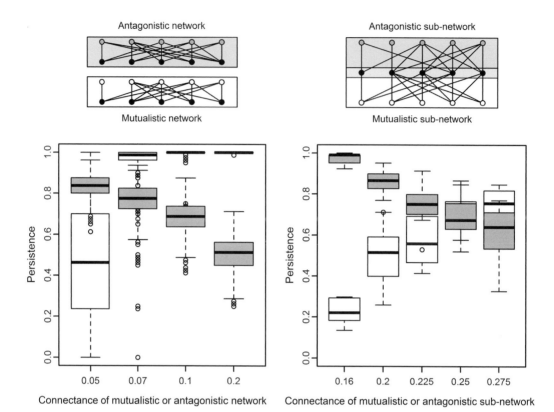

Figure 5.2 Species persistence as a function of connectance when mutualistic and antagonistic networks are studied separately (left panels) or linked together (right panels). Effects of mutualistic connectance are represented in white, while effects of antagonistic connectance are in gray. For the left panels, model parameter values are defined as in Thébault and Fontaine (2010) with networks of 80 species. For the right panels, model parameter values are set as in Sauve *et al.* (2014) with networks of 104 species and connectance of mutualistic (resp. antagonistic) sub-network is set constant to 0.25 (resp. 0.25) when connectance of antagonistic (resp. mutualistic) sub-network varies.

persistence of mutualistic networks while nestedness decreases the persistence of antagonistic networks (Figure 5.4; Thébault and Fontaine, 2010).

The lower persistence of highly modular mutualistic networks reflects the extinction of entire modules: the four levels of persistence at high modularity observed in Figure 5.4a correspond to persistence of either one, two, three, or four out of the four modules of the simulated networks. In interlinked networks, mutualistic modularity and antagonistic nestedness both decrease species persistence but their effects are weaker than in isolated networks (Figure 5.4; Sauve *et al.*, 2014).

Overall, our results show that the consequences of the structure of antagonistic and mutualistic sub-networks are qualitatively the same whether we consider networks with one single type of interaction or both types of interactions. However, the effects of mutualistic modularity and antagonistic nestedness are strongly weakened in interlinked

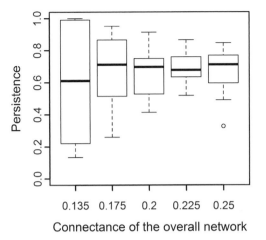

Figure 5.3 Species persistence as a function of global connectance (i.e., ignoring interaction types) of the interlinked networks. Model parameter values are set as in Sauve *et al.* (2014) with networks of 104 species.

networks. The loss of effects of sub-network modularity could be explained by the fact that the modules of one sub-network are now linked via the interactions in the other sub-network. The negative effects of nestedness in antagonistic networks arise from strong negative indirect interactions among species due to interaction overlap. In interlinked networks, these negative indirect interactions might be partly canceled out by positive indirect interactions generated through the mutualistic sub-network (Sauve *et al.*, 2016b). These last results call for the development of new network metrics that consider the diversity of interaction types. More specifically, as the effects of modularity and nestedness of each sub-network might depend on the way the networks are combined, we need new metrics to describe and quantify the way the sub-networks are linked.

5.4   Consequences of Interlinked Patterns Between Antagonistic and Mutualistic Sub-Networks on Stability

The linking species connect both sub-networks and should thus channel the perturbations from one sub-network to the other. To study the structure and dynamics of interlinked networks, we can focus on these linking species and on the way they connect the two sub-networks, hereafter referred as "the interlinked pattern." Two interlinked networks can indeed have identical sub-network structures but still differ in their interlinked patterns: for example, the linking species can have a similar, opposite or unrelated degree, in the two sub-networks. They can also be part of same or different modules in each of the two sub-networks (Figure 5.1).

Focusing on the impact of the correlation between the sub-network degrees of the linking species (Figure 5.1d, e) and the similarity between their module affiliations in the two sub-networks (Figure 5.1f, g), we extended the analysis of Sauve *et al.* (2014) to

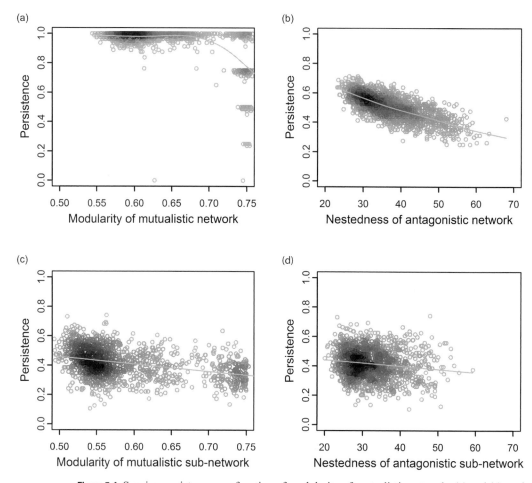

Figure 5.4 Species persistence as a function of modularity of mutualistic networks (a) and (c), and of nestedness of antagonistic networks (b) and (d), when both network types are studied in isolation (top panels) or when the two networks are linked together (bottom panels). Same model parameter values as in Figure 5.2, with connectance = 0.07 in (a), connectance = 0.2 in (b), and connectance of mutualistic sub-network = 0.1 and of antagonistic sub-network = 0.2 in (c) and (d).

study the impacts of these interlinked patterns on the stability of tripartite networks that combine an antagonistic and a mutualistic bipartite network (Figure 5.1c). Using the same model as described in the previous section, we simulated the dynamics of inter-linked networks for varying interlinked patterns while keeping the structure of the sub-networks constant. The correlation of degrees is measured as the Spearman correlation coefficient between the mutualistic and the antagonistic degrees of the linking species. When positive, linking species that are generalists in one sub-network tend to be generalists in the other sub-network. When negative, linking species that are generalists in one sub-network tend to be specialists in the other sub-network. The similarity between module affiliations is measured by the normalized mutual information between

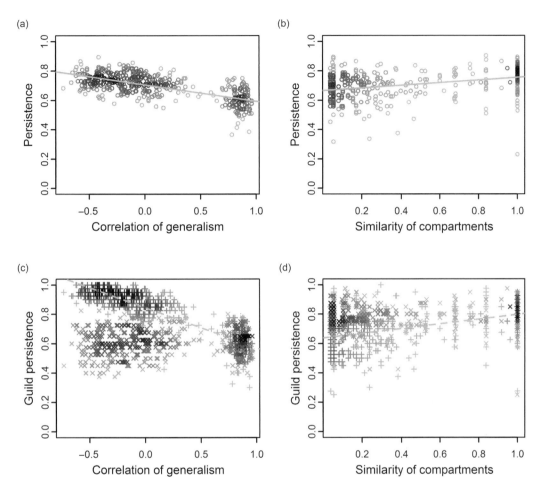

Figure 5.5 Species persistence as a function of the correlation between mutualistic and antagonistic degrees of linking species (a) and (c) and of the similarity of modules between both sub-networks (b) and (d). Overall species persistence is represented by circles (a) and (b) while mutualist and antagonist persistence are represented respectively by crosses and exes. Same model parameter values as in Figure 5.4c, d, except for connectance = 0.15 in both sub-networks.

the partitions of linking species in the antagonistic and mutualistic sub-networks (Thébault, 2013). This metric ranges from 0 to 1 and it is at its highest if the linking species that belong to the same module in one sub-network also belong to the same module in the other sub-network. As done in the previous section, we analyze here the impact of these structural patterns on species persistence.

Our results show that a positive correlation between the degrees of the linking species decreases species persistence whereas a negative correlation, meaning that linking species that are generalists in one sub-network are specialists in the other, increases persistence (Figure 5.5a).

On the other hand, a high similarity between module composition of linking species in the two sub-webs slightly favors species persistence (Figure 5.5b). It is interesting to note that persistence is decreased in the two configurations that would favor perturbation spread: when there is a positive correlation between the degrees of linking species, and when module affiliations of linking species in the two sub-networks do not fully overlap. In the first case, most species from the two sub-networks are linked together by the same generalist species. In the second case, perturbations can spread from one module to another as the modules of one sub-network are linked together by those of the other sub-network. One can notice that it is the persistence of the mutualist guild that is most affected by interlinked patterns here (Figure 5.5c, d). This suggests a propagation of the negative effects of the antagonist guild in the mutualistic sub-network. These results are coherent with those of a previous study on a similar type of interlinked network: Melián *et al.* (2009) found that the presence of linking species with high ratios of mutualistic to antagonistic interactions (i.e., species that would support a negative correlation between degrees) favored species persistence in the network. Nonetheless, our study requires further development to assess whether these results can be generalized to a broader range of conditions (for example for different strengths of mutualistic and antagonistic interactions).

5.5 Discussion and Conclusion

The chapter proposes a way to describe the structural features of networks with multiple interaction types, and then evaluates the effect of these structures on the dynamics of networks combining antagonistic and mutualistic bipartite webs. Results show that: (1) Structures focusing on one type of interaction at a time (i.e., sub-network structures) explain species persistence better than global network structures, which ignore interaction types. (2) The consequences of the structure of antagonistic and mutualistic sub-networks are qualitatively similar to those obtained for networks with only antagonistic or mutualistic interactions, but the effects of sub-network modularity and nestedness are weakened in interlinked networks. (3) The way the two sub-networks are linked (i.e., interlinked patterns) has consequences on species persistence.

We focused here on particular interlinked networks where the linking species are negatively affected by antagonists and positively affected by mutualists. Such networks could correspond, for example, to plants interacting with both herbivores and pollinators (or seed dispersers) as described by Pocock *et al.* (2012) and Melián *et al.* (2009). Our analyses will need to be performed on other networks combining different sub-networks as structure–stability relationships will likely depend on the type of interlinked networks considered. For instance, one could consider interlinked networks of mutualistic and antagonistic sub-webs where the linking species are both mutualists and predators or herbivores (such as interactions between plants and Lepidoptera; Altermatt and Pearse, 2011), or interlinked networks combining either two types of mutualistic or antagonistic interactions (e.g., plants with pollinators and seed dispersers: Albrecht *et al.*, 2007; networks of plants, aphids, and their parasitoids: Elias *et al.*, 2013).

In this chapter, we have limited the study of interlinked networks to cases where sub-networks of each type of interaction are described by a bipartite web. However, for the interlinked patterns analyzed here, the correlation in degree and the similarity in module affiliations could be applied to unipartite networks, such as food webs described at ecosystem level (the type of network that is most studied in ecology). Indeed, the relation between trophic degree and parasitic degree has started to be investigated in food webs including parasites (Chen *et al.*, 2008; Amundsen *et al.*, 2009). The proposed framework is also well suited to link different ecosystem services, such as pollination and biological pest control, or pollination and seed dispersal (Albrecht *et al.*, 2007), because the focus of studies in agro-ecosystems is often on one particular guild (generally plants) that is at the core of these issues. Future work on interlinked networks should thus give interesting insights on applied issues, such as ecosystem management and ecological restoration (Pocock *et al.*, 2012; Evans *et al.*, 2013).

References

Albrecht, M., Duelli, P., Schmid, B., and Muller, C. B. (2007). Interaction diversity within quantified insect food webs in restored and adjacent intensively managed meadows. *Journal of Animal Ecology*, **76**, 1015–1025.

Altermatt, F. and Pearse, I. S. (2011). Similarity and specialization of the larval versus adult diet of European butterflies and moths. *American Naturalist*, **178**, 372–382.

Amundsen, P.-A., Lafferty, K. D., Knudsen, R. *et al.* (2009). Food web topology and parasites in the pelagic zone of a subarctic lake. *Journal of Animal Ecology*, **78**, 563–572.

Bascompte, J., Jordano, P., Melián, C. J., and Olesen, J. M. (2003). The nested assembly of plant–animal mutualistic networks. *Proceedings of the National Academy of Sciences of the United States of America*, **100**, 9383–9387.

Bersier, L., Banašek-Richter, C., and Cattin, M. (2002). Quantitative descriptors of food-web matrices. *Ecology*, **83**, 2394–2407.

Cagnolo, L., Salvo, A., and Valladares, G. (2011). Network topology: patterns and mechanisms in plant–herbivore and host–parasitoid food webs. *Journal of Animal Ecology*, **80**, 342–351.

Chen, H.-W., Liu, W.-C., Davis, A. J., *et al.* (2008). Network position of hosts in food webs and their parasite diversity. *Oikos*, **117**, 1847–1855.

Elias, M., Fontaine, C., and van Veen, F. J. F. (2013). Evolutionary history and ecological processes shape a local multilevel antagonistic network. *Current Biology*, **23**, 1355–1359.

Evans, D. M., Pocock, M. J. O., and Memmott, J. (2013). The robustness of a network of ecological networks to habitat loss. *Ecology Letters*, **16**, 844–852.

Fontaine, C., Guimarães, P. R., Kéfi, S., *et al.* (2011). The ecological and evolutionary implications of merging different types of networks. *Ecology Letters*, **14**, 1170–1181.

Jordano, P., Bascompte, J., and Olesen, J. M. (2003). Invariant properties in coevolutionary networks of plant-animal interactions. *Ecology Letters*, **6**, 69–81.

Kéfi, S., Berlow, E. L., Wieters, E., *et al.* (2012). More than a meal… integrating non-feeding interactions into food webs. *Ecology Letters*, **15**, 291–300.

Krasnov, B. R., Fortuna, M. A., Mouillot, D., *et al.* (2012). Phylogenetic signal in module composition and species connectivity in compartmentalized host–parasite networks. *American Naturalist*, **179**, 501–511.

Leibold, M. (1996). A graphical model of keystone predators in food webs: trophic regulation of abundance, incidence, and diversity patterns in communities. *American Naturalist*, **147**, 784–812.

McCann, K., Hastings, A., and Huxel, G. (1998). Weak trophic interactions and the balance of nature. *Nature*, **395**, 794–798.

Melián, C. J., Bascompte, J., Jordano, P., and Krivan, V. (2009). Diversity in a complex ecological network with two interaction types. *Oikos*, **118**, 122–130.

Memmott, J., Waser, N. M., and Price, M. V. (2004). Tolerance of pollination networks to species extinctions. *Proceedings of the Royal Society B: Biological Sciences*, **271**, 2605–2611.

Montesinos-Navarro, A., Segarra-Moragues, J. G., Valiente-Banuet, A., and Verdú, M. (2012). The network structure of plant–arbuscular mycorrhizal fungi. *New Phytologist*, **194**, 536–547.

Mougi, A. and Kondoh, M. (2012). Diversity of interaction types and ecological community stability. *Science*, **337**, 349–351.

Olff, H., Alonso, D., Berg, M. P., *et al.* (2009). Parallel ecological networks in ecosystems. *Philosophical Transactions of the Royal Society B: Biological Sciences*, **364**, 1755–1779.

Pocock, M. J. O., Evans, D. M., and Memmott, J. (2012). The robustness and restoration of a network of ecological networks. *Science*, **335**, 973–977.

Sanders, D., Jones, C. G., Thébault, E., *et al.* (2014). Integrating ecosystem engineering and food webs. *Oikos*, **123**, 513–524.

Sauve, A. M. C., Fontaine, C., and Thébault, E. (2014). Structure–stability relationships in networks combining mutualistic and antagonistic interactions. *Oikos*, **123**, 378–384.

Sauve, A. M. C., Thébault, E., Pocock, M. J. O., and Fontaine, C. (2016a). How plants connect pollination and herbivory networks and their contribution to community stability. *Ecology*, **97**, 908–918.

Sauve, A. M. C., Fontaine, C., and Thébault, E. (2016b). Stability of a diamond-shaped module with multiple interaction types. *Theoretical Ecology*, **9**, 27–37.

Thébault, E. (2013). Identifying compartments in presence-absence matrices and bipartite networks: insights into modularity measures. *Journal of Biogeography*, **40**, 759–768.

Thébault, E. and Fontaine, C. (2010). Stability of ecological communities and the architecture of mutualistic and trophic networks. *Science*, **329**, 853–856.

Van Veen, F. J. F., Müller, C. B., Pell, J. K., and Godfray, H. C. J. (2008). Food web structure of three guilds of natural enemies: predators, parasitoids and pathogens of aphids. *Journal of Animal Ecology*, **77**, 191–200.

Vazquez, D. P., Poulin, R., Krasnov, B. R., and Shenbrot, G. I. (2005). Species abundance and the distribution of specialization in host–parasite interaction networks. *Journal of Animal Ecology*, **74**, 946–955.

Verdú, M. and Valiente-Banuet, A. (2008). The nested assembly of plant facilitation networks prevents species extinctions. *American Naturalist*, **172**, 751–760.

6 Toward Multiplex Ecological Networks: Accounting for Multiple Interaction Types to Understand Community Structure and Dynamics

Sonia Kéfi, Elisa Thébault, Anna Eklöf, Miguel Lurgi, Andrew J. Davis, Michio Kondoh, and Jennifer Adams Krumins

6.1 Introduction

In drylands, there is often not enough water for the whole land to be covered by vegetation. Instead, vegetation occurs in patches (Figure 6.1a), where established plants provide a favorable environment for the recruitment of new individuals, for example, by creating shading and locally favoring soil and resource retention (Aguiar and Sala, 1999). This facilitation mechanism is well documented in drylands and known to be of great importance for dryland plant communities (Soliveres *et al.*, 2015). At the same time, plant species compete with each other for water, the main limiting resource in those ecosystems, and are consumed by herbivores. This is only one example of the variety of ecological interactions that co-occur in ecological communities (Figure 6.1b, c). As early as 1859, Charles Darwin highlighted the diverse interaction types that link species in nature. One of his famous examples on how cats influence seed set in red clover (Darwin, 1859) illustrates how indirect effects can percolate through a variety of ecological interactions that co-occur in ecological communities. Cats eating mice is a trophic interaction, mice building nests that are later used by bumble bees is an ecological engineering interaction, and these bees in turn increasing seed set in clover is a mutualistic pollination interaction.

Despite the recognized importance of this variety of interaction types in nature, ecological research has largely focused on analyzing one single type of interaction at a time, for example trophic networks or food webs (Pimm, 1982; Cohen *et al.*, 1993; de Ruiter *et al.*, 1995; Brose *et al.*, 2005, 2006; Neutel *et al.*, 2007), mutualistic communities (Jordano *et al.*, 2003; Blüthgen *et al.*, 2007), host–parasite and host–parasitoids webs (Vázquez *et al.*, 2005; Krasnov *et al.*, 2012), and facilitation networks (Verdú and Valiente-Banuet, 2008). Studies on networks of single interaction types – greatly dominated by food-web studies (e.g., Ings *et al.*, 2009) – have suggested that single-interaction ecological networks exhibit predictable structural regularities with important consequences for their dynamics (Williams and Martinez, 2000; Bascompte *et al.*, 2003; Montoya *et al.*, 2006; Verdú and Valiente-Banuet, 2008;

(a)   (b)   (c)

Figure 6.1 Diverse types of interspecific interactions in ecology. (a) Vegetation patches in Cabo de Gata Natural Park, Spain, due to facilitation between plants (photo: S. Kéfi). (b) *Eristalis* pollinating a butterfly bush (*Buddleja*) (photo: C. Fontaine). (c) Kelp serving as a habitat for crabs, Chile (photo: E. Wieters).

Thébault and Fontaine, 2010). For example, mutualistic (Bascompte *et al.*, 2003) and facilitation networks (Verdú and Valiente-Banuet, 2008) tend to be more nested than expected by chance, a feature that has been shown to give rise to positive complexity–stability relationships in dynamic models (Okuyama and Holland, 2008; Thébault and Fontaine, 2010).

Recently, the joint importance of different types of species interactions for structure, stability, and functioning in ecological communities has gained increased attention. Indeed, the stability of ecological networks may depend on the way different interaction types are combined in real communities (Berlow *et al.*, 2004; Olff *et al.*, 2009; Fontaine *et al.*, 2011; Kéfi *et al.*, 2012). This consideration of networks of multiple interaction types prompts several questions. These include: To what extent can we understand community stability by looking exclusively at the properties of a single interaction type? How may combining different interactions in ecological networks affect the stability–complexity relationship? Do networks with multiple interaction types exhibit particular structures? In this chapter, we provide an overview of the approaches that have attempted to describe and model ecological networks in which many different types of ecological interactions operate simultaneously. We will refer to these networks as "multiplex" networks (Cardillo *et al.*, 2013; Boccaletti *et al.*, 2014; Kivelä *et al.*, 2014).

6.2   The Structure of Multiplex Ecological Networks: Data and Indices Used

Pioneering studies have explored ecological networks that simultaneously include different types of trophic interactions, for example predation and parasitism (Lafferty *et al.*, 2006, 2008). These studies show, for instance, that food webs that also take parasites into account are substantially different in structure from food webs that do not. They have increased connectance, longer food chains, and the parasite–host links dominate numerically over predator–prey links (Lafferty *et al.*, 2006, 2008).

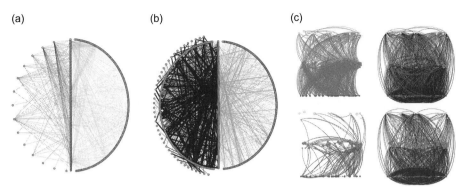

Figure 6.2 (a) Doñana Ecological Network based on data from Melián *et al.* (2009) showing herbivores (left), plants (middle), and pollinators and seed dispersers (right). Antagonistic links are in blue and mutualistic ones in orange. (b) Species interaction network at Norwood Farm, Somerset, UK based on data from Pocock *et al.* (2012). Antagonistic links are in blue and mutualistic ones in orange plants in the middle. (c) Chilean interaction web based on data from Kéfi *et al.* (2015). Trophic network in yellow (top left), positive non-trophic network in green (bottom left), negative non-trophic network in blue (bottom right), and the top-right network shows all trophic (in yellow) and all non-trophic links (positive and negative) in purple; between the 104 species of the community. (A black and white version of this figure will appear in some formats. For the color version, please refer to the plate section.)

Melián *et al.* (2009) compiled an ecological network, the Doñana Biological Reserve network, including both mutualistic (pollinator–plant and seed disperser–plant) and antagonistic (plant–herbivore) interactions. They associated two bipartite webs by letting the plants be the joining nodes between both networks (Figure 6.2a). They found a non-random distribution of mutualistic to antagonistic ratios among species, with high ratios concentrated in a few plant species. Moreover, most strong interactions were concentrated in those few plants with the highest mutualistic to antagonistic ratio. Two alternative hypotheses have been proposed to explain the observed high mutualistic to antagonistic ratios in a few plant species (Melián *et al.*, 2009). The first one, the neutral hypothesis, relies on the patterns of species abundance. Indeed, four of the plant species with the highest mutualistic to antagonistic ratios are also the most abundant species in the dataset, which could contribute to those species having more links overall. In addition, a sampling bias could be the reason why those species have proportionally more mutualistic than antagonistic links. There are indeed more mutualistic than consumer species in the network because most sampling of consumers was done on large-body consumers and not on invertebrates (C. Melián, personal communication, 2016). Under neutrality, such a sampling bias would lead to the most abundant plants having proportionally more mutualistic than antagonistic links. It is unclear, however, whether this pattern would remain if the sampling was more balanced. More data would be needed to test this neutral hypothesis, i.e., to evaluate whether abundance alone can explain the concentration of high mutualistic to antagonistic ratios in a few plant species. The second hypothesis, the evolutionary one, suggests that species with high mutualistic to antagonistic ratios have developed reward and defense systems (e.g., production of compounds to attract pollinators) that concentrate most of the strong interactions around them. Whether this is actually the case in that community and whether this would also induce the concentration of strong interactions in a few plant species requires further investigation.

Missing at this point, however, were more comprehensive ecological networks that included the multi-trophic architecture of natural food webs as well as other types of non-trophic interactions (Borer *et al.*, 2002; Berlow *et al.*, 2004; Ings *et al.*, 2009; Olff *et al.*, 2009; Fontaine *et al.*, 2011). A multiplex ecological network was published by Pocock *et al.* (2012), who assembled seven bi- or tripartite networks of an agro-ecosystem in the UK, each of the networks including plants (i.e., seven sub-networks; Figure 6.2b). In the same way as Melián *et al.* (2009), plants thereby formed the linking nodes between all these webs, which include mutualistic (flower visitors–plants) and antagonistic interactions (plants–herbivores, herbivores–parasitoids, plants–seed preda-tors, and parasitic interactions). Pocock *et al.* (2012) evaluated the robustness of the animal groups by simulating the sequential removal of plant taxa. Following the removal of a plant species (primary extinction), animal taxa became extinct (secondary extinc-tion) when all their resources were extinct. Robustness was calculated for each of the sub-networks as the average area under the curve of the secondary extinctions against primary extinctions across all simulations. The robustness of some of the sub-networks covaried, but the covariance was less than expected as compared to a null model. This suggests that, for example, positive or negative effects of management on plants for one animal group would not necessarily ripple through the network of networks to affect other animal groups. Sauve *et al.* (2016) further analyzed the patterns of connection between the antagonistic (plant–herbivore, plant–seed predator) and the mutualistic (plant–flower visitor) sub-networks in this dataset. The analysis showed that, when considering the binary version of the Pocock *et al.* (2012) network, plants' generalism with regard to flower visitors was positively correlated to their generalism with regard to herbivores, a pattern that was partly related to the higher degree of abundant plant species in both sub-networks. However, when considering the quantified version of the Pocock *et al.* (2012) network (i.e., including interaction strengths), the correlation between plant generalism in the mutualistic and in the antagonistic sub-networks did not differ from random expectations (Sauve *et al.*, 2016).

More recently, a comprehensive unipartite network including all known interactions between 104 species in the rocky shore community of central Chile was published (Kéfi *et al.*, 2015; Figure 6.2c). The dataset includes all reported interactions between species pairs, but these interactions were categorized as trophic, positive non-trophic, and negative non-trophic (i.e., three sub-networks). In this dataset, non-trophic interactions are more than twice as abundant as trophic ones, with 3296 non-trophic interactions and 1458 trophic interactions. Among non-trophic interactions, negative ones are 20 times more abundant than positive ones. In this community, by far the most abundant non-trophic interactions are competition for space and positive interactions related to the provision of habitat or refuges by sessile species or primary producers. The three sub-networks were analyzed separately, and the results suggest that the structure of the non-trophic links is non-random both alone and with respect to the trophic structure, with many more non-trophic interactions at the base of the food web than expected by chance (compared against randomized networks keeping only the connectance and the number of species fixed). These basal non-trophic interactions consist primarily of negative interactions due to competition for space among sessile primary producers. Furthermore, the occurrence of different types of interactions, relative to all possible links, is predicted well by trophic structure and simple attributes of the source and target species.

Similarly, Wootton *et al.* (2015) recently published a detailed multiplex network from the middle intertidal zone in Tatoosh Island, Washington, USA. Tatoosh Island was originally described as an intertidal food web dominated by the mussel *Mytilus californianus* and was complemented with additional interaction types based on observed interactions and natural history information. The Tatoosh network contains 110 species and 1898 interactions, which are feeding (+/–), competitive (–/–), mutualist (+/+), commensal (+/0), or amensal (–/0). Sander *et al.* (2015) recently analyzed this network together with the Doñana Biological Reserve network (Melián *et al.*, 2009) and the Norwood Farm network (Pocock *et al.*, 2012). They compared how the structural characteristics of these three networks, primarily regarding the grouping of the species, changed depending on which type of interactions were included in the analysis. In the Tatoosh network, Sander *et al.* (2015) showed that the groupings obtained using all interactions or only trophic interactions were overall very similar. However, the inclusion of additional interaction types could reveal novel interacting groups of species, with possible implications for ecological functioning. For example, with the addition of non-trophic interactions, a group of algae species differentiates into two groups, suggesting that these two groups differ in how they interact non-trophically with the rest of the network.

Besides observational and experimental studies recording the type of interaction between a given set of species, a number of studies have used proxies to infer species interactions from temporal or spatial data. For instance, studies have built ecological networks on the assumption that correlated changes in abundance represent the combined effects of all interaction types. The correlations have been identified by applying logical rules to simple and partial Mantel correlations (Davis *et al.*, 2004) or other techniques such as the SparCC methodology (Jordán *et al.*, 2015). Patterns of spatial co-occurrence have also been used to infer the sign of interactions between species in plant communities (Saiz and Alados, 2011).

This brief overview shows that multiplex networks have been obtained in two different ways in the literature so far:

- by attaching single interaction type networks together (Figure 6.2a, b); this requires some of the nodes to belong to different sub-networks for the sub-networks to be linked in the multiplex network;
- by making unipartite networks including links of different nature between a given set of species (Figure 6.2c).

Independently of the approach adopted, these empirical studies of multiplex ecological networks suggest that the structure of the networks is not random and raise questions about the consequences of these observed structures for community stability and ecosystem functioning. Integrating interaction diversity into theoretical models could help investigate that question (Kéfi *et al.*, 2012).

6.3   The Dynamic Properties of Multiplex Ecological Networks: Theory and Models

There have been several attempts to incorporate more than one interaction type in a combined model network (Table 6.1). Robert May's pioneering work in the early 1970s already included several interaction types (1972). In his work (and many others after him) an ecological community was summarized by a "community matrix," containing the net effect of a species on another. This net effect can be positive, null, or negative. More recent extensions of May's work, focusing on the role of the diversity of interaction types, shed new light on the complexity–stability debate (Allesina and Tang, 2012; Mougi and Kondoh, 2012).

Table 6.1 Examples of studies of dynamic models including several interaction types in an ecological network (we discarded the studies on modules of few species). This list is not exhaustive.

Reference	Type of model	Type of multiplex network	Type of links
Stability and complexity in model ecosystems May (1972)	Community matrix based on Lotka–Volterra-like models.	One network with a continuum of interactions ranging from competition to facilitation.	Only one link between species A and B, which summarizes the net effect of A on B. Can be negative, null, or positive (continuum).
The stability of real ecosystems Yodzis (1981)	Community matrix whose food-web structure is based on real food webs. The intensity and positions of intraspecific interference for primary producers and interspecific interference among consumers that share at least a prey are randomly chosen for different fractions of those links.	Same as May (1972).	Three types of links: – trophic – intraspecific competition – interspecific competition.
Rheagogies: modeling non-trophic effects in food webs Arditi *et al.* (2005)	Ordinary differential equations (ODE), food web.	Food web = structure into which the non-trophic interactions are transplanted.	Two types of links: – trophic – non-trophic interactions as modifiers of trophic links.
Non-trophic interactions, biodiversity, and ecosystem functioning:	Same as Arditi *et al.* (2005).	Same as Arditi *et al.* (2005).	Same as Arditi *et al.* (2005).

Table 6.1 (cont.)

Reference	Type of model	Type of multiplex network	Type of links
an interaction web model Goudard and Loreau (2008)			
Diversity in a complex ecological network with two interaction types Melián *et al.* (2009)	ODE ecological network with mutualistic and antagonistic (e.g., predation) interactions.	Combination of a bipartite mutualistic network and a bipartite antagonistic network, linked by one group of species (plants interacting with both pollinators and herbivores).	Two types of interactions: – antagonistic – mutualistic.
Stability criteria for complex ecosystems Allesina and Tang (2012)	Same as May (1972) (random) but with structures reflecting predator–prey, mutualistic, and competitive networks.	Same as May (1972).	Same as May (1972) (continuum).
Diversity of interaction types and ecological community stability Mougi and Kondoh (2012)	Same as May (1972) with Holling type I and II. (Cascade and bipartite webs.)	Same as May (1972).	Same as May (1972).
Positive interactions among competitors can produce species-rich communities Gross (2008)	ODE, multiple consumer species feeding on one resource.	Facilitation affects plant mortality in a competitive plant community.	Two types of interactions: – trophic – facilitative.
More than a meal. . . integrating non-feeding interactions into food webs Kéfi *et al.* (2012)	Bioenergetic model. (Structure: niche model.)	Unipartite web including all known interactions between species of the community.	Three link types: – trophic – facilitation between plants – interference between predators of the same prey.
Color and degree of interspecific synchrony of environmental noise affect the variability of complex ecological networks Lin and Sutherland (2013)	Bioenergetic model. (Structure: niche model + environmental noise.)	Unipartite web.	Four types of interactions: – trophic – facilitation – parasitism – competition.
Structure–stability relationships in networks combining	ODE, ecological network with mutualistic and	Same as Melián *et al.* (2009).	Same as Melián *et al.* (2009).

Table 6.1 (cont.)

Reference	Type of model	Type of multiplex network	Type of links
mutualistic and antagonistic interactions Sauve *et al.* (2014)	antagonistic (e.g., predation) interactions.		
The effects of space and diversity of interaction types on the stability of complex ecological networks Lurgi *et al.* (2016)	Cellular automata-like, individual-based spatially explicit model. (Structure: niche model with a fraction of herbivore links becoming mutualistic.)	Unipartite web.	Two types of interactions: – trophic – mutualistic.

Allesina and Tang (2012) showed that matrices including a mixture of competition and mutualism were less likely to be locally stable than predator–prey matrices. Going further in the analysis of the role of mixing interaction types for local stability of ecological communities, Mougi and Kondoh (2012) showed that introducing even a small proportion of mutualistic links can completely destabilize an otherwise stable food web. Stability in this case is evaluated as the proportion of locally stable systems (estimated based on the Jacobian matrix following May) among simulation replicates (each corresponding to random sampling of parameters in a uniform distribution). As the proportion of mutualistic links further increases, stability reaches a peak (i.e., the case where the highest proportion of stable communities was achieved among the simulation replicates) at a moderate mixture of both interaction types (but see Lin and Sutherland, 2013, who did not find such a stabilizing effect of mutualism in a model including environmental stochasticity). In other words, this study suggests that there could be an "optimal" way of mixing different interaction types (antagonism, mutualism, or competition) that maximizes community stability. Moreover, in those communities with a moderate mixture of interaction types, Mougi and Kondoh (2012) observed a consistent positive complexity–stability relationship, irrespective of network structure or functional response. These results suggest that interaction diversity can be an essential component for the maintenance of biodiversity.

Some studies have focused on how the particular structure of mutualistic and antagonistic interactions affect community stability (Melián *et al.*, 2009; Sauve *et al.*, 2014). They studied the dynamics of multiplex networks merging bipartite antagonistic and mutualistic networks, showing that the way these two bipartite networks are combined affects their stability (Melián *et al.*, 2009; Chapter 5). Melián *et al.* (2009) showed that the observed combination of strong interactions in a few plant species with high mutualistic to antagonistic ratios generated significantly more species persistence than found in the randomized networks. Similarly, Sauve *et al.* (2016) suggested that the way plants connected pollination and herbivory webs in the dataset by Pocock *et al.* (2012) promoted local stability. Sauve *et al.* (2014) also showed that structures known to promote stability in networks made of a single interaction type, such as nestedness and modularity, had weaker effects in networks merging mutualistic and antagonistic interactions.

Other studies adopted a different approach by integrating non-trophic interactions as modifications of trophic interactions or "rheagogies," i.e., a change in the direct interaction between two species due to the density of a third species (Arditi *et al.*, 2005; Goudard and Loreau, 2008; Ohgushi *et al.*, 2012). This modification can be positive or negative. They studied the role of non-trophic interactions in complex food-web models, showing that interaction modifications (their connectance, magnitude, and nature) affect ecosystem functioning and the relationship between biodiversity and biomass.

More recently, Kéfi *et al.* (2012) suggested that besides the modifications of trophic interactions, there were two other ways non-trophic interactions could be included in models of ecological networks: the modification of non-trophic species traits (e.g., mortality, reproduction traits) or the modification of flows of matter and biomass across system boundaries (e.g., effects on immigration and emigration rates of some species or indirect effects through the modification of input or loss rate of an abiotic resource in the system, such as shading, which decreases evaporative water loss). Simulations of a bioenergetic food web showed that adding plant facilitation and predator interference affected species diversity and could result in higher species diversity, in agreement with the positive effect of facilitation found by Gross *et al.* (2008) in a resource-competition model (Kéfi *et al.*, 2012).

Sanders *et al.* (2014) built on the approach of Kéfi *et al.* (2012) by studying more specifically the incorporation into food webs of ecosystem engineering, i.e., the physical modification of the environment by organisms. Combining trophic and engineering interactions may result in complex feedbacks (both trophic feedbacks on engineers and engineering feedbacks on consumers). Those feedbacks depend on the direction and strength of the engineering effect, the trophic position of the engineer, and the trophic levels affected by the engineering effect. For example, engineers such as ants have been shown to have contrasting effects on a grassland food web: mound-building by ants leads to a positive bottom–up engineering effect, which results in increased primary productivity; while ant predation induces a negative top–down effect, which counteracts the positive bottom–up effect (Sanders and van Veen, 2011). Using a simple three-species food chain model, Sanders *et al.* (2014) showed that the complex feedbacks resulting from the combination of engineering and feeding interactions could affect community dynamics and responses to environmental perturbations. For example, when the engineer is the intermediate consumer and when engineering activity increases species ingestion rates, cyclic dynamics occur whose amplitude is larger than when the engineer is external to the food chain (i.e., without trophic feedback to the engineer). On the contrary, when the engineer is a primary producer, feedbacks between the engineer and the rest of the food web tend to stabilize the species dynamics (Sanders *et al.*, 2014).

A diversity of interaction types can also have important consequences for the spatial organization of ecological communities. The spatial coupling of food webs has been shown to have a stabilizing potential, and to affect the persistence and diversity of local communities (Holt, 2002; Amarasekare, 2008), with important implications for community dynamics (McCann *et al.*, 2005). The incorporation of other (in addition to trophic relationships) interaction types within a spatial ecology framework is thus important if we want to fully understand the implications of a mixture of interaction types for community organization and stability. Using an individual-based approach, Lurgi *et al.* (2016) showed that an increasing proportion of mutualistic species in a food web positively affected not only the dynamic stability of the model communities (sensu May, i.e., evaluated as the average interaction strength), but also enhanced the spatial aggregation of species populations (i.e., the degree of clustering of individuals of each species in space). Lurgi *et al.* (2016) argued that spatial aggregation was linked to reproductive stability because it increased the likelihood of finding a reproductive partner in the neighborhood, enhancing in this way the ability of certain species to persist.

These recent advances in theory and modeling reflect various ways of incorporating different interaction types into ecological networks. Some consider one link between each pair of species whose intensity reflects the net interaction between the two species (Allesina and Tang, 2012; "May-type" networks; May, 1972; Mougi and Kondoh, 2012; Sauve *et al.*, 2014). The sign of this net effect determines the nature of the interaction, which is thereby a continuum between competition and facilitation. Others include explicitly different interaction types, which are modeled in different ways – reflecting different effects from one species to another often due to an explicit underlying ecological mechanism (Arditi *et al.*, 2005; Goudard and Loreau, 2008; Kéfi *et al.*, 2012; Sanders *et al.*, 2014).

Taken together, these initial investigations of the dynamics of multiplex ecological networks highlight that taking a variety of species interactions into account can affect predictions about species diversity and ecosystem functioning as well as the complexity–stability relationship. This suggests that accounting for different interaction types in ecological networks may be essential to our understanding of complex ecological communities.

6.4   Limits of Current Approaches and Challenges Ahead for Studying Multiplex Ecological Networks

This brief overview of the state of the art of empirical and theoretical approaches to complex ecological networks incorporating a diversity of interaction types shows the variety of approaches currently used in the literature.

Empirical studies have taken advantage of data available on single-interaction type networks by trying to merge these sub-networks into multiplex webs, when species are common to several sub-networks (Melián *et al.*, 2009; Fontaine *et al.*, 2011; Pocock *et al.*, 2012). Others have tried to record all interaction types known between a given set of species (Kéfi *et al.*, 2015). The suggested non-random structure of these multiplex networks opens new avenues of research: (1) How are the sub-networks composed by different interaction types intertwined (Fontaine *et al.*, 2011)? (2) Do multiplex ecological networks show common structural properties, and what factors explain these structures? (3) What are the dynamic and functional consequences of accounting for several interaction types in ecological networks? Finally, (4) What are the implications of looking at multiplex networks in space?

Likewise, model approaches have been diverse, from approaches based on local stability analysis on community matrices looking at the net interaction effect between a pair of species (May, 1972; Allesina and Tang, 2012; Mougi and Kondoh, 2012) to more mechanistic approaches focusing on sets of non-trophic interactions (Arditi et al., 2005; Goudard and Loreau, 2008; Kéfi et al., 2012; Sanders et al., 2014; Lurgi et al., 2016). In this second family of models, trophic networks are usually considered the backbone of the ecological network on which the non-trophic links are transplanted. These models have investigated the consequences of the presence (intensity, connectance, localization...) of these non-trophic interactions for species persistence, biomass, productivity, and stability. The importance of considering these ecologically diverse communities within a spatial context has also been highlighted (Lurgi et al., 2016). This is particularly important for understanding the effects of inherently spatial perturbations (such as habitat loss and fragmentation) on these communities. However, as this area of ecological research is still in its infancy, there is no consensus yet about how different types of interactions affect the complexity–stability relationship, and further research in that direction is needed. Additional emphasis should be put on what is the role of a mixture of interaction types in different aspects of community stability (Pimm, 1984; Ives and Carpenter, 2007), and how these are related to each other (Donohue et al., 2013), since the majority of studies so far have focused on only one particular type of stability (local stability analysis). A comprehensive understanding of the role that different kinds of ecological interactions play in the stability of complex ecological communities is essential to study the effects that anthropogenic and other kinds of perturbations have on these communities.

Overall, these approaches to multiplex ecological networks remain recent and scarce. Other multiplex ecological network data would help investigate whether the structural regularities observed in the datasets currently available are general and whether they can be extrapolated to different ecosystem types. Moreover, these data could be used to impose realistic structure in network models. Measures of the strength of the interactions between species would also be extremely useful for model calibration (Wood et al., 2010). Obtaining such data is a significant challenge. Many types of interactions require specific experiments to quantify their effects. For example, estimating the costs and benefits of pollination interactions for plants is still a challenge because pollinator efficiency strongly differs between species and can involve cheating (Genini et al., 2010). In addition, the construction of fully quantified multiplex networks requires that the strengths of the different interaction types are expressed in common units (for example in terms of flow of carbon or nitrogen, or by the net effects on species fitness).

Ecological network studies can furthermore benefit from the recent advances in multilayer network theory (Boccaletti et al., 2014; Kivelä et al., 2014). A set of associated metrics and tools are progressively becoming available in the literature, and will allow ecologists to move beyond the study of monolayer networks (Pilosof et al., 2017). For example, metrics of community detection (where "communities" are defined as groups of nodes that are tightly linked to each other) have been adapted for multilayer networks (Mucha *et al.*, 2010). A first application of a clustering method to a multiplex ecological network allowed grouping species that are similar in their combined trophic, non-trophic positive, and non-trophic negative interactions, thereby revealing their multidimensional interaction niche (Kéfi *et al.*, 2016). A mathematical model suggests that the specific combination of the three interaction types observed in that ecological network seemed to improve species persistence, overall productivity, and community resilience to extinction.

One of the greatest challenges of environmental biology is to predict how human impacts propagate through the complex network of interactions among organisms in natural communities. Incorporating the diversity of species interactions observed in nature in both empirical datasets and theoretical models is an important step toward improving our understanding of the resilience of complex ecological communities to the changes to come.

References

Aguiar, M. R. and Sala, O. E. (1999). Patch structure, dynamics and implications for the functioning of arid ecosystems. *Trends in Ecology and Evolution*, **14**(7), 273–277.

Allesina, S. and Tang, S. (2012). Stability criteria for complex ecosystems. *Nature*, **483** (7388), 205–208.

Amarasekare, P. (2008). Spatial dynamics of food webs. *Annual Review of Ecology, Evolution, and Systematics*, **39**(1), 479–500.

Arditi, R., Michalski, J., and Hirzel, A. H. (2005). Rheagogies: modelling non-trophic effects in food webs. *Ecological Complexity*, **2**(3), 249–258.

Bascompte, J., Jordano, P., Melián, C. J., and Olesen, J. M. (2003). The nested assembly of plant–animal mutualistic networks. *Proceedings of the National Academy of Sciences of the United States of America*, **100**(16), 9383–9387.

Berlow, E. L., Neutel, A.-M., Cohen, J. E., *et al.* (2004). Interaction strengths in food webs: issues and opportunities. *Journal of Animal Ecology*, **73**(3), 585–598.

Blüthgen, N., Menzel, F., Hovestadt, T., Fiala, B., and Blüthgen, N. (2007). Specialization, constraints, and conflicting interests in mutualistic networks. *Current Biology*, **17**(4), 341–346.

Boccaletti, S., Bianconi, G., Criado, R., *et al.* (2014). The structure and dynamics of multilayer networks. *Physics Reports*, **544**(1), 1–122.

Borer, E. T., Anderson, K., Blanchette, C. A., *et al.* (2002). Topological approaches to food web analyses: a few modifications may improve our insights. *Oikos*, **99**(2), 397–401.

Brose, U., Cushing, L., Berlow, E. L., *et al.* (2005). Body sizes of consumers and their resources. *Ecology*, **86**(9), 2545.

Brose, U., Jonsson, T., Berlow, E. L., *et al.* (2006). Consumer–resource body-size relationships in natural food webs. *Ecology*, **87**(10), 2411–2417.

Cardillo, A., Gómez-Gardeñes, J., Zanin, M., *et al.* (2013). Emergence of network features from multiplexity. *Scientific Reports*, **3**, 1344.

Cohen, J. E., Pimm, S. L., Yodzis, P., and Saldana, J. (1993). Body sizes of animal predators and animal prey in food webs. *Journal of Animal Ecology*, **62**(1), 67–78.

Darwin, C. (1859). *The Origin of Species by Means of Natural Selection, or the Preservation of Favoured Races in the Struggle for Life*. London: John Murray, Albemarle Street.

Davis, A. J., Liu, W., Perner, J., and Voigt, W. (2004). Reliability characteristics of natural functional group interaction webs. *Evolutionary Ecology Research*, **6**(8), 1145–1166.

de Ruiter, P. C., Neutel, A.-M., and Moore, J. C. (1995). Energetics, patterns of interaction strengths, and stability in real ecosystems. *Science*, **269**(5228), 1257–1260.

Donohue, I., Petchey, O. L., Montoya, J. M., *et al.* (2013). On the dimensionality of ecological stability. *Ecology Letters*, **16**, 421–429.

Fontaine, C., Guimarães, P. R., Kéfi, S., *et al.* (2011). The ecological and evolutionary implications of merging different types of networks. *Ecology Letters*, **14**(11), 1170–1181.

Genini, J., Morellato, L. P. C., Guimarães, P. R., and Olesen, J. M. (2010). Cheaters in mutualism networks. *Biology Letters*, rsbl20091021.

Goudard, A. and Loreau, M. (2008). Nontrophic interactions, biodiversity, and ecosystem functioning: an interaction web model. *American Naturalist*, **171**(1), 91–106.

Gross, K. (2008). Positive interactions among competitors can produce species-rich communities. *Ecology Letters*, **11**(9), 929–936.

Holt, R. D. (2002). Food webs in space: on the interplay of dynamic instability and spatial processes. *Ecological Research*, **17**(2), 261–273.

Ings, T. C., Montoya, J. M., Bascompte, J., *et al.* (2009). Ecological networks: beyond food webs. *Journal of Animal Ecology*, **78**(1), 253–269.

Ives, A. R. and Carpenter, S. R. (2007). Stability and diversity of ecosystems. *Science*, **317**(5834), 58–62.

Jordán, F., Lauria, M., Scotti, M., *et al.* (2015). Diversity of key players in the microbial ecosystems of the human body. *Scientific Reports*, **5**, 15920.

Jordano, P., Bascompte, J., and Olesen, J. M. (2003). Invariant properties in coevolutionary networks of plant–animal interactions. *Ecology Letters*, **6**(1), 69–81.

Kéfi, S., Berlow, E. L., Wieters, E. A., *et al.* (2012). More than a meal… integrating non-feeding interactions into food webs. *Ecology Letters*, **15**, 291–300.

Kéfi, S., Berlow, E. L., Wieters, E. A., *et al.* (2015). Network structure beyond food webs: mapping non-trophic and trophic interactions on Chilean rocky shores. *Ecology*, **96**(1), 291–303.

Kéfi, S., Miele, V., Wieters, E. A., Navarrete, S. A., and Berlow, E. L. (2016). How structured is the entangled bank? The surprisingly simple organization of multiplex ecological networks leads to increased persistence and resilience. *PLoS Biology*, **14**(8), e1002527.

Kivelä, M., Arenas, A., Barthelemy, M., *et al.* (2014). Multilayer networks. *Journal of Complex Networks*, **2**(3), 203–271.

Krasnov, B. R., Fortuna, M. A., Mouillot, D., *et al.* (2012). Phylogenetic signal in module composition and species connectivity in compartmentalized host–parasite networks. *American Naturalist*, **179**(4), 501–511.

Lafferty, K. D., Dobson, A. P., and Kuris, A. M. (2006). Parasites dominate food web links. *Proceedings of the National Academy of Sciences of the United States of America*, **103**(30), 11211–11216.

Lafferty, K. D., Allesina, S., Arim, M., *et al.* (2008). Parasites in food webs: the ultimate missing links. *Ecology Letters*, **11**(6), 533–546.

Lin, Y. and Sutherland, W. J. (2013). Color and degree of interspecific synchrony of environmental noise affect the variability of complex ecological networks. *Ecological Modelling*, **263**, 162–173.

Lurgi, M., Montoya, D., and Montoya, J. M. (2016). The effects of space and diversity of interaction types on the stability of complex ecological networks. *Theoretical Ecology*, **9**(1), 3–13.

May, R. M. (1972). Will a large complex system be stable? *Nature*, **238**(5364), 413–414.

McCann, K. S., Rasmussen, J. B., and Umbanhowar, J. (2005). The dynamics of spatially coupled food webs. *Ecology Letters*, **8**(5), 513–523.

Melián, C. J., Bascompte, J., Jordano, P., and Krivan, V. (2009). Diversity in a complex ecological network with two interaction types. *Oikos*, **118**(1), 122–130.

Montoya, J. M., Pimm, S. L., and Solé, R. V. (2006). Ecological networks and their fragility. *Nature*, **442**(7100), 259–264.

Mougi, A. and Kondoh, M. (2012). Diversity of interaction types and ecological community stability. *Science*, **337**(6092), 349–351.

Mucha, P. J., Richardson, T., Macon, K., Porter, M. A., and Onnela, J.-P. (2010). Community structure in time-dependent, multiscale, and multiplex networks. *Science*, **328**(5980), 876–878.

Neutel, A.-M., Heesterbeek, J. A. P., van de Koppel, J., *et al.* (2007). Reconciling complexity with stability in naturally assembling food webs. *Nature*, **449**(7162), 599–602.

Ohgushi, T., Schmitz, O., and Holt, R. D. (2012). *Trait-Mediated Indirect Interactions*. Cambridge, UK: Cambridge University Press.

Okuyama, T. and Holland, J. N. (2008). Network structural properties mediate the stability of mutualistic communities. *Ecology Letters*, **11**(3), 208–216.

Olff, H., Alonso, D., Berg, M. P., *et al.* (2009). Parallel ecological networks in ecosystems. *Philosophical Transactions of the Royal Society B: Biological Sciences*, **364**(1524), 1755–1779.

Pilosof, S., Porter, M. A., Pascual, M., and Kéfi, S. (2017). The multilayer nature of ecological networks. *Nature in Ecology and Evolution*, **1**, 101.

Pimm, S. L. (1982). *Food Webs*. Chicago, IL: University of Chicago Press.

Pimm, S. L. (1984). The complexity and stability of ecosystems. *Nature*, **307**(5949), 321–326.

Pocock, M. J. O., Evans, D. M., and Memmott, J. (2012). The robustness and restoration of a network of ecological networks. *Science*, **335**(6071), 973–977.

Saiz, H., and Alados, C. L. (2011). Effect of *Stipa tenacissima* L. on the structure of plant co-occurrence networks in a semi-arid community. *Ecological Research*, **26**(3), 595–603.

Sander, E. L., Wootton, J. T., and Allesina, S. (2015). What can interaction webs tell us about species roles? *PLOS Computational Biology*, **11**(7), e1004330.

Sanders, D. and van Veen, F. J. F. (2011). Ecosystem engineering and predation: the multi-trophic impact of two ant species. *Journal of Animal Ecology*, **80**, 569–576.

Sanders, D., Jones, C. G., Thébault, E., *et al.* (2014). Integrating ecosystem engineering and food webs. *Oikos*, **123**(5), 513–524.

Sauve, A. M. C., Fontaine, C., and Thébault, E. (2014). Structure–stability relationships in networks combining mutualistic and antagonistic interactions. *Oikos*, **123**(3), 378–384.

Sauve, A. M. C., Thébault, E., Pocock, M. J. O., and Fontaine, C. (2016). How plants connect pollination and herbivory networks and their contribution to community stability. *Ecology*, **97**(4), 908–917.

Soliveres, S., Smit, C., and Maestre, F. T. (2015). Moving forward on facilitation research: response to changing environments and effects on the diversity, functioning, and evolution of plant communities. *Biological Reviews of the Cambridge Philosophical Society*, **90**(1), 297–313.

Thébault, E. and Fontaine, C. (2010). Stability of ecological communities and the architecture of mutualistic and trophic networks. *Science*, **329**(5993), 853–856.

Vázquez, D. P., Poulin, R., Krasnov, B. R., and Shenbrot, G. I. (2005). Species abundance and the distribution of specialization in host–parasite interaction networks. *Journal of Animal Ecology*, **74**(5), 946–955.

Verdú, M., and Valiente-Banuet, A. (2008). The nested assembly of plant facilitation networks prevents species extinctions. *American Naturalist*, **172**(6), 751–760.

Williams, R. J. and Martinez, N. D. (2000). Simple rules yield complex food webs. *Nature*, **404**(6774), 180–183.

Wood, S. A., Lilley, S. A., Schiel, D. R., and Shurin, J. B. (2010). Organismal traits are more important than environment for species interactions in the intertidal zone. *Ecology Letters*, **13**(9), 1160–1171.

Wootton, J. T., Sander, E. L., and Allesina, S. (2015). Data from: What can interaction webs tell us about species roles? *Dryad Digital Repository*, http://dx.doi.org/10.5061/dryad.39jv1 [Accessed March 22, 2017].

Yodzis, P. (1981). The stability of real ecosystems. *Nature*, **289**, 674–676.

7 Unpacking Resilience in Food Webs: An Emergent Property or a Sum of the Parts?

Ross M. Thompson and Richard Williams

7.1 Introduction

Disturbance is a pervasive force in ecology. Understanding how disturbance influences natural systems is of growing importance as human impacts become increasingly widespread and drive changes in the magnitude and frequency of disturbance events (e.g., Archibald *et al.*, 2012; Bellard *et al.*, 2012). Critical to managing the effects of disturbance is gaining an understanding of what characteristics of natural communities allow them to persist in the face of disturbance. This endeavor has a long history. Early ecologists concluded that diversity begat stability based on observations of naturally fluctuating systems (e.g., Lindeman, 1942, see Rooney and McCann, 2012). These views were challenged by modeling work in the 1970s (e.g., May, 1972) which found that in highly simplified ecological models, diversity resulted in dynamic instability. Models in later years, which incorporated realistic distributions of link strengths and non-equilibrium dynamics, have yielded the hypothesis that diverse systems are stabilized by a complex net of weak interactions (e.g., Yodzis, 1981; McCann, 2000).

Key to the concept of "stability" in ecological systems is the idea of dynamic responses that allow systems to "rebound" to their previous state after disturbance. Holling (1973) called this "ecological resilience." Fundamental to the idea of resilience is that there are multiple stable states in which an ecosystem can exist, and which have characteristics that maintain those states (Walker *et al.*, 2004). In food-web ecology, this is considered as the tendency of a food web to return to its original topology after a disturbance event (McCann, 2000).

The mechanisms that underlie resilience in food webs can be divided into three groups (Figure 7.1). The first is that the individual nodes (populations of species) within the food web have resilient traits, such that when the populations are subject to disturbance, they are able to persist and recover. We term this "nodal resilience" (Figure 7.1a). Nodal resilience of a taxon is not affected by the impacts of disturbance on its resources or consumers, but rather is a direct consequence of the taxon's traits. Resilient traits of taxa can be extremely diverse but include the ability to seek refugia, resting stages that

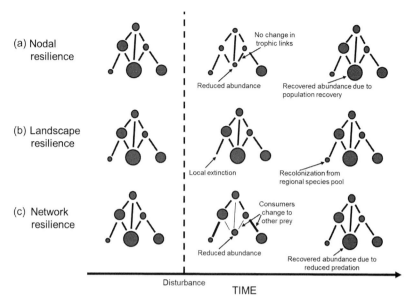

(a) Nodal resilience

No change in trophic links

Reduced abundance

Recovered abundance due to population recovery

(b) Landscape resilience

Local extinction

Recolonization from regional species pool

(c) Network resilience

Consumers change to other prey

Reduced abundance

Recovered abundance due to reduced predation

Disturbance

TIME

Figure 7.1 Conceptual model for mechanisms that may underlie resilience of food-web structure in response to disturbance. Food webs in the left panel are exposed to a disturbance that results in changes in food webs (middle panel) including reductions in abundance (smaller circles) and loss of taxa (circles absent), with consequences for the strength of trophic links to other taxa (weights of lines). Food webs then recover to their original structure (right panel) as a result of: (a) recovery in abundance within the nodes (nodal resilience); (b) recolonization from other locations (landscape resilience); or (c) food-web processes that reduce consumption rates of a node and allow recovery (network resilience).

protect against disturbance, and life-history characteristics favoring fast reproduction (Townsend *et al.*, 1997; Bolnick *et al.*, 2011; De Lange *et al.*, 2013).

Resilience of local populations may also be a function of "landscape resilience" – where local reductions in abundance or local extinctions are "rescued" by landscape processes. For example, where a taxon becomes locally rare or extinct due to disturbance, the population may exhibit resilience due to immigration from neighboring populations (Figure 7.1b) (Amarasekare, 2008). Landscape resilience may also arise when consumers move in the landscape to exploit other patches when a resource taxon becomes rare in a patch, for example after being impacted by disturbance (McCann *et al.*, 2005). This then reduces the consumptive pressure on the resource taxon and allows it to recover.

Finally, food webs may be resilient as a consequence of "network resilience" – where interactions with competitors, resources, or consumers facilitate resilience (Figure 7.1c). The idea that network characteristics could influence resilience is implicit in the definition presented by Holling (1973), and the mechanisms that might underlie this have been developed by several authors (e.g., Chesson and Huntly, 1997; McCann, 2000). For example, if disturbance reduces the abundance of a taxon, and its consumer then switches to an alternate prey source, this will facilitate the recovery of the taxon and constitutes network resilience. Similarly, if a resource responds favorably to disturbance and facilitates a consumer's recovery, this is also resilience as a consequence of network processes. If network resilience is important to the overall resilience of a food web, then we would expect particular attributes of food webs to be associated with a tendency of a food web to return to its previous state after disturbance. It has been hypothesized that food webs characterized by abundant weak links and high diversity will be more resilient to disturbance (McCann et al., 1998). Other attributes have been developed to explicitly assess the potential resilience of food webs. Network robustness (Dunne et al., 2002) models the potential impacts of extinctions on food webs and provides a proposed index of resilience for a food web.

Clear-cut harvesting of trees in the riparian margins of streams dramatically impacts the aquatic environment through alteration of both the physical environment and biological processes. Clear-cutting catchments can also increase the "peakiness" of hydrological responses to rainfall and alter the physical nature of the stream channel (Tetzlaff et al., 2007; Thompson et al., 2009). Introduction of large amounts of organic matter into the channel as a consequence of clear-cutting, in combination with increased light, alters basal resource availability (algae and detritus) (Göthe et al., 2009; Clapcott and Barmuta, 2010). These impacts on organic matter processing result in increased abundances of the invertebrates that utilize those resources (Noel et al., 1986; Growns and Davis, 1994; Trayler and Davis, 1998). Collectively these studies show that clear-cut forestry in stream catchments represents a significant disturbance (Thompson et al., 2009; Zhang et al., 2009).

Here we explore whether initial food-web structure and community traits determine the resilience of stream food webs to the effects of a forestry disturbance. The purpose is not to describe in detail the effects of the clear-cutting disturbance on physical conditions, basal resources, or invertebrate community composition, as this is done elsewhere (Thompson et al., 2009). Instead, this analysis specifically addresses what attributes of the pre-harvest food webs might determine the resilience of the food webs to the disturbance. High resilience in this context was defined as the post-disturbance food web having similar topological structure to the pre-disturbance food web. Specifically, we were interested in the following questions:

1. Were food webs with more taxa with resilient traits before the disturbance more resilient in terms of food-web structure (nodal resilience)?
2. Were there topological attributes of the food webs before the disturbance that made particular food webs more resilient (network resilience)?

As the disturbance was occurring at whole-of-landscape scales, it was assumed that landscape resilience was not a relevant process. We expected food webs that pre-disturbance were characterized by high connectivity and diversity (McCann et al., 1998), and had high network robustness values (Dunne et al., 2002), would exhibit greater resilience and "rebound" more closely to their pre-disturbance structure.

Table 7.1 Catchment details for the six streams in the study. Each study site was a 30 m reach. The stream catchment in this table is defined as the catchment of the stream upstream of the study site. The years in which the sites were sampled for food webs are shown, with brackets indicating where sampling occurred after the disturbance event (catchment clear-cutting) (after Thompson et al., 2009).

Catchment	Site	Code	Altitude (m asl)	Lithology	Catchment area (ha)	Years sampled (clear-cut)
Meggatburn	Berwick	b	140	Schist	130.2	1, (2)
Akatore	Akatore A	aa	300	Schist	255.3	1, (2)
Akatore	Akatore B	ab	300	Schist	132.8	1, (2)
Craggy Tor	Catlins	ca	500	Sandstone	129.0	1, (2)
Narrowdale	Narrowdale	co	300	Schist	376.7	1, 2
Mimihau	Venlaw	v	220	Sandstone	643.7	1, 2, (3)

7.2   Methods: Descriptions, Metrics, and Analytical Approaches

7.2.1   Description of Study Streams and Design

Six representative 30 m reaches in separate 2nd order streams in the South Island of New Zealand (Table 7.1) were chosen as study sites. At the beginning of the study all streams had catchments dominated by mature ($\approx$ 30 year old), first-rotation plantations of Monterey Pine (*Pinus radiata*) planted to the stream's edge. Detailed descriptions of the study sites and sampling of the food webs can be found elsewhere (Thompson and Townsend, 2003; Thompson et al., 2009). Briefly here, all streams were perennial, with substrate dominated by cobbles and had circum-neutral pH. Four of the streams (Akatore A, Akatore B, Catlins, and Berwick) had their catchments harvested between year 1 and 2 of the study. Two additional sites (Venlaw and Narrowdale) were not disturbed by catchment harvesting in year 1 or 2 and act as controls. The Venlaw site was then harvested at the end of year 2, and was sampled in year 3 to provide an additional post-disturbance food web. Therefore sampling across the six sites provided food webs not affected by disturbance and controlling for an effect of time, and five "pre-disturbance/ post-disturbance" pairs of food webs.

7.2.2   Description of Food Webs

The benthic food web in each stream was sampled as described elsewhere (Thompson and Townsend, 2003). For algal species composition, biofilm at each site was measured from 20 cobbles gathered at random along each stream reach, scrubbed clean of biofilm into a known volume of distilled water and homogenized. Five 0.5 mL samples were taken from the rock scrubbings from each site and used to construct wet mounts. These were inspected under 200–400× magnification to identify the algal species present. Ten randomly located Surber samples (area 0.06 m^2, mesh size 250 μm) were taken at each site and searched for macroinvertebrates, excluding those less than 1.5 mm in length, terrestrial invertebrates, partial invertebrates, pupae, and empty insect cases.

Macroinvertebrates were identified to the highest degree of taxonomic resolution possible, attributed a functional feeding group (Thompson and Townsend, 2005, and taxonomic references therein) and counted to provide abundance data. Of the 136 invertebrate taxa, 53% were identified to species and 86% to genera. Fish were sampled using electro-fishing and crayfish by electro-fishing and searching under stones. Ten individuals of each species were selected to represent the full size spectrum collected. Where fewer than 20 individuals of a species were sampled, half were retained for gut analysis. These were euthanazed and the remainder returned live to the stream.

Ten individuals of each fish taxon and fifteen of each invertebrate taxon were selected to represent the full size range present, and processed for gut analysis (Thompson and Townsend, 2003, 2005). Although ten and fifteen individuals can be considered a minimum effort, it is sufficient to estimate most food-web attributes and represents a logistically challenging but achievable sample size for food webs as rich in species as those considered here (Thompson and Townsend, 2000). Invertebrate gut contents were filtered onto 0.45 μm filters and mounted on permanent slides. Contents were determined at 200–400× magnification. The gut contents of fish and crayfish were identified under 40× magnification. Most gut contents could be identified to either species or genus for both animal and algal taxa. Terrestrial invertebrates in the gut were grouped into a single basal category. Unidentifiable material was either classified as organic detritus and included as a single basal category, or inorganic material and excluded from the food web. Each item was attributed a trophic interaction score of either 1 (present) or 0 (not detected).

7.2.3   Calculation of Resilience-Trait Scores

Resilience traits of taxa were allocated based on the New Zealand Macroinvertebrate Community Index (MCI) (Stark, 1985, 1993). This approach gives each taxon a score based on expert opinion of its tolerance to disturbance, with low scores indicating high tolerance and high scores indicating low tolerance. For each site and sampling occasion the taxa found were attributed their MCI tolerance score (Stark and Maxted, 2007). For each site this score was then averaged across the community for each sampling occasion. High site values indicate a community dominated by taxa lacking resilience traits, and low values a resilient community dominated by tolerant taxa with resilience traits.

7.2.4   Calculation of Food-Web Attributes

Food-web attributes (Table 7.2) were calculated from the binary food-web matrices. Network robustness was calculated for each pre-disturbance food web using the approach described in Dunne *et al.* (2002). Robustness provides a measure of the vulnerability of a food web to secondary extinctions as a result of taxa loss, and in this case was the percentage of non-basal taxa that needed to be removed in order to generate 50% loss of all non-basal taxa. Other food-web attributes were calculated using Foodweb3D (Yoon *et al.*, 2004) and included: number of taxa, percentage of taxa that were top (taxa with no consumers), intermediate (taxa that both consume and are

Table 7.2 Food-web attributes and resilience-trait score (average taxon MCI) of the six streams over the study period. For site codes see Table 7.1. Bold values indicate post-harvesting samples.

Site	aa	aa	ab	ab	b	b	ca	ca	v	v	v	co	co
Sampling occasion	0	**1**	0	**1**	0	**1**	0	**1**	0	0	**1**	0	0
No. of taxa	75	**56**	55	**54**	74	**60**	42	**50**	53	51	**69**	50	48
% Top	40.00	**41.07**	32.73	**33.33**	37.84	**16.67**	40.48	**40.00**	16.98	15.69	**21.74**	42.00	33.57
% Intermediate	12.00	**5.36**	20.00	**16.67**	14.86	**36.67**	26.19	**22.00**	36.98	41.18	**31.88**	18.00	23.57
% Basal	48.00	**53.57**	47.27	**50.00**	47.30	**46.67**	33.33	**38.00**	46.03	43.14	**46.38**	40.00	42.86
Connectance	0.04	**0.05**	0.05	**0.05**	0.05	**0.06**	0.06	**0.06**	0.05	0.05	**0.05**	0.05	0.06
Average links/taxon	3.00	**2.98**	2.53	**2.67**	3.85	**3.43**	2.48	**2.80**	3.16	2.82	**3.68**	2.50	2.57
Average generality	1.28	**1.46**	1.24	**1.46**	1.30	**1.36**	0.99	**1.20**	1.32	1.35	**1.31**	1.11	1.19
Average vulnerability	1.90	**1.66**	1.68	**1.63**	1.64	**1.58**	1.89	**1.78**	1.61	1.56	**1.66**	1.88	1.51
Average trophic level	1.57	**1.51**	1.63	**1.59**	1.62	**1.61**	1.77	**1.71**	1.65	1.66	**1.61**	1.73	1.71
Max. trophic level	3.04	**3.00**	3.20	**3.00**	3.08	**3.07**	3.00	**3.00**	3.11	3.02	**3.07**	3.38	3.22
Network diameter	2.60	**2.55**	2.62	**2.57**	2.47	**2.42**	2.38	**2.38**	2.49	2.43	**2.51**	2.44	2.48
Clustering coefficient	0.02	**0.01**	0.02	**0.02**	0.02	**0.03**	0.02	**0.02**	0.02	0.01	**0.01**	0.06	0.05
Network robustness	0.06		0.07		0.03		0.10		0.05				
Av. taxa MCI score	5.17		4.91		5.66		5.03		5.33				

consumed), and basal (taxa that do not consume other taxa), connectance (Warren, 1994), average number of links per taxa, average generality and vulnerability (Memmott *et al.*, 2000), average and maximum trophic level (Williams and Martinez, 2004a), diameter (Martinez *et al.*, 2002), clustering coefficient (Dunne, 2006), and dietary discontinuity (Cattin *et al.*, 2004).

7.2.5   Analyses

Food-web attributes were analyzed multivariately by normalizing a site by attribute matrix and subjecting it to principal components analysis to generate a two-dimensional ordination (Primer, Plymouth Marine Laboratories). To provide an index of resilience, the amount of change in food-web structure for each stream was summarized as vectors within the principal components ordination space. This meant that a site with high resilience had a short vector length (food-web structure changed a small amount from pre-disturbance to post-disturbance). Resilience values were regressed against the average taxon MCI score for each pre-disturbance food web to assess the influence of traits on the resilience of the food webs. The resilience values were then regressed against attributes of the pre-disturbance food webs that have been hypothesized to affect resilience: number of taxa, connectance, links per taxon, and network robustness. A hypothesis-driven approach was taken because the number of replicates was relatively low, and with the large number of potential predictors, the chance of detecting false correlations was high.

7.3   Results: Patterns in the Responses of Food Webs to Disturbances

7.3.1   Summary of the Food Webs

The food webs and study sites are described in some detail in Thompson and Townsend (2003) and Thompson *et al.* (2009). In brief here, the study streams were generally small (1st/2nd order, 1 to 2 m in width) but permanently flowing, with pre-harvest substrate comprising pebbles and cobbles. The basal resources underpinning the food webs were a combination of algae (predominantly diatoms) and terrestrially derived organic matter. The communities were dominated pre-harvest by insect larvae, including mayflies, stoneflies, blackflies, and caddis-flies. The top consumers in the streams were fish, either native (galaxiids) or exotic (salmonids).

7.3.2   Nature of the Disturbance

Detailed data on physical, chemical, and biological effects of the harvesting disturbance can be found in Thompson *et al.* (2009). Effects varied depending on the topography, and the harvesting practices used, for example sites that were incised tended to have large logs left suspended over the channel but had riparian vegetation removed, flatter sites tended to retain some riparian vegetation. At all sites there was evidence of considerable disturbance of the stream banks and channel after disturbance, with an increase in organic material in the channel and the addition of large amounts of fine sediment. There was evidence of hydrological disturbance as a consequence of upstream land clearance.

7.3.3   Changes in Food-Web Structure

Changes in community composition after harvest are described in Thompson *et al.* (2009). In general, taxonomic composition of the streams did not alter dramatically after disturbance, but patterns of abundance changed. After disturbance, several taxa that were not present beforehand were sampled at high abundances. These included the chironomid *Chironomus zelandicus*, the amphipod *Phreatogammarus* sp., and the snail *Potamopyrgus antipodarum*. A number of taxa increased in abundance after the disturbance: oligochaetes, chironomids, the mayfly *Coloburiscus humeralis*, and the stoneflies *Austroperla cyrene* and *Cristaperla fimbriae*. Fish were present at three of the four sites, but were at low densities.

The food-web matrices are available from the GlobalWeb database (www.globalwebdb .com). The food webs were diverse (42 to 75 taxa), with a well resolved basal algal flora (approximately 50% of taxa in each web), low levels of connectance, short food chains, and relatively simple food-web structure (Table 7.2). The pre-disturbance food webs varied considerably in terms of how tolerant the taxa were (range: 1, very tolerant, to 10, very intolerant), with average taxon MCI values ranging from 4.9 to 5.7. Network robustness for the pre-disturbance food webs ranged from 0.03 to 0.10.

Food webs that were not affected by disturbance changed relatively little from year to year and vectors of change were generally shorter than food webs that were impacted by disturbance (Figure 7.2). The exception to this was the Akatore B site, where food-web structure changed minimally despite disturbance, and less than the two control sites.

Table 7.3 Loadings on principal component axes (see Figure 7.2) for food-web structure

	PC1	PC2
	59.4%	33.6%
Number of taxa	0.181	0.882
% Top	0.654	−0.316
% Intermediate	−0.730	−0.029
% Basal	0.076	0.346
Connectance	0.000	0.000
Average links/taxon	−0.005	0.034
Average generality	−0.001	0.007
Average vulnerability	0.005	−0.003
Average trophic level	−0.001	−0.005
Maximum trophic level	0.000	−0.002
Network diameter	0.002	0.004
Clustering coefficient	0.000	−0.001

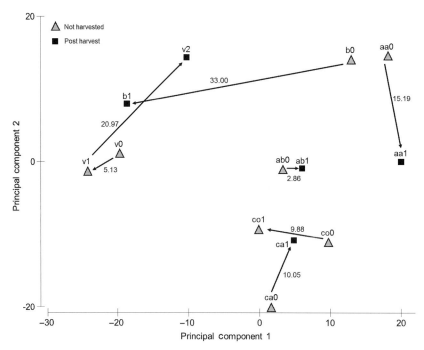

Figure 7.2 Principal components analysis ordinations of food-web attributes pre- and post-disturbance. Letter codes indicate streams (see Table 7.1), with symbols indicating pre-harvesting disturbance (triangles) or post-disturbance (squares). Where no disturbance occurred between the two samplings, numbers indicating time of sampling (e.g., 1 = +1 year). Arrows show the magnitude and trajectory of change. For loadings on principal component axes and explanatory power see Table 7.3.

For the other sites, food-web structure was changed by the disturbance, but the magnitude of that change and the attributes affected varied depending on the stream (Tables 7.2 and 7.3). Berwick and Venlaw changed the most, and in a similar way, in response to disturbance. Both exhibited a reduction in the percentage of top taxa, and an increase in the number of links per taxon (Table 7.2). Food-web structure at the Catlins site and Akatore A sites both changed, but the nature of that change was very different (Figure 7.2). Akatore A had a large decrease in clustering coefficient and in the percentage of intermediate species. Using the length of the vector of change as a measure of resilience, the Akatore B food web was the most resilient, followed by Catlins, Akatore A, Venlaw, and Berwick.

7.3.4   Relationship Between Pre-Disturbance Traits and Food-Web Resilience

There was an extremely strong positive relationship between the resilience of the food webs and the average MCI score, indicating the number of taxa with tolerant traits in the pre-disturbance food web predicted the resilience of the food web to disturbance (Figure 7.3a).

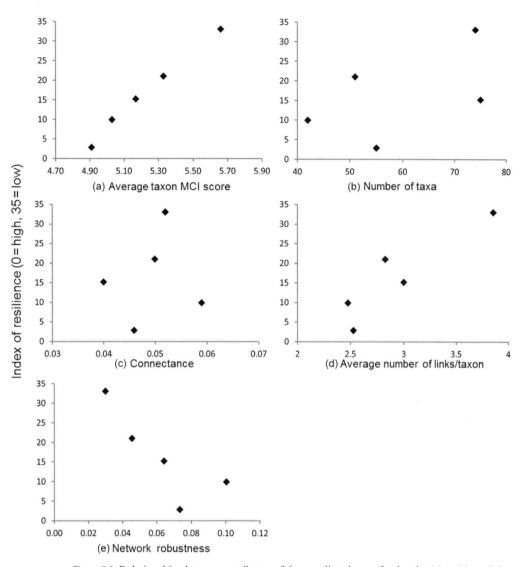

Figure 7.3 Relationships between attributes of the pre-disturbance food webs (a) to (e), and the degree to which food webs recovered to their original topology after being exposed to disturbance (resilience). A high value for the resilience index indicates low resilience.

7.3.5 Relationship Between Pre-Disturbance Food-Web Attributes and Food-Web Resilience

There was a weak negative relationship between the number of taxa in the pre-resilience food web and the resilience of that food web – that is, more diverse food webs appeared to be less resilient to the effects of disturbance (Figure 7.3b). There was no relationship evident between pre-disturbance connectance and food-web resilience (Figure 7.3c).

Pre-disturbance food webs with more links per taxa were less resilient (Figure 7.3d). There was some evidence that food webs with higher network robustness values were more resilient to disturbance (Figure 7.3e).

7.4   Discussion and Summary

Understanding what characteristics of natural systems make them resilient to the effects of disturbance is a key ecological question of both fundamental and applied importance. However, the logistic difficulty of addressing this problem has led to progress being limited to conceptual work, modeling, and experiments that consider a single trophic level or a small part of a food web. Here, relatively complex food webs were described before and after disturbance, and in a design that controlled for any confounding effects of time. While sample size was low, there were clear patterns suggesting that both nodal resilience (the presence of tolerance traits in the pre-disturbance community) and network resilience (characteristics of the pre-disturbance food webs) were contributing to the overall resilience of the food webs.

Clear-cut harvesting of the catchment represented a significant disturbance. Effects of the disturbance on in-stream habitat were considerable and consistent between streams, altering the physical nature of the channel and changing the availability of food resources, with flow on effects to the invertebrate communities (Thompson *et al.*, 2009). The disturbance did alter food-web structure, but effects were not consistent across streams. Some of these differences in responses may be due to differences in the physical characteristics of the streams and differences in harvesting technique (Thompson *et al.*, 2009). However, it is also possible the same disturbance may result in different topological outcomes, due to complex direct and indirect effects being propagated through the food web (Menge, 1995). Previous studies have shown that disturbance of similar food webs can generate diverse outcomes (Wardle, 1995). The absence of dramatic changes in the food web despite radical changes in habitat and in community composition is somewhat surprising, given that previous work from similar environments has shown that food-web analysis is capable of revealing profound differences in stream communities along environmental gradients (Lancaster and Robertson, 1995; Tavares-Cromar and Williams, 1996; Townsend *et al.*, 1998). These food webs are likely to be based largely on detritus, which may act to stabilize food webs by providing a large and stable energy base (DeAngelis, 1975; Closs and Lake, 1994).

While the topology of the food webs post-disturbance was variable, all sites exposed to disturbance exhibited changes in food-web structure. The magnitude of the changes varied sufficiently between sites to allow exploration of what attributes of the pre-disturbance food webs might have acted to influence resilience. There was an extremely strong relationship between the traits of the constituent taxa and food-web resilience. Communities that had more taxa with tolerant traits exhibited greater resilience in food-web structure. Townsend *et al.* (1997) described clear trait relationships with fluvial disturbance in streams close to those in the current study. Here it appears that one factor that can determine a food web's resilience is simply to have constituent taxa that are resilient, rather than the food web *per se* having resilient characteristics. We call this contribution to overall resilience of a community "nodal resilience," reflecting the fact that the resilience of the food web is a product of the characteristics of the nodes, not of the structure of the food web. In these systems it is possible that sites that have a community with a large component of tolerant taxa may have been exposed to previous disturbances, for example hydrologic disturbance, or may even be reflecting the disturbance of the original land clearance and conversion (Harding *et al.*, 1998). Previous disturbance history may influence current community structure and "pre-adapt" a community to be resilient to future disturbances. Nodal resilience is an emergent property of the resilience traits of taxa, which are analogous to the "segments of species" that were a component of early community stability models (Levin, 1970).

Implicit in the definition of resilience by Holling (1973) was that it is a product of characteristics of the network and not just of the constituent nodes. This perspective on stability speaks to the earlier multispecies stability modeling of May (1971). We explored this concept of "network resilience" by seeing whether attributes of the pre-disturbance food web resulted in greater resilience. One of the most evocative arguments for biodiversity protection has been the assertion that systems that have more species are more resilient (Hooper *et al.*, 2005). The analogies that have been used to support this assertion have been intuitively appealing, for example the "Rivet Hypothesis" (Ehrlich and Ehrlich, 1983) and the "Portfolio Effect" (Doak *et al.*, 1998). Both of those analogies imagine extra species as building redundancy into systems, increasing their ability to maintain function and structure in the face of disturbance. However, meta-analyses that have looked for a simple relationship between diversity and stability have not found one (Hooper *et al.*, 2005; Thompson and Starzomski, 2007). In the current study, there was weak evidence that the food webs that were more diverse before the disturbance were *less* resilient. This pattern may reflect the fact that dynamic modeling suggests that having more species is insufficient – those species must also be embedded in a web of weak interactions to be stable (McCann *et al.*, 1998; Berlow, 1999).

Numerous studies have suggested that connectance may infer resilience to food webs (McCann *et al.*, 2000; Dunne *et al.*, 2002; Gilbert, 2009). There was not a strong relationship between pre-disturbance food-web connectance and resilience, although the two most highly connected food webs were the two that exhibited the greatest resilience. Connectance was low and relatively invariant in these food webs, and it is possible that some of these food webs were not sufficiently connected to generate the stabilizing mechanisms that have been proposed. Interestingly another measure of connectivity, links per taxon, showed a strong negative relationship with resilience. Pre-disturbance food webs with more links per taxon were less resilient. While this could be interpreted as support of the conclusions made by May (1972), the more likely scenario here is that taxa with tolerant traits also had relatively few trophic links. This is as a result of a group of tolerant taxa (including chironomids) feeding only on detritus (Thompson *et al.*, 2009), which is represented in this food web as a single basal resource. Average taxa MCI and links per species were strongly correlated ($R^2 = 0.97$), suggesting in this system that the taxa with the most trophic links lacked tolerant traits. However, it is not possible to rule out the potential for a network resilience mechanism being in operation.

Network robustness (Dunne *et al.*, 2002) measures the number of secondary extinctions that result from taxa being removed from the network. It is expressed here as the proportion of non-basal taxa that can be removed before 50% of all non-basal taxa become extinct in the food web. There was a positive relationship between pre-disturbance food-web robustness and resilience in this dataset. This is despite the fact that the disturbance resulted in very few extinctions. The mechanisms whereby robustness would indicate resilience in the absence of extinction are unclear and worthy of future research. Interestingly, in this dataset there was no support for the previously described relationship between connectance and robustness (Dunne *et al.*, 2002), although this may again be a consequence of the relatively small range of connectance values.

The characterization here of mechanisms that may contribute to overall resilience is a simplification, in that it treats the mechanisms as independent of each other. This may not be true, particularly where, for example, life-history traits associated with resilience also affect food-web role. For example, in some aquatic systems species with high dispersal capabilities (high landscape resilience) are the dominant top predators, and therefore associated with particular food-web attributes (such as longer food chains) (Morán-Ordóñez *et al.*, 2015). In other cases, it may be expected that taxa with high nodal resilience (e.g., generalist taxa) may be more likely to have particular food-web roles (e.g., omnivory; Fagan, 1997). There are also potential feedbacks that may affect multiple aspects of resilience, for example disturbance effects on productivity that then drive changes in composition and food-web structure. Ideally, it would be possible to study food webs in a factorial design that disentangled these different mechanisms, but any such experiment is likely to be intractable for larger food webs.

It is important to acknowledge that this study has a number of limitations. Despite the considerable effort spent describing food-web topology, the power of the study is limited by low replication. While no previous study has had a dataset of this nature or size to explore these questions, a larger number of food webs would allow a more definitive test of the hypotheses. Trophic connections in these food webs were based on a snapshot of the trophic connections between a small number of individuals in a single season. Describing all trophic interactions can require sampling many thousands of individuals over time (Woodward *et al.*, 2005). Nevertheless, the effort expended here has been shown to be sufficient to meaningfully summarize food-web structure. The gut contents methodology used here also has some limitations in that it describes a dietary snapshot rather than an integrated diet through time. Stable isotope analysis provides a time-integrated assessment of diet, but for large food webs the mixing models required are intractable (Lubetkin and Simenstad, 2004). Describing all interactions between species (both trophic and non-trophic) represents an unrealistic task in systems as diverse as these. However, it is possible that non-trophic interactions, such as competition or facilitation, stabilize the most resilient food webs in this study.

Using an unusually large and complete set of food webs collected in an identical manner, it was possible to explore the relationships between pre-disturbance food-web structure and resilience to disturbance. Conceptually, a novel framework for "unpacking" resilience into a series of components was proposed. This provides the basis for further work to uncover the mechanistic basis for resilience. Such studies should include a range of disturbance regimes and ecosystem types. In the systems studied here it seems that the tolerant nature of the constituent taxa provides "nodal resilience" by being able to persist and recover despite exposure to a major disturbance. However, there was also some evidence that network characteristics may be important. Taken collectively these results emphasize the need for an understanding of both the traits of constituent taxa and the structure of the food web in predicting which systems may be vulnerable to the effects of disturbance.

Acknowledgments

We would like to acknowledge the University of Otago Stream Team 1998–2001 and the technical staff of the Departments of Zoology and Surveying. ITT Rayonier Ltd. and Wenita Forest Products Ltd. allowed us access to the study sites. Gerry Closs, the STARR Lab at the University of British Columbia, the Thompson Lab at Monash University, the Institute for Applied Ecology at the University of Canberra, and a very large number of anonymous reviewers provided useful comments on earlier versions. RT would like to acknowledge the support of the Biodiversity Research Centre at the University of British Columbia, and support from two Australian Research Council grants (FT110100957 and LP120200217).

References

Amarasekare, P. (2008). Spatial dynamics of food webs. *Annual Review of Ecology, Evolution, and Systematics*, **39**(1), 479–500.

Archibald, S., Staver, A. C., and Levin, S. A. (2012). Evolution of human-driven fire regimes in Africa. *Proceedings of the National Academy of Sciences of the United States of America*, **109**, 847–852.

Bellard, C., Bertelsmeier, C., Leadley, P., Thuiller, W., and Courchamp, F. (2012). Impacts of climate change on the future of biodiversity. *Ecology Letters*, **15**, 365–377.

Berlow, E. L. (1999). Strong effects of weak interactions in ecological communities. *Nature*, **398**, 330–334.

Bolnick, D. I., Amarasekare, P., Araújo, M. S., *et al.* (2011). Why intraspecific trait variation matters in community ecology. *Trends in Ecology and Evolution*, **26**(4), 183–192.

Chesson, P. and Huntly, N. (1997). The roles of harsh and fluctuating conditions in the dynamics of ecological communities. *American Naturalist*, **150**, 519–553.

Clapcott, J. E. and Barmuta, L. A. (2010). Forest clearance increases metabolism and organic matter processes in small headwater streams. *Journal of the North American Benthological Society*, **29**, 546–561.

Closs, G. P. and Lake, P. S. (1994). Spatial and temporal variation in the structure of an intermittent-stream food web. *Ecological Monographs*, **64**, 1–21.

DeAngelis, D. L. (1975). Stability and connectance in food web models. *Ecology*, **56**, 238–243.

De Lange, H. J., Kramer, K., and Faber, J. H. (2013). Two approaches using traits to assess ecological resilience: a case study on earthworm communities. *Basic and Applied Ecology*, **14**, 64–73.

Doak, D. F., Bigger, D., Harding, E. K., *et al.* (1998). The statistical inevitability of stability–diversity relationships in community ecology. *American Naturalist*, **151**, 264–276.

Dunne, J. A. (2006). The network structure of food webs. In *Ecological Networks: Linking Structure to Dynamics in Food Webs*, ed. M. Pascual and J. A. Dunne, Oxford, UK: Oxford University Press, pp. 27–86.

Dunne, J. A., Williams, R. J., and Martinez, N. D. (2002). Network structure and biodiversity loss in food webs: robustness increases with connectance. *Ecology Letters*, **5**, 558–567.

Ehrlich, P. R. and Ehrlich, A. H. (1983). *Extinction: The Causes and Consequences of the Disappearance of Species*. New York: Random House.

Fagan, W. F. (1997). Omnivory as a stabilizing feature of natural communities. *American Naturalist*, **150**, 554–567.

Gilbert, A. J. (2009). Connectance indicates the robustness of food webs when subjected to species loss. *Ecological Indicators*, **9**, 72–80.

Göthe, E., Lepori, F., and Malmqvist, B. (2009). Forestry affects food webs in northern Swedish coastal streams. *Fundamental and Applied Limnology/Archiv Für Hydrobiologie*, **175**, 281–294.

Growns, I. and Davis, J. (1994). Effects of forestry activities (clearfelling) on stream macroinvertebrate fauna in south-western Australia. *Marine and Freshwater Research*, **45**, 963–975.

Harding, J. S., Benfield, E. F., Bolstad, P. V., Helfman, G. S., and Jones, E. B. (1998). Stream biodiversity: the ghost of land use past. *Proceedings of the National Academy of Sciences of the United States of America*, **95**, 14843–14847.

Holling, C. S. (1973). Resilience and stability of ecological systems. *Annual Review of Ecology and Systematics*, **4**, 1–23.

Hooper, D. U., Chapin, F. S. I., Ewel, J. J., *et al.* (2005). Effects of biodiversity on ecosystem functioning: a consensus of current knowledge. *Ecological Monographs*, **75**, 3–35.

Lancaster, A. L. and Robertson, J. R. (1995). Microcrustacean prey and macroinvertebrate predators in a stream food web. *Freshwater Biology*, **34**, 123–134.

Levin, S. A. (1970). Community equilibria and stability, and an extension of the competitive exclusion principle. *American Naturalist*, **104**, 413–423.

Lindeman, R. L. (1942). The trophic dynamics aspect of ecology. *Ecology*, **23**, 399–418.

Lubetkin, S. C. and Simenstad, C. A. (2004). Multi-source mixing models to quantify food web sources and pathways. *Journal of Applied Ecology*, **41**, 996–1008.

Martinez, N. D., Williams, R. J., Berlow, E. L., and Dunne, J. A. (2002). Two degrees of separation in complex food webs. *Proceedings of the National Academy of Sciences of the United States of America*, **99**, 12913–12916.

May, R. M. (1971). Stability in multispecies community models. *Mathematical Biosciences*, **12**, 59–79.

May, R. M. (1972). Will a large complex system be stable? *Nature*, **238**, 413–414.

McCann, K. S. (2000). The diversity–stability debate, *Nature*, **405**, 227–233.

McCann, K., Hastings, A., and Huxel, G. (1998). Weak trophic interactions and the balance of nature. *Nature*, **395**, 794–798.

McCann, K. S., Rasmussen, J. B., and Umbanhowar, J. (2005). The dynamics of spatially coupled food webs. *Ecology Letters*, **8**, 513–523.

Memmott, J., Martinez, N. D., and Cohen, J. E. (2000). Predators, parasitoids and pathogens: species richness, trophic generality and body sizes in a natural food web. *Journal of Animal Ecology*, **69**, 1–15.

Menge, B. A. (1995). Indirect effects in marine rocky intertidal interaction webs: patterns and importance. *Ecological Monographs*, **65**, 21–74.

Morán-Ordóñez, A., Pavlova, A., Pinder, A., *et al.* (2015). Aquatic communities in arid landscapes: local conditions, dispersal-traits and landscape configuration determine local biodiversity. *Diversity and Distributions*, **21**, 1230–1241.

Noel, D. S., Martin, C. W., and Federer, C. A. (1986). Effects of forest clearcutting in New England on stream macroinvertebrates and periphyton. *Environmental Management*, **10**, 661–670.

Rooney, N. and McCann, K. S. (2012). Integrating food web diversity, structure and stability. *Trends in Ecology and Evolution*, **27**, 40–46.

Stark J. D. (1985). A macroinvertebrate community index of water quality for stony streams. *Water & Soil Miscellaneous Publication* **87**. Wellington, New Zealand: National Water and Soil Conservation Authority.

Stark, J. D. (1993). Performance of the Macroinvertebrate Community Index: effects of sampling method, sample replication, water depth, current velocity, and substratum on index values. *New Zealand Journal of Marine and Freshwater Research*, **27**, 463–478.

Stark, J. D. and Maxted, J. R. (2007). A user guide for the Macroinvertebrate Community Index. Wellington, New Zealand: Ministry for the Environment.

Tavares-Cromar, A. F. and Williams, D. D. (1996). The importance of temporal resolution in food web analysis: evidence from a detritus-based stream. *Ecological Monographs*, **66**, 91–113.

Tetzlaff, D., Malcolm, I. A., and Soulsby, C. (2007). Influence of forestry, environmental change and climatic variability on the hydrology, hydrochemistry and residence times of upland catchments. *Journal of Hydrology*, **346**, 93–111.

Thompson, R. and Starzomski, B. M. (2007). What does biodiversity actually do? A review for managers and policy makers. *Biodiversity and Conservation*, **16**, 1359–1378.

Thompson, R. M. and Townsend, C. R. (2000). Is resolution the solution? The effect of taxonomic resolution on the calculated properties of three stream food webs. *Freshwater Biology*, **44**, 413–422.

Thompson, R. M. and Townsend, C. R. (2003). Impacts on stream food webs of native and exotic forest: an intercontinental comparison. *Ecology*, **84**, 145–161.

Thompson, R. M. and Townsend, C. R. (2005). Energy availability, spatial heterogeneity and ecosystem size predict food-web structure in streams. *Oikos*, **108**, 137–148.

Thompson, R. M., Phillips, N. R., and Townsend, C. R. (2009). Biological consequences of clear-cut logging around streams: moderating effects of management. *Forest Ecology and Management*, **257**, 931–940.

Townsend, C., Doledec, S., and Scarsbrook, M. (1997). Species traits in relation to temporal and spatial heterogeneity in streams: a test of habitat templet theory. *Freshwater Biology*, **37**, 367–387.

Townsend, C. R., Thompson, R. M., and Mcintosh, A. R. (1998). Disturbance, resource supply, and food web architecture in streams. *Ecology Letters*, **1**, 200–209.

Trayler, K. M. and Davis, J. A. (1998). Forestry impacts and the vertical distribution of stream invertebrates in south-western Australia. *Freshwater Biology*, **40**, 331–342.

Walker, B., Holling, C. S., Carpenter, S. R., and Kinzig, A. (2004). Resilience, adaptability and transformability in social-ecological systems. *Ecology and Society*, **9**, 5.

Wardle, D. A. (1995). Impacts of disturbance on detritus food webs in agro-ecosystems of contrasting tillage and weed management practices. *Advances in Ecological Research*, **26**, 105–185.

Warren, P. H. (1994). Making connections in food webs. *Trends in Ecology and Evolution*, **9**, 136–141.

Williams, R. J. and Martinez, N. D. (2004). Limits to trophic levels and omnivory in complex food webs: theory and data. *American Naturalist*, **163**, 458–468.

Woodward, G., Speirs, D. C., and Hildrew, A. G. (2005). Quantification and resolution of a complex, size-structured food web. *Advances in Ecological Research*, **36**, 85–135.

Yodzis, P. (1981). The stability of real ecosystems. *Nature*, **289**, 674–676.

Yoon, I., Williams, R., Levine, E., *et al.* (2004). Webs on the web (WOW): 3D visualization of ecological networks on the WWW for collaborative research and education. *Proceedings of SPIE – The International Society for Optical Engineering*, **5295**, 124–132.

Zhang, Y., Richardson, J. S., and Pinto, X. (2009). Catchment-scale effects of forestry practices on benthic invertebrate communities in Pacific coastal streams. *Journal of Applied Ecology*, **46**, 1292–1303.

Part II

Food Webs: From Traits to Ecosystem Functioning

8   Integrating Food-Web and Trait-Based Ecology to Investigate Biomass-Trait Feedbacks

Ursula Gaedke and Toni Klauschies

8.1   General Concept

Biodiversity is rapidly declining while the frequency and strength of anthropogenically influenced changes in climate and land use is increasing (Chapin *et al.*, 2000; Butchart *et al.*, 2010). A diminished biodiversity leads to a reduced capability of ecological systems, such as individuals, populations, communities, and food webs, to buffer environmental changes and to maintain ecosystem functions and ecosystem services, leading, in turn, to a further decline in biodiversity. This profoundly impacts human well-being and our economy on a global scale (Naeem *et al.*, 2009). Thus mechanistic understanding and models predicting future ecosystem responses to climate changes are an important basis for management decisions.

We aim to understand how an understudied aspect of biodiversity, the biodiversity-related flexibility of ecological systems, allows ecosystems to adjust their properties to altered abiotic and biotic conditions. Depending on the different facets of biodiversity (e.g., genetic, phenotypic, and species diversity) individuals, populations, and communities possess an inherent flexibility, which influences their dynamics and, consequently, those of the entire food web (Figure 8.1).

For example, enhanced grazing may lead to a higher proportion of less edible plants. This dampens the reduction of plant biomass, which likely will have a feedback on the biomass and community composition of the herbivores, e.g., the share of herbivorous species able to exploit less edible plants may increase. As a result, the advantage of being less edible is reduced and the edible plants may recover with positive consequences for their consumers. This promotes the coexistence of different plant types and hence biodiversity. Given such feedback loops, the responses of large, non-linear, and intricately interconnected networks, such as food webs, to altered conditions are very difficult to understand and to predict, but of outstanding importance for fundamental and applied ecology. Thus we must broaden our very limited quantitative knowledge and predictive power on how biodiversity affects ecological dynamics and responses to environmental changes.

One way to put this idea into practice is to combine approaches from food-web and trait-based ecology. For example, previous more descriptive considerations of biodiversity effects on overall temporal variability focused mainly on the effects of the sheer number of

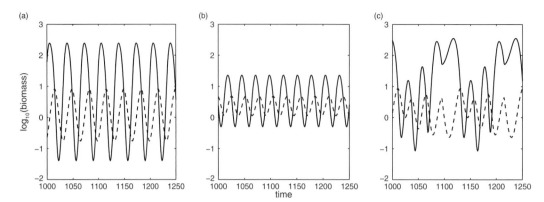

Figure 8.1 Simulations of a predator (dashed line) prey (solid line) system in: (a) a constant environment without flexibility to adjust to altered conditions; (b) with potential for adjustment only for the prey; and (c) with potential for adjustment in both predator and prey. The latter results in more complex dynamic patterns than classical predator–prey cycles (modified after Tirok *et al.*, 2011). For the sake of simplicity and brevity, the examples given here concern trophic interactions (e.g., predation and competition) but, in principle, this holds true for all other kinds of interactions, such as parasite–host interactions, mutualism, facilitation, and syntrophy.

species rather than their functional differences and the actual dynamic patterns (e.g., the shapes of population fluctuations; Hooper *et al.*, 2005). In addition, food webs are traditionally considered as (ecological) networks consisting of nodes (e.g., species) with rigid and constant properties, which holds neither in ecology nor in many other disciplines (e.g., social science, transportation networks). To overcome such limitations classical food-web theory may incorporate an innovative, flexible, trait-based approach. It explicitly considers functional traits as measurable properties of individuals (e.g., growth rate, edibility of prey, selectivity of consumers) affecting their performance and response to environmental change that may change over time depending on the prevailing conditions (for details see Chapter 10). Such a trait-based approach accounts for the fact that natural food webs continuously alter the properties of their components, which may translate into more complex system dynamics (Figure 8.1). This is in line with current research on general dynamic interaction networks, where structure and interaction strengths are dynamic and can coevolve (Gross and Blasius, 2008) enabling adjustment of the entire system to environmental changes. This approach also advances the increasingly recognized field of trait-based ecology by investigating the role of feedbacks between different groups of organisms, i.e., traits are not only considered as a suitable tool to describe the properties of (groups of) organisms but are used to study how trait changes are intertwined with the continuously changing environment.

Adjustment in functional trait values may arise at the level of individuals, populations, or communities, by different mechanisms such as:

- adaptive phenotypic plasticity of individual organisms, e.g., by reversible changes in morphology, life history, behavior, or physiological acclimation; these may or may not be much faster than the subsequently mentioned mechanisms

- evolution by mutation, recombination, and selection, changing the genetic composition; it also includes shifts in the frequency of different clones or genotypes within a population of the same species or inherited up- or down-regulation of genes implying changes in genotype frequencies
- species sorting, i.e., shifts in the species composition within a community.

That is, changes in trait values may or may not be reversible and arise at different scales by evolutionary or ecological mechanisms. Adaptive phenotypic plasticity generally enables a more flexible, generalistic behavior, e.g., with respect to defense, but may be associated with costs (e.g., DeWitt *et al.*, 1998). In metazoans ontogenetic growth and metamorphosis may also give rise to substantial trait changes within a population, which affects its dynamics. Here densities of individuals with different trait values are directly linked to each other (e.g., no juveniles without adults and vice versa). However, most models and experiments conducted in the present context so far consider unicellular organisms for operational reasons. Hence consequences of ontogenetic trait changes are beyond the scope of this chapter (but see Chapter 10). We consider a population or trophic level as adaptive if its trait values may change independent of whether the change is inherited or due to phenotypic plasticity. The environment is any property outside the organism under consideration, i.e., all biological and physicochemical conditions external to the organism. This implies that the organism can shape its environment and the environment shapes the trait values of the organism.

Previously, evolutionary processes have been mostly assumed to occur over such long time scales that they would hardly affect ecological processes. However, recent empirical and theoretical results show that evolutionary and ecological dynamics may occur on similar time scales and that they may co-determine the dynamical behavior of ecological systems i.e., giving rise to eco-evolutionary dynamics (e.g., Fussmann *et al.*, 2007; Urban *et al.*, 2008; Bolnick *et al.*, 2011; Miner *et al.*, 2012; Norberg *et al.*, 2012) (Figure 8.2).

The study of eco-evolutionary feedback dynamics has greatly benefited from theoretical modeling and experimental approaches. Eco-evolutionary feedback dynamics were investigated for example in predator–prey communities, because the strong selection exerted by predation drives evolution rapidly enough to synchronize evolutionary and ecological dynamics (Abrams, 2000). Abrams and Matsuda (1997) used a model to show that when a prey species evolves a defended genotype at a cost of a lower growth rate, classical predator–prey dynamics exhibiting a typical quarter-phase lag are shifted toward longer cycles, where predator and total prey biomass cycle are out of phase. Yoshida *et al.* (2003) and Becks *et al.* (2010) presented empirical evidence for such eco-evolutionary feedback dynamics from corresponding laboratory experiments. Rapid evolution within their prey population, i.e., changes in the frequencies of defended and undefended prey clones (Figure 8.2) as a response to predation, determined the dynamics of these predator–prey systems and whether or not the different clones coexisted, i.e., trait variation was maintained. In addition to such eco-evolutionary dynamics, adaptive phenotypic plasticity may change trait values often at faster time scales and may substantially contribute to the trait variation and to the stability of

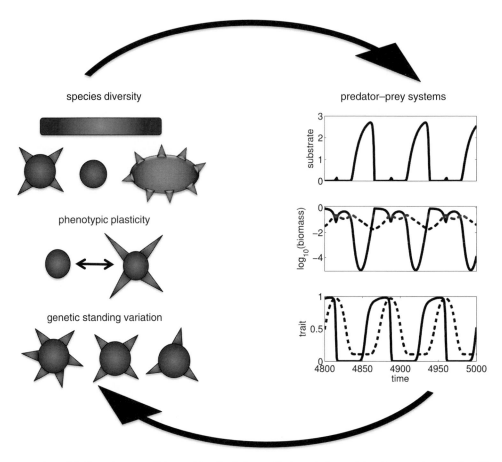

Figure 8.2 Complex causalities and feedback loops. Trait variation, arising from evolution, clonal/
species shifts, or phenotypic plasticity, represents the biodiversity present and influences the
dynamics of the ecological system in respect to biomasses, trait values, and trait variation.
The latter has a feedback on the maintenance of trait variation, which depends on the trade-off(s)
among traits and their shape.

population and community dynamics (e.g., Vos *et al.*, 2004). The integration of these
processes at the system level is still in its infancy (Cortez, 2011; Yamamichi *et al.*, 2011).

To fully understand the role of temporal and standing trait variation we have to
consider the ubiquitous and inevitable trade-offs among the different functional traits
of an organism. They constrain the temporal or standing variation in individual traits
since it is inherently impossible for an organism to maximize all trait values simultaneously ("A jack of all trades is a master of none"). The existence of such trade-offs
among traits is widely recognized, based on energetic constraints, but little empirical
knowledge exists on their magnitude and shape although simulation models and first
experimental evidence suggest that these properties are decisive for system dynamics
and the maintenance of trait variation (e.g., Becks *et al.*, 2010; Tirok *et al.*, 2011; Kasada
et al., 2014). The trade-offs quantify the costs involved in realizing distinct trait values.

A typical example is that a defense against predation (1st trait) has certain costs (e.g., the energy to grow spines or hiding without access to food), which subsequently reduce the growth rate (2nd trait). Given the various ways that ecological interactions and food webs are shaped within natural systems different types of costs may arise that affect predator–prey interactions differently (van Velzen and Etienne, 2015). For example, direct costs, such as reduced growth or enhanced mortality rates, always affect a species performance independent of the presence or absence of the resident community. They may arise from the ability of a generalist to use a broader range of prey or to adopt two metabolic strategies (e.g., mixotrophs). In contrast, ecological costs are only expressed in the context of ecological interactions and thus in the presence of (part of) the resident community, e.g., a predator or competitor.

Direct costs promoted the coexistence of specialist and generalist species in consumer–resource (Abrams, 2006; van Velzen and Etienne, 2013) and predator–prey models (Yamamichi *et al.*, 2011). Because ecological costs can have very different effects from direct costs, it is important to know how direct and ecological costs compare and jointly influence the coexistence of generalists and specialists in predator–prey systems and under which circumstances a potential for trait variation pays off – since little is known about the interaction between direct and ecological costs.

8.2   Costs for Trait Changes Determine Coexistence of Generalists and Specialists

We present a model study that elucidates consequences of direct and ecological costs arising from a generalist's potential for trait changes on the coexistence of generalist and specialist prey. To facilitate comparisons with prospective experimental studies we consider a flow-through chemostat model comprising three types of prey and two specialist predators (Figure 8.3).

The substrate concentration, prey, and predator biomasses and the generalist's mean trait value change according to the following equations:

$$\frac{dC}{dt} = \delta \cdot (C_0 - C) - \left(\frac{r \cdot C \cdot \left(S_1 + S_2 + (1-c) \cdot G \right)}{C + H} \right) \tag{8.1}$$

$$\frac{dS_1}{dt} = e \cdot \left(\frac{r \cdot C}{C + H} - \delta \right) \cdot S_1 - \frac{g \cdot S_1 \cdot P_1}{S_1 + (1-p)^n \cdot G + K} \tag{8.2}$$

$$\frac{dS_2}{dt} = e \cdot \left(\frac{r \cdot C}{C + H} - \delta \right) \cdot S_2 - \frac{g \cdot S_2 \cdot P_2}{S_2 + p^n \cdot G + K} \tag{8.3}$$

$$\frac{dG}{dt} = e \cdot \left(\frac{r \cdot (1-c) \cdot C}{C + H} - \delta \right) \cdot G - \frac{g \cdot (1-p)^n \cdot G \cdot P_1}{S_1 + (1-p)^n \cdot G + K} - \frac{g \cdot p^n \cdot G \cdot P_2}{S_2 + p^n \cdot G + K} \tag{8.4}$$

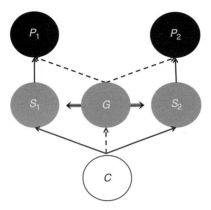

Figure 8.3 Sketch of the food web described by Eqs. (8.1) to (8.7). The specialist and generalist prey, S_1, S_2, and G, compete for the substrate, C. The specialist predators, P_1 and P_2, feed on their specific prey, S_1 and S_2, respectively, and on G depending on its actual trait value. The overall strength of nutrient uptake and feeding interaction is denoted by solid (strong) and dashed (weak) lines. The double arrow indicates that the generalist prey is able to adapt its trait value in response to the presence of P_1 and P_2.

$$\frac{dP_1}{dt} = \left(e \cdot g \cdot \frac{S_1 + (1-p)^n \cdot G}{S_1 + (1-p)^n \cdot G + K} - \delta \right) \cdot P_1 \tag{8.5}$$

$$\frac{dP_2}{dt} = \left(e \cdot g \cdot \frac{S_2 + p^n \cdot G}{S_2 + p^n \cdot G + K} - \delta \right) \cdot P_2 \tag{8.6}$$

$$\frac{dp}{dt} = v \cdot \left(\frac{\partial \left(\frac{1}{G} \frac{dG}{dt} \right)}{\partial p} \right) + \frac{w}{p^2} - \frac{w}{(1-p)^2} \tag{8.7}$$

C_0 and C are the substrate concentrations in the inflowing medium and the chemostat vessel. Substrate, prey, and predator biomasses are washed out with the dilution rate δ. The specialist prey S_1 and S_2 are only grazed by their specialist predators P_1 and P_2, respectively, implying a bidirectional trait axis. In contrast, the generalist prey G can adopt the trait values of S_1 or S_2 or any value in between, and is preyed upon by P_1 and P_2 accordingly, depending on its mean trait value p. Hence a defense of G against one of the predators increases its susceptibility against predation by the other one. This implies that the costs for the original defense depend on the density of the other predator. We use a common half-saturation constant H, conversion efficiency e, and maximum growth rate r for all prey but the maximum growth rate of G is reduced by a factor of $(1-c)$ accounting for the direct costs c arising from its flexibility in defending itself against P_1 or P_2. For the predators we assumed a Holling type II functional response with an extension to multiple prey types (Murdoch, 1973) with a common half-saturation constant K and maximum grazing rate g.

The generalist's mean trait value p defines its susceptibility to grazing by P_1 (high and identical with S_1 for $p = 0$) and P_2 (high and identical with S_2 for $p = 1$). It is linearly interpolated between these extremes for $n = 1$, which implies that the grazing pressures of

P_1 and P_2 on G linearly declines and increases, respectively, with p. For $n < 1$ the grazing pressure of P_1 decreases slower than that of P_2 increases with p. Hence decreasing n increases the overall ecological costs of G by an enhanced top–down control, since G experiences a higher per-capita grazing pressure than the mean of the two specialist prey for intermediate values of p. We model temporal changes of p with an approach first developed in quantitative genetics (e.g., Lande, 1976; Abrams, 2006, 2010) where p changes according to the local fitness gradient increasing the per-capita net-growth rate of G (dG/Gdt, Eq. [8.7]). The parameter v scales the rate of mean trait changes relative to changes in the biomasses. Increasing v increases the speed of adjustment and thus the ability of G to track the environmental fluctuations and to adopt the currently favored trait value. Hence v influences the ecological costs of G in a temporally variable environment, since it influences the degree of maladaptation to the ambient conditions. In addition, we introduced with the right hand-side of Eq. (8.7) a boundary function, representing non-adaptive trait changes to prevent p from exceeding the ecologically feasible extremes of being perfectly edible or entirely inedible (Abrams, 2006, 2010; Mougi, 2012). The magnitude of this trait change is scaled by the parameter w.

We parameterized our model according to a microbial plankton food web comprising bacteria and protozoans (for details see Table 8.1). In such webs, trophic interactions strongly influence biomass dynamics and the lower trophic levels that are most relevant for the systems' metabolism are dominated by short living, highly diverse, and abundant organisms with a large potential for rapid trait changes. Hence feedbacks between biomass and trait dynamics are expected to be highly relevant. For example, S_1 may represent a bacterial strain growing as single cells, whereas S_2 forms colonies that are too large to be consumed by P_1, whereas P_2 cannot take up single-celled bacteria but only colonies. G may belong to the frequently occurring bacteria that may form both single cells and colonies. We conducted numerical simulations to investigate the combined effects of direct (c) and ecological (n) costs of the generalist's potential for trait changes on the coexistence of the three prey species. Specifically, we performed simulations

Table 8.1 Parameter description

Parameter	Value	Unit	Description
δ	0.1	d^{-1}	Dilution rate
C_0	3	mg C L^{-1}	Substrate concentration in the inflowing medium
r	1	d^{-1}	Maximum growth rate of S_1 and S_2
c	0–0.9	–	Relative reduction of the maximum growth rate of G compared to the specialists
H	0.01	mg C L^{-1}	Half-saturation constant for substrate uptake
g	1.2	d^{-1}	Maximum grazing rate of P_1 and P_2
n	0.2–1	–	Shape of the trade-off between the susceptibilities to grazing by P_1 and P_2
K	0.1	mg C L^{-1}	Half-saturation constant for prey ingestion
e	0.3		Conversion efficiency
v	0.01, 0.1, 1	–	Scaling factor of adaptive trait adjustment
w	0.001	d^{-1}	Scaling factor of non-adaptive trait adjustment

comprising a full-factorial combination of 19 different values of c and 17 different values of n. In addition, we varied the generalist's speed of adaptation to mimic slow ($v = 0.01$), intermediate ($v = 0.1$), and fast ($v = 1$) trait changes. Simulations were terminated after 5000 time steps. Species extinction was assumed if average biomasses were below 10^{-10} during the last 1000 time steps. For all simulations we used $C = 1$, $S_1 = G = 0.11$, $S_2 = P_1 = P_2 = 0.1$, and $p = 0.5$ as initial conditions.

The generalist prey, G, and specialist prey, S_1 and S_2, coexisted in the presence of both predators within a certain range of direct (c) and ecological (n) costs of G (for details see below). This system exhibited cyclic dynamics where bottom–up and top–down control of the prey species alternated as did the dominance of the prey and the predators, P_1 and P_2 (Figure 8.4). A high substrate concentration and a low grazing pressure by P_1 promoted a strong increase in the biomasses of S_1 and G exhibiting similar susceptibilities to predation ($p \approx 0$; Figure 8.4, t_1). In contrast, S_2 and subsequently also P_2 decreased because of a persistingly high grazing pressure by P_2 on S_2.

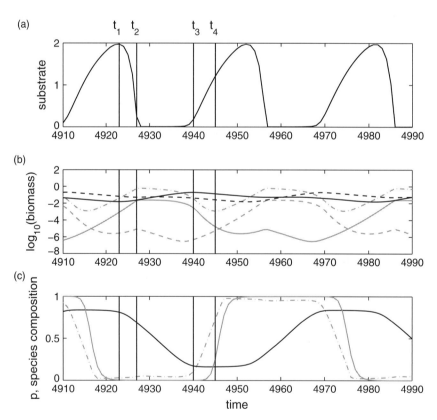

Figure 8.4 (a) Substrate, (b) biomasses (S_1, solid gray line; S_2, dotted gray line; G, dashed dotted gray line; P_1, solid black line; P_2, dotted black line), and (c) composition of the specialized prey community (solid gray, $S_2/(S_1 + S_2)$) and the predator community (black, $P_2/(P_1 + P_2)$) and trait dynamics (dashed dotted grey, p) of G for intermediate direct and ecological costs ($c = 0.1$, $n = 0.8$) and intermediate trait velocity ($v = 0.1$). A high value of p implies a low predation pressure by P_1 and a high one by P_2 (for details see text).

The biomass of G reached its maximum after substrate depletion and started to decline in response to the increase in the biomass of P_1 (Figure 8.4, t_2). Although S_1 experienced the same enhanced mortality by P_1 its biomass still slightly increased due to its higher growth rate. As a consequence of the ongoing shift in the relative grazing pressure by P_2 and P_1, G started to re-increase its mean trait value (Figure 8.4c, t_3) which allowed G to escape from predation by P_1 into an almost predator-free space as the absolute grazing pressure by P_2 was still decreasing and low. This opened a window of opportunity for G allowing it to compensate for its lower maximum growth rate because the increase in the relative abundance of S_2 lagged behind the increase in the generalist's trait value. This growth advantage depended on the speed of trait adjustment (v), reduced the biomass variability of G, and promoted its coexistence with S_1 and S_2 (Figure 8.4, t_4). Overall, changes in the generalist's trait value depended on the weighted sum of the per-capita grazing rates by P_1 and P_2. Hence both predator (numerical response) and prey biomasses (functional response) influenced the magnitude and direction of trait changes. This explains why differences in the functional responses of the two predators arising from differences in their prey biomasses caused a time lag between the time when the contribution of P_2 to the total predator community (numerical response) dropped below 0.5 and the increase of p (Figure 8.4c, t_{3-4}). In contrast to G, S_1 strongly declined, experiencing an increasingly high grazing pressure. The latter resulted from the predators' functional type II response implying an increased per-capita death rate when total prey biomass ($S_1 + G$) is reduced. The concomitant competitive release leading to higher substrate concentrations and the low grazing pressure allowed S_2 to recover. After reaching very high trait values ($p \approx 1$), G and S_2 jointly increased in biomass. The cycle repeats as described above but this time with S_2 and G dominating the system.

In general, coexistence of different prey species depended on the costs for the generalist's potential to change its trait value (Figure 8.5). In the absence of direct ($c = 0$) and ecological costs ($n = 1$, i.e., a linear trade-off between the generalist's susceptibilities against predation by P_1 and P_2, Figure 8.5d) a perfect generalist ($p = 0.5$) experiences the same growth rates and grazing losses as the mean of the two specialist prey. That is, G has never any disadvantage compared to the specialist prey. However, since G is able to change its trait value in response to the current selection pressure it experiences temporal benefits compared to the specialist prey. This reduces the overall grazing pressure on G, leading to the exclusion of the specialist prey when the generalist's speed of trait adjustment is sufficiently high ($v \gg 0$, e.g., $v = 0.1$; Figure 8.5b).

Increasing the direct or ecological costs for the generalist's potential for trait changes increased the likelihood of the specialist prey to persist. In the presence of low direct costs ($0 < c < 0.2$, i.e., a growth rate reduction of G by up to 20% compared to the specialists) G suffered from increased resource competition with the specialist prey which themselves experienced competitive release from G. However, at low or no additional ecological costs (n close to 1), G experienced a lower overall grazing pressure than the specialist prey because of its ability to temporally escape grazing, overcompensating the reduction in its growth rate and excluding the specialist prey. The latter

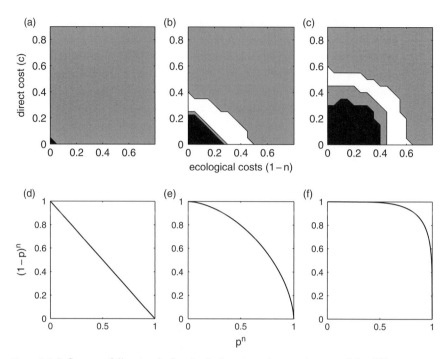

Figure 8.5 Influence of direct and of ecological costs on the coexistence of the different prey types (black: one prey species (generalist) survived; gray: two prey species coexist; white: all three prey species coexist) with (a) low ($v = 0.01$), (b) intermediate ($v = 0.1$), and (c) high speed of adjustment ($v = 1$) of the generalist G. (d)–(f) Trade-off between the generalist's sensitivities against grazing by P_2 and P_1, p^n and $(1 - p)^n$, respectively, for $n = 1$ (linear trade-off) (d); $n = 0.6$ (strong trade-off) (e); and $n = 0.2$ (very strong trade-off) (f).

was prevented by a further increase in the direct or ecological costs. This resulted either from a further competitive release of the specialist prey from resource competition ($c > 0.2$) or an increased overall grazing pressure on G ($n < 1$) equalizing the performance of the different prey types. Indeed, decreasing n increased the strength of the trade-off between the generalist's susceptibilities for the two predators leading to a higher grazing pressure on G when exhibiting intermediate trait values. As a consequence G and one or two specialist prey coexisted (Figure 8.5b). When all three prey species persisted, G was predominantly resource limited whereas the specialist prey were mainly limited by their respective specialist predators. A further increase in the direct and ecological costs destabilized the coexistence once again but promoted this time the exclusion of G. In case of enhanced direct costs ($c > 0.4$) the flexibility of G could not compensate anymore for the strong resource competition with the specialist prey whereas an increase in the ecological costs increased the overall grazing pressure on G when exhibiting intermediate trait values. This weakened its overall advantage of being flexible, thus making G more vulnerable to resource competition (Figure 8.5b).

Decreasing the generalist's speed of adaptation also reduced its advantage of being flexible. At very low speed of adaptation ($v = 0.01$, Figure 8.5a) the coexistence of generalist and specialist prey became very unlikely as G only persisted in the system in the absence of significant costs and was easily outcompeted by the specialist prey. In contrast, increasing the generalist's speed of adaptation ($v = 1$, Figure 8.5c) increased its ability to track changes in the environment reducing the amount of time of being maladapted. Indeed, higher trait velocities increased the generalist's window of opportunity allowing it to compensate higher costs. In addition, realized ecological costs were reduced since G was able to quickly overcome intermediate trait values associated with relatively high grazing pressure.

Frequently an increase in defense against one predator comes at the direct cost of a reduced growth rate or the ecological cost of enhanced mortality by another predator. Our model results show that such direct and ecological costs of a generalist's potential for trait changes are important for the coexistence of specialists and generalists. In accordance with Abrams (2006) and van Velzen and Etienne (2013) we show that under asynchronous fluctuations a generalist may coexist with two specialists when its costs are not too high and its trait adjustments are effective ($v \gg 0$, e.g., $v = 0.1$) and adaptive, i.e., trait changes follow the local fitness gradient. In contrast to the former studies, we considered a chemostat model with an explicit equation for the substrate and the generalist and specialists being at the prey trophic level. They differ in their susceptibilities to the two specialist predators (top–down perspective), whereas Abrams (2006) and van Velzen and Etienne (2013) considered generalist and specialist consumers showing differences in their abilities to use two different logistically growing prey species. We complement their results by considering the joint effects of direct and ecological costs on the coexistence of specialists and generalists. Furthermore, we show that an increase in the speed of adaptation not only increases the likelihood of a generalist to coexist with two specialists but also increases the range of parameters where the generalist excludes the specialists. This results from the fact that the generalist is able to counteract larger costs when adapting very fast to the current selection pressure because it can then minimize the ecological costs by shortening periods of maladaptation. This allows the generalist to rapidly pass intermediate trait values associated with relatively high grazing losses.

Similar to our model, Yamamichi *et al.* (2011) considered a chemostat system with two specialists and one generalistic, phenotypically plastic prey but only one selective predator mainly feeding on the undefended prey type. In contrast to our model they assumed a unidirectional trait axis, a trade-off between the growth rate and level of defense, and that the generalist's trait changes follow the equations for induced defenses. That is, the trait changes are only influenced by predator but not prey densities and are thus not necessarily adaptive since trait changes may lead to maladaptation of the generalist. In contrast, our approach assumed that the trait changes of the generalist are adaptive depending on both prey (resource mediated) and predator biomasses acting in the direction of increased fitness. The non-adaptive trait changes may explain the low potential of the generalist prey to exclude the specialists even when exhibiting low direct costs (Yamamichi *et al.*, 2011). In contrast, we observed a substantial range of parameter values where the generalist excluded the specialist prey especially when the speed of adaptation was high.

8.3   Relevance and Outreach

Previous work and the results presented above show that incorporating traits as dynamical properties in food-web models will stimulate a reconsideration of established concepts in fundamental ecology. Above all this, it strongly contributes to applied ecology as we gain a deeper insight into how some of the most important anthropogenic threats (biodiversity loss, climate change, and biological invasions which greatly impact trait variation) interact with ecosystem function, services, and biogeochemical cycles. The type of ecological dynamics (e.g., static, oscillating, transient, extinctions) is not only decisive for fundamental questions but influences the management of species extinction risks, the control of climate-change related outbreaks of pests, and the reliability of other ecosystem services. Further combining food-web and trait-based ecology will enable us to identify and experimentally test mechanisms maintaining biodiversity that can then be implemented in (applied, forecasting) models to improve their validity. A distinct example includes the advanced understanding and modeling of microbial food webs in the activated sludge of wastewater treatment plants, which helps to optimize their performance (e.g., removal of pathogenic bacteria, organic carbon, and nutrients) under fluctuating conditions. Such food webs are mostly composed of bacteria, phages, and protozoa and exhibit a very large potential for rapid adjustment due to the high abundances, short generation times, and large trait variation of the component organisms.

Furthermore, predictive management models (e.g., large-scale vegetation or ocean and limnetic "N–P–Z" models) are increasingly employed to advise policy and to guide appropriate actions to mitigate consequences of land-use and climate change. In such models we urgently need the ability to consider the consequences of the varying degrees of flexibility arising from changing levels of biodiversity with only a modest increase in complexity and computational effort (Mooij *et al.*, 2010; Göthlich and Oschlies, 2012), which is, however, so far in its infancy. Already rather simple multispecies predator–prey systems show a very rich behavior depending on specific model assumptions (Ellner and Becks, 2011; Bauer *et al.*, 2014). For example, increasing the number of species within a trophic level altered the type of dynamics depending on assumption details concerning, e.g., the concomitant increase in trait range, species packing, and minimum feeding thresholds (Bauer *et al.*, 2014). Similarly, intra- and interspecific trait changes may jointly influence species coexistence and population and community dynamics. In accordance, long-term observation of plankton dynamics in Lake Constance could only be reproduced by a highly resolved plankton food-web model (20 guilds including bacteria and different fast growing protozoans) when accounting for the potential of short-living guilds to adjust to ambient predator or prey conditions (Boit *et al.*, 2012). In contrast to the model of Tirok and Gaedke (2010), this could only be done in a non-mechanistic way, which reduced the generality of the model. Hence we need to find better solutions also for complex models tailored to specific systems to account for the ubiquitous flexibility of ecological systems.

Acknowledgments

We thank Jacob Hauschildt for work on a preliminary version of the model, and Ellen van Velzen and Peter de Ruiter for their constructive comments on the manuscript.

References

Abrams, P. A. (2000). The evolution of predator–prey interactions: theory and evidence. *Annual Review of Ecology and Systematics*, **31**, 79–105.

Abrams, P. A. (2006). The prerequisites for and likelihood of generalist-specialist coexistence. *American Naturalist*, **167**, 329–342.

Abrams, P. A. (2010). Quantitative descriptions of resource choice in ecological models. *Population Ecology*, **52**, 47–58.

Abrams, P. A. and Matsuda, H. (1997). Prey adaptation as a cause of predator–prey cycles. *Evolution*, **51**, 1742–1750.

Bauer, B., Vos, M., Klauschies, T., and Gaedke, U. (2014). Diversity, functional similarity and top–down control drive synchronization and the reliability of ecosystem function. *American Naturalist*, **183**, 394–409.

Becks, L., Ellner, S. P., Jones, L. E., and Hairston, N. G. (2010). Reduction of adaptive genetic diversity radically alters eco-evolutionary community dynamics. *Ecology Letters*, **13**, 989–997.

Boit, A., Martinez, N. D., Williams, R. J., and Gaedke, U. (2012). Mechanistic theory and modeling of complex food web dynamics in Lake Constance. *Ecology Letters*, **15**, 594–602.

Bolnick, D. I., Amarasekare, P., Araujo, M. S., *et al.* (2011). Why intraspecific trait variation matters in community ecology. *Trends in Ecology and Evolution*, **26**, 183–192.

Butchart, S. H. M., Walpole, M., Collen, B., *et al.* (2010). Global biodiversity: indicators of recent declines. *Science*, **328**, 1164–1168.

Chapin, F. S., Zavaleta, E. S., Eviner, V. T., *et al.* (2000). Consequences of changing biodiversity. *Nature*, **405**, 234–242.

Cortez, M. H. (2011). Comparing the qualitatively different effects rapidly evolving and rapidly induced defences have on predator–prey interactions. *Ecology Letters*, **14**, 202–209.

DeWitt, T. J., Sih, A., and Wilson, D. S. (1998). Costs and limits of phenotypic plasticity. *Trends in Ecology and Evolution*, **13**, 77–81.

Ellner, S. P. and Becks, L. (2011). Rapid prey evolution and the dynamics of two-predator food webs. *Theoretical Ecology*, **4**, 133–152.

Fussmann, G. F., Loreau, M., and Abrams, P. A. (2007). Eco-evolutionary dynamics of communities and ecosystems. *Functional Ecology*, **21**, 465–477.

Göthlich, L. and Oschlies, A. (2012). Phytoplankton niche generation by interspecific stoichiometric variation. *Global Biogeochemical Cycles*, **26**. doi:10.1029/2011GB004042.

Gross, T. and Blasius, B. (2008). Adaptive coevolutionary networks: a review. *Journal of the Royal Society Interface*, **5**, 259–271.

Hooper, D. U., Chapin, F. S., Ewel, J. J., *et al.* (2005). Effects of biodiversity on ecosystem functioning: a consensus of current knowledge. *Ecological Monographs*, **75**, 3–35.

Kasada, M., Yamamichi, M., and Yoshida, T. (2014). Form of an evolutionary tradeoff affects eco-evolutionary dynamics in a predator–prey system. *Proceedings of the National Academy of Sciences of the United States of America*, **111**, 16035–16040.

Lande, R. (1976). Natural-selection and random genetic drift in phenotypic evolution. *Evolution*, **30**, 314–334.

Miner, B. E., De Meester, L., Pfrender, M. E., Lampert, W., and Hairston, Jr., N. G. (2012). Linking genes to communities and ecosystems: *Daphnia* as an ecogenomic model. *Proceeding of the Royal Society B: Biological Sciences*, **279**, 1873–1882.

Mooij, W. M., Trolle, D., Jeppesen, E., *et al.* (2010). Challenges and opportunities for integrating lake ecosystem modelling approaches. *Aquatic Ecology*, **44**, 633–667. doi:10.1007/s10452-010–9339-3 ER.

Mougi, A. (2012). Unusual predator–prey dynamics under reciprocal phenotypic plasticity. *Journal of Theoretical Biology*, **305**, 96–102.

Murdoch, W. W. (1973). Functional response of predators. *Journal of Applied Ecology*, **10**, 335–342.

Naeem, S., Bunker, D. E., Hector, A., *et al.* (eds.) (2009). *Biodiversity, Ecosystem Functioning, and Human Wellbeing: An Ecological and Economic Perspective.* Oxford, UK: Oxford University Press.

Norberg, J., Urban, M. C., Vellend, M., Klausmeier, C. A., and Loeuille, N. (2012). Eco-evolutionary responses of biodiversity to climate change. *Nature Climate Change*, **2**, 747–751.

Tirok, K. and Gaedke, U. (2010). Internally driven alternation of functional traits in a multi-species predator–prey system. *Ecology*, **91**, 1748–1762.

Tirok, K., Bauer, B., Wirtz, K., and Gaedke, U. (2011). Predator–prey dynamics driven by feedback between functionally diverse trophic levels. *PLoS ONE*, **6**, e27357. doi:10.1371/journal.pone.0027357.

Urban, M. C., Leibold, M. A., Amarasekare, P., *et al.* (2008). The evolutionary ecology of metacommunities. *Trends in Ecology and Evolution*, **23**(6), 311–317.

van Velzen, E. and Etienne, R. S. (2013). The evolution and coexistence of generalist and specialist herbivores under between-plant competition. *Theoretical Ecology*, **6**, 87–98.

van Velzen, E. and Etienne, R. S. (2015). The importance of ecological costs for the evolution of plant defense against herbivory. *Journal of Theoretical Biology*, **372**, 89–99.

Vos, M., Kooi, B. W., DeAngelis, D. L., and Mooij, W. M. (2004). Inducible defences and the paradox of enrichment. *Oikos*, **105**, 471–480.

Yamamichi, M., Yoshida, T., and Sasaki, A. (2011). Comparing the effects of rapid evolution and phenotypic plasticity on predator–prey dynamics. *American Naturalist*, **178**, 287–304.

Yoshida, T., Jones, L. E., Ellner, S. P., Fussmann, G. F., and Hairston, Jr., N. G. (2003). Rapid evolution drives ecological dynamics in a predator–prey system. *Nature*, **424**, 303–306.

9   Including the Life Cycle in Food Webs

Karin A. Nilsson, Amanda L. Caskenette, Christian Guill, Martin Hartvig, and
Floor H. Soudijn

9.1   Introduction

To grow and reproduce is fundamental for living organisms. In essence all organisms go
through a life cycle with ontogenetically driven changes in their physiological rates and
trophic interactions (Figure 9.1; Box 9.1).

This ontogenetic development occurs even in unicellular organisms but is more
striking in other groups. For example: dragonflies undergo metamorphoses that span
several habitats, Atlantic marlin increase up to 500 times their length, and the cod-worm
has different host requirements for each life-history stage. All of these ontogenetic
changes correspond to large shifts in the ecological role of an individual.

In spite of the drastic changes many individuals undergo over their life history,
classical ecological theory typically assumes that all individuals within a population
are identical. As a consequence, a large part of our ecological understanding relies
on this assumption. This is surprising considering that ecological theory strongly
links to evolution, which is critically dependent on variation among individuals.
Acknowledging ecological variation of individuals within the species is relatively
recent to food-web ecology. While individual variation can arise from genetic or
stochastic processes, this chapter focuses on individual variation that relates to
ontogenetic development.

Biological interactions that are susceptible to ontogenetic variation include: resource
use, vulnerability to predators and parasites, mutualistic interactions, cannibalism, and
commensalism. Therefore the consideration of ontogeny has major implications for the
way we consider food-web topology (Box 9.2). In a broader sense, the function of an
organism, such as the nutrient fluxes it contributes to and the ecosystem services it takes
part in, may also change over ontogeny. By ignoring the individual life history, ecologists focus on interactions between populations rather than between individuals, an
abstraction that may be biologically inaccurate. In fact, differences between individuals
within species can exceed, and have larger effects on food-web dynamics, than differences between individuals of different species. This suggests that the consideration of
differences between life stages within populations is essential for our understanding of
food-web structure and ecosystem functioning.

An essential dynamical aspect of the consideration of life history is that somatic
growth and development precedes the reproductive part of the life cycle. Thus only
mature individuals take part in producing new offspring. In contrast, classical models

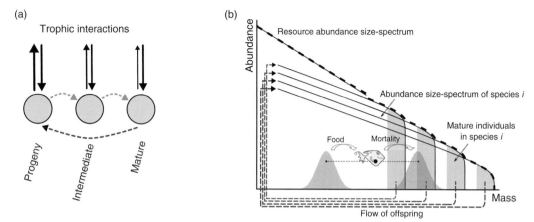

Figure 9.1 Two approaches for modeling ontogenetic variation. (a) *Stage structure*: individuals within a species develop through different stages. Each of the nodes could be trophically connected to other stage-structured species (stage transition via development shown in light-gray dashed arrows, and reproduction in dark-gray dashed arrows). (b) *Size spectra*: individuals within a species develop through continuous growth in body size. The abundance distribution as a function of body size is modeled explicitly using somatic growth, reproduction, and mortality from (typically) larger individuals in the community. The community size spectrum is the sum of all species size spectra including any additional resource size spectra. Common for both approaches is that trophic interactions and physiological rates are likely to vary between stages/body sizes, and that only adults are capable of producing offspring.

Box 9.1 Modeling the Individual and Its Life Cycle

A structured model describes the distribution of individuals across life stages within the population (Figure 9.1). These models come in different forms, age-structured, stage-structured, and size-structured, and can all be based on a broad set of different methodologies. In age-structured models transitions between stages typically depend only on time and the dynamic coupling between the food environment and somatic growth is not explicitly considered (Persson *et al.*, 2014). On the other hand, size- and stage-structured models can incorporate mechanistic and process-driven narratives of how individuals move between stages and grow in size by explicitly taking into account the feedback on the resource.

When incorporating food-dependent transitions in models, additional practices are required to describe how individuals allocate energy to reproduction and somatic growth. Most mechanistic approaches rely on bioenergetic models of individual-level processes as a function of body size to describe food intake and somatic growth (Kitchell *et al.*, 1977; Kooijman, 2000). These bioenergetic models in turn are based on size-dependent relationships of physiological rates and processes (Peters, 1983). Many of these rates, such as the metabolic rate, can be described by allometric scaling relationships both intra- and interspecifically (Glazier, 2005). Other rates, such as the intake and attack rate, follow allometric interspecific scaling relationships (Rall *et al.*, 2012) and hump-shaped intraspecific scaling relationships (Ursin, 1973;

Byström and Garcíia-Berthou, 1999; Byström and Andersson, 2005; Rall *et al.*, 2012). Models structured to incorporate physiological rates and processes are required to allow for this variety in parameter relationships with body size.

Physiologically structured population models (PSPM; Metz and Diekmann, 1986; de Roos and Persson, 2001, 2013) allow for a full consideration of size-dependent characteristics. The framework for PSPM is built around a bioenergetics model that describes the energy use and food intake of individuals, while dynamically keeping track of the abundance and size distribution. The inclusion of population structure in a dynamic framework in this way also allows for food-dependent growth as individual somatic growth depends on food availability, which is shaped by consumption. The bioenergetics of each species are described using intraspecific allometric relationships along with size-dependent processes describing food intake and maturation. Trait-based species characterization can be incorporated within the PSPM framework to characterize a species and its individuals using both intra- and interspecific scaling relationships (Hartvig *et al.*, 2011). This trait-based approach additionally allows for a natural inclusion of the life-history trade-offs (Stearns, 1989; Charnov *et al.*, 2001; Hartvig *et al.*, 2011).

The PSPM framework allows for a continuous flow of individuals through their ontogeny. This framework is, however, computationally draining and hard to analyze for more than two-species systems. In order to make more computationally efficient models that still characterize the dynamics of PSPMs, one option is to simplify the model by dividing the population into stages (de Roos *et al.*, 2008b). We give an example of such a model later, in Box 9.3, where we also describe how differences in competitiveness between different stages can lead to biomass overcompensation. Alternatively, one may also do a fine, but finite, discretization of the size distribution and use standard numerical techniques to model size-structured population dynamics (Hartvig *et al.*, 2011).

Explicit dynamic modeling of the community size distribution without resolving individual species identities (often referred to as the community size spectrum) has been used as a crude characterization of the community-level dynamics of an entire food web (e.g., Silvert and Platt, 1980; Benoît and Rochet, 2004; Blanchard *et al.*, 2009; Law *et al.*, 2009; Zhang *et al.*, 2013). In these community models trophic interactions are assumed to occur solely on the basis of individual body size. Finally, we want to emphasize that the biomass-based stage-structured model we use (Box 9.3), the PSPMs, and the size spectra models are based on the same underlying individual level assumptions (concerning energy intake, growth, metabolism, etc.) and that the effects observed at both the population and community level arise from these.

typically assume that reproduction starts at birth. To include this aspect of life history in dynamical food-web models it is necessary to describe the way individuals grow and transition between life stages. This can be accomplished using a stage-structured model (see Box 9.1: Modeling the Individual and Its Life Cycle). The focus of this chapter lies

Box 9.2 Topological Characteristics of Food Webs With Stage-Structured Populations

Network characteristic	Examples: ontogenetic structured resolved vs. aggregated

Number of nodes in the network, N
Classically, the number of nodes is just the number of species: N = S. With Z ontogenetic stages per species, the number of nodes becomes N = S × Z.

(dotted arrows denote ontogenetic links)

Edges (trophic links)
In aggregated populations, a single link defines a consumer–resource pair. This simple situation cannot be translated unambiguously to structured populations. Anything from 1 to (2 × Z)^2 links is possible.

Connectance, C
C = L/N^2. Calculations based on aggregated populations yield higher values than those based on stages as nodes and links between stages unless in all predator–prey pairs in the network all predator stages feed on all prey stages.

C=1/12   C=2/9

Trophic level, TL
Ontogenetic stages of the same species can feed on different TLs. Defining the TL of a species as a whole is therefore problematic, and would also confound the true TL of stage-specific predators.

TL 3   TL 3
TL 2   TL 2
TL 1   TL 1

Ecological role (top, intermediate, or basal species)
In aggregated populations the ecological role is uniquely defined for each species, but in stage-structured populations the ontogenetic stages may have different roles. This is most obvious for top species (or stages). Differences in ecological role can have implications for the biomass distribution between stages.

T   T
I, T   I
B   B

Cannibalism
In aggregated populations cannibalism only has the function of density-dependent mortality. In stage-structured populations cannibalism between stages can occur (most likely more often than cannibalism within single stages), which shapes the distribution of biomass between the stages.

Omnivory

Depending on the situation, resolving the ontogenetic structure of populations can either increase or decrease the number of omnivores in a food web. In the example to the right, omnivory of the top predator is not detected if the stage structure of its prey is not resolved.

Generality, vulnerability

Generalist predators can be ontogenetic specialists, which makes them more vulnerable to the loss of prey species (cf. Rudolf and Lafferty, 2011). Similar for vulnerability.

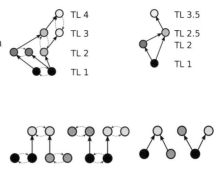

on stage- and size-structured approaches, as these, as opposed to age-structured models, directly link to the vast literature on, e.g., body size and energetic rates (Box 9.1). While individual-based models (IBMs) can account for variation among individuals at a given life stage, these are not considered here.

In this chapter we address the implications of ontogenetic development for food-web structure and food-web dynamics. We do this by reviewing recently developed examples and putting the effects of ontogeny in a food-web perspective. We start by explaining the consequences of considering ontogeny for food-web topology (Section 9.2 and Box 9.2) and dynamical properties of food webs such as stability (Section 9.4), coexistence (Section 9.3), and alternative stable states (Section 9.5). We then address the consequences of these dynamical properties for how the food-web dynamics respond to environmental gradients (Section 9.5 and Box 9.4) and harvest (Section 9.6). As we will show, there are already several examples of how ontogeny affects all these aspect of food-web ecology, and the implication can indeed be major. The full extent of these effects is yet to be determined, but it is nevertheless clear that there is a large scope for implications in a broad range of food webs, considering the prevalence of ontogenetic development. Finally, we propose some future directions regarding the necessary data and theoretical tools to further investigate the importance of ontogenetic development for food-web ecology (Section 9.7).

9.2   Food-Web Structure

We first address the basic consequences of considering ontogeny and changes in body size over life for food-web structure and functioning. By adding ontogeny, the definitions normally used for food webs change in many aspects, starting from the definition of nodes and extending to higher level concepts, such as trophic levels, omnivory, and cannibalism (Box 9.2; Pimm and Rice, 1987).

To manage the complexity of food webs, species are often lumped together as trophic species (Pimm *et al.*, 1991) or functional groups. However, for most species, individuals in different life stages cannot be categorized in the same way (Persson, 1999). Due to growth in body size, metamorphosis, or differences in habitat use, individuals may travel through several trophic levels or functional groups over their life (Werner and Gilliam, 1984; Rudolf and Lafferty, 2011; Hartvig and Andersen, 2013; Rudolf and Rasmussen, 2013a). Thus the topology of food webs may change drastically when taking different life stages into consideration.

Often, the links and position of a species in the food web are based on the average size and behavior of adult individuals in a population (Digel *et al.*, 2014) instead of being specific to life stage. This focus on averages naturally alters how a predator with multiple changes in resource use over its life cycle appears in the eventual food web (Box 9.2 panels: *Ecological role* and *Generality, vulnerability*). Similarly, the vulnerability of species to predators can change over their life. What starts as a highly preyed upon juvenile might turn out to be a top predator once it is mature: making the common practice of averaging a species' vulnerability perplexing. The resolution of food webs also depends on how data collection handles the population-level heterogeneity (Martinez, 1991). Due to the changes in body size related to ontogenetic development, differences in the ecological role can be larger within species than between species (Polis, 1984; Rudolf and Rasmussen, 2013b). Individual body size tends to be a better predictor for the trophic position and trophic links than species identity (Jennings *et al.*, 2001; Woodward *et al.*, 2010). One could therefore suggest that measuring individuals may be as, or more, important than identifying them to species (see e.g., Gilljam *et al.*, 2011).

Resolving the size or stage structure of species inevitably increases the complexity of the food web, but it also increases understanding of the true structure of interactions between species and improves the predictability of the effect of changes in topology, for example species extinctions. When ontogenetic stages instead of species (or even trophic species) are considered as the relevant unit of organization, the number of network nodes and the number of trophic links can in general be expected to increase (Box 9.2). The connectance of the extended network, on the other hand, will decrease unless there is a complete diet overlap between the stages (Box 9.2).

Beyond the consequences of including ontogeny for the food-web complexity in terms of number of nodes and links, ontogeny also adds different types of links: ontogenetic links that represent development and reproduction. While reproduction and somatic growth connect life stages within species rather than forming a network between species, they still contribute to the distribution of biomass in the food web. The somatic growth and reproduction energy channels are regulated by different processes than the typical trophic channel, which may have important consequences for the food web in terms of the flux pattern of energy in a food web.

For example, natural predator–prey body mass ratios support species coexistence by generating a stabilizing flux pattern of biomass (Brose *et al.*, 2006a, 2006b; Kartascheff *et al.*, 2010; Heckman *et al.*, 2012). However, juveniles often operate at lower trophic levels than adults, which opens alternative, ontogenetic pathways for biomass that have a different flux pattern than trophic links, i.e., juveniles move biomass up the food chain by growing. Adults, in contrast, reverse the general flux direction and move biomass via reproduction to a lower trophic level. How this affects the stability in a food web has not been explored yet. Currently, even reliable estimates of the quantitative importance of biomass flux via ontogenetic versus trophic links are lacking.

In addition to these novel fluxes between stages, the size or stage structure of a population becomes an integrated component of the topology of the web, which may have implications for ecosystem dynamics and functioning. For example, dramatic shifts in perch population structure have been shown to result in trophic cascades, where subsequent changes in zooplankton and phytoplankton abundance resulted in remarkably different food-web compositions (Persson *et al.*, 2003). Also, recent experimental studies with insect populations have shown that different population structures may cause differences in the abundances of other species, community structure, production, and respiration (Rudolf and Rasmussen, 2013a, b). Importantly, these effects did not scale with total biomass of the manipulated species, but depended on the identity of the different life stages, reinforcing the notion that the different ecological roles of life stages are significant for food-web dynamics. These results imply that the consideration of differences between life stages within populations is essential for our understanding of food-web structure, dynamics, and ecosystem functioning. The consequences of differences between individuals in different life stages for community and food-web dynamics will be discussed in the following sections.

9.3 Coexistence

A fundamental problem in ecology is the coexistence of many species in a system on few resources when, theoretically, this coexistence should be impossible (Tilman, 1994). In this section we consider how ontogenetic development results in mechanisms that might support or limit species coexistence.

Ontogenetic diet shifts can induce niche partitioning between consumer species that compete for the same resources (Schellekens *et al.*, 2010). In contrast to the competitive exclusion principle (Tilman, 1994), two predators with strong size or stage specificity can coexist on the same prey if they prefer different life stages of the prey. Moreover, two predators feeding on the same prey may facilitate each other's existence (de Roos *et al.*, 2008a). This effect results from the fact that one predator induces biomass overcompensation in the prey (i.e., an increase in biomass of a certain stage; see Box 9.3: Biomass Overcompensation) that favors another predator. Thus by incorporating stage-structure and size-specific predation, the interaction between two predators sharing a prey may change from competitive exclusion to co-dependence. There is experimental evidence for predator-to-predator facilitation (de Roos *et al.*, 2008a; Huss and Nilsson, 2011) and also an example of how a pathogen may facilitate a predator in a natural system

Box 9.3 Biomass Overcompensation

Biomass overcompensation is defined as an increase in biomass of a certain life stage of a population in response to increased mortality. This biomass overcompensation depends on an asymmetry in the resource acquisition for different life stages, for example between juveniles and adults. When juveniles are superior competitors compared to adults, the population will be regulated through reproduction, and biomass overcompensation will occur in the juvenile stage (Figure 9.2a). On the other hand, developmental regulation occurs when adults are superior competitors, and biomass overcompensation takes place in the adult stage (Figure 9.2b; de Roos *et al.*, 2007, 2008b). Increased mortality results in higher resource levels for the survivors, leading to both higher reproduction and developmental rates (until the mortality becomes too severe). Due to the differences in competitiveness (or resource availability) between the different life stages, these rates will increase to a different extent, resulting in a redistribution of biomass between the life stages. This response occurs irrespectively of which size is targeted by mortality, but can differ in magnitude depending on the stage selectivity of the mortality. Biomass overcompensation is the underlying mechanism for several phenomena, such as the emergent Allee effect and emergent facilitation, which are discussed in this chapter.

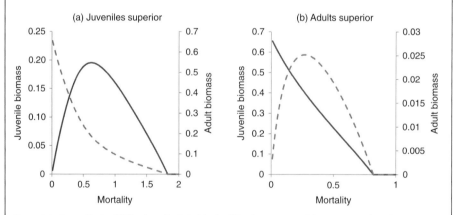

Figure 9.2 Juvenile (solid line) and adult (dashed line) consumer biomass as a function of stage independent mortality (m), for when (a) juveniles and (b) adults are superior competitors. (Figure modified from de Roos *et al.* 2007: Figure 1).

Model Equations and Description

The model consists of a stage-structured consumer, with a juvenile (J) and an adult (A) stage, feeding on an unstructured resource (R) that follows semichemostatic resource growth. The turnover rate of the resource is denoted with r (default $r = 1$) and the carrying capacity with K (default $K = 2$). The consumer attack rate is a_{cr} (default $a = 10$) and the functional response is a type II with the half-saturation constant H (default $H = 1$). Importantly, the parameters q_j and q_a scale the competitiveness of the stages, when juveniles are superior competitors $q_j = 1$ and $q_a = 0.5$, and when adults are superior competitors $q_j = 0.5$ and $q_a = 1$. Conversion efficiency c (default $c = 0.5$), metabolic cost T (default $T = 1$), and mortality m (default $m = 0.1$) are assumed to be equal for juveniles and adults. The adults allocate all their net energy production to reproduction and hence the term for adult feeding shows up in the juvenile equation (term 2, Eq. [9.2]). The juvenile feeding is represented by term 1, Eq. (9.2). The transition from the juvenile to the adult state is determined by the maturation rate, where z (default $z = 0.01$) defines the duration of the juvenile stage (term 3, Eq. [9.2] and term 1 [Eq. 9.3]). The relatively complicated maturation function is what assures the approximation of a continuously size-structured population model by the much simpler stage-structured model. De Roos et al. (2007, 2008b) described the model and its derivation in detail.

$$\frac{dR}{dt} = r(K - R) - \frac{q_j aRJ}{H + R} - \frac{q_a aRA}{H + R} \tag{9.1}$$

$$\frac{dJ}{dt} = \left(\frac{cq_j aR}{H + R} - T\right)J + \left(\frac{cq_a aR}{H + R} - T\right)A - \left(\frac{\left(\frac{cq_j aR}{H + R} - T\right) - m}{1 - \left(z\right)^{1 - \frac{cq_j aR}{H + R} - T}{m}}\right)J - mJ \tag{9.2}$$

$$\frac{dA}{dt} = \left(\frac{\left(\frac{cq_j aR}{H + R} - T\right) - m}{1 - \left(z\right)^{1 - \frac{cq_j aR}{H + R} - T}{m}}\right)J - mA \tag{9.3}$$

The model was also modified to represent the scenario in Box 9.3 by adding another resource (N) and allowing the adult stage to feed exclusively on that resource (Eqs. [9.4]–[9.7]).

$$\frac{dR}{dt} = r(K - R) - \frac{q_j aRJ}{H + R} \tag{9.4}$$

$$\frac{dN}{dt} = r(K - R) - \frac{q_a aNA}{H + N} \tag{9.5}$$

$$\frac{dJ}{dt} = \left(\frac{cq_j aR}{H + R} - T\right)J + \left(\frac{cq_a aN}{H + N} - T\right)A - \left(\frac{\left(\frac{cq_j aR}{H + R} - T\right) - m}{1 - \left(z^{1 - \frac{cq_j aR}{H + R} - T}\right)}\right)J - mJ \tag{9.6}$$

$$\frac{dA}{dt} = \left(\frac{\left(\frac{cq_j aR}{H + R} - T\right) - m}{1 - \left(z^{1 - \frac{cq_j aR}{H + R} - T}\right)}\right)J - mA \tag{9.7}$$

(Ohlberger et al., 2011). However, it is still unexplored as to what extent emergent facilitation occurs in natural food webs. This is important to determine since, while it promotes coexistence, it makes species more dependent on each other and food webs thus more sensitive to secondary extinctions.

Omnivory is generally thought to decrease the scope for species coexistence (Pimm and Rice, 1987). While life-history omnivory was originally claimed to increase coexistence compared to classical, unstructured omnivory models (Pimm and Rice, 1987; Mylius et al., 2001), the actual effect of life-history omnivory on coexistence has later been shown to depend on how the diet of the intraguild predator changes over its life. While diet broadening decreases possibilities for coexistence (van der Wolfshaar et al., 2006), an abrupt diet shift increases coexistence between prey and predator (Abrams, 2011; Hin et al., 2011). Furthermore, life-history cannibalism and omnivory have been shown to make coexistence possible at all resource levels (Hartvig and Andersen, 2013).

While few studies have considered the implications of ontogenetic development and associated changes in community interactions for coexistence in large food webs, it is recognized that species loss affect webs differently depending on how the food web is conceptualized (Brose et al., 2016). In a theoretical study that took only the binary network structure of food webs into account, i.e., the presence or absence of species and links, Rudolf and Lafferty (2011) investigated how the occurrence of ontogenetic specialists (Section 9.2) affects the resilience of food webs against species loss. It was found that the extinction risk of an ontogenetic specialist was much higher than if its ontogenetic structure was ignored, related to that species-level extinction occurred even when just one stage lost its resources. It was also shown that if higher network complexity (higher connectance) means more ontogenetic stages per species, higher complexity would lead to lower resistance of the food web against secondary extinctions.

In general, species richness may be increased through reduction of niche overlap between species (Chesson and Kuang, 2008). This may be achieved by increasing the number of traits used to characterize a species (Rossberg *et al.*, 2010a, b; Eklöf *et al.*, 2013). Ontogeny might simply allow for more ways of being different, as trait variation within species creates more niches (Bolnick *et al.*, 2011), thereby providing the basis for higher diversity. Mechanisms of differentiation between species are size-specific trophic interactions (ontogenetic niche shifts and differential vulnerability to predators), the size at maturation, or the size of offspring produced (size at birth). Moreover, intraspecific scaling of food-ingestion rates and metabolic rates with body size might be considered (Glazier, 2005). How different ontogeny-related sources of variation between species opens the potential for increased diversity is yet to be explored, but it has been demonstrated that increased trait dimensionality opens the possibility of assembling large and stable size-structured food webs (Hartvig, 2011; Hartvig *et al.*, 2011).

In summary, while the consideration of ontogeny may increase the potential for coexistence by providing more ways of "being different," theoretical results regarding complex food webs and food-web modules of two or three species show mixed results regarding the potential of coexistence of ontogenetically structured populations. The occurrence of ontogenetic specialists or the effect of emergent facilitation can increase the scope for coexistence but also enhance the vulnerability of food webs to secondary extinctions. Also size-dependent trophic interactions that structure competitive, omnivorous, or cannibalistic feeding relations can under certain circumstances enhance coexistence possibilities. Yet, to what extent the latter mechanisms that have been analyzed in the context of small food-web modules actually affect coexistence has so far not been rigorously tested in the context of complex food-web models or food-web data.

9.4 Dynamic Stability

Differences between individuals occurring over the life cycle do not only affect the coexistence or extinction probability of species, but are also important drivers of population dynamics. They can determine whether a population settles at a stable equilibrium or whether population fluctuations occur. In this section we will mainly describe the consequences of ontogenetic structure for predictions regarding a classical stability measure, linear stability of equilibriums in population or community dynamics (Rosenzweig and MacArthur, 1963; May, 1972).

The pattern of interaction strengths in food webs is a major determinant of food-web stability (Berlow, 1999; May, 2006). Specifically, in unstructured populations weak interactions stabilize population cycles driven by strong interactions (McCann *et al.*, 1998). This stabilizing effect of weak interactions is found in stage-structured populations as well (Caskenette and McCann, 2017). Yet, interestingly, weak interactions may also be destabilizing in stage-structured populations due to the re-routing of energy between different stages through reproduction and development. Size-selective feeding (that is characteristic for size-structured models) may stabilize predator–prey cycles, when one stage is unsusceptible to predation and acts as a refuge (Hastings, 1983; Abrams and Walters, 1996; de Roos *et al.*, 2003a). In addition to classical predator–prey cycles, structured populations may display another form of instability, i.e., generation cycles or cohort cycles. These cycles are driven by the differences in competitiveness between different stages in the same population (Gurney and Nisbet, 1985; de Roos and Persson, 2003; de Roos *et al.*, 2003a) and have been shown to be common in nature, at least among generalist species (Murdoch *et al.*, 2002). Predator–prey cycles are often seen as something destructive that leads to population densities traveling dangerously close to zero. On the other hand, temporal variability may increase competitive coexistence (e.g., Chesson, 2000). A recent study shows that temporal variability may increase predator persistence when stage structure is considered in their prey.

The question of linear stability also arises in studies using a community size-spectrum approach (Box 9.1; Figure 9.1). Compared to the studies cited above, they offer a more holistic view as they aim to describe full ecological communities instead of only two or three interacting species. Size-spectrum models, however, simplify by lumping all species together into a single spectrum that recognizes only individual body size and hence ignores species identity. Studies using this approach show that the community size spectrum can either have a stable stationary shape or exhibit traveling waves. Factors that promote stability include decreasing predator–prey mass ratios, increasing diet breadth, or coupling of slow benthic and fast pelagic energy pathways (Law *et al.*, 2009; Blanchard *et al.*, 2011; Plank and Law, 2012). The latter is similar to the stabilizing effect of coupling slow and fast energy channels by top predators in classical food web models (Rooney *et al.*, 2006).

If the community size spectrum is not modeled directly, but is an emergent property that results from explicitly modeling each species and summing up the species size spectra, its stability is enhanced (Zhang *et al.*, 2013). This enhanced stability comes from the difference in growth rates across species at a given body size in the community spectrum, which ensures a more steady and stable flow of biomass along the mass axis in the community size spectrum. Thus this shows that species richness plays an important role for stability in food webs of size-structured populations, and why a species-based framework may often be needed (Hartvig *et al.*, 2011).

In the classical picture of a food web as a network of unstructured populations, increasing diet breadth (i.e., higher connectance) and increasing species diversity are associated with lower probability of stability (May, 1972). Resolving the ontogenetic structure of all species in a network without lumping them together in a single size spectrum greatly enhances the complexity of the network, even if only two stages per species are considered (Box 9.2). From a dynamical system's point of view this doubles the degrees of freedom, which, following May (1972), has a negative impact on stability.

However, the additional complexity due to considering ontogenetic stages is clearly added in a non-random way and it is possible that the intrinsic order of the additional degrees of freedom will act in stabilizing, in the same way that consumer–resource interaction strengths in real food webs are organized to promote stability (Yodzis, 1981). If, and under which conditions, either one of the two possibilities occurs has to be determined by future studies.

9.5   Alternative States and Environmental Gradients

9.5.1   Catastrophes and Alternative Stable States

Alternative stable ecosystem states have received a lot of attention in the last decades. Predator collapses and consequent trophic cascades have been observed to change entire landscapes (Estes *et al.*, 2011). Often, when alternative stable states exist, so-called tipping points occur where a relatively small perturbation may push the system from one state into another, drastically different, state (Scheffer *et al.*, 2001). Causes for these catastrophic changes are gradual changes in, for example, harvesting pressures or environmental conditions that drive a system toward a tipping point. The sudden shifts are hard to handle by ecosystem management as they are often unexpected and can, as the sudden crash of cod populations in the North-West Atlantic (Hutchings and Myers, 1994), have major socio-economic impacts.

In models that resolve the ontogenetic structure of populations and include stage-specific feeding interactions, the existence of alternative stable states has been reported frequently. Given the prevalence of ontogenetic niche shifts (Werner and Gilliam, 1984) this may point at a certain susceptibility of ecological systems to this phenomenon. Moreover, while the occurrence of early-warning signals has been reported for systems that are close to a catastrophic transition (Scheffer *et al.*, 2009), early-warning signals may be absent in ontogenetically structured populations (Boerlijst *et al.*, 2013). Here we give a few examples of cases where ontogenetic development is a crucial part of the mechanisms that maintain alternative states. However, the switching between states may still be driven by environmental factors. It is also worth noting that changes in population structure may permeate to a larger part of the food web, for example by inducing a trophic cascade.

One way by which processes within populations can create the potential for alternative stable states, is a complete diet shift between the juvenile and adult stage (Schreiber and Rudolf, 2008; Guill, 2009; Nakazawa, 2011). Depending on, for example, the relative amounts of resources that are available in the system for the different life stages, either small individuals or large individuals are dominant in the population. In this scenario a change in, for example, the productivity of the juvenile resource may cause a shift to an alternative state (further explained in Section 9.5.2: Environmental Gradients).

Second, when a predator induces biomass overcompensation (Box 9.3) in its prey, alternative stable community states with and without predators may occur. These alternative states may occur because size- or stage-specific predation mortality affects the prey size distribution such that more prey of the size fed on by the predator is produced through predation. Under these circumstances, a decrease in the abundance of the predator may force the system into a state where the predator loses control of the prey distribution, and the predator does not have sufficient prey available in its preferred size range to persist (de Roos et al., 2003b; van Leeuwen et al., 2008). This is usually referred to as an emergent Allee effect. An empirical example of the emergent Allee effect comes from a lake in northern Norway, where harvesting changed the prey (Arctic char) population's size distribution and allowed for the establishment of the predator (brown trout). Once present at a high density, brown trout kept the community in its new state (Persson et al., 2007). In this example it was harvest that induced the shift to the alternative state that was subsequently maintained by predation. This kind of state shift may, however, also be induced by changes in environmental factors (van Leeuwen et al., 2008).

Alternative stable states have also been found in theoretical studies on intraguild predation with a stage-structured intraguild predator. If the predator has an ontogenetic niche shift and switches from preying on a resource shared with its prey as a juvenile to preying on the intraguild consumer as an adult, decreasing adult predator density can enable the consumer to outperform juvenile predators and either drive the predator population to extinction (Hin et al., 2011) or trap it in a low-density state (Walters and Kitchell, 2001). Also cannibalism in size-structured models can result in alternative stable community states through a multitude of mechanisms (Claessen et al., 2004). The alternative states can simply be different shapes of the species' size distribution or, more drastically, presence or absence of the cannibalistic species. In either case strong effects on other species in the food web may occur.

It is unclear how common alternative stable community states as a result of ontogenetic niche shifts, the emergent Allee effect, cannibalism, or life history omnivory and intraguild predation are. There is clear empirical evidence for some of these phenomena, and theoretical evidence indicating they might be common, but see Hartvig and Andersen (2013) for a study that identifies alternative stable states as a rather exotic phenomenon. Hence before they can be called general phenomena, more empirical evidence of their existence is needed. As they are relatively newly described, other examples may (or may not) follow when more studies in this area are conducted. The potential for drastic changes throughout the food web as a result of trophic cascades in response to changes in population structure emphasizes the importance of including ontogenetic development. In the following section we consider some specific examples of alternative stable states in more detail.

9.5.2　Environmental Gradients

A survey of the natural world shows us that the environment in which food webs exist varies across space and time and, furthermore, that food-web dynamics are affected by these changes in environmental conditions. In this section we give a number of examples of how some ontogenetically driven phenomena play out under different environmental conditions, including switches between alternative stable states.

The classical exploitation ecosystem hypothesis (EEH) predicts that increasing productivity results in an increase in the number of trophic levels in a food chain. In more detail, the EEH predicts a monotonic increase of biomass of the highest trophic level with increasing productivity and that when productivity reaches a certain threshold level, a new trophic level can invade the system (Oksanen et al., 1981). This general prediction can also be made based on stage-structured models (de Roos and Persson, 2013). However, when ontogenetic development is considered, abrupt regime shifts can occur and for the same productivity levels alternative stable community states may occur. These may, for example, be due to the aforementioned emergent Allee effect. Hence a smooth response of the system to a smooth change of environmental conditions is unlikely when ontogenetic development is accounted for in the population dynamics.

As mentioned above, individuals in different life stages in a population readily utilize different resources, due to ontogenetic changes in habitat use and feeding efficiency for example. An increase in the productivity of the resource that individuals in one life stage are feeding on may lead to a change in the regulation and the structure of the population (de Roos et al., 2007; Guill, 2009; van de Wolfshaar et al., 2011; Box 9.4). For example, if the productivity of the resource for adults is increased, this may lead to a shift in the population from a competitive bottleneck in the adult stage to a bottleneck in the juvenile stage (this is analogous to changes in the relative competitiveness and population regulation, as in Box 9.3; see also Box 9.4).

As a result, a shift in the stage distribution occurs and biomass increases in the juvenile stage and decreases in the adult stage. Due to the changes in population regulation that may occur over a productivity gradient (in a stage-specific resource), predictions that are opposite to the predictions of the EEH may arise. For example, in a food-chain scenario with a size-specific predator feeding on a consumer that undergoes an ontogenetic niche-shift, increasing the productivity of a stage-specific resource for the consumer may lead to the extinction of the predator due to a shift in the population structure of the prey (emergent predator exclusion; de Roos and Persson, 2013).

We have discussed productivity of stage-specific resources as an example of an environmental gradient. However, in essence, any factor that affects the competitiveness between juveniles and adults can cause changes in population regulation, thereby altering both biomass stage distributions and how the population will respond to harvesting. To demonstrate this, in Box 9.4 we give an example of how a turbidity gradient in lakes could affect the stage structure and regulation of a consumer population. In some

---

Box 9.4 Turbidity Gradient

An example of the effects of an environmental gradient on the structure, the regulation, and the response to harvesting of a population.

The results are based on a consumer–resource biomass model with one juvenile and one adult consumer stage, feeding on two different resources. The model and parameters are as described in Box 9.1, but with separate resources for juveniles and adults (carrying capacity for the juvenile and adult resource respectively is 0.5 and q is now incorporated into the stage-specific attack rate).

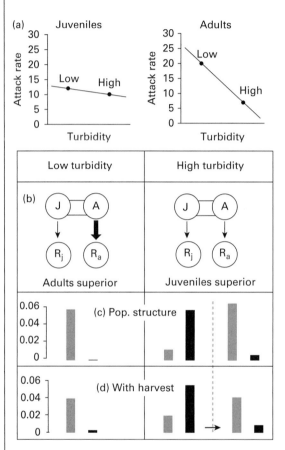

Figure 9.3 In panel (a) the relationships between the level of turbidity and the attack rate of juvenile and adult consumers are shown. Panel (b) shows the module configuration where the thickness of arrows represents the relative strength of the attack rates for the two different levels of turbidity. Panel (c) shows the population structure without harvesting ($m = 0.1$); when adults (black) are superior competitors as a result of low turbidity levels (left column) and the two alternative states when juveniles (gray) are superior, as a result of high turbidity levels (right columns). Panel (d) shows the population's stage structures at a high mortality level, for both a situation when adults (black) are superior competitors (left column, $m = 0.3$), and for the two alternative states when juveniles (gray) are superior, (right columns, $m = 0.14$ and 0.3). Note that these alternative states depend on the initial conditions. When starting from the adult-dominated state, increasing the mortality rate will result in a switch to the juvenile-dominated state (indicated by an arrow in panel d). Juveniles are shown in gray and adults in black.

areas, turbidity and organic matter input in lakes are predicted to increase as a consequence of increasing rainfall following global warming, which links our example to ongoing climate change.

Finally, we would like to point out that these changes in environmental factors might occur in both spatial and temporal contexts. Gradients in productivity, temperature, as well many other environmental factors follow a predictable latitudinal pattern, but also occur due to human disturbance, such as eutrophication. We argue that it is useful to consider ontogenetically driven phenomena and changes in population regulation and stage structure in the context of global change and human impact.

9.6   Harvest

Humans are extremely successful at altering ecosystems through harvesting. Harvest mortality on commercially important species (e.g., sockeye or pink salmon) can reach up to 70% in some years (English *et al.*, 2011) and can cause cascading extinctions (Baum and Worm, 2009). This harvest mortality is often not equal across sizes and stages; in fact harvest is often concentrated on a small part of the total species size spectrum. Further, even with equal harvest, there are stage-specific food-web consequences of exploitation; making harvest a human disturbance that we feel deserves its own section in this chapter.

There is a long history of including age and size structure in harvesting literature, especially in forestry and fisheries research (e.g., Murphy and Smith, 1990; Newton, 1997). Size limitations to harvesting are often imposed, especially in aquatic food webs, based on the desire to protect juveniles, allowing them to reach maturity in order to add to the fitness of the population. This, along with preferences for larger, more lucrative resources, has led to harvesting regimes that target particular sizes and, often correspondingly, particular stages. While most empirical and theoretical research focuses on the impact of size-specific harvesting on sustainability and optimal yield of target species (e.g., Rijnsdorp, 1993; Hidalgo *et al.*, 2011; Brunel and Piet, 2013), size-selective harvesting has implications for the dynamics of the entire food web. Size-selective harvest may induce damped trophic cascades that propagate both up and down in the community size spectrum (Andersen and Pedersen, 2010), and can amplify oscillations and cause traveling waves in the community size spectrum (Rochet and Benoît, 2012).

Harvesting, whether it be selective or not, can also cause sudden changes in population structure through the emergent Allee effect that, as shown in the global stability section, can cascade through the entire food web. In the Baltic Sea, for example, harvesting of adult cod led to a shift in size structure of cod as well as their prey (van Leeuwen *et al.*, 2008; Casini *et al.*, 2008, 2009; Gårdmark *et al.*, 2014). Subsequently, these changes affected lower trophic levels, resulting in a decrease in zooplankton and increase in phytoplankton biomass.

Beyond the implications for the topology of the food web, harvest can have important consequences for stability. For example, stability is more likely to be lost when targeting the juvenile stage in cannibalistic populations (Magnusson, 1999). Moreover, the magnitude of these dynamical consequences of size-selective harvesting can change depending on the targeted size or stage (van Kooten *et al.*, 2007).

Based on the necessity to improve fish-stock management there has been a recent emerging push toward taking an ecosystem-based approach to fisheries management that includes multiple species (Pikitch *et al.*, 2004). Food-web ecology plays a central role in this new approach, as optimal and maximal yield depend on indirect effects of harvesting that permeate food webs through species interactions (Matsuda and Abrams, 2006). Several types of models have been implemented for multispecies management; however, they often do not include food-dependent growth, size-dependent interactions, or dynamic energy budgets (Persson *et al.*, 2014). The inclusion of these ontogenetic traits is essential for an assessment of the effect of harvesting on populations and even more so for the effect of harvesting on ensembles of interacting species (see Persson *et al.*, 2014 for further reading).

9.7   Above and Beyond the Food-Web Module

While we point out some important consequences of incorporating the life cycle, there are some outcomes that have not been explored yet. Including ontogeny adds a new type of link that changes the way that biomass flows through the community, and while this has been explored in combination with trophic interactions, other types of links should be considered. Researchers are starting to create models that mix interaction types (Kéfi *et al.*, 2012; Mougi and Kondoh, 2012) and including ontogenetic structure will increase the possibility that a species will exhibit multiple types of interactions with other species in the food web.

The insights that were gained by systematically adding stage and size structure generally come from food-web modules (de Roos and Persson, 2013). Scaling up from the food-web module to complex ecological networks will allow us to determine how the life cycle impacts more complex webs. For example, it is important to determine how ontogenetically driven phenomena, such as biomass overcompensation and emergent facilitation, play a role in large food webs. However, when adding the life cycle to complex ecological networks, it should be considered "when" and "where" this is done, as adding ontogeny will necessarily add more complexity. Unfortunately, there is so far no simple answer to the question: When does ontogenetic complexity matter?

Beyond the question of when complexity should be added, it remains so far unresolved at which level of organization unstructured models should be compared with structured models (e.g., individual, population, trait, community, food web, landscape) and which or how many levels need to correspond such that the system is sufficiently realistically represented by a model.

While it is useful to compare structured models with models that do not include ontogeny, the biggest test of models still remains their ability to explain the real world. An effective way to determine the required amount of detail in a model is by confronting model predictions with real data. However, the full species life cycle is often not covered by data collection; at times, sampling of certain life stages is even actively avoided. Data collection tends to be focused on adult individuals because they are generally larger and to reduce stage-specific variation in datasets. This bias leaves us with a minimal amount of information regarding the role of young individuals in food webs (with the exception of some forestry datasets). In order for empirical data to be used to test structured models, it needs to be collected with species ontogeny in mind.

We have demonstrated the importance and far-reaching consequences of including ontogenetic development and, we hope, provided compelling arguments in support of the notion that ontogenetic development is of importance to many research questions in food-web ecology. We covered a broad range of topics, including food-web topology and structure, stability and coexistence, alternative states, environmental gradients, and the effects of harvesting. All these aspects have major implications for the management of food webs, understanding and maintaining biodiversity, and for predicting the responses of food webs to changing environmental conditions.

References

Abrams, P. A. (2011). Simple life-history omnivory: responses to enrichment and harvesting in systems with intraguild predation. *American Naturalist*, **178**(3), 305–319.

Abrams, P. A. and Walters, C. J. (1996). Invulnerable prey and the paradox of enrichment. *Ecology*, **77**(4), 1125–1133.

Andersen, K. H. and Pedersen, M. (2010). Damped trophic cascades driven by fishing in model marine ecosystems. *Proceedings of the Royal Society B: Biological Sciences*, **277**, 795–802.

Baum, J. K. and Worm, B. (2009). Cascading top–down effects of changing oceanic predator abundances. *Journal of Animal Ecology*, **78**(4), 699–714.

Benoît, E. and Rochet, M. J. (2004). A continuous model of biomass size spectra governed by predation and the effects of fishing on them. *Journal of Theoretical Biology*, **226**(1), 9–21.

Berlow, E. L. (1999). Strong effects of weak interactions in ecological communities. *Nature*, **398**(6725), 330.

Blanchard, J. L., Jennings, S., Law, R., *et al.* (2009). Size-spectra dynamics from stochastic predation and growth of individuals. *Journal of Animal Ecology*, **78**(1), 270–280.

Blanchard, J. L., Law, R., Castle, M. D., and Jennings, S. (2011). Coupled energy pathways and the resilience of size-structured food webs. *Theoretical Ecology*, **4**(3), 289–300.

Boerlijst, M. C., Oudman, T., and de Roos, A. M. (2013). Catastrophic collapse can occur without early warning: examples of silent catastrophes in structured ecological models. *PloS One*, **8**(4).

Bolnick, D. I., Amarasekare, P., Araújo, M. S., *et al.* (2011). Why intraspecific trait variation matters in community ecology. *Trends in Ecology & Evolution*, **26**(4), 183–192.

Brose, U., Williams, R. J., and Martinez, N. D. (2006a). Allometric scaling enhances stability in complex food webs. *Ecology Letters*, **9**(11), 1228–1236.

Brose, U., Jonsson, T., Berlow, E., *et al.* (2006b). Consumer-resource body-size relationships in natural food webs. *Ecology*, **87**(10), 2411–2417.

Brose, U., Blanchard, J. L., Eklöf, A., *et al.* (2016). Predicting the consequences of species loss using size-structured biodiversity approaches. *Biological Reviews*. doi:10.1111/brv.12250.

Brunel, T. and Piet, G. J. (2013). Is age structure a relevant criterion for the health of fish stocks? *ICES Journal of Marine Science*, **70**, 270–283.

Byström, P. and Andersson, J. (2005). Size-dependent foraging capacities and intercohort competition in an ontogenetic omnivore (Arctic char). *Oikos*, **3**, 523–536.

Byström, P. and Garcia-Berthou, E. (1999). Density dependent growth and size specific competitive interactions in young fish. *Oikos*, **86**, 217–232.

Casini, M., Lövgren, J., Hjelm, J., *et al.* (2008). Multi-level trophic cascades in a heavily exploited open marine ecosystem. *Proceedings of the Royal Society B: Biological Sciences*, **275**, 1793–1801.

Casini, M., Hjelm, J., Molinero, J.-C., *et al.* (2009). Trophic cascades promote threshold-like shifts in pelagic marine ecosystems. *Proceedings of the National Academy of Sciences of the United States of America*, **106**, 197–202.

Caskenette, A. L. and McCann, K. S. (2017). Biomass reallocation between juveniles and adults mediates food web stability by distributing energy away from strong interactions. *PLoS ONE*, **12**(1), e0170725. doi:10.1371/journal.pone.0170725.

Charnov, E. L. (2001). Reproductive efficiencies in the evolution of life histories. *Evolutionary Ecology Research*, **3**(7), 873–876.

Chesson, P. (2000). Mechanisms of maintenance of species diversity. *Annual Review of Ecology and Systematics*, **31**(2000), 343–358.

Chesson, P. and Kuang, J. J. (2008). Competition and biodiversity in spatially structured habitats. *Nature*, **456**(7219), 235–238.

Claessen, D., de Roos, A. M., and Persson, L. (2004). Population dynamic theory of size-dependent cannibalism. *Proceedings of the Royal Society B: Biological Sciences*, **271**(1537), 333–340.

de Roos, A. M. and Persson, L. (2001). Physiologically structured models: from versatile technique to ecological theory. *Oikos*, **94**(1), 51–71.

de Roos, A. M. and Persson, L. (2003). Competition in size-structured populations: mechanisms inducing cohort formation and population cycles. *Theoretical Population Biology*, **63**(1), 1–16.

de Roos, A. M. and Persson, L. (2013). *Population and Community Ecology of Ontogenetic Development*. Princeton, NJ: Princeton University Press.

de Roos, A. M., Persson, L., and McCauley, E. (2003a). The influence of size-dependent life-history traits on the structure and dynamics of populations and communities. *Ecology Letters*, **6**(5), 473–487.

de Roos, A. M., Persson, L., and Thieme, H. R. (2003b). Emergent Allee effects in top predators feeding on structured prey populations. *Proceedings of the Royal Society B: Biological Sciences*, **270**(1515), 611–618.

de Roos, A. M., Schellekens, T., van Kooten, T., van de Wolfshaar, K. E., Claessen, D., and Persson, L. (2007). Food-dependent growth leads to overcompensation in stage-specific biomass when mortality increases: the influence of maturation versus reproduction regulation. *American Naturalist*, **170**(3), E59–E76.

de Roos, A. M., Schellekens, T., van Kooten, T., and Persson, L. (2008a). Stage-specific predator species help each other to persist while competing for a single prey. *Proceedings of the National Academy of Sciences of the United States of America*, **105**(37), 13930–13935.

de Roos, A. M., Schellekens, T., van Kooten, T., van de Wolfshaar, K. E., Claessen, D., and Persson, L. (2008b). Simplifying a physiologically structured population model to a stage-structured biomass model. *Theoretical Population Biology*, **73**(1), 47–62.

Digel, C., Curtsdotter, A., Riede, J., Klarner, B., and Brose, U. (2014). Unravelling the complex structure of forest soil food webs: higher omnivory and more trophic levels. *Oikos*, **123**(10), 1157–1172.

Eklöf, A., Jacob, U., Kopp, J., *et al.* (2013). The dimensionality of ecological networks. *Ecology Letters*, **16**(5), 577–583.

English, K. K., Edgell, T. C., Bocking, R. C., Link, M., and Raborn, S. (2011). Fraser River sockeye fisheries and fisheries management and comparison with Bristol Bay sockeye fisheries. *LGL Ltd. Cohen Commission Technical Report* 7, 190.

Estes, J. A., Terborgh, J., Brashares, J. S., *et al.* (2011). Trophic downgrading of planet Earth. *Science*, **333**(6040), 301–306.

Gårdmark, A., Casini, M., Huss, M., *et al.* (2014). Regime shifts in exploited marine food-webs: detecting mechanisms underlying alternative stable states using size-structured community dynamics theory. *Philosophical Transactions of the Royal Society B: Biological Sciences*, **370**, DOI: 10.1098/rstb.2013.0262.

Gilljam, D., Thierry, A., Edwards, F. K., *et al.* (2011). Seeing double: size-based and taxonomic views of food web structure. *Advances in Ecological Research*, **45**, 67–133.

Glazier, D. S. (2005). Beyond the "3/4-power law": variation in the intra- and inter-specific scaling of metabolic rate in animals. *Biological Reviews of the Cambridge Philosophical Society*, **80**(4), 611–662.

Guill, C. (2009). Alternative dynamical states in stage-structured consumer populations. *Theoretical Population Biology*, **76**(3), 168–178.

Gurney, W. and Nisbet, R. (1985). Fluctuation periodicity, generation separation, and the expression of larval competition. *Theoretical Population Biology*, **180**, 150–180.

Hartvig, M. (2011). *Food Web Ecology: Individual Life-Histories and Ecological Processes Shape Complex Communities*. Ph.D. thesis, Department of Biology, Lund University, Sweden.

Hartvig, M. and Andersen, K. H. (2013). Coexistence of structured populations with size-based prey selection. *Theoretical Population Biology*, **89**, 24–33.

Hartvig, M., Andersen, K. H., and Beyer, J. E. (2011). Food web framework for size-structured populations. *Journal of Theoretical Biology*, **272**(1), 113–122.

Hastings, A. (1983). Age-dependent predation is not a simple process. 1. Continuous – time models. *Theoretical Population Biology*, **23**(3), 347–362.

Heckmann, L., Drossel, B., Brose, U., and Guill, C. (2012). Interactive effects of body-size structure and adaptive foraging on food-web stability. *Ecology Letters*, **15**(3), 243–250.

Hidalgo, M., Rouyer, T., Molinero, J. C., *et al.* (2011). Synergistic effects of fishing-induced demographic changes and climate variation on fish population dynamics. *Marine Ecology Progress Series*, **2011**(426), 1–12.

Hin, V., Schellekens, T., Persson, L., and de Roos, A. M. (2011). Coexistence of predator and prey in intraguild predation systems with ontogenetic niche shifts. *American Naturalist*, **178**(6), 701–714.

Huss, M. and Nilsson, K. A. (2011). Experimental evidence for emergent facilitation: promoting the existence of an invertebrate predator by killing its prey. *Journal of Animal Ecology*, **80**(3), 615–621.

Hutchings, J. and Myers, R. (1994). What can be learned from the collapse of a renewable resource? Atlantic cod, *Gadus morhua*, of Newfoundland and Labrador. *Canadian Journal of Fisheries and Aquatic Sciences*, **51**, 2126–2146.

Jennings, S., Pinnegar, J. K., Polunin, N. V. C., and Boon, T. W. (2001). Weak cross-species relationships between body size and trophic level belie powerful size-based trophic structuring in fish communities. *Ecology Letters*, **70**(6), 934–944.

Kartascheff, B., Heckmann, L., Drossel, B., and Guill, C. (2010). Why allometric scaling enhances stability in food web models. *Theoretical Ecology*, **3**(3), 195–208.

Kéfi, S., Berlow, E. L., Wieters, E. A., *et al.* (2012). More than a meal. . . integrating non-feeding interactions into food webs. *Ecology Letters*, **15**(4), 291–300.

Kitchell, J. F., Stewart, D. J., and Weininger, D. (1977). Applications of a bioenergetics model to yellow perch (*Perca flavescens*) and walleye (*Stizostedion vitreum vitreum*). *Journal of the Fisheries Research Board of Canada*, **34**(10), 1922–1935.

Kooijman, S. A. L. M. (2000). *Dynamic Energy and Mass Budgets in Biological Systems*. Cambridge, UK: Cambridge University Press.

Law, R., Plank, M. J., James, A., and Blanchard, J. L. (2009). Size-spectra dynamics from stochastic predation and growth of individuals. *Ecology*, **90**(3), 802–811.

Magnússon, K. G. (1999). Destabilizing effect of cannibalism on a structured predator–prey system. *Mathematical Biosciences*, **155**, 61–75.

Martinez, N. (1991). Artifacts or attributes? Effects of resolution on the Little Rock Lake food web. *Ecological Monographs*, **61**(4), 367–392.

Matsuda, H. and Abrams, P. A. (2006). Maximal yields from multispecies fisheries systems: rules for systems with multiple trophic levels. *Ecological Applications*, **16**(1), 225–237.

May, R. M. (1972). Will a large complex system be stable? *Nature*, **238**, 413–414.

May, R. M. (2006). Network structure and the biology of populations. *Trends in Ecology and Evolution*, **21**(7), 394–399.

McCann, K., Hastings, A., and Huxel, G. (1998). Weak trophic interactions and the balance of nature. *Nature*, **395**, 794–798.

Metz, J. A. J. and Diekmann, O. (1986). *The Dynamics of Physiologically Structured Populations*. Lecture Notes in Biomathematics, vol. 68. Amsterdam: Springer-Verlag.

Mougi, A. and Kondoh, M. (2012). Diversity of interaction types and ecological community stability. *Science*, **349**(2012), 337.

Murdoch, W. W., Kendall, B. E., Nisbet, R. M., *et al.* (2002). Single-species models for many-species food webs. *Nature*, **417**(6888), 541–543.

Murphy, L. F. and Smith, S. J. (1990). Optimal harvesting of an age-structured population. *Journal of Mathematical Biology*, **29**, 77–90.

Mylius, S. D., Klumpers, K., de Roos, A. M., and Persson, L. (2001). Impact of intraguild predation and stage structure on simple communities along a productivity gradient. *American Naturalist*, **158**(3), 259–276.

Nakazawa, T. (2011). Ontogenetic niche shift, food-web coupling, and alternative stable states. *Theoretical Ecology*, **4**(4), 479–494.

Newton, P. F. (1997). Stand density management diagrams: review of their development and utility in stand-level management planning. *Forest Ecology and Management*, **98**, 251–265.

Ohlberger, J., Langangen, Ø., Edeline, E., *et al.* (2011). Stage-specific biomass over-compensation by juveniles in response to increased adult mortality in a wild fish population. *Ecology*, **92**(12), 2175–2182.

Oksanen, L., Fretwell, S. D., Arruda, J., and Niemela, P. (1981). Exploitation ecosystems in gradients of primary productivity. *American Naturalist*, **118**(2), 240–261.

Persson, L. (1999). Trophic cascades: abiding heterogeneity and the trophic level concept at the end of the road. *Oikos*, **85**(3), 385–397.

Persson, L., de Roos, A. M., Claessen, D., *et al.* (2003). Gigantic cannibals driving a whole-lake trophic cascade. *Proceedings of the National Academy of Sciences of the United States of America*, **100**(7), 4035–4039.

Persson, L., Amundsen, P.-A., de Roos, A. M., Klemetsen, A., Knudsen, R., and Primicerio, R. (2007). Culling prey promotes predator recovery: alternative states in a whole-lake experiment. *Science*, **316**(5832), 1743–1746.

Persson, L., van Leeuwen, A., and de Roos, A. M. (2014). The ecological foundation for ecosystem-based management of fisheries: mechanistic linkages between the individual-, population-, and community-level dynamics. *ICES Journal of Marine Science*, **71**(8), 2268–2280.

Peters, R. H. (1983). *The Ecological Implications of Body Size*. New York: Cambridge University Press.

Pikitch, E., Santora, C., Babcock, E., *et al.* (2004). Policy forum: ecosystem-based fishery management. *Science*, **305**, 346–347.

Pimm, S. L. and Rice, J. C. (1987). The dynamics of multispecies, multi-life-stage models of aquatic food webs. *Theoretical Population Biology*, **32**(3), 303–325.

Pimm, S., Lawton, J., and Cohen, J. (1991). Food web patterns and their consequences. *Nature*, **350**, 669–674.

Plank, M. J. and Law, R. (2012). Size-spectra dynamics from stochastic predation and growth of individuals. *Theoretical Ecology*, **5**(4), 465–480.

Polis, G. (1984). Age structure component of niche width and intraspecific resource partitioning: can age groups function as ecological species? *American Naturalist*, **123**(4), 541–564.

Rall, B. C., Brose, U., Hartvig, M., *et al.* (2012). Universal temperature and body-mass scaling of feeding rates. *Philosophical Transactions of the Royal Society B: Biological Sciences*, **1605**, 2923–2934.

Rijnsdorp, A. D. (1993). Fisheries as a large-scale experiment on life-history evolution: disentangling phenotypic and genetic effects in changes in maturation and reproduction of North Sea plaice, *Pleuronectes platessa* L. *Oecologia*, **96**(3), 391–401.

Rochet, M.-J. and Benoit, E. (2012). Fishing destabilizes the biomass flow in the marine size spectrum. *Proceedings of the Royal Society B: Biological Sciences*, **279**(1727), 284–292.

Rooney, N., McCann, K., Gellner, G., and Moore, J. C. (2006). Structural asymmetry and the stability of diverse food webs. *Nature*, **442**(7100), 265–269.

Rosenzweig, M. and MacArthur, R. (1963). Graphical representation and stability conditions of predator–prey interactions. *American Naturalist*, **97**(895), 209.

Rossberg, A. G., Brännström, A., and Dieckmann, U. (2010a). Food-web structure in low- and high-dimensional trophic niche spaces. *Journal of the Royal Society Interface*, **7**(53), 1735–1743.

Rossberg, A. G., Brännström, Å., and Dieckmann, U. (2010b). How trophic interaction strength depends on traits. *Theoretical Ecology*, **3**(1), 13–24.

Rudolf, V. H. W. and Lafferty, K. D. (2011). Stage structure alters how complexity affects stability of ecological networks. *Ecology Letters*, **14**(1), 75–79.

Rudolf, V. H. W. and Rasmussen, N. L. (2013a). Ontogenetic functional diversity: size structure of a keystone predator drives functioning of a complex ecosystem. *Ecology*, **94**(5), 1046–1056.

Rudolf, V. H. W. and Rasmussen, N. L. (2013b). Population structure determines functional differences among species and ecosystem processes. *Nature Communications*, **4**, 2318.

Scheffer, M., Carpenter, S., Foley, J. A., Folke, C., and Walker, B. (2001). Catastrophic shifts in ecosystems. *Nature*, **413**(6856), 591–596.

Scheffer, M., Bascompte, J., Brock, W. A., *et al.* (2009). Early-warning signals for critical transitions. *Nature*, **461**(7260), 53–59.

Schellekens, T., de Roos, A. M., and Persson, L. (2010). Ontogenetic diet shifts result in niche partitioning between two consumer species irrespective of competitive abilities. *American Naturalist*, **176**(5), 625–637.

Schreiber, S. and Rudolf, V. H. W. (2008). Crossing habitat boundaries: coupling dynamics of ecosystems through complex life cycles. *Ecology Letters*, **11**(6), 576–587.

Silvert, W. and Platt, T. (1980). Dynamic energy-flow model of the particle size distribution in pelagic ecosystems. In *Evolution and Ecology of Zooplankton Communities*, ed. W. C. Kerfoot, Illanover, NH: University Press of New England, pp. 754–763.

Stearns, S. (1989). Trade-offs in life-history evolution. *Functional Ecology*, **3**(3), 259–268.

Tilman, D. (1994). Competition and biodiversity in spatially structured habitats. *Ecology*, **75**(1), 1228–1236.

Ursin, E. (1973). On the prey size preferences of cod and dab. *Meddelelser fra Danmarks Fiskeri-og Havundersgelser*, **7**, 85–98.

van de Wolfshaar, K., de Roos, A., and Persson, L. (2006). Size-dependent interactions inhibit coexistence in intraguild predation systems with life-history omnivory. *American Naturalist*, **168**(1), 62–75.

van de Wolfshaar, K. E., HilleRisLambers, R., and Gårdmark, A. (2011). Effect of habitat productivity and exploitation on populations with complex life cycles. *Marine Ecology Progress Series*, **438**, 175–184.

van Kooten, T., Persson, L., and de Roos, A. M. (2007). Size-dependent mortality induces life-history changes mediated through population dynamical feedbacks. *American Naturalist*, **170**, 258–270.

van Leeuwen, A., de Roos, A. M., and Persson, L. (2008). How cod shapes its world. *Journal of Sea Research*, **60**(1–2), 89–104.

Walters, C. and Kitchell, J. F. (2001). Cultivation/depensation effects on juvenile survival and recruitment: implications for the theory of fishing. *Canadian Journal of Fisheries and Aquatic Sciences*, **58**(1), 39–50.

Werner, E. and Gilliam, J. (1984). The ontogenetic niche and species interactions in size-structured populations. *Annual Review of Ecology and Systematics*, **15**, 393–425.

Woodward, G., Blanchard, J., Lauridsen, R. B., *et al.* (2010). Individual-based food webs: species identity, body size and sampling effects. *Advances in Ecological Research*, **43**, 211–266.

Yodzis, P. (1981). The stability of real ecosystems. *Nature*, **289**, 674–676.

Zhang, L., Thygesen, U. H., Knudsen, K., and Andersen, K. H. (2013). Trait diversity promotes stability of community dynamics. *Theoretical Ecology*, **6**(1), 57–69.

10 Importance of Trait-Related Flexibility for Food-Web Dynamics and the Maintenance of Biodiversity

Ursula Gaedke, Beatrix E. Beisner, Amrei Binzer, Amy Downing, Christian Guill, Toni Klauschies, Jan J. Kuiper, Floor H. Soudijn, and Wolf M. Mooij

10.1 Introduction

Although the ubiquitous biodiversity-related flexibility of ecological systems is qualitatively well established, most empirical and theoretical studies regard ecological systems so far as units with rigid, predefined properties. The reason for this static approach is that incorporating the tremendous diversity and flexibility of natural systems into empirical and theoretical studies has been extremely challenging in terms of developing consistent mathematical frameworks and designing appropriate experiments. This approach has also been necessary owing to the lack of empirical data on the ability of species to change properties over time. A recent approach to solve this problem is to move from a species- to a trait-based perspective. This is not just a change in terminology but in concept, providing a mechanistic basis for biodiversity–ecosystem function relationships and improving our potential to identify general rules in community ecology (McGill *et al.*, 2006; Savage *et al.*, 2007; Hillebrand and Matthiessen, 2009). Functional traits are used to link species to their function in the ecosystem. They are well defined, measurable properties of individuals (e.g., edibility or diet selectivity) affecting their performance and responses to environmental changes and hence population and community dynamics as well as trophic interactions. The frequency distribution of functional traits (Figure 10.1a) enables a quantification of functional diversity. Large variation in trait values (e.g., a full range from highly edible, fast growing to almost inedible, slow growing species) implies a high functional diversity and vice versa. This trait distribution may be described by its shape and central tendency (Figure 10.1b) and may change in response to altered abiotic (e.g., temperature) and biotic conditions (e.g., predator density) and thus characterize the milieu with which individual organisms interact (McGill *et al.*, 2006) (Figure 10.1c).

10.1.1 Scientific Background

Maintaining the different kinds of ecosystem services in a way that optimizes human well-being and economy is one of the most urgent tasks of our century, which challenges policy-makers as well as scientists. The frequency and intensity of land use, climate

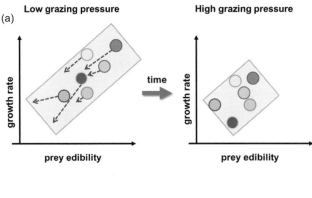

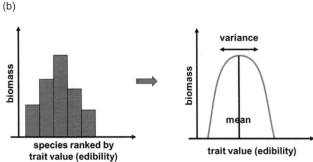

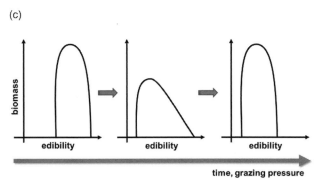

Figure 10.1 Individuals are not entirely alike, but differ – to varying degrees – in their trait values. (a) Example displaying the location of six individuals (or populations) in a two-dimensional trait space with the two functional traits "growth rate" and "prey edibility" under low grazing pressure (left). The trait values are correlated with each other due to a trade-off between the growth rate and edibility. In the upper right corner of the trait space, growth but also vulnerability to grazing are maximal (and minimal in the lower left corner). Under high grazing pressure, the individuals (or populations) move in the trait space (dashed arrows) to re-adjust their trade-off between growth and grazing losses depending on their specific flexibility (right). (b) The trait distribution is the biomass-weighted frequency with which a functional trait is represented in a population or community and reflects the functional diversity present (left discrete, right continuous form). (c) Responding to altered abiotic or biotic conditions, trait distributions change in time, which in turn alters system dynamics.

change, and other anthropogenically induced environmental disturbances are accelerating biodiversity declines worldwide. The negative impact of these processes on ecological systems (e.g., individuals, populations, communities, and food webs) may amplify each other: environmental changes can accelerate biodiversity loss and a reduced biodiversity may increase the sensitivity of ecological systems to environmental changes. So far, comprehensive research has revealed population and community responses to altered environmental conditions (e.g., Wagner and Benndorf, 2007), but have mostly ignored the potential effects of simultaneously altered biodiversity influencing future potential to adjust to these changed conditions. Intensive biodiversity research led to new insights into the effects of biodiversity on separate ecosystem properties, including average plant biomass and its overall variability (Hooper *et al.*, 2005). Yet, rarely has the way in which biodiversity itself impacts the dynamic behavior and organism responses to altered environmental conditions been studied in large-scale terrestrial, marine, or freshwater biodiversity experiments. Recent evidence shows, however, that biodiversity very likely influences ecological dynamics given that changes in one part of the ecosystem are not immediately, proportionally, or directly transmitted to the whole system (Bolnick *et al.*, 2011). Rather, the inherent biodiversity of ecological systems may promote adjustments at lower hierarchical levels that may buffer the response at a higher level (Verschoor *et al.*, 2004; Tirok and Gaedke, 2010). With respect to the different facets of biodiversity (e.g., phenotypic, genetic, and species diversity) individuals, populations, and communities may change their properties to adjust to ambient conditions and improve their fitness, thereby influencing food-web dynamics (Norberg, 2004; Tirok and Gaedke, 2010; Tirok *et al.*, 2011).

Consider an example of increasing herbivory leading to a higher proportion of less edible plants (algae) possessing chemical or structural defense mechanisms. Plant biomass losses are thus dampened as herbivory declines (Sommer *et al.*, 2003), which could feed back on the herbivore biomass and community composition through an increase in the proportion of herbivorous species able to exploit less edible algae. As a result, the relative advantage possessed by less edible algae compared to edible ones is reduced. Overall, this dynamic promotes the coexistence of different algal types, enhancing biodiversity. Such ongoing feedbacks are based on the biodiversity and functional characteristics of interacting communities (e.g., mean edibility of algae and selectivity of the herbivores) and the properties of individual species (e.g., formation of morphological defenses against predation, altered behavior in the presence of predators, prey switching, etc.). This biodiversity-dependent potential for adjustment leads to complex feedbacks and simultaneous changes at different levels of ecological organization. For example, a diverse group of predators (species level) may adjust its diet composition to altered prey availability (environmental change), which, in turn, reduces the consequences of the environmental change for the overall energy flow in the food web (system level). As a consequence, the responses of complex and highly interconnected networks such as food webs to altered conditions can be very difficult to understand and predict (Duffy *et al.*, 2007), despite their importance in fundamental and applied ecology.

10.1.2   Theory Development and Mathematical Modeling Approaches

Mathematical models provide a powerful tool to elucidate the consequences of trait variation for ecological systems of different complexity. Current community models generally use static process parameterizations ignoring the natural potential to adjust to changing environmental conditions. Current guiding principles for formulating an improved generation of model equations are characterized by "adaptive optimization-based models considering the adaptive capacity of life" (Smith and Yamanaka, 2007). The pivotal role of trait variation mediating the interaction with the environment and other organisms is made explicit by trait-based modeling approaches. At least three types of approaches are currently discussed that have complementary strengths. Trait variation may be accounted for by considering explicitly a number of functionally different entities such as several prey species of different susceptibility to grazing (e.g., species sorting models, Figure 10.1b left; Yoshida *et al.*, 2007; Tirok and Gaedke, 2010), which allows direct links to empirical work. Alternatively, functionally different entities are represented by a continuous trait value distribution (Figure 10.1b right), which reduces model complexity (Wirtz and Eckhardt, 1996; Norberg, 2004; Tirok *et al.*, 2011). Some studies used this so-called dynamic trait or gradient-dynamics approach to describe evolution (Abrams and Matsuda, 1997) and coevolution of predator and prey (Dieckmann and Law, 1996), adaptive behavioral dynamics, and community dynamics (Savage *et al.*, 2007; Merico *et al.*, 2014). Both approaches are based on ordinary differential equations describing biomass dynamics in time. The second one also incorporates partial differential equations to describe changes of trait distributions with selection pressure. In combination, they allow considering of, for example, simultaneously intra- and interspecific trait variation (see Chapter 8 in this book). The advantages of these approaches (e.g., tractability, straightforward links to well-known theory) may be combined with individual-based modeling, which enables us to represent the dynamics of the shape of the trait distribution explicitly (Grimm *et al.*, 2005; May *et al.*, 2009; dos Santos *et al.*, 2011) and to include individual-level theories accounting for trade-offs between traits (Kooijman, 2010).

In addition to phenotypic plasticity and evolution, ontogenetic growth may provide a substantial source of intraspecific trait variation. In population models it is possible to account for the full, continuous ontogenetic trait distribution of populations. So-called physiologically structured population models keep track of several characteristics of individuals in a population simultaneously (e.g., body size and age) (Metz and Diekmann, 1986). However, when modeling large ensembles of species, such as food webs, this approach becomes exceedingly complex and an approach that approximates the trait distribution with discrete stages can be more appropriate (De Roos *et al.*, 2008). For further and more detailed information please refer to Chapter 9 in this book.

The different types of models also differ with respect to the implementation of mechanisms to avoid species extinctions or to maintain trait variation. In natural systems trait variation may be locally supported by factors such as sexual reproduction, mutations, immigration, refuges (spatial heterogeneity), seed banks, dormant stages, and predator behavior (Holling type III functional response). It has been speculated (Jones *et al.*, 2009) or shown (Tirok *et al.*, 2011; Merico *et al.*, 2014) that model results may be sensitive to such assumptions. Hence the robustness of model results against the implicitly or explicitly stated underlying assumptions must be carefully tested. Additionally, the type of dynamics revealed might critically depend on the nature of trade-offs assumed in the model, which can differ among ecosystem types. Furthermore, the mechanisms giving rise to trait variation, such as phenotypic plasticity, inherited changes, or species sorting, may be relevant as well (Cortez, 2011).

10.2   Endogenously Driven Feedbacks Between Traits and Biomass Dynamics That Maintain Biodiversity

Feedbacks between trait and biomass dynamics may be relevant in nature, as indicated by evidence from highly controlled chemostat experiments with fast living microorganisms (e.g., Becks *et al.*, 2010) and some field observations (Tirok and Gaedke, 2007; Post and Palkovacs, 2009). For example, an attempt to model the empirically well-studied seasonal dynamics of the species-rich phytoplankton and ciliate communities in the large deep Lake Constance failed when using established types of models based on ordinary differential equations. The models recurrently failed to reproduce the observed spring dynamics characterized by rather high and constant biomasses of both predator and prey when using realistic parameter values and assumptions concerning ciliate mortality. This stimulated a more profound data analysis revealing that during the spring bloom, the relative contribution of the selective ciliated interception feeders was positively related to that of the highly edible algae constituting most of their potential diet, whereas the relative contribution of generalist filter feeders correlated positively with that of the less edible algae that encompassed part of their diet spectrum (Tirok and Gaedke, 2007). In contrast to the rather constant community biomasses, pronounced variability in population biomasses and species shifts were observed in both algae and ciliates, but were unrelated to seasonal changes. They were presumably internally forced: selective feeders alternated with generalist consumers in their relative importance, as did their edible and less edible prey leading to recurrent coupled biomass-trait dynamics owing to partly coupled predator–prey cycles.

When including these insights in a standard Rosenzweig–MacArthur-type model of three interacting phytoplankton and three ciliate species differing in their edibility and selectivity, the model reproduced the aforementioned patterns, i.e., rather constant community biomass, highly variable population biomasses, and recurrent changes in the trait values of the predator and prey communities (Tirok and Gaedke, 2010). This model assumed two simple trade-offs between trait values: edible algae had higher growth rates than less edible ones, and generalists had higher half-saturation constants than specialists, thus requiring higher overall food concentrations. The model enabled the simultaneous adjustment of traits on both trophic levels for the first time in such species- or clonal-sorting models. It resulted in dynamics strongly deviating from the classical predator–prey dynamics. Results included intermittent patterns where dampened classical predator–prey cycles alternated with more complex dynamics, including out-of-phase cycles between predator and prey community biomasses. They originated – among others – from the temporally variable dominance of individual species and reflected the patterns observed in the Lake Constance data (Tirok and Gaedke, 2010). In line with these findings, the observed seasonal biomass and production patterns in a pelagic food web with 20 plankton guilds of the large, well-studied Lake Constance could only be reproduced by a model that allowed for trait variation (Boit et al., 2012).

Using the same basic model structure but incorporating the dynamic trait approach revealed that the maintenance of trait variation critically depended on the number, type, and shape of the trade-off curves (Jones et al., 2009; Tirok et al., 2011). Previous models using this approach for describing community dynamics restricted the potential for trait variation to the first trophic level (Abrams and Matsuda, 1997; Cortez and Ellner, 2010). However, because the data suggested that adjacent trophic levels strongly influenced each other's biomass and trait dynamics, species compositions of the two trophic levels were described by the mean functional traits (i.e., prey edibility and predator food-selectivity), and functional diversities by their variances (Tirok et al., 2011). As in Tirok and Gaedke (2010), altered edibility triggered shifts in food selectivity so that consumers continuously responded to the present prey composition, and vice versa. This trait-mediated feedback mechanism resulted in a complex dynamic behavior with ongoing oscillations in the mean trait values, reflecting a continuous reorganization on both trophic levels. Trait variation was internally maintained among the prey because no ideal trait value existed for any composition of the predator trophic level due to the trade-offs between edibility, growth, and carrying capacity. The predators were only subject to a single trade-off between food-selectivity and grazing ability and in the absence of trait variation in the prey, one predator type became dominant and thus trait variation declined to zero. Ignoring substantial trait variation, the system showed the same dynamics as conventional predator–prey models. Hence these studies suggested that resolving variation in traits together with the respective trade-offs on different trophic levels can strongly shape the outcome of mathematical models including the maintenance of trait variation. To conclude, accounting for trait variation at two adjacent trophic levels reproduced the observed Lake Constance dynamics at the population and community levels, but the field observation did not permit for a rigorous test of the underlying mechanisms. Thus a more in-depth analysis of the field data, and the analysis of simpler, more artificial but highly controlled laboratory systems such as algae–rotifer chemostats are required.

Pioneering experimental work using a defended and an undefended algal clone and a non-adaptive predator supplemented by respective modeling revealed that adaptive variation in a prey trait relevant for the trophic interactions can lead to eco-evolutionary cycles, i.e., changes in the ecological interactions drive changes in the frequency of genotypes (evolution) feeding back on the ecological interactions (Post and Palkovacs, 2009). Backed-up by empirical evidence the corresponding models assumed that the edible clone had a higher growth rate or a lower half-saturation constant for the limiting nutrient than the less edible one. These experiments stimulated further modeling studies, which suggested among others that eco-evolutionary cycles arise when the defense against grazing is effective but involves low costs (Jones and Ellner, 2007). If defense is very cheap so-called cryptic cycles were found, which were characterized by relatively pronounced predator cycles and low variability in total prey biomass. The latter was due to almost exact out-of-phase cycles of the defended and the undefended algal type (Jones and Ellner, 2007; Jones *et al.*, 2009). In the specific chemostat setting, coexistence of algal clones and thus the maintenance of trait variation enabling eco-evolutionary cycles typically demanded large differences in trait values. Thus relatively small differences in the initial trait range could change not only the type of dynamics, e.g., stasis, usual quarter-lagged or out-of-phase, or cryptic predator–prey cycles (e.g., Yoshida *et al.*, 2003; 2007; Becks *et al.*, 2010; 2012), but also the amount of trait variation maintained (coexistence of both prey types or competitive exclusion) in this chemostat system. To summarize, substantial steps toward creating a general theory on eco-evolutionary dynamics have been achieved for simple predator–prey systems.

In these chemostat systems out-of-phase predator–prey cycles were already found when only the prey were adaptive contrasting with the above-mentioned model results based on observations in Lake Constance. Here out-of-phase predator–prey cycles only emerged when both trophic levels were adaptive (Tirok *et al.*, 2011). This may be explained by the following differences. In the models reflecting the chemostat conditions the edible prey was competitively superior to the less edible one and the dilution caused continuous losses, but also potential for re-growth and thus shifts in clonal composition. As a consequence, the edible clone started to replace the less edible one at low predator densities leading to a new predator–prey cycle as this stimulated the subsequent recovery of the non-adaptive selective predator. In the models based on the Rosenzweig–MacArthur equations, the prey clones/species experienced no non-grazing mortality and were either equal competitors (Tirok and Gaedke, 2010) or – following the growth-rate hypothesis – the fast growing edible prey species were competitively inferior (Tirok *et al.*, 2011). This implies – keeping everything else equal (cf. below) – that with a non-adaptive selective predator the prey community composition gets locked in the less edible highly competitive prey state, or with a non-adaptive, non-selective predator grazing on the entire prey spectrum the different prey species coexist and get synchronized as they experience similar environmental conditions (see also Bauer *et al.*, 2014). The latter implies regular quarter-phase lagged predator–prey cycles. Recurrent out-of-phase cycles between prey and predator community biomasses were only realized when both trophic levels can exhibit trait variation: the grazing of selective predators on highly edible prey initiated the out-of-phase cycles whereas the non-selective grazing promoted the recurrent trait changes as it re-opened the possibility of edible prey increase again. The latter, in turn, provoked a change in the predator community toward more selective grazers. Thus achieving general results demands inclusion of the effect of the food-web structure.

In addition, the potential occurrence of out-of-phase dynamics between predator and prey community biomasses seemingly may not only depend on the number of trophic levels with trait variation and associated trade-offs, but also on the source of trait variation. In contrast to Cortez (2011), Mougi (2012) modeled phenotypic plasticity at two rather than one trophic level and assumed that trait changes depended on both predator and prey densities rather than only on predator densities. This may explain, respectively, the absence (Cortez, 2011) or the presence (Mougi, 2012) of out-of-phase cycles but further comparative studies are required also accounting systematically for the source of trait variation.

10.3   Consequences of Trait Variation Arising from Ontogenetic Growth Rather Than Species Sorting or Phenotypic Plasticity

Body mass is often considered as a master trait as it influences fundamental physiological rates of individuals such as metabolism and digestion (Peters, 1993) but also their foraging abilities and vulnerability to predation (Kalinkat et al., 2013). Consequently, body mass largely determines an organism's energy requirement, biomass uptake, and thus the interactions and dynamics among species (Binzer et al., 2012). Due to morphological constraints, predators are usually larger than their prey (Brose et al., 2006a). This body-mass dependence of trophic interactions leads to a body-size architecture of food webs that is favorable for the stability of multispecies communities as a whole (Brose et al., 2006b). The stabilizing effect of the body-mass scaling of populations' energetics is actually so strong that model food webs with a randomized body-size architecture evolve dynamically toward a state in which predators are on average larger than their prey (Heckmann et al., 2012).

Consequently, models of food-web structure are often based on species body size, either implicitly (Williams and Martinez, 2000) or explicitly (Petchey et al., 2008). Usually, a rigid, constant body mass is used that represents an average individual of the population independent of its life stage or age (Digel et al., 2014). However, there can be significant intraspecific variation in body mass. For unicellular organisms, which were the focus in most of the studies mentioned above, size variation may arise from phenotypic plasticity (e.g., the formation of spines) or clonal sorting. However, typically, the largest source of variation in body mass within species is ontogenetic growth of individuals (Werner and Gilliam, 1984): before they can start reproducing, individuals of any given species need to grow and at least double from their size at birth. Within-species variation in body size often matches or exceeds between-species variation in body size (Woodward and Hildrew, 2002). This holds, for example, for fish upon which many models based on the master trait size have been based. This inherent biological principle implies that when considering body mass as a trait, no additional mechanisms are necessary to maintain trait variation in the population. In summary, variation in body size due to ontogeny can be substantial, and is inevitably present in all biological populations.

To better understand the functioning and stability of food webs, we must recognize that the dynamics of a single species with ontogenetic trait diversity (e.g., the range of body sizes spanned by the life stages) is fundamentally different from the dynamics of separate species with the same trait diversity. This emphasizes that we need to understand the source of trait variation (e.g., changes in clonal or species composition, phenotypic plasticity, ontogenetic growth) if we are to understand the consequences of trait variation for system dynamics (see Chapter 8; Gaedke and Klauschies, 2016). The

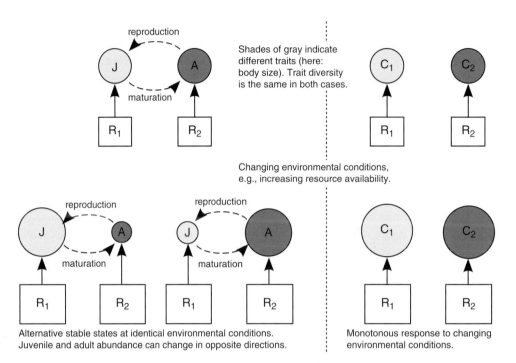

Figure 10.2 Response of consumer-resource systems (left) with and (right) without ontogenetic resource shift to changing environmental conditions. J – juvenile consumers, A – adult consumers, R_i – resources, C_i – consumers. While separate consumer species that feed on different resources will react similarly to changing environmental conditions (e.g., increasing consumer biomass as resource availability increases), consumer stages that are coupled by ontogenetic links (reproduction and maturation) can change in opposite directions even if the availability of both resources increases.

life stages of species are coupled by the ontogenetic links of maturation and reproduction. This means that unlike separate species, the net biomass production of a life stage does not only affect its own biomass, but also feeds into the biomass of another life stage (i.e., another node in the food web). This results in strikingly different effects of changing environmental conditions on the structured population compared to an ensemble of separate, unstructured populations (Figure 10.2).

If trait variation between the stages leads to an ontogenetic resource shift, alternative stable states can occur between populations dominated by small or large individuals (Guill, 2009) due to positive feedback loops that trap the population either in a juvenile-dominated or adult-dominated state. Perhaps contrary to expectations, a high biomass density of an ontogenetic stage is usually a sign of low productivity of the respective stage, as accumulating biomass limits per-capita resource availability and increases competition in the dominating stage. This leads to a low biomass density of the other stage, which in turn has a high per-capita resource availability. The resulting high per-capita production rates of biomass in the low-abundance stage consequently reinforce the strong intrastage competition in the high-abundance, low-productivity stage.

Another consequence of a size-structured population is that increased mortality at one size can lead to positive effects on the biomass of the total population (De Roos *et al.*, 2007; Guill, 2009, see also Chapter 9 in this book). The mechanism behind this biomass overcompensation works similarly to the positive feedback loop just described. In a population characterized by two stages, the overall productivity of the population will, in general, be limited by only one stage, either the juveniles and their ability to grow and mature, or the adults and their ability to produce offspring. As accumulating biomass in one stage increases competition and limits per-capita resource availability, a bottleneck is created that limits overall productivity. Increasing mortality of either stage will widen this bottleneck and increase the biomass of the more productive stage. Finally, because stage-specific mortality and resource availability are usually determined by other species in the community (predators, competitors, and resources), ontogenetic trait diversity consequently influences community dynamics and ultimately food-web stability (De Roos *et al.*, 2003; Nakazawa, 2011; Zhang *et al.*, 2013).

10.4   Maintenance of Biodiversity and Ecosystem Functioning in Variable Environments Considering a Trait-Based Approach

The experimental and theoretical studies presented so far mostly operate under constant abiotic conditions. However, trait-based approaches hold great potential for exploring the response of biodiversity and ecosystem functioning in ecosystems exposed to environmental variability. Environmental variability can occur both temporally and spatially as a result of natural processes or human-induced changes. Below we discuss three examples of how trait-based approaches have increased our understanding of how biodiversity, in the form of trait variation, and ecosystem functioning are maintained in variable environments.

10.4.1   Trait-Based Approaches Used to Understand Food-Web Composition in Environments with Natural, Temporally Recurring Fluctuations

For species that reside in environments exposed to repeated fluctuations, three main classes of life-history strategies have been identified, along which trait variation occurs. Species can take advantage of long periods of relative stability between fluctuations, and are good competitors (C-strategists or "affinity adapted" or "gleaner" species) because they have high affinity for resources, removing access to them by other species. Alternatively, species may grow quickly after a perturbation, when per-capita resources have increased. These are variously called "ruderals" (R-strategists), "velocity-adapted," "growth strategists," or "opportunists." Finally, storing resources when times are good to survive resource-scarce or otherwise harsh periods is the strategy adopted by "storage adapted" (S-strategists) species. These three main strategies define critical traits of species in fluctuating environments, and appear to be pervasive enough so that this classification scheme (C–R–S) has formed the base of functional group classifications for primary producers ranging from terrestrial plants (Grime, 1977) to aquatic phytoplankton (Grover, 1991; Sommer *et al.*, 1993). Recent studies have shown that populations face strong trade-offs in these traits (e.g., Litchman *et al.*, 2007; Edwards *et al.*, 2013a). This suggests that once environmental variation in a habitat is characterized, knowledge of species traits as they relate to the C–R–S scheme can be used to predict which species are expected to occupy a food web (e.g., Edwards *et al.*, 2013b). This trait-based approach can provide advantages over species-based approaches in at least two ways. First, trait-based approaches allow general predictions about changes in food-web structure as the rate and size of environmental fluctuations change. Second, trait-based approaches could improve efficiency in restoring food webs in conservation efforts. Much of the work to date has focused on primary producers and on single trophic level interactions, considering the trade-offs in C–R–S strategies and dispersal abilities between competitors. Future frameworks should consider how to expand this view to food webs, including animal communities.

10.4.2   Trait-Based Approaches Used to Predict Ecosystem Functioning in Environments with Human-Induced Variability

Trait-based approaches can also be used to predict the response of biodiversity and ecosystem functioning to human-induced environmental change (Violle *et al.*, 2012; Tomimatsu *et al.*, 2013). A particularly useful trait-based approach has been to describe species in terms of functional response traits and functional effect traits (e.g., Naeem and Wright, 2003; Violle *et al.*, 2007; Hillebrand and Matthiessen, 2009). Although not always mutually exclusive, functional response traits determine a species' sensitivity and response to a perturbation (e.g., drought or chemical tolerance, behavioral responses), while functional effect traits are expected to directly affect ecosystem functioning (e.g., nutrient uptake or excretion rates). Together, functional response and effect traits can be used to predict changes in biodiversity and ecosystem functioning in response to disturbance or perturbation that go beyond the C–R–S strategies. For example, functional response traits have been used to predict changes in freshwater invertebrate community composition after chemical disturbances (Feio and Doledec, 2012) and in response to future climate changes (Conti *et al.*, 2014). Similarly, Solan *et al.* (2004) predicted future scenarios for the composition of benthic marine food webs and ecosystem functioning in response to pollution, and Gallagher *et al.* (2013) predicted changes in plant functional diversity and ecosystem functioning in response to climate change.

10.4.3   Trait-Based Approaches to Predict Food-Web Composition in Environments That Are Spatially Variable or Patchy

When environments are spatially variable or patchy, trait-based approaches that incorporate traits related to dispersal and colonization ability are important for understanding how local biodiversity is maintained (Urban *et al.*, 2008). Species are predicted to successfully colonize new habitats if they possess traits that allow them to survive periods of scarce resources, move easily between patches, pass through hostile environments, and compete strongly when they enter a new community. For example, species composition in local cloud forest patches depends on species traits related to dispersal ability, home range, and fecundity (Ponce-Reyes *et al.*, 2013). Freshwater systems are also by nature patchy environments, existing within a terrestrial landscape. Thus freshwater species must have mechanisms such as desiccant-resistant dormant stages that allow them to survive through the harsh terrestrial environment. Trait-based approaches indicate that dispersal limitation is a critical factor in the very slow recovery of lake food webs following industrial acidification in the 1970s (Gray and Arnott, 2011).

10.5   Perspectives for Representing Ongoing Trait Changes in Ecosystem Models

Ecosystem models aim to capture the fundamental dynamics of ecosystems in response to various external forcing factors (Jørgensen, 1994). Food-web dynamics, nutrient and energy cycles, and spatial exchange processes across boundaries are important concepts in the design of ecosystem models (Mooij *et al.*, 2010). As a trade-off of their wide scope, ecosystem models necessarily lack a detailed description of the comprising components. Often the biotic components of the system consist of a single number representing the amount of carbon stored in a functional group, thereby ignoring the traits of the comprising species and individuals. The only trait dynamics that need to be incorporated are those that have a measurable effect on the state of the ecosystem as a whole. Although there certainly is some subjectivity in making these choices, analysis of the structure of well-tested ecosystem models may provide a list of traits whose ongoing dynamics clearly scale up to the level of the ecosystem.

Here we focus on a specific ecosystem model, namely the PCLake model for shallow lake ecosystems (Janse *et al.*, 2008). The current version of the model emerged following multi-decadal development and has been successfully confronted with data for a wide range of lake ecosystems (Janse *et al.*, 2010). The core of the model consists of six functional groups in the water phase and another four in the sediments. Each functional group is described in terms of carbon, nitrogen, and phosphorus. The abiotic pools of nitrogen and phosphorus complement a mass-balanced description of these essential nutrients. Exchange processes between the pelagic and benthic spatial compartment as well as with the surrounding environment (either adjacent waters or shorelines) are essential for both the functioning of the system as well as for defining scenarios of external perturbation of the system (Sollie *et al.*, 2008). While PCLake stands out among aquatic ecosystem models for its attention to the food web, it shares many properties with other aquatic ecosystem models in the way in which lower trophic levels and biogeochemical processes are dealt with (Mooij *et al.*, 2010).

10.5.1   Trait-Based Approaches in the Ecosystem Model for Shallow Lakes PCLake

Most of the trait dynamics incorporated in PCLake relate to the primary producers: phytoplankton and macrophytes. The first trait included deals with the edibility of the phytoplankton so as to mimic the typical seasonal succession from diatoms to cyanobacteria in eutrophic lakes, with a clear-water phase in between (Sommer *et al.*, 1993, 2012). A second trait that proved to be essential in the functioning of lake ecosystems is the stoichiometric composition of the phytoplankton, depending on the supply rates of nitrogen and phosphorus. Stoichiometric composition of the phytoplankton is tightly linked with traits determining edibility (Sterner and Elser, 2002). A third set of traits included in PCLake and its twin model for linear waters is the growth form of the macrophytes (e.g., floating, submerged, etc.). Growth form has a direct impact on the competition of macrophyte species for light and nutrients and a resulting shift in species dominance has consequences for the ecosystem as a whole (Van Gerven *et al.*, 2015). Finally, a key trait necessary to understand food-web dynamics is fish size (De Roos and Persson, 2013). Through its ontogenetic development, large shifts occur in the position individual fish occupy in the food web. Given the importance of fish for the aquatic ecosystem as a whole, through top–down effects, these ontogenetic niche shifts need to be represented in the ecosystem model, even if in a very simple form of distinguishing between small and large fish (De Roos and Persson, 2013). A more thorough approach would be to deal with fish in the model with physiologically structured formulations (De Roos and Persson, 2001).

10.5.2   The Future of Trait-Based Approaches in Ecosystem Models

New insights regarding the importance of trait-based approaches combined with accelerating anthropogenic environmental change and global biodiversity loss will likely lead to the inclusion of a richer array of trait dynamics in ecosystem models. However, including more detail in ecosystem models is not without cost, as the model may become too complex for the correct interpretation of its outcomes. As a result, only those traits (effect traits) for which a substantial impact on ecosystem functioning can be proven should be included. In fact, models such as PCLake are an interesting playground for developing hypotheses regarding how trait-based processes are linked to ecosystem functioning. Now that some aspects of trait dynamics are already included for phytoplankton, macrophytes, and fish (see previous paragraph), it is logical that future trait-based expansions of the model will focus on zooplankton and the microbial community. While a large literature exists on traits of zooplankton in relation to its food and predators (Van der Stap *et al.*, 2007) none of this is represented in PCLake. The microbial community is not even dealt with as a compartment in PCLake but only as mineralization flux from detritus to nutrients. From this perspective, it is remarkable that acceptable fits can be produced with the current version of PCLake for ecosystem parameters such as chlorophyll-a concentration, the standing stock of macrophytes, or turbidity (Janse *et al.*, 2010). It will be very interesting to see how including more trait-based dynamics in the model will improve its fit and thereby provide even more evidence for the importance of trait dynamics for the understanding and prediction of ecosystem functioning.

References

Abrams P. A. and Matsuda H. (1997). Prey adaptation as a cause of predator–prey cycles. *Evolution*, **51**, 1742–1750.

Bauer, B., Vos, M., Klauschies, T., and Gaedke, U. (2014). Diversity, functional similarity and top–down control drive synchronization and the reliability of ecosystem function. *American Naturalist*, **183**, 394–409.

Becks, L., Ellner, S. P., Jones, L. E., and Hairston, N. G. (2010). Reduction of adaptive genetic diversity radically alters eco-evolutionary community dynamics. *Ecology Letters*, **13**, 989–997.

Becks, L., Ellner, S. P., Jones, L. E., and Hairston, Jr., N. G. (2012). The functional genomics of an eco-evolutionary feedback loop, linking gene expression, trait evolution, and community dynamics. *Ecology Letters*, **15**, 492–501.

Binzer, A., Guill, C., Brose, U., and Rall, B. C. (2012). The dynamics of food chains under climate change and nutrient enrichment. *Philosophical Transactions of the Royal Society B: Biological Sciences*, **367**, 2935–2944.

Boit, A., Martinez, N. D., Williams, R. J., and Gaedke, U. (2012). Mechanistic theory and modeling of complex food web dynamics in Lake Constance. *Ecology Letters*, **15**, 594–602.

Bolnick, D. I., Amarasekare, P., Araujo, M. S., *et al.* (2011). Why intraspecific trait variation matters in community ecology. *Trends in Ecology and Evolution*, **26**, 183–192.

Brose, U., Jonsson, T., Berlow, E. L., *et al.* (2006a). Consumer-resource body-size relationships in natural food webs. *Ecology*, **87**, 2411–2417.

Brose, U., Williams, R. J., and Martinez, N. D. (2006b). Allometric scaling enhances stability in complex food webs. *Ecology Letters*, **9**, 1228–1236.

Conti, L., Schmidt-Kloibe, A., Grenouillet, G., and Graf, W. (2014). A trait-based approach to assess the vulnerability of European aquatic insects to climate change. *Hydrobiologia*, **721**(1), 297–315.

Cortez, M. H. (2011). Comparing the qualitatively different effects rapidly evolving and rapidly induced defences have on predator–prey interactions. *Ecology Letters*, **14**, 202–209.

Cortez, M. H. and Ellner, S. P. (2010). Understanding rapid evolution in predator–prey interactions using the theory of fast–slow dynamical systems. *American Naturalist*, **176**, 109–127.

De Roos, A. M. and Persson, L. (2001). Physiologically structured models: from versatile technique to ecological theory. *Oikos*, **94**(1), 51–71.

De Roos, A. M., Persson, L., and McCauley, E. (2003). The influence of size-dependent life-history traits on the structure and dynamics of populations and communities. *Ecology Letters*, **6**, 473–487.

De Roos, A. M., Schellekens, T., van Kooten, T., *et al.* (2007). Food-dependent growth leads to overcompensation in stage-specific biomass when mortality increases: the influence of maturation versus reproduction regulation. *American Naturalist*, **170**, 59–76.

De Roos, A. M., Schellekens, T., van Kooten, T., *et al.* (2008). Simplifying a physiologically structured population model to a stage-structured biomass model. *Theoretical Population Biology*, **73**, 47–62.

De Roos, A. M. and Persson, P. (2013). *Population and Community Ecology of Ontogenetic Development*. Princeton, NJ: Princeton University Press.

Dieckmann, U. and Law, R. (1996). The dynamical theory of coevolution: a derivation from stochastic ecological processes. *Journal of Mathematical Biology*, **34**, 579–612.

Digel, C., Curtsdotter, A., Riede, J., Klarner, B., and Brose, U. (2014). Unravelling the complex structure of forest soil food webs: higher omnivore and more trophic levels. *Oikos*, **123**, 1157–1172.

dos Santos, F. A. S., Johst, K., and Grimm, V. (2011). Neutral communities may lead to decreasing diversity–disturbance relationships: insights from a generic simulation model. *Ecology Letters*, **14**, 653–660.

Duffy, J. E., Cardinale, B. J., France, K. E., *et al.* (2007). The functional role of biodiversity in ecosystems: incorporating trophic complexity. *Ecology Letters*, **10**, 522–538.

Edwards, K. F., Klausmeier, C. A., and Litchman, E. (2013a). A three-way tradeoff maintains functional diversity under variable resource supply. *American Naturalist*, **182**, 786–800.

Edwards, K. F., Klausmeier, C. A., and Litchman, E. (2013b). Functional traits predict variation in phytoplankton community structure across lakes of the United States. *Ecology*, **94**, 1626–1635.

Feio, M. J. and Doledec, S. (2012). Integration of invertebrate traits into predictive models for indirect assessment of stream functional integrity: a case study in Portugal. *Ecological Indicators*, **15**, 236–247.

Gallagher, R. V., Hughes, L., and Leishman, M. R. (2013). Species loss and gain in communities under future climate change: consequences for functional diversity. *Ecography*, **36**(5), 531–540.

Gray, D. K. and Arnott, S. E. (2011). Does dispersal limitation impact the recovery of zooplankton communities damaged by a regional stressor? *Ecological Applications*, **21**, 1241–1256.

Grime, J. P. (1977). Evidence for existence of three primary strategies in plants and its relevance to ecological and evolutionary theory. *American Naturalist*, **111**, 1169–1194.

Grimm, V., Revilla, E., Berger, U., *et al.* (2005). Pattern-oriented modeling of agent-based complex systems: lessons from ecology. *Science*, **310**, 987–991.

Grover, J. P. (1991). Resource competition in a variable environment: phytoplankton growing according to the Variable–Internal–Stores model. *American Naturalist*, **138**, 811–835.

Guill, C. (2009). Alternative dynamical states in stage-structured consumer populations. *Theoretical Population Biology*, **76**, 168–178.

Heckmann, L., Drossel, B., Brose, U., and Guill, C. (2012). Interactive effects of body-size structure and adaptive foraging on food-web stability. *Ecology Letters*, **15**, 243–250.

Hillebrand, H. and Matthiessen, B. (2009). Biodiversity in a complex world: consolidation and progress in functional biodiversity research. *Ecology Letters*, **12**, 1405–1419.

Hooper, D. U., Chapin, F. S., Ewel, J. J., *et al.* (2005). Effects of biodiversity on ecosystem functioning: a consensus of current knowledge. *Ecological Monographs*, **75**, 3–35.

Janse, J. H., De Senerpont Domis, L. N., Scheffer, M., *et al.* (2008). Critical phosphorus loading of different types of shallow lakes and the consequences for management estimated with the ecosystem model PCLake. *Limnologica – Ecology and Management of Inland Waters*, **38**(3), 203–219.

Janse, J. H., Scheffer, M., Lijklema, L., *et al.* (2010). Estimating the critical phosphorus loading of shallow lakes with the ecosystem model PCLake: sensitivity, calibration and uncertainty. *Ecological Modelling*, **221**(4), 654–665. doi:10.1016/j.ecolmodel.2009.07.023 ER.

Jones, L. E. and Ellner, S. P. (2007). Effects of rapid prey evolution on predator–prey cycles. *Journal of Mathematical Biology*, **55**, 541–573.

Jones, L. E., Becks, L., Ellner, S. P., *et al.* (2009). Rapid contemporary evolution and clonal food web dynamics. *Philosophical Transactions of the Royal Society B: Biological Sciences*, **364**, 1579–1591.

Jørgensen, S. E. (1994). Models as instruments for combination of ecological theory and environmental practice. *Ecological Modelling*, **75**, 5–20.

Kalinkat, G., Schneider, F. D., Digel, C., *et al.* (2013). Body masses, functional responses and predator–prey stability. *Ecology Letters*, **16**, 1126–1134.

Kooijman, S. A. L. M. (2010). *Dynamic Energy Budget Theory for Metabolic Organisation*. Cambridge, UK: Cambridge University Press.

Litchman, E., Klausmeier, C. A., Schofield, O. M., and Falkowski, P. G. (2007). The role of functional traits and trade-offs in structuring phytoplankton communities: scaling from cellular to ecosystem level. *Ecology Letters*, **10**, 1170–1181.

May, F., Grimm, V., and Jeltsch, F. (2009). Reversed effects of grazing on plant diversity: the role of below-ground competition and size symmetry. *Oikos*, **118**, 1830–1843.

McGill, B. J., Enquist, B. J., Weiher, E., and Westoby, M. (2006). Rebuilding community ecology from functional traits. *Trends in Ecology and Evolution*, **21**, 178–185.

Merico, A., Brandt, G., Smith, S. L., and Oliver, M. (2014). Sustaining diversity in trait-based models of phytoplankton communities. *Frontiers in Ecology and Evolution*, **2**, 59, 1–8.

Metz, J. A. J. and Diekmann, O. (1986). *The Dynamics of Physiologically Structured Populations*. Springer-Verlag.

Mooij, W. M., Trolle, D., Jeppesen, E., *et al.* (2010). Challenges and opportunities for integrating lake ecosystem modelling approaches. *Aquatic Ecology*, **44**(3), 633–667. doi:10.1007/s10452-010-9339-3 ER.

Mougi, A. (2012). Unusual predator–prey dynamics under reciprocal phenotypic plasticity. *Journal of Theoretical Biology*, **305**, 96–102.

Naeem, S. and Wright J. P. (2003). Disentangling biodiversity effects on ecosystem functioning: deriving solutions to a seemingly insurmountable problem. *Ecology Letters*, **6**, 567–579.

Nakazawa, T. (2011). Ontogenetic niche shift, food-web coupling, and alternative stable states. *Theoretical Ecology*, **4**, 479–494.

Norberg, J. (2004). Biodiversity and ecosystem functioning: a complex adaptive systems approach. *Limnology and Oceanography*, **49**, 1269–1277.

Petchey, O. L., Beckerman, A. P., Riede, J. O., and Warren, P. H. (2008). Size, foraging, and food-web structure. *Proceeding of the National Academy of Sciences*, **105**, 4191–4196.

Peters, R. H. (1993). *The Ecological Implications of Body Size*. Cambridge University Press.

Ponce-Reyes, R., Nicholson, E., Baxter, P. W. J., Fuller, R. A., and Possingham, H. (2013). Extinction risk in cloud forest fragments under climate change and habitat loss. *Biodiversity Research*, **19**, 518–529.

Post, D. M. and Palkovacs, E. P. (2009). Eco-evolutionary feedbacks in community and ecosystem ecology: interactions between the ecological theatre and the evolutionary play. *Philosophical Transactions of the Royal Society B: Biological Sciences*, **364**, 1629–1640.

Savage, V. M. and Norberg, J. (2007). A general multi-trait-based framework for studying the effects of biodiversity on ecosystem functioning. *Journal of Theoretical Biology*, **247**, 213–229.

Smith, S. L. and Yamanaka, Y. (2007). Optimization-based model of multinutrient uptake kinetics. *Limnology and Oceanography*, **52**, 1545–1558.

Solan, M., Cardinale, B. J., Downing, A. L., *et al.* (2004). Extinction and ecosystem function in the marine benthos. *Science*, **306**, 1177–1180.

Sollie, S., Janse, J. H., Mooij, W. M., Coops, H., and Verhoeven, J. T. A. (2008). The contribution of marsh zones to water quality in Dutch shallow lakes: a modeling study. *Environmental Management*, **42**(6), 1002–1016. doi:10.1007/s00267-008-9121-7.

Sommer, U., Padisak, J., Reynolds, C. S., and Juhasz-Hagy, P. (1993). Hutchinson's heritage: the diversity–disturbance relationship in phytoplankton. *Hydrobiologia*, **249**, 1–7.

Sommer, U., Sommer, F., Santer, B., *et al.* (2003). *Daphnia* versus copepod impact on summer phytoplankton: functional compensation at both trophic levels. *Oecologia*, **135**, 639–647.

Sommer, U., Adrian, R., Domis, L. D. S., *et al.* (2012). Beyond the Plankton Ecology Group (PEG) model: mechanisms driving plankton succession. *Annual Review of Ecology, Evolution and Systematics*, **43**, 429–448. doi:10.1146/annurev-ecolsys-110411-160251 ER.

Sterner, R. W. and Elser, J. J. (2002). *Ecological Stoichiometry: The Biology of Elements from Molecules to the Biosphere*. Princeton, NJ: Princeton University Press.

Tirok, K. and Gaedke, U. (2007). Regulation of planktonic ciliate dynamics and functional composition during spring in Lake Constance. *Aquatic Microbial Ecology*, **49**, 87–100.

Tirok, K. and Gaedke, U. (2010). Internally driven alternation of functional traits in a multi-species predator–prey system. *Ecology*, **91**, 1748–1762.

Tirok, K., Bauer, B., Wirtz, K., and Gaedke, U. (2011). Predator–prey dynamics driven by feedback between functionally diverse trophic levels. *PLoS ONE*, **6** (11), e27357. doi:10.1371/journal.pone.0027357.

Tomimatsu, H., Sasaki, T., Kurokawa, H., *et al.* (2013). Sustaining ecosystem functions in a changing world: a call for an integrated approach. *Journal of Applied Ecology*, **50** (5), 1124–1130.

Urban, M. C., Leibold, M. A., Amarasekare, P., *et al.* (2008). The evolutionary ecology of metacommunities. *Trends in Ecology and Evolution*, **23**(6) 311–317.

van der Stap, I., Vos, M., and Mooij, W. M. (2007). Induced defenses in herbivores and plants differentially modulate a trophic cascade. *Ecology*, **88**, 2474–2481.

Van Gerven, L. P. A., de Klein, J. J. M., Gerla, D. J., *et al.* (2015). Competition for light and nutrients in layered communities of aquatic plants. *American Naturalist*, **186**(1), 72–83.

Verschoor, A. M., Vos, M., and van der Stap, I. (2004). Inducible defences prevent strong population fluctuations in bi- and tritrophic food chains. *Ecology Letters*, **7**, 1143–1148.

Violle, C., Navas, M.-L., Vile, D., *et al.* (2007). Let the concept of trait be functional! *Oikos*, **116**, 882–892.

Violle, C., Enquist, B. J., McGill, B. J., *et al.* (2012). The return of the variance: intraspecific variability in community ecology. *Trends in Ecology and Evolution*, **27**, 244–252.

Wagner, A. and Benndorf, J. (2007). Climate-driven warming during spring destabilises a *Daphnia* population: a mechanistic food-web approach. *Oecologia*, **151**, 351–364.

Werner, E. E. and Gilliam, J. F. (1984). The ontogenetic niche and species interactions in size-structured populations. *Annual Review of Ecology and Systematics*, **15**, 393–425.

Williams, R. J. and Martinez, N. D. (2000). Simple rules yield complex food webs. *Nature*, **404**, 180–183.

Wirtz, K. W. and Eckhardt, B. (1996). Effective variables in ecosystem models with an application to phytoplankton succession. *Ecological Modelling*, **92**, 33–53.

Woodward, G. and Hildrew, A. G. (2002). Body-size determinants of niche overlap and intraguild predation within a complex food web. *Journal of Animal Ecology*, **71**, 1063–1074.

Yoshida, T., Jones, L. E., Ellner, S. P., Fussmann, G. F., and Hairston, Jr., N. G. (2003). Rapid evolution drives ecological dynamics in a predator–prey system. *Nature*, **424**, 303–306.

Yoshida, T., Ellner, S. P., Jones L. E., *et al.* (2007). Cryptic population dynamics: rapid evolution masks trophic interactions. *PLoS Biology*, **5**, 1868–1879.

Zhang, L., Thygesen, U. H., Knudsen, K., and Andersen, K. H. (2013). Trait diversity promotes stability of community dynamics. *Theoretical Ecology*, **6**, 57–69.

11 Ecological Succession Investigated Through Food-Web Flow Networks

Antonio Bodini, Cristina Bondavalli, and Giampaolo Rossetti

11.1 Introduction

Ecosystems change in time following dynamics that scholars have characterized as a true developmental process (Ulanowicz, 1986). Seen through the lenses of ecological community these changes are known as an overall progression called succession (McMahon, 1980). During this process species in an environment are gradually substituted by other, more adapted, associations until a final arrangement called climax emerges, with a well-established community (Clements, 1936; Whittaker, 1953; Odum, 1969). Although individual successional processes appear to be unique and dependent on timing, initial conditions, types of disturbance, and other factors (Sousa, 1984), if the development is orderly and directional it descends that some predictive capability can be achieved from its analysis (Noble and Slatyer, 1977). According to this, similarities and differences in processes and patterns have been investigated to extract generalizations that could help in shaping a general model for ecological succession (for a review of the various findings see Drury and Nisbet, 1973). Prominent among the many attempts to single out those changes is the summary proposed by Odum (1969). He predicted patterns for 24 features grouped in six main categories such as community energetic, life history, nutrient cycling, overall homeostasis, selection pressure, and community structure. This scheme has been extremely influential on ecological thought but it has also been criticized (Drury and Nisbet, 1973; Oksanen, 1991). The ideas that pervade the entire construction, such as balance, homeostasis, stability, and equilibrium, were challenged by a viewpoint that stresses disturbance, chance, and non-equilibrium as important drivers of ecosystem dynamics (Archer and Stokes, 2000).

If succession would take place along defined pathways, then it should be possible to mark their progress in quantitative fashion. Schindler (1990) tested Odum's hypotheses by experimentally analyzing the effects of nutrient enrichment and acidification on whole-lake ecosystems. He found that acidified lakes showed progressively increasing P/R ratio, dominance of r-strategists among zooplankton, food-chain shortening, and decreased species diversity; this evidence confirmed that ecosystems under stress may show reversal trends with respect to Odum's (1985) scenario.

Ulanowicz approached the question using the apparatus of network analysis, which offers the opportunity to quantify ecosystem growth and development (Ulanowicz,

1986, 1997). He performed his testing by contrasting two marsh creek ecosystems: one undisturbed and the other impacted by thermal stress (Ulanowicz, 1986). In another study (Mageau et al., 1998) he developed a model that simulated outputs for an ecosystem advancing through the various stages of succession. In both exercises he tested the ability of certain systems-level information indices to quantitatively grasp ecosystem developmental trends. In those studies Odum's hypotheses seemed to hold. Two main drawbacks affect these investigations. First, limited data availability allowed the comparison of only one snapshot representation for every ecosystem (perturbed and unperturbed). This precluded any possibility to track temporal sequences. Second, the simulation approach allowed including time in the analysis but there were no field data to assess and corroborate model outcomes.

   To overcome these two problems we applied network analysis to a mountain lake ecosystem for which long-term data were available. Such an enlarged dataset was compiled during intensive sampling campaigns that were conducted during the period 1989 to 1994. The data allowed us to construct a suite of ecological flow networks representative of the ecosystem. Using the apparatus of network analysis we computed system-level indices that quantify ecosystem growth and development (Ulanowicz, 1986, 1997). These indices were first confronted with the ecosystem attributes that Odum listed as indicators of ecosystem development, in search for correspondences. Next, the trends that emerged for system-level indices were compared with the tendencies that Odum predicted for the attributes he selected. We could define network counterparts unambiguously only for a limited set of Odum's 24 indices. We provide here a short preview of the correspondence between the two types of attributes, as obtained by exploring the original meanings of the indices (Odum, 1969; Drury and Nisbet, 1973 and references therein; Ulanowicz, 1997) and the main concepts of the information theory upon which system-level indices of network analysis are based (Ulanowicz, 2001; Latham and Scully, 2002; Allesina and Bodini, 2008). Odum's entropy, stability, and information correspond to development capacity, total system overhead, and average mutual information, respectively. The dissipative overhead can be the natural counterpart for Odum's respiration. Redundancy, a network attribute that quantifies the multiplicity of links, can be a proxy for the linearity of food chains. Cycling activity and internalization of medium can be approximated by the Finn cycling index (FCI, Ulanowicz, 1986) and by the overhead on import.

11.2   Lake Scuro: Food Web and Quantitative Indices of Development

Lake Scuro Parmense is an oligotrophic dimictic lake located in the Northern Apennines (Italy) at an altitude of 1527 meters above sea level (m asl) (longitude: 10.046° E; latitude: 44.382° N). During the spring thaw, when the water volume is at its maximum (44×10^3 m^3), the area is about 1.2 ha and the maximum depth is 10.4 m. The water is weakly buffered and its nutrient and chlorophyll-a contents are rather low and qualify the water body as oligotrophic (Antonietti et al., 1988). The food web is mainly composed of zooplankton species and it is resolved at the species level except for those species that

Table 11.1 List of the zooplankton species included in the Lake Scuro food web.

Zooplankton compartments

ROTIFERS

3 *Keratella cochlearis*
4 *Ascomorpha* **spp.**
   Ascomorpha ecaudis, Ascomorpha saltans
5 *Synchaeta* **sp.**
6 *Polyarthra* **sp.**
7 **Predator rotifers**
   Ploesoma hudsoni, Asplanchna priodonta
8 *Conochilus unicornis-hippocrepis*
9 **Other rotifers**
   Cephalodella sp., *Collotheca mutabilis, Eothinia elongata, Erignatha clastopis, Euchlanis* sp., *Gastropus stylifer, Keratella* gr. *quadrata, Lecane closterocerca, Lecane flexilis, Lecane luna, Lecane* gr. *lunaris, Lepadella* sp., *Lindia* sp., *Mytilina ventralis, Notholca squamula, Notommata* sp., *Pleurata* sp., *Resticula nyssa, Scaridium longicaudum, Taphrocampa annulosa, Testudinella parva, Trichocerca* spp., *Trichotria tetractis*

CLADOCERANS

10 *Diaphanosoma brachyurum*
11 *Daphnia longispina*
12 **Other cladocerans**
   Acroperus harpae, Alona guttata, Alonella nana, Biapertura affinis, Graptoleberis testudinaria

COPEPODS

13 **Eudiaptomus intermedius nauplii**
14 **Eudiaptomus intermedius copepodites (CI-CII-CIII)**
15 **Eudiaptomus intermedius copepodites (CIV-CV)**
16 **Eudiaptomus intermedius adults**
17 **Cyclopoid copepods nauplii**
   Eucyclops serrulatus, Macrocyclops albidus, Mesocyclops leucktarti
18 **Cyclopoid copepods copepodites**
19 **Cyclopoid copepods adults**

were only represented by few individuals or found sporadically in the lake. These latter were grouped in aggregated nodes of microfilter feeders, macrofilter feeders, and predators. This latter group is composed of predatory planktonic invertebrates as the lake is naturally fishless. Phytoplankton species included Chlorophyceae, with 12 taxa, Chrysophyceae (ten taxa), Diatomeae (six taxa), Cryptophyceae (four taxa), Dynophyceae (three taxa), and Cyanophyceae (one taxon). All the algal species were grouped in a unique node for the food web (phytoplankton). Zooplankton species as they are included in the food webs are listed in Table 11.1.

Limnological surveys were carried out in the open-water season (May to October) of the years: 1989 to 1994. Samples were collected at weekly to monthly intervals from the point of maximum lake depth. This long-term research effort allowed characterizing of most of the ecological relationships. Predator–prey interactions, in particular, were

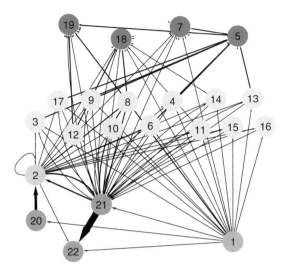

Figure 11.1 The food web of the Lake Scuro ecosystem. Keys are: 1 phytoplankton; 2 microbial loop; 3 *Keratella cochlearis*; 4 *Ascomorpha* spp.; 5 *Synchaeta* sp.; 6 *Polyarthra* sp.; 7 predator rotifers; 8 *Conochilus unicornis-hippocrepis*; 9 other rotifers; 10 *Diaphanosoma brachyurum*; 11 *Daphnia longispina*; 12 other cladocerans; 13 *Eudiaptomus nauplii*; 14 *Eudiaptomus* CI-CII-CIII; 15 *Eudiaptomus* CIV-CV; 16 *Eudiaptomus intermedius* adults; 17 Cyclopoid copepods nauplii; 18 Cyclopoid copepods copepodites; 19 Cyclopoid copepods adults; 20 WDOC; 21 WPOC; 22 BPOC. (A black and white version of this figure will appear in some formats. For the color version, please refer to the plate section.)

detailed through the analysis of stomach contents (Rossetti *et al.*, 1997), and by enclosure experiments (Paris *et al.*, 1995). With the support of the vast literature on the subject (Kerfoot, 1980; Lynch and Shapiro, 1981; Kerfoot and Sih, 1987) a complete description of the resource–consumer interactions became possible. The network description of the 22-component food web is given in Figure 11.1.

All links were quantified as carbon flows and measured in g C m^{-2} yr^{-1}. A technical report compiled for a twin lake (Lake Santo, Northern Appennines, 10°00'38" E, 44°24'06" N, 1507 m asl, Bondavalli *et al.*, 2006) provides detailed information on how flows were calculated using abundance data (number of individuals for zooplankton, and cells for phytoplankton) per unit volume. The quantification of flows allowed for the calculation of system-level indices. A brief sketch of the indices that were essential for the analysis is provided below. Since all the measures are based on flow values, an essential description of how indices are obtained from flows is also provided. The term T_{ij} identifies a flow from node i to node j. Summations are extended to include the external environment as: (a) source of input (compartment 0) for the system; (b) receiver of usable medium (compartment $N + 1$) from the system; and (c) sink of medium (compartment $N + 2$) dissipated by the system. Labels 1, 2, ..., N identify the system's components. Contractions are used when convenient to shorten equations: $T.$ stands for summation across all rows (first dot) and columns (second dot).

11.2.1   Entropy (H), Average Mutual Information (AMI), and Overhead (O)

MacArthur (1955) applied Shannon's information index to the flows in ecosystem networks, and its expression becomes

$$H = k \sum_{i=1}^{N+2} \sum_{j=1}^{N+2} \frac{T_{ij}}{T_{..}} log_2 \left(\frac{T_{ij}}{T_{..}} \right) \qquad (11.1)$$

where H is the diversity of flows in the network, k is a scalar constant, T_{ij} denotes the flow from node i to node j. $T_{..}$ is the total system throughput (TST), the sum of all flows in the network and quantifies the level of ecosystem activity:

$$TST \sum_{i=1}^{N+2} \sum_{j=1}^{N+2} T_{ij} = T_{..} \qquad (11.2)$$

H, also called entropy (Ulanowicz, 2001; Allesina and Bodini, 2008), can be interpreted as the overall complexity of the system or the potential the system has at its disposal for improving its structure of flows to a completely coherent structure. Rutledge $et~al.$ (1976) used the notion of conditional probability to decompose MacArthur's index in two complementary terms, amending the measure of total flow diversity: $H = AMI + H_c$ where the average mutual information (AMI) quantifies the amount of flow diversity that is encumbered by structural constraints. Average mutual information depends on flows in the following way:

$$AMI = k \sum_{i=1}^{N+2} \sum_{j=1}^{N+2} \frac{T_{ij}}{T_{..}} log_2 \left(\frac{T_{ij} T_{..}}{T_{i.} T_{.j}} \right) \qquad (11.3)$$

H_c represents the amount of "choice" (residual diversity/freedom) pertaining to both the inputs and outputs of an average node in the network:

$$H_c = k \sum_{i=1}^{N+2} \sum_{j=1}^{N+2} \frac{T_{ij}}{T_{..}} log_2 \left(\frac{T_{ij}^2}{T_{i.} T_{.j}} \right) \qquad (11.4)$$

Therefore the overall complexity of the flow structure, as measured by the MacArthur's index (entropy), can be resolved into two components: (a) AMI, which estimates how orderly and coherently flows are connected; and (b) H_c, which gauges the disorder and freedom that is preserved. If the flows are constrained to follow a given path, then knowing that a particle is leaving compartment i also tells us that it will enter compartment j. In this case there is no uncertainty about the fate of the particle and the information we possess is the highest. Therefore AMI acquires the maximum possible value. If, on the other hand, each compartment of a network communicates with all the others through equally efficient channels (fluxes of the same magnitude) the uncertainty about which route the particle will take is maximal and AMI equals to zero. An example of maximally and minimally constrained structures is provided in Figure 11.2. Intermediate configurations yield AMI values between 0 and a maximum that depends

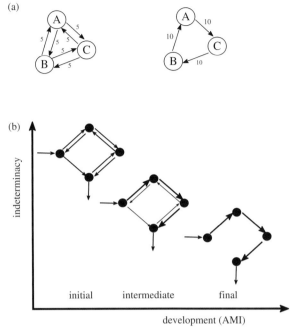

Figure 11.2 (a) Minimal (left) and maximal (right) AMI for a simple network composed by four nodes (modified from Allesina and Bodini, 2008). (b) Hypothetical mechanism for ecosystem development. More efficient connections become dominant in channeling medium throughout the system. The final configuration is less redundant and more articulated (modified from Allesina and Bodini, 2008).

on the system's structure and not on system size, which is preserved throughout the example graphs.

Higher values for AMI identify structures that are tightly constrained in respect to the movement of the currency, and are highly organized. Currency distribution, in fact, would take place along few, efficient routes with a lower cost of maintenance for the system. Highly redundant networks possess lower AMI values, and are less organized. For example, in the upper left network of Figure 11.2b the four components exchange matter using numerous equiponderant links. If it happens that certain links are more efficient in transferring medium, at every cycle they will become more important so that an ever-increasing amount of currency will pass through them, and they will progressively predominate over less efficient connections. The ultimate consequence is that this positive feedback will select few routes, pruning away less efficient links yielding a final configuration similar to the lower right graph in Figure 11.2b. Positive feedback cycles, in synthesis, would force ecological networks toward less redundant, more efficient configurations. According to this, ecosystems would develop in the direction of more organized structure of exchange. AMI, thus, quantifies ecosystem development and it provides detail about how the ecosystem distributes medium to its components. H_c gauges the disorder and freedom that is preserved in the system, and has been suggested by Rutledge *et al.* (1976) as a proper measure of stability (MacArthur, 1955).

Scaling H by the total sum of network flows (TST), we obtain the maximum development capacity of the system:

$$DC = T_{..} \sum_{i=1}^{N+2} \sum_{j=1}^{N+2} \frac{T_{ij}}{T_{..}} log_2 \left(\frac{T_{ij}}{T_{..}} \right) = \sum_{i=1}^{N+2} \sum_{j=1}^{N+2} T_{ij} log_2 \left(\frac{T_{ij}}{T_{..}} \right) \qquad (11.5)$$

The development capacity (DC) quantifies the maximum potential that a system has at its disposal to achieve further development and it serves as an upper boundary for ecosystem organization. Also the term H_c that measures the residual diversity of flows can be scaled by TST yielding the total system overhead (O), which pertains to redundant flows (Ulanowicz, 1986). We define the overhead as:

$$\sum O = \sum_{i=1}^{N+2} \sum_{j=1}^{N+2} T_{ij} log_2 \left(\frac{T_{ij}}{T_{..}} \right) - \sum_{i=1}^{N+2} \sum_{j=1}^{N+2} \frac{T_{ij}}{T_{..}} log_2 \left(\frac{T_{ij} T_{..}}{T_{i.} T_{.j}} \right)$$

$$= - \sum_{i=2}^{N+2} \sum_{j=1}^{N+2} T_{ij} log_2 \left(\frac{T_{ij}^2}{T_{i.} T_{.j}} \right) \qquad (11.6)$$

It is also made of different contributions: overhead on import: $O_I = - \sum_{j=1}^{N} T_{0j}$ $log_2 \left(\frac{T_{0j}^2}{T_{0.} T_{.j}} \right)$; on export: $O_E = - \sum_{j=1}^{N} T_{i,N+1} log_2 \left(\frac{T_{i,N+1}^2}{T_{i.} T_{.N+1}} \right)$; dissipation: $DO = - \sum_{j=1}^{N} T_{i,N+2} log_2 \left(\frac{T_{i,N+2}^2}{T_{i.} T_{.N+2}} \right)$; and redundancy $R = - \sum_{i=1}^{N} \sum_{j=1}^{N} T_{ij} log_2 \left(\frac{T_{ij}^2}{T_{i.} T_{.j}} \right)$ of internal exchanges. While dissipative overhead is naturally related to what any ecosystem dissipates through internal processes (i.e., respiration), the other terms are strongly tied to the effective multiplicity of parallel flows by which energy is imported (overhead on import), exported as usable medium (overhead on export), or exchanged between the compartments (redundancy).

Overhead has been defined as a proxy for stability (Mageau et al., 1995). Finally, scaling AMI by the TST one obtains an index called ascendency, whose expression is as follows:

$$A = AMI \times TST = T_{..} \sum_i \sum_j \frac{T_{ij}}{T_{..}} log \left(\frac{T_{ij} T_{..}}{T_{i.} T_{.j}} \right) = \sum_i \sum_j T_{ij} log \left(\frac{T_{ij} T_{..}}{T_{i.} T_{.j}} \right) \qquad (11.7)$$

Ascendency quantifies the fraction of the TST that is managed along efficient connections (Ulanowicz, 1997, 2004).

11.3   Observed and Expected Trends: A Comparative Analysis

Some indices of the ecological flow networks directly correspond to attributes that appear in Odum's list as indices of homeostasis: entropy, stability, and information. Entropy was said to decrease from early stages of succession to more mature

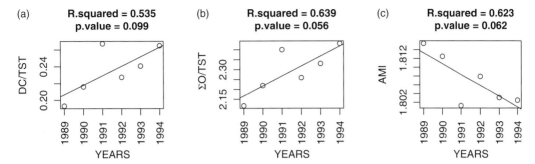

Figure 11.3 (a) Trend shown by entropy (DC/TST), (b) for stability ($\left(\sum O\right)/TST$), and (c) for information (AMI), for the Lake Scuro ecosystem during the period 1989 to 1994. Entropy is given as development capacity divided by TST because the software gives the value scaled by the total system throughput. The same applies to total overhead.

configurations but information and stability would be increasing. Figure 11.3 shows trends observed in Lake Scuro for entropy, total overhead, and information, as obtained from the food-web networks.

The trends of the three indices are significant at the 0.1 probability level. Overhead seems to be in agreement with Odum's expectation but this is not so for entropy and information decrease. Figure 11.4 portrays the same indices but scaled (multiplied) by the TST, that is the overall quantity of currency that the system handles (sum of all flows).

Multiplying by the TST redefines their meaning in terms of the total energy (matter) that flows throughout the system. So, for example, ascendency, which is the product of system information (AMI) and TST, can be interpreted as the fraction of the total currency that flows along constrained pathways, that is the amount of currency that is handled efficiently by the system. At the same time overhead quantifies the fraction of matter/energy that the system has at its disposal to overcome the effects of perturbation. Trends for entropy and overhead shown in Figure 11.4 coincide with those displayed in Figure 11.3. Information, instead, shows an opposite behavior in respect to its intensive component. The trend for ascendency is not significant but it is close to the 0.1 probability level. TST (which is increasing but not significantly, R-squared = 0.518, p = 0.107, chart not shown here) plays a disproportionate effect upon these indices. Yet, as seen in Figure 11.4, entropy contrasts with Odum's expectation whereas stability meets his prediction.

The pattern of these three indices seems to mirror that expected by Ulanowicz (1986, 2004) who described entropy, overhead, and information as all increasing during ecosystem development, but only in the earlier stages. According to Ulanowicz also AMI should increase during succession, but this is not confirmed for Lake Scuro (Figure 11.3).

Ulanowicz (1986) and Mageau *et al.* (1995) broadened the scenario hypothesizing that during succession information would continue to increase relative to entropy, hence ascendency relative to development capacity, and this would occur at the expense of the

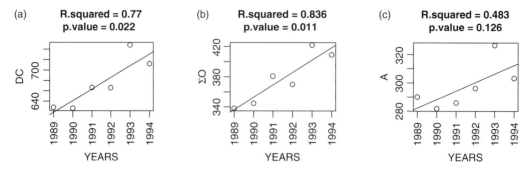

Figure 11.4 Trends shown by entropy (a), stability (b), and information (c) as scaled by the TST.

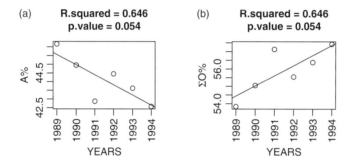

Figure 11.5 (a) Ascendency and (b) overhead as fraction of the development capacity.

overhead. Figure 11.5 is useful to explore this hypothesis as it shows trends for ascendency and overhead as fractions of the development capacity (entropy).

Both trends are significant at the 0.1 probability level but contrast with the normal course of succession hypothesized by Ulanowicz: overhead, in fact, increases at the expense of information.

Other features included in Odum's scenario can be addressed by the components of the overhead. Respiration is predicted to increase as the ecosystem moves to late stages of development because energy fixed tends to be balanced by the energy cost of maintenance (that is, total community respiration). The dissipative overhead (DO, Figure 11.6) shows a positive significant trend ($p < 0.05$), in agreement with Odum's expectation.

Odum asserted that food chains tend to be linear in earlier stages of development and to become more web-like as succession progresses. In networks this feature is quantified by redundancy (R). Its trend is portrayed in Figure 11.6. This index increased significantly, showing that Lake Scuro seems to follow the predicted evolution toward more ramified configurations. Ulanowicz (1986) found it problematic to reconcile this view with one in which development would be accompanied by increasing ascendency, as this latter index generally favors linear, chain-like transfers. Ascendency in Lake Scuro increases due to the disproportionate contribution of TST. But if we consider

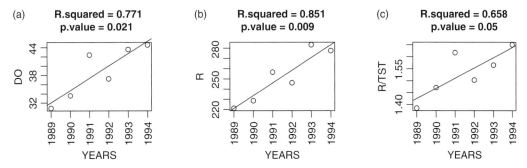

Figure 11.6 (a) Dissipative overhead (DO), which quantifies respiration; and the two forms of the redundancy index: (b) R, and (c) R/TST.

information only in its intensive component (AMI) our data show that this index decreased (Figure 11.3) coherently with increased redundancy. This holds even if we consider the intensive component of redundancy (R/TST, Figure 11.6), which shows a significant increasing trend. Thus during the period of investigation the lake showed a propensity to become more web-like.

Mineral cycles would increase in importance as the ecosystem develops; this would be accompanied by greater internalization of medium, and the ecosystem would rely less on imported matter as cycles become more and more important. The Finn cycling index (FCI, Ulanowicz, 1986), which quantifies the fraction of matter that takes part in cycling, and the overhead on import (OI), the redundancy of channels that import matter in the system, can serve to ascertain how Lake Scuro behaved in respect to these features. Figure 11.7 shows the trends for these indices along the period of investigation.

The Finn cycling index is increasing as expected although its trends is not significant. The overhead on import shows a complex behavior. If we consider its intensive part, that pertaining to the topology of import channels (OI/TST), the trend is decreasing but is not significant. With the necessary circumspection that statistical significance imposes in this case, it is likely that during the period of investigation the structure of imports became less redundant. However OI is increasing (Figure 11.7). Thus throughout the period 1989 to 1994, cycling seems to become more important as the overall amount of cycling activity increased. This latter evidence would suggest that the lake increased its dependence from imported medium meanwhile its cycling activity was increasing. Nonetheless the intensive component of the overhead on import was decreasing; this indicates that Lake Scuro was evolving toward a configuration with fewer channels to import medium, although more currency traveled along these channels. This suggests that we rethink about the role of cycling during succession and the way we define it.

Finally we computed also classical indices for ecosystem development, such as bioenergetic indices and the Shannon index of community diversity. In particular, we computed the ratio of production over respiration (P/R), production over biomass (P/B),

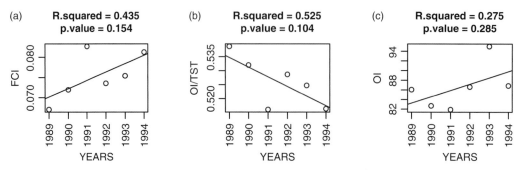

Figure 11.7 (a) Finn cycling index (FCI) and (c) overhead on import (OI). (b) OI/TST is the intensive component of the overhead, which refers to the topology of import channels. OI measures the amount of matter that is conveyed to the system from the outside through multiple import channels.

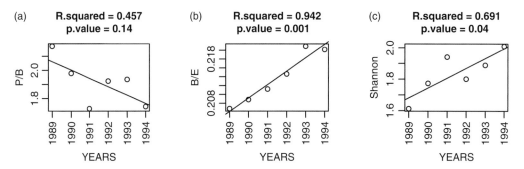

Figure 11.8 Bioenergetic indices (P/B, B/E) and the Shannon index of diversity.

and that of biomass over the sum of production and respiration (B/E, E = P + R). The first did not show any significant trend and it is not shown in Figure 11.8, which presents the trends for P/B and B/E. The former is decreasing, although close to the 0.1 probability level; the latter index increased significantly along the period considered. Both trends are thus in agreement with Odum's scenario. So is the Shannon index, which shows a significant increasing trend (Figure 11.8).

In his hypothesis for development Odum divided diversity into two components: that pertaining to the variety of species and that referring to the apportionment of individuals among the species (equitability or evenness). In Lake Scuro only this latter component contributes to the index as the number of species did not change during the whole period of investigation.

11.4   Disturbed Ecosystem, Earlier Stages of Succession or What?

Contrasting evidence emerges from this analysis. The majority of the indices considered in the study would speak in favor of a developmental pathway that mirrors Odum's generalization. These are indices that capture bioenergetics, diversity (e.g., Shannon index), cycling, community respiration, food chains, and stability. By contrast, entropy and information show reversal trends in comparison with the expectation. The trend for entropy (both as H and DC) can be accommodated in the general scheme if we consider that Ulanowicz's simulation yielded that information initially increased toward a maximum and then leveled off and slowly decreased (Mageau *et al.*, 1995). So Lake Scuro could still be at some early stage of succession. Clearly this cannot be demonstrated with the dataset at our disposal. However, as a speculative argument one can posit that during the period of investigation Lake Scuro could be at early stages of a developmental pathway following some perturbation or stressful event that had disrupted previous configurations. This hypothesis is, however, hard to accommodate in the picture that emerges for ecological investigations on the lake. In fact this water body has always been described as a rather undisturbed ecosystem (Bertani *et al.*, 2016). Nonetheless, information, quantified by the AMI, and overhead would indicate that some stressful conditions could have affected the lake. Overhead has been indicated as a proxy for stability because it may quantify a system's ability to absorb stress without dramatic loss of function, due to the multiplicity of connections (both internal and with the outside environment, Mageau *et al.*, 1995). In Ulanowicz's scenario this index should be decreasing along with ecosystem development, because redundant, less efficient connections would be pruned away in favor of fewer efficient links as the ecosystem develops. Only a reversal of successional pathways would produce an increase in overhead, a sign that the system would require protection to face stressful conditions. High overhead, however, renders the system more stable, according to Odum's picture. Thus Ulanowicz's scenario seems to contradict the idea that an ecosystem would tend to be more stable in the course of succession. However, in Odum's generalization stability and information seem not to be in relation with one another, whereas in Ulanowicz's perspective they are. In particular, information would increase at the expenses of overhead, exactly as observed in Lake Scuro. To possibly reconcile the two views an in-depth re-examination of the theoretical foundations of the successional paradigm seems necessary. This would require further long-term investigations targeted at pristine ecosystems as compared to their disturbed counterparts.

References

Allesina, S. and Bodini, A. (2008). Ascendency. In *Encyclopedia of Ecology Vol. 1*, ed. S. E. Jørgensen and B. D. Fath, Oxford: Elsevier, pp. 254–263.

Antonietti, R., Ferrari, I., Rossetti, G., Tarozzi, L., and Viaroli, P. (1988). Zooplankton structure in an oligotrophic mountain lake in Northern Italy. *Verhandlungen des Internationalen Verein Limnologie.*, **23**, 545–552.

Archer, S. and Stokes, C. (2000). Stress, disturbance and change in rangeland ecosystems: rangeland desertification. *Advances in Vegetation Science*, **19**, 17–38.

Bertani, I., Primicerio, R., and Rossetti, G. (2016). Extreme climatic event triggers a lake regime shift that propagates across multiple trophic levels. *Ecosystems*, **19**, 16–31.

Bondavalli, C., Bodini, A., Rossetti, G., and Allesina, S. (2006). Detecting stress at the whole ecosystem level. The case of a mountain lake: Lake Santo (Italy). *Ecosystems*, **9**, 768–787.

Clements, F. E. (1936). Nature and structure of the climax. *Journal of Ecology*, **24**, 252–284.

Drury, W. H. and Nisbet, I. C. T. (1973). Succession. *Journal of the Arnold Arboretum*, **54**, 331–368.

Kerfoot, W. C. (ed.) (1980). *The Evolution and Ecology of Zooplankton Communities*. Hanover, NH: University Press of New England.

Kerfoot, W. C. and Sih, A. (eds.) (1987). *Predation: Direct and Indirect Impacts on Aquatic Communities*. Hanover, NH: University Press of New England.

Latham, L. G. and Scully, E. P. (2002). Quantifying constraint to assess development in ecological networks. *Ecological Modelling*, **154**, 25–44.

Lynch, M. and Shapiro, J. (1981). Predation, enrichment, and phytoplankton community structure. *Limnology and Oceanography*, **26**, 86–102.

MacArthur, R. (1955). Fluctuation of animal populations and a measure of community stability. *Ecology*, **36**, 533–536.

MacMahon, J. A. (1980). Ecosystems over time: succession and other types of change. In *Forests: Fresh Perspectives from Ecosystem Analysis*, ed. R. H. Waring and O. R. Corvalis, Oregon State University Press, pp. 27–58.

Mageau, M. T., Costanza, R., and Ulanowicz, R. E. (1995). The development and initial testing of a quantitative assessment of ecosystem health. *Ecosystem Health*, **1**, 201–213.

Mageau, M. T., Costanza, R., and Ulanowicz, R. E. (1998). Quantifying the trends expected in developing ecosystems. *Ecological Modelling*, **112**, 1–22.

Noble, I. R. and Slatyer, R. O. (1977). Post-fire succession of plants in Mediterranean ecosystems. In *Proceedings of the symposium on the environmental consequences of fire and fuel management in Mediterranean climate ecosystems*. USDA Forest Service General Technical Report. WO-3, pp. 27–36.

Odum, E. P. (1969). The strategy of ecosystem development. *Science*, **164**, 262–270.

Odum, E. P. (1985). Trends expected in stressed ecosystems. *BioScience*, **35**, 419–422.

Oksanen, L. (1991). Trophic levels or trophic dynamics: a consensus emerging? *Trends in Ecology and Evolution*, **6**(2), 58–60.

Paris, G., Rossetti, G., Cattadori, M., and Giordani, G. (1995). Phytoplankton–zooplankton interactions in a small mountain lake (Lake Scuro, Parma Appennines): results from enclosure experiments. *Proceedings of the Italian Society of Ecology*, **18**, 147–150.

Rossetti, G., Hamzah, W., and Paris, G. (1997). Zooplankton grazing activity and algal food electivity in a mountain lake. *Proceedings of the Italian Society of Ecology*, **18**, 147–150.

Rutledge, R. W., Basorre, B. L., and Mulholland, R. J. (1976). Ecological stability: an information theory viewpoint. *Journal of Theoretical Biology*, **57**, 355–371.

Schindler, D. W. (1990). Experimental perturbations of whole lakes as tests of hypotheses concerning ecosystem structure and function. *Oikos*, **57**, 25–41.

Sousa, W. P. (1984). Intertidal mosaics: patch size, propagule availability, and spatially variable patterns of succession. *Ecology*, **65**, 1918–1935.

Ulanowicz, R. E. (1986). *Growth and Development: Ecosystems Phenomenology.* New York: Springer-Verlag.

Ulanowicz, R. E. (1997). *Ecology: The Ascendent Perspective.* New York: Columbia University Press.

Ulanowicz, R. E. (2001). Information theory in ecology. *Computers and Chemistry*, **25**, 393–399.

Ulanowicz, R. E. (2004). Quantitative methods for ecological network analysis. *Computational Biology and Chemistry*, **28**, 321–339.

Whittaker, R. H. (1953). A consideration of climax theory: the climax as a population and pattern. *Ecological Monographs*, **23**, 41–78.

12 Statistical Approaches for Inferring and Predicting Food-Web Architecture

Rudolf P. Rohr, Russell E. Naisbit, Christian Mazza, and Louis-Félix Bersier

12.1 Introduction

Food webs are complex networks of trophic interactions (Cohen, 1978). Identification of the factors underlying the architecture of these networks remains a key question in ecology, in the hope that this will reveal how communities may be conserved in the face of species loss and climate change. Until now, this question has mainly been tackled with two different approaches: "a priori" (Ross, 1911), through stochastic or evolutionary models aimed at reproducing the essence of the system (Cohen and Newman, 1985; Williams and Martinez, 2000; Drossel et al., 2001; Cattin et al., 2004; Stouffer et al., 2005; Rossberg et al., 2006; Allesina et al., 2008; Capitán et al., 2013), and "a posteriori," through mechanistic or statistical models aimed at inferring observed networks (Petchey et al., 2008; Allesina and Pascual, 2009; Rohr et al., 2010, 2016).

In the first approach, assumptions about the principles underlying food-web structure are used to construct models. Following the assumptions, model food webs are generated and compared with observed data. In almost all of these models, the input parameters are the numbers of species and the numbers of trophic links. For each species, some "abstract traits" are generated randomly ("ranks" for the cascade model, "niche values" for the niche model, etc.). The comparison with observed food webs is typically achieved indirectly, by generating a large number of networks and comparing them with observed webs through statistical descriptors, e.g., the proportion of top species, or the average chain length (Cohen et al., 1990; Williams and Martinez, 2008). A limitation of this approach is that different models can yield very similar results, a well-known problem with a-priori models (e.g., Cohen, 1968).

In the second approach, observed food webs are fitted using models with the objective of identifying the underlying structure. In food-web ecology, this approach is based on mechanistic (Petchey et al., 2008), or statistical (also called probabilistic) models that use biological traits (usually body size) as explanatory variables or latent traits (Rohr et al., 2010; Williams et al., 2010; Williams and Purves, 2011; Rohr et al., 2016). Direct comparison of the inferred and observed food webs is possible, e.g., by using the percentage of correctly fitted trophic links.

The statistical approach works in general as follows. Let A_{ij} be the adjacency matrix of a food web, i.e., $A_{ij} = 1$ when prey i is eaten by predator j and $A_{ij} = 0$ otherwise. The aim of a statistical model is to infer the probability of existence of trophic links between pairs of species denoted by $P(A_{ij} = 1)$. A standard approach with such binary data is to write a model for the logit of those probabilities (Kolaczyk, 2009), which takes the following general form:

$$\text{logit}\left(P(A_{ij} = 1)\right) = log\left(\frac{P(A_{ij} = 1)}{1 - P(A_{ij} = 1)}\right) = f(\theta, X, Z), \qquad (12.1)$$

with θ the parameters, X the observed species traits as explanatory variables, and/or Z the latent traits (if present in the model). Then the likelihood of observing the network A_{ij} is given by the formula

$$L(A|\theta, X, Z) = \prod_{ij} P(A_{ij} = 1)^{A_{ij}} (1 - P(A_{ij} = 1))^{(1 - A_{ij})}. \qquad (12.2)$$

Here, we first present four statistical models: the body-size, latent-traits, niche, and matching-centrality models. Second, we compare their performance in inferring empirical food webs. Third, we explain how they can be used for exploring the biological factors underlying food-web architecture. Finally, we show how the matching-centrality model can be used to infer partially observed food webs, and to predict their architecture (i.e., the trophic links that a new species would form in a network).

12.2   Body-Size Model

Our first model, the body-size model, uses species body size as explanatory variables (Rohr *et al.*, 2010). It is based on the assumption that there exists an optimal ratio between the predator and its prey body sizes. The formulation for the probability of existence of a trophic link between a predator of body-size m_j and a prey of body-size m_i is given by

$$\text{logit}\left(P(A_{ij} = 1)\right) = \alpha + \beta log\left(\frac{m_j}{m_i}\right) + \gamma log^2\left(\frac{m_j}{m_i}\right). \qquad (12.3)$$

The parameters α, β, and γ are estimated using the maximum likelihood technique. Indeed, this model is simply a generalized linear model with a binomial distribution, and with $log(m_j/m_i)$ and $log^2(m_j/m_i)$ as explanatory variables. The quadratic term is used to capture the optimum in the body-size ratio. This model can also be viewed as a simplified statistical version of the allometric diet breadth model (ADBM) of Petchey *et al.* (2008).

12.3   Latent-Traits Model

The latent-traits model is an extension of the body-size model, which aims at quantifying the structure that is left unexplained by the optimal body-size ratio. The idea is to add, for each species, parameters that quantify their behavior as prey or as predator. These parameters are called latent traits (or latent variables): they are considered as important characteristics of species, but are not measured; however, they can be estimated from the data. The model is as follows:

$$\text{logit}\left(P\left(A_{ij} = 1\right)\right) = \alpha + \beta log\left(\frac{m_j}{m_i}\right) + \gamma log^2\left(\frac{m_j}{m_i}\right) + v_i \delta f_j. \qquad (12.4)$$

Compared to the body-size model, the additional parameters are v_i, the vulnerability traits of prey species i; f_j, the foraging trait of predator j; and δ, a scaling parameter proportional to the relative importance of the latent term. We constrain the scale of the latent traits as follows: $\sum_i (v_i)^2 = \sum_i (f_j)^2 = 1$. The latent term can be thought of as the dominant component of a singular value decomposition (SVD) applied to an analog of the matrix of residuals once controlling for body-size ratios, with the v_i the left (or exit) dominant singular vector, the f_j the right (or entry) dominant singular vector, and δ the dominant singular value (Hoff, 2009; Rohr *et al.*, 2010). All parameters and latent traits are fitted at the same time by maximum likelihood using a simulated annealing algorithm.

12.4   Niche Model

The probabilistic niche model (Fournier *et al.*, 2009; Williams *et al.*, 2010; Williams and Purves, 2011) aims to be a statistical counterpart of the original stochastic niche model of Williams and Martinez (2000). The original niche model assumes that each species in a food web is characterized by a niche center (v_i), a diet center (f_i), and a diet breadth (r_i), and that predators j consume all prey i whose v_i values are within an interval of size r_j, centered on f_j. In the statistical version, the niche parameters are not anymore randomly drawn, but estimated directly from the food web itself. In this sense, these parameters are latent traits. We define the model as follows:

$$\text{logit}\left(P\left(A_{ij} = 1\right)\right) = -\frac{1}{r_j}(v_i - f_j)^2 + m \qquad (12.5)$$

with m the common intercept. The rationale behind the equation is the following: the closer the diet center of a predator to a niche value of a prey, the larger the probability of a trophic link; the larger the diet breadth of a predator, the higher the probability of forming a trophic link. As for the latent-traits model, the parameters (m, v_i, f_i, and r_i) are estimated by maximum likelihood using a simulated annealing algorithm. Note that top species have no v_i terms, and basal species have no f_j and r_j terms. It is worth mentioning that this equation does not perfectly match the formulation of the original niche model, where all prey species within an interval are consumed with probability 1. This constraint is incompatible with the non-intervality of diet intervals of observed food webs (Cattin *et al.*, 2004; Bersier *et al.*, 2006; Stouffer *et al.*, 2006), and would result in a statistical model with a likelihood of zero.

12.5 Matching-Centrality Model

The matching-centrality model (Rohr *et al.*, 2016) aims to be a "generalization" of the previous models. As for the niche model, species are characterized by latent traits only. Each species as prey is described by a latent trait of centrality v_i^*, and by d latent traits of matching v_i^k ($k = 1, \ldots, d$). The parameter d represents the number of matching dimensions. Similarly, each species as predator is described by a latent trait of centrality f_j^*, and by d latent traits of matching f_j^k (Rossberg *et al.*, 2006, 2010, 2013). The model is mathematically defined as follows:

$$\text{logit}\left(P(A_{ij} = 1) \right) = -\sum_{k=1}^{d} \lambda^k (v_i^k - f_j^k)^2 + \delta_1 v_i^* + \delta_2 f_j^* + m, \qquad (12.6)$$

where the parameter $\lambda^k > 0$ is the relative importance of the matching terms of dimension k, δ_1 the relative importance of the prey centrality term, δ_2 the relative importance of the predator centrality term, and m the common intercept. The model is based on the following ideas: the smaller the difference between the matching traits of a prey and of a predator, the higher the probability that a trophic link is formed; the larger the centrality trait of a prey (of a predator), the higher the expected number of predators (of prey). As in the two previous models, the latent traits of matching and centrality are estimated by maximum likelihood using a simulated annealing algorithm. Note that top species have no v_i^* and v_i^k terms, and similarly basal species have no f_j^* and f_j^k terms; for $d > 1$, the v_i^k vectors are pairwise orthogonal, and similarly the f_j^k vectors.

12.6 Performance of the Models

We judged the performance of the models based on two criteria: the fraction of correctly fitted trophic links (Ω), and the Akaike information criterion (AIC). The fraction of correctly fitted trophic links is computed as follows: from the fitted probabilities given by the model, we construct a fitted adjacency matrix by setting to 1 the L elements corresponding to the pairs of species having the highest linking probability ($L =$ number of observed trophic interactions). Then we compute the fraction of trophic links that have been correctly fitted. This measure of performance is independent of the complexity of the model. In contrast, the equation of the AIC includes a penalty for the number of parameters of the model. In our case, the body-size model has three parameters, while all

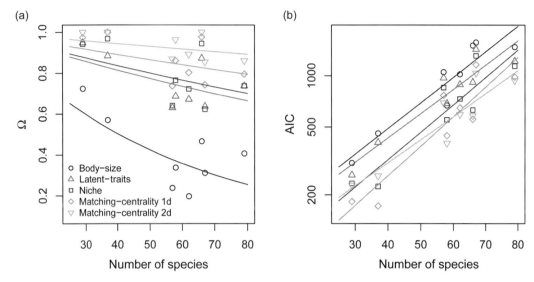

Figure 12.1 Performance of the five statistical models. Panel (a) shows the fraction of correctly fitted trophic links as a function of the number of species in the eight fitted food webs. More complex models in terms of the number of latent traits provide better fits. Panel (b) gives the AIC as a function of the number of species. It indicates that, based on this criterion, the increase in the complexity of the models is justified. (A black and white version of this figure will appear in some formats. For the color version, please refer to the plate section.)

other models involving latent traits are much more complex, with the number of parameters scaling with the number of species in the food web.

We fitted the models to eight aquatic food webs from the published dataset of Brose *et al.* (2005): Sierra Lakes (Harper-Smith *et al.*, 2005), Tuesday Lake (Jonsson *et al.*, 2005), Mill Stream (Ledger *et al.*, unpublished data), Celtic Sea (Pinnegar *et al.*, 2003), Mulgrave River (Rayner, unpublished data), Skipwith Pond (Warren, 1989), Sheffield (Warren, unpublished data), and Broadstone Stream (Woodward *et al.*, 2005); the number of species in these systems vary from 29 to 79.

The results of fitting the body-size model, latent-traits model, niche model, and matching-centrality model in one and two dimensions are shown in Figure 12.1. Regarding the fraction of correctly fitted trophic links, we expectedly find that more complex models do perform better (Figure 12.1a). Expectedly, the decline in performance with food-web size is less pronounced for more complex models, with the two-dimensional matching-centrality being the less affected. Comparison of the models based on AIC provides the same ranking (Figure 12.1b). This is an interesting result given the large number of parameters of the models involving latent traits compared to the body-size model. It indicates that the improvement in the goodness-of-fit is not a mere consequence of increasing the complexity. Again, the matching-centrality model largely outperforms the other ones. In food webs with less than 60 species, two matching dimensions are superfluous, but provide a clear improvement for larger food webs. It is likely that more dimensions would be needed for very large webs.

12.7 Linking Latent Traits to Biological Information

Figure 12.2 represents the food web of Tuesday Lake fitted with the different models. The dots are the observed trophic interactions ($A_{ij} = 1$); the species in their role of prey (rows) and of predator (columns) are ordered according to body size or to the different

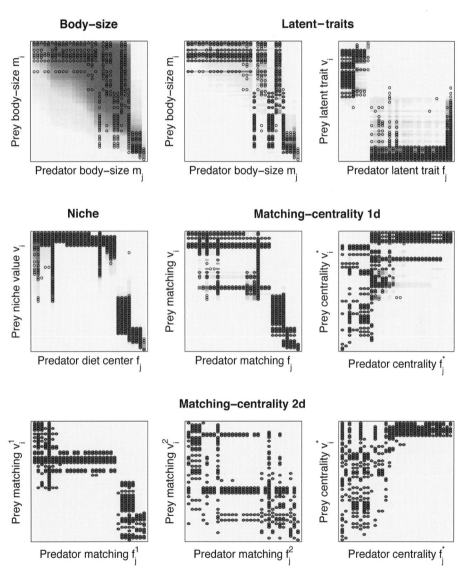

Figure 12.2 Representation of the food web of Tuesday Lake in the latent-traits space. Each panel represents the adjacency matrix, with the dots indicating a trophic interaction between a predator (columns) and a prey (row). The color, from yellow to red, indicates increasing fitted linking probability of the respective models. Species are ordered according to the relevant variable or latent trait (see axis legends). (A black and white version of this figure will appear in some formats. For the color version, please refer to the plate section.)

latent traits; the background color gives the fitted probability that a trophic link exists between species i and j. Once the latent traits are estimated for each species (latent traits of vulnerability and of foraging in the latent-traits model; niche values, diet center, and diet breadth in the niche model; matching and centrality traits of vulnerability and of foraging in the matching-centrality model), it can be tested if they are related to biological information about the species. This can be achieved with different statistical approaches depending on the type of biological information available (e.g., simple non-parametric correlation with a quantitative variable describing the species).

In our case, information on body size and on taxonomy (as a proxy for phylogeny) is available for the eight food webs, but we present the results only for the food web of Tuesday Lake. Because taxonomic information is a qualitative variable, we choose partial Mantel tests (Legendre and Legendre, 1998) for this analysis; in this way, we can compare the relative importance of body size and of taxonomy in explaining the latent traits (Naisbit *et al.*, 2012). We compute distance matrices for the biological variables and the relevant latent traits. The distance in body size between two species i and j is computed as the absolute value in log differences, i.e., $d_{ij}^{BS} = \text{abs}(log(m_i) - log(m_j))$. For the distance in phylogeny, we used as a proxy the proportional number of taxonomic levels for which two species differ (Naisbit *et al.*, 2012). The distance in latent traits is simply given by the absolute value of the differences (e.g., for the latent traits of vulnerability, $d_{ij} = \text{abs}(v_i - v_j)$). We relate these three distance matrices using partial Mantel tests with Pearson correlation.

For the latent-traits model, we find that latent traits of vulnerability and foraging are both strongly related to the phylogeny (Table 12.1). This indicates that phylogeny has an effect on network structure that is independent of the optimal body-size ratio between the predators and prey. Interestingly, body size of the species in their role of prey is still correlated with the latent trait of vulnerability (once phylogeny has been accounted for); thus the optimal ratio does not fully capture the effect of body size on network structure for the species in the role of prey (Rohr *et al.*, 2010). The reason underlying this result is difficult to unravel; one possibility is that the distribution of body-mass ratios is strongly skewed, generating a correlation between body mass and latent traits. However, why this effect is apparent only for species in their role of prey requires further investigation.

The results are consistent for the other models based on latent traits only. In the niche model, the niche trait is related to the position of the species on an abstract axis depicting their position as prey, while the diet center depicts their position as predator. In the matching-centrality model, this corresponds to the vulnerability and to the foraging traits, respectively. We find that body size is correlated to all these latent traits once phylogeny is accounted for (since we use partial Mantel tests; see Table 12.1). Globally, this observation indicates that species of similar body size tend to have similar trophic roles, which is expected in aquatic systems (e.g., Cohen *et al.*, 2003).

Another important result is the presence of a strong phylogenetic signal in almost all latent traits, which indicates that phylogenetic constraints are an important determinant of the trophic structure of communities (Cattin *et al.*, 2004; Naisbit *et al.*, 2012). Thus taxonomically similar prey species tend to be eaten by taxonomically similar predators,

Table 12.1 Results of partial Mantel tests between the latent traits and body size and phylogeny for the food web of Tuesday Lake.

Type of latent trait	Body size		Phylogeny	
	partial-r	p-value	partial-r	p-value
Latent-traits model				
Latent traits of vulnerability	**0.545**	**0.001**	**0.272**	**0.001**
Latent traits of foraging	−0.010	0.472	**0.341**	**0.003**
Niche model				
Niche traits	**0.785**	**0.001**	**0.192**	**0.001**
Diet center	**0.714**	**0.001**	−0.070	0.720
Diet range	0.074	0.205	−0.056	0.772
Matching-centrality 1D model				
Vulnerability matching	**0.590**	**0.001**	**0.267**	**0.001**
Foraging matching	**0.810**	**0.001**	−0.014	0.550
Vulnerability centrality	**0.294**	**0.004**	**0.074**	**0.031**
Foraging centrality	**0.649**	**0.001**	0.064	0.183
Matching-centrality 2D model				
Vulnerability matching 1st dim.	**0.330**	**0.001**	**0.070**	**0.010**
Foraging matching 1st dim.	**0.221**	**0.040**	**0.149**	**0.031**
Vulnerability matching 2nd dim.	**0.194**	**0.003**	**0.267**	**0.001**
Foraging matching 2nd dim.	**0.255**	**0.014**	**0.402**	**0.001**
Vulnerability centrality	0.075	0.161	**0.098**	**0.002**
Foraging centrality	0.058	0.203	**0.522**	**0.001**

Legend: dim. = dimension.
Bold values indicate significant correlation at the $p \leq 0.05$ level.

and similar predator species tend to consume similar prey. This result can be seen as unsurprising, as phylogeny is a powerful integrator of the ecological characteristics of species (body size included). However, since we use partial Mantel tests, this result indicates that ecological traits uncorrelated to body size are also necessary to account for the structure of trophic interactions of prey species (e.g., in terrestrial systems, similar secondary compounds in phylogenetically related plant species; Price, 2003).

12.8   Reconstruction of Partially Observed Food Webs

One of the advantages of statistical models for food-web architecture is their ability of inferring and reconstructing partially observed food webs. In such a case, no information is available for the existence of a link for a fraction of species pairs (for those species pairs, $A_{ij} = $ NA). A statistical model can be fitted on a partially observed adjacency matrix: the likelihood function (Eq. 12.2) is simply computed only on the part that is known. Mathematically this likelihood L^P is given by

$$L^P(A|\theta, X, Z) = \prod_{A_{ij} \neq \text{NA}} P(A_{ij} = 1)^{A_{ij}}(1 - P(A_{ij} = 1))^{(1-A_{ij})} \tag{12.7}$$

The inference of all the parameters and all the latent traits can be achieved with only the known part of the adjacency matrix. Using the equation of the models (Eqs. 12.3 to 12.6), we can estimate the linking probabilities for all pairs of species, in particular for the pairs of species set to NA. With these estimated linking probabilities, we can finally reconstruct the food web by predicting the presence and absence of trophic links for all pairs of species. The prediction works as follows: first, the estimated total number of trophic interactions L_P is computed as the sum over all species pairs of the linking probabilities ($L_P = \sum_{ij} P(A_{ij} = 1)$); second, the L_P pairs with the highest linking probability are set to 1 in the adjacency matrix.

We test the reconstruction of the Tuesday Lake food web using the matching-centrality model with two dimensions of matching. We simulate partially observed networks by setting to NA a given fraction of the elements of the adjacency matrix. Then, we reconstruct these generated partially observed food webs and compare the outputs with the observed matrix. As a measure of the performance of the method, we use the fraction of 1s and 0s correctly predicted. Figure 12.3a shows the results for 10% and 30% of the matrix elements set to NA. In general, the ability of the model to reconstruct Tuesday Lake's food web is quite high. Note, however, that these values have to be considered against the baseline given by the fraction of non-trophic links (one minus the connectance; the horizontal dashed line in Figure 12.3): trivially predicting an absence of trophic link for all pairs of species would result in a fraction of correct predictions equal to the fraction of non-trophic links.

12.9   Forecasting Trophic Interactions

A very interesting feature of statistical models for food-web architecture is the possibility to forecast the trophic links that a new incoming species will make when joining an existing community. For such a forecast to be sensible, it is first necessary to have a model able to faithfully infer the network. For this reason, we choose the 2D matching-centrality model. The core of the methodology resides in the use of latent traits as intermediate between the linking probabilities and observed species biological traits. We explain the method using body size and species phylogeny as biological information. The procedure is as follows: the first step consists of fitting the matching-centrality model to the observed food web. From the result, we extract the estimated latent traits of matching and of centrality for each species. The second step consists of relating each matching and centrality trait of all species to the available biological traits. In our case, this is achieved by a phylogenetic regression (Grafen, 1989; Freckleton *et al.*, 2002): we assume each latent trait to be linearly related to the log of the body size $\vec{m}$, and the phylogeny to induce a correlation structure. This linear model (here for a matching trait of vulnerability) is mathematically given by

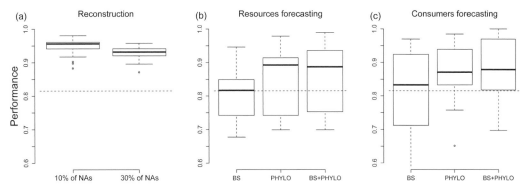

Figure 12.3 Performance of the 2D matching-centrality model in reconstructing and forecasting the Tuesday Lake food web. In all panels, the y axis gives the proportion of correctly predicted presence and absence of trophic links (0s and 1s in the adjacency matrix); the dashed lines give the baseline performance given by the proportion of 0s in the adjacency matrix (see text). Panel (a) shows the performance of reconstructing the food web after having set to NA (absence of information) a given percentage of the adjacency matrix. Panels (b) and (c) give the performance of forecasting the set of prey and the set of predators, respectively. We perform an out-of-sample test by removing triplets of species.

$$\vec{v} \sim N\left(\alpha + \beta \cdot log(\vec{m}),\ \sum(\lambda) \right) \qquad (12.8)$$

with α and β the intercept and slope, respectively, and $\Sigma(\lambda)$ the variance–covariance matrix induced by the phylogenetic relatedness. We use Pagel's λ correlation structure (Freckleton *et al.*, 2002): the elements of the variance–covariance matrix are then given by

$$\sum(r)_{ij} = \sigma^2 \begin{cases} 1 & \text{if} \quad i = j \\ \lambda t_{ij} & \text{if} \quad i \neq j \end{cases} \qquad (12.9)$$

where t_{ij} is the fraction of common time between species i and j on the phylogenetic tree, λ quantifies the strength of the correlation induced by the phylogeny, and σ^2 is the common variance. The third step consists of forecasting the latent traits of matching and of centrality of the new species k. Using the estimated parameters of the phylogenetic regressions of step two, we can use the conditional expectation to forecast these values, which are given by

$$\hat{v}_k = \hat{\alpha} + \hat{\beta} \cdot log(m_k) + \sum(\hat{r})_{-k,k} \sum(\hat{r})_{k,k}^{-1} \left(\vec{v}_{-k} - \hat{v}_{-k} \right) \qquad (12.10)$$

where $\hat{\alpha}, \hat{\beta}, \hat{\gamma}$ are the fitted parameters from the second step; $\sum(\hat{r})_{-k,k}$ is the kth column without the kth row (indicated by subscript $-k$) of the variance–covariance matrix; $\sum(\hat{r})_{k,k}$ is the (k,k) element of the variance–covariance matrix; $\left(\vec{v}_{-k} - \hat{v}_{-k} \right)$ is the row vector of residuals obtained from the phylogenetic regression of the second step. The final step consists of estimating the linking probabilities for the new species and then forecasting the presence or absence of trophic links. Based on the forecasted matching and centrality traits and using Eq. (12.6), we forecast the linking probabilities between the new species and the species already present; then using the same technique as in the reconstruction method, we forecast the presence or absence of trophic links.

To test our forecasting method, we perform an out-of-sample test by removing three species at the same time and then trying to forecast their trophic links in the food web of Tuesday Lake. As a measure of performance, we use the fraction of correctly predicted elements of the adjacency matrix for each triplet of test species. Figures 12.3b and c show that using body size alone as biological information provides only poor predictions; phylogeny is necessary to attain good performance. With this information, the model performs in general very well in forecasting trophic links for species, both in their role of prey and of predator.

12.10   Discussion

The motivation behind the recent development of statistical models for food-web structure was to offer a simple and intuitive tool to explore the factors underlying the architecture of ecological networks. They represent a complement to stochastic models, such as the cascade model (Cohen and Newman, 1985) and its successors (Williams and Martinez, 2000; Cattin *et al.*, 2004; Stouffer *et al.*, 2005; Rossberg *et al.*, 2006; Allesina *et al.*, 2008; Capitán *et al.*, 2013), and are more versatile than the ADBM of Petchey *et al.* (2008), which was developed specifically to explore how allometry and optimal foraging can be used to infer trophic interactions. A problem with stochastic models, which generate families of networks intended to reproduce the structure of real food webs, is that different assumptions used to build the models can lead to very similar results, a problem well-known for example with species-abundance models (Cohen, 1968). Statistical models can provide a more direct assessment of the relationships between putative underlying factors and network structure. Here we use body size and phylogeny as factors, but it would be very interesting to include other ecological variables in the models. Such information is becoming available for food webs, for example in Eklöf *et al.* (2013).

Another key advantage of statistical models is that they can be used to reconstruct partially observed systems, or to forecast the links produced by a new species. Any statistical model could theoretically be used for these purposes. For example, a version of the latent-trait model has been used to infer the linking probabilities for a plant–herbivore network (Pellissier *et al.*, 2013). Here we base our analyses on the matching-centrality model (Rohr *et al.*, 2016) since it offers by far the best fitting capabilities. We believe that these features could be very useful, for example to build so-called metawebs (Gravel *et al.*, 2013) from partial information on a system, or to estimate the trophic role of an introduced species. The approach also has its limits. For the reconstruction of food webs, the predicted trophic interactions should of course not be taken for granted, but could serve as a guide to direct sampling effort on the system in a cost-efficient way. For forecasting the links that a new species would create, the approach is meaningful only if the new species shares enough ecological characteristics with some of the species present in the system.

The first model including latent variables in food-web ecology is the latent-trait model (Rohr *et al.*, 2010). This model requires the estimation of a large number of parameters, which is achieved with Morkov Chain Monte Carlo (aka MCMC) or simulated annealing methods (as in the present contribution). As with the other models based on latent variables, the goodness-of-fit is impressive, which may not seem surprising given the large number of parameters. However, one important outcome of our analyses is that these parameters are not merely abstract values, but are bearing meaningful biological information. Another key aspect is the possibility to formulate much more complex (and hopefully sensible) models than classical generalized linear models (e.g., the body-size model). The probabilistic-niche model (Fournier *et al.*, 2009; Williams *et al.*, 2010; Williams and Purves, 2011) is an excellent example: it is possible from the food-web matrix to infer the parameters of the classical niche model. The niche model takes the point of view of the predators to constrain the possible prey that enter their diet (Williams and Martinez, 2000); however, there is no constraint for the species in their role of prey. The matching-centrality model was developed to circumvent this limitation, with a formulation that is symmetric for prey and predators. The first step was to separate the species in their role of prey and in their role of predator in two sets. In doing so, the food web is expressed as a bipartite network, with the intermediate species appearing in both sets. The traits of vulnerability and of foraging are estimated for intermediate species, while basal species have no foraging traits, and similarly top species have no vulnerability traits. The main difference with the niche model is that prey species have a centrality trait in the matching-centrality model, which is related to the number of species they prey upon (the centrality trait for the predators is akin to the range of the niche model). Interestingly, it appeared that the simplest formulation accounting for these desiderata yielded an equation similar to that found in Rossberg *et al.* (2010), from which we named the latent traits of our model. We found out that the matching-centrality model is very versatile in decomposing any adjacency matrix into several quantitative traits for the nodes, from which the adjacency matrix can then be reconstructed. We applied this model to the analysis, reconstructing and forecasting networks as diverse as social, terrorist-association, aggression between countries, genetic, and neuronal (Rohr *et al.*, 2016). We hope that this model that emanates from food-web ecology will be useful in a wide range of scientific domains.

Acknowledgments

This work was supported by the National Centre of Competence in Research "Plant Survival" to LFB, by SystemsX.ch, the Swiss Initiative in Systems Biology to LFB and CM, and by the Swiss National Science Foundation grant 31003A_138489 to LFB.

References

Allesina, S. and Pascual, M. (2009). Food web models: a plea for groups. *Ecology Letters*, **12**, 652–662.

Allesina, S., Alonso, D., and Pascual, M. (2008). A general model for food web structure. *Science*, **320**, 658–661.

Bersier, L. F., Cattin, M. F., Banasek-Richter, C., Baltensperger, R., and Gabriel, J. P. (2006). Box B: Reply to Martinez and Cushing. In *Ecological Networks: Linking Structure to Dynamics in Food Webs*, ed. M. Pascual and J. A. Dunne, New York: Oxford University Press, pp. 91–92.

Brose, U., Cushing, L., Berlow, E. L., *et al.* (2005). Body sizes of consumers and their resources. *Ecology*, **86**, 2545.

Capitán, J. A., Arenas, A., and Guimerà, R. (2013). Degree of intervality of food webs: from body-size data to models. *Journal of Theoretical Biology*, **334**, 35–44.

Cattin, M. F., Bersier, L. F., Banasek-Richter, C., Baltensperger, R., and Gabriel, J. P. (2004). Phylogenetic constraints and adaptation explain food-web structure. *Nature*, **427**, 835–839.

Cohen, J. E. (1968). Alternate derivations of a species-abundance relation. *American Naturalist*, **102**, 165–172.

Cohen, J. E. (1978). *Food Webs and Niche Space*. Princeton, NJ: Princeton University Press.

Cohen, J. E. and Newman, C. M. (1985). A stochastic theory of community food webs. 1. Models and aggregated data. *Proceedings of the Royal Society B: Biological Sciences*, **224**, 421–448.

Cohen, J. E., Briand, F., and Newman, C. M. (1990). *Community Food Webs, Data and Theory*. Berlin: Springer-Verlag.

Cohen, J. E., Jonsson, T., and Carpenter, S. R. (2003). Ecological community description using the food web, species abundance, and body size. *Proceedings of the National Academy of Sciences of the United States of America*, **100**, 1781–1786.

Drossel, B., Higgs, P. G., and McKane, A. J. (2001). The influence of predator–prey population dynamics on the long-term evolution of food web structure. *Journal of Theoretical Biology*, **208**, 91–107.

Eklöf, A., Jacob, U., Kopp, J., *et al.* (2013). The dimensionality of ecological networks. *Ecology Letters*, **16**, 577–583.

Fournier, T., Rohr, R. P., Scherer, H., Mazza, C., and Bersier, L. F. (2009). How to estimate the niche values in a food web? The 94th ESA Annual Meeting (August 2–7, 2009), abstract: http://eco.confex.com/eco/2009/techprogram/P20894.HTM.

Freckleton, R. P., Harvey, P. H., and Pagel, M. (2002). Phylogenetic analysis and comparative data: a test and review of evidence. *American Naturalist*, **160**, 712–726.

Grafen, A. (1989). The phylogenetic regression. *Philosophical Transactions of the Royal Society B: Biological Sciences*, **326**, 119–157.

Gravel, D., Poisot, T., Albouy, C., Velez, L., and Mouillot, D. (2013). Inferring food web structure from predator–prey body size relationships. *Methods in Ecology and Evolution*, **4**, 1083–1090.

Harper-Smith, S., Berlow, E. L., Knapp, R. A., Williams, R. J., and Martinez, N. D. (2005). Communicating ecology through food webs: visualizing and quantifying the effects of stocking alpine lakes with trout. In *Dynamic Food Webs: Multispecies*

Assemblages, Ecosystem Development, and Environmental Change, ed. P C. de Ruiter, V. Wolters, and J. C. Moore, Elsevier Academic Press, pp. 407–423.

Hoff, P. D. (2009). Multiplicative latent factors models for description and prediction of social networks. *Computational and Mathematicals Organization*, **15**, 261–272.

Jonsson, T., Cohen, J. E., and Carpenter, S. R. (2005). Food webs, body size, and species abundance in ecological community description. In *Advances in Ecological Research, Vol. 36*, ed. H. Caswell, Elsevier Academic Press, pp. 1–84.

Kolaczyk, E. D. (2009). *Statistical Analysis of Network Data*. New York: Springer.

Legendre, P. and Legendre, L. (1998). *Numerical Ecology*. Amsterdam: Elsevier.

Naisbit, R. E., Rohr, R. P., Rossberg, A. G., Kehrli, P., and Bersier, L.-F. (2012). Phylogeny versus body size as determinants of food web structure. *Proceedings of the Royal Society B: Biological Sciences*, **279**, 3291–3297.

Pellissier, L., Rohr, R. P., Ndiribe, C., *et al.* (2013). Combining food web and species distribution models for improved community projections. *Ecology and Evolution*, **3**, 4572–4583.

Petchey, O. L., Beckerman, A. P., Riede, J. O., and Warren, P. H. (2008). Size, foraging, and food web structure. *Proceedings of the National Academy of Sciences of the United States of America*, **105**, 4191–4196.

Pinnegar, J. K., Trenkel, V. M., Tidd, A. N., Dawson, W. A., and Du Buit, M. H. (2003). Does diet in Celtic Sea fishes reflect prey availability? In *Annual Symposium of the Fisheries Society of the British Isles*, Norwich, pp. 197–212.

Price, P. W. (2003). *Macroevolutionary Theory on Macroecological Patterns*. Cambridge, UK: Cambridge University Press.

Rohr, R. P., Scherer, H., Kehrli, P., Mazza, C., and Bersier, L. F. (2010). Modeling food webs: exploring unexplained structure using latent traits. *American Naturalist*, **173**, 170–177.

Rohr, R. P., Naisbit, R. E., Mazza, C., and Bersier, L. F. (2016). Matching-centrality decomposition and the forecasting of new links in networks. *Proceedings of the Royal Society B: Biological Sciences*, **283**, 20152702.

Ross, R. (1911). Some quantitative studies in epidemiology. *Nature*, **87**, 466–467.

Rossberg, A. G. (2013). *Food Webs and Biodiversity: Foundations, Models, Data*. Wiley.

Rossberg, A. G., Matsuda, H., Amemiya, T., and Itoh, K. (2006). Food webs: experts consuming families of experts. *Journal of Theoretical Biology*, **241**, 552–563.

Rossberg, A. G., Brannstrom, A., and Dieckmann, U. (2010). How trophic interaction strength depends on traits. *Theoretical Ecology*, **3**, 13–24.

Stouffer, D. B., Camacho, J., Guimera, R., Ng, C. A., and Nunes Amaral, L. A. (2005). Quantitative patterns in the structure of model and empirical food webs. *Ecology*, **86**, 1301–1311.

Stouffer, D. B., Camacho, J., and Amaral, L. A. N. (2006). A robust measure of food web intervality. *Proceedings of the National Academy of Sciences of the United States of America*, **103**, 19015–19020.

Warren, P. H. (1989). Spatial and temporal variation in the structure of a freshwater food web. *Oikos*, **55**, 299–311.

Williams, R. J. and Martinez, N. D. (2000). Simple rules yield complex food webs. *Nature*, **404**, 180–183.

Williams, R. J. and Martinez, N. D. (2008). Success and its limits among structural models of complex food webs. *Journal of Animal Ecology*, **77**, 512–519.

Williams, R. J. and Purves, D. W. (2011). The probabilistic niche model reveals substantial variation in the niche structure of empirical food webs. *Ecology*, **92**, 1849–1857.

Williams, R. J., Anandanadesan, A., and Purves, D. (2010). The probabilistic niche model reveals the niche structure and role of body size in a complex food web. *PLoS ONE*, **5**, e12092.

Woodward, G., Speirs, D. C., and Hildrew, A. G. (2005). Quantification and resolution of a complex, size-structured food web. In *Advances in Ecological Research, Vol. 36*, ed. H. Caswell, Elsevier Academic Press, pp. 85–135.

13 Global Metawebs of Spider Predation Highlight Consequences of Land-Use Change for Terrestrial Predator–Prey Networks

Klaus Birkhofer, Eva Diehl, Volkmar Wolters, and Henrik G. Smith

13.1 Land-Use Change and Terrestrial Predator–Prey Networks

Land-use change, here defined as the conversion of one land-use type into another (e.g., forest to arable land), affects biodiversity and biotic interactions worldwide (Sala *et al.*, 2000). Although there has been large regional variation in the extent of agricultural expansion and abandonment in Europe in the past 50 years (Rabbinge and van Diepen, 2000), there has been a general trend that forest has expanded at the expense of agricultural land (Kankaanpää and Carter, 2004; Rounsevell *et al.*, 2006). The patterns have been similar in North America the past decades (Smith *et al.*, 2010). Globally, particularly in developing countries, the general pattern has instead been agricultural expansion (Smith *et al.*, 2010), threatening forest ecosystems (DeFries *et al.*, 2010; Lambin and Meyfroidt, 2011; but see Angelsen, 2010). In fact, models predict an increase of cropland between 10 and 25% up to 2050, mainly due to agricultural expansion in developing countries (Schmitz *et al.*, 2014). Natural and semi-natural non-forest ecosystems are primarily threatened by a conversion to pasture land (Schmitz *et al.*, 2014) or by cultivation with biofuels (Havlík *et al.*, 2011). Climate change is an important additional driver of land-use conversion, as the range of crop species contract or expand (Olesen and Bindi, 2002) and as forests adapt to changing climatic conditions (Spittlehouse and Stewart, 2004).

In the past, effects of land-use change were often exclusively assessed by their impact on species richness (Tilman *et al.*, 2001). Today, it has increasingly become evident that we need metrics that capture additional features of biological communities to understand consequences of land-use change on ecosystem functions and the provision of ecosystem services (Tylianakis *et al.*, 2007; Diehl *et al.*, 2013). Trophic interactions that link species in food webs are important components that modulate functions provided by biological communities (Laliberté and Tylianakis, 2010; Tylianakis *et al.*, 2010; Thompson *et al.*, 2012). For example, the loss of large apex predators from an ecosystem due to anthropogenic disturbance may cascade through the food chain and lead to drastic effects on primary producers (Estes *et al.*, 2011). Predator populations are often severely affected by anthropogenic disturbances (Attwood *et al.*, 2008). Consequently, the conversion from one major land-use type into another may alter predator–prey interactions

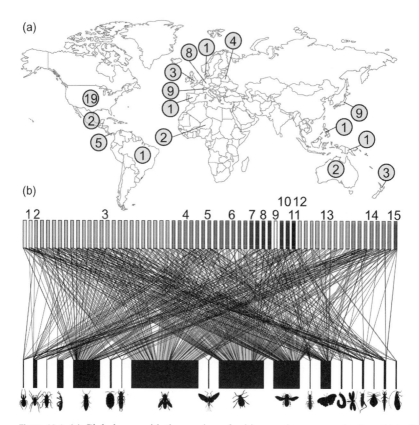

Figure 13.1 (a) Global map with the number of spider species per country for which diets were reported with sufficient level of detail to be included here. All data come from visual searches for web-building spiders and their prey in primary studies. For spider species that occurred more than once in the database only the dataset with most prey items was included in the global metaweb. (b) Global metaweb of web-building spider species (N = 63) and the relative predation on prey orders (N = 19). Spider species are color coded according to family identity and numbered: 1 Agelenidae, 2 Amaurobiidae, 3 Araneidae, 4 Dictynidae, 5 Eresidae, 6 Linyphiidae, 7 Mimetidae, 8 Nephilidae, 9 Pholcidae, 10 Pisauridae, 11 Scytodidae, 12 Sicariidae, 13 Tetragnathidae, 14 Theridiidae, 15 Uloboridae. Prey orders from left to right are: Acari, Araneae, Blattodea, Collembola, Coleoptera, Isopoda, Dermaptera, Diptera, Ephemeroptera, Hemiptera, Hymenoptera, Isoptera, Lepidoptera, Myriapoda, Neuroptera, Orthopteroidea, Psocoptera, Thysanoptera, and Trichoptera. (A black and white version of this figure will appear in some formats. For the color version, please refer to the plate section.)

(Ives *et al.*, 2005) and the provision of predator-mediated ecosystem services such as the control of agricultural pests (conservation biological control; Barbosa, 1998).

The complex interactions between predators and prey can be analyzed and visualized by using interaction networks, in which nodes are predator or prey taxa and links connect nodes, i.e., represent a predator taxon feeding on a prey taxon (antagonistic predator–prey networks; see Figure 13.1b; Dunne, 2006).

Quantitative antagonistic networks include information about the strength of the realized predator–prey links, i.e., how many prey individuals have been consumed by

Box 13.1 Definition of Network Metrics Used to Compare Metawebs Between Spider Species and Prey Orders in Agricultural, Semi-Natural/Natural and Forest Ecosystems Worldwide

(1) Connectivity

(a) Connectance ("connectance") is the realized proportion of possible links in a network (Dunne et al., 2002).

(b) Links per species ("links per species") is the mean number of links per species in a network.

(2) Nestedness

(a) Nestedness ("weighted NODF") describes to what extent predator and prey specialists interact preferentially with generalists. The nestedness metric NODF is based on overlap and decreasing fill of interactions in a matrix of interactions between predators and prey (Almeida-Neto et al., 2008) and the weighted version considers the interactions' strength between predators and prey (Almeida-Neto and Ulrich, 2011).

(3) Specialization

(a) Specialization ("H2") characterizes the degree of specialization among predators and prey in the entire network (Bluthgen et al., 2006).

(b) Generality/Vulnerability ("generality" and "vulnerability") are mean effective numbers of prey orders per predator species (generality) or predator species per prey order (vulnerability) weighted by their marginal totals in an interaction matrix.

(4) Niche overlap

(a) Niche overlap ("niche.overlap.HL") describes the similarity in interactions between predator species (HL, higher trophic level) based on Horn's index of dispersion.

(5) Clustering

(a) Modularity ("Q"; Dormann and Strauss, 2014) characterizes subgroups of predator species and prey orders with within-module interactions that are more prevalent than between-module interactions based on quantitative interaction data.

(6) Interactions

(a) Dissimilarity of interactions between networks originates from differences in composition or because shared species may interact with prey differently (Poisot et al., 2012). The latter component is used here.

Note: Major categories (1–6) follow the definition in Miranda et al. (2013). Terms in parentheses give the respective function in the package bipartite (Dormann et al., 2008) and betalink (Poisot et al., 2012) in the software R (R Development Core Team, 2008) that was used to calculate the respective network metric.

---

a predator. Hence quantitative networks add important details to the understanding of trophic interactions in food-web compartments. Network approaches have the advantage that they produce metrics from trophic interactions that are independent of the taxonomy of species (Box 13.1) and it is therefore possible to compare networks across regions or ecosystems over large scales (Gray et al., 2014).

Our understanding of how anthropogenic changes affect trophic interactions in antagonistic networks of terrestrial invertebrate communities is very limited (van der Putten et al., 2004; Bohan and Woodward, 2013), particularly if compared to mutualistic or parasitoid–host networks (Miranda et al., 2013). Predator–prey interactions are often difficult to observe and quantify (Ings et al., 2009; Tixier et al., 2013). Even though recent advances such as molecular gut content analyses are very promising, they also suffer from methodological problems (e.g., that secondary predation or cannibalism are not detectable; Traugott *et al.*, 2013). To construct networks it is therefore a common practice to infer information about interactions from databases or scientific literature, e.g., that predator A has been previously observed to feed on prey B and C (Gray *et al.*, 2014).

In this chapter, we present antagonistic networks between abundant terrestrial predators (63 web-building spider species) and their prey (19 arthropod orders) that have been constructed from published records of primarily single-species diets around the world. This global metaweb of spider predation is divided into three metawebs for major ecosystem classes affected by land-use change: (1) forest ecosystems such as deciduous forests; (2) semi-natural/natural ecosystems such as heathland, hedges, and unmanaged grassland; and (3) agricultural ecosystems including crops such as maize, wheat, and soybean, but also meadows and pastures. We aim to identify major differences in prey composition and network metrics between these three ecosystem types to better understand the potential consequences of land-use conversions from one major ecosystem type into another for the role of these predators in terrestrial ecosystems and the structure of antagonistic networks.

13.2   A Global Metaweb of Spider Predation

Certain groups of terrestrial, aboveground predators are ideal study objects for the purpose of observing predation events under field conditions (Sunderland *et al.*, 2005). Spiders are the most abundant and predominantly predaceous arthropod order in terrestrial biomes (Wise, 1993). Ecologists have long been interested in the feeding ecology and diet of spiders (e.g., Bilsing, 1920). Given the rich data from these individual studies, we are now able to compare diet composition and network structure for web-building spiders between major terrestrial ecosystem types. Originally studies were descriptive, but the focus became more applied later in the twentieth century by studying the role of spiders in population control of agricultural pests (e.g., Kiritani *et al.*, 1972), human disease vectors or parasites (e.g., Dąbrowska-Prot *et al.*, 1968), or general aspects of foraging ecology (Wise and Barata, 1983). Comparing the diets of individual spider species between different habitats was a popular research field among ecologists with a particularly notable number of contributions by Elchin Fizuli oglu Huseynov (Guseinov *et al.*, 2004; Guseinov, 2005; Huseynov *et al.*, 2005; Huseynov, 2005, 2006a, b, 2007a, b, c, 2008), Wolfgang Nentwig (Nentwig, 1982, 1983a, 1993b, 1985), Martin Nyffeler (Nyffeler and Benz, 1978, 1979, 1981a, b, 1988a, b, c; Nyffeler *et al.*, 1986, 1987, 1988, 1992), and George W. Uetz (Uetz *et al.*, 1978; Uetz and Hartsock, 1987). However, studies about the diet composition of whole spider communities in single habitats are relatively rare (Nyffeler and Sterling, 1994; Bardwell and Averill, 1997; Pérez-de la Cruz *et al.*, 2007). Systematic comparisons between the diet composition of spider communities along a land-use gradient (Diehl *et al.*, 2013) and studies in replicated designs (Birkhofer *et al.*, 2015) are to our knowledge limited to two recent studies.

Research has moved on from single species toward synthesizing individual studies focusing on hunting strategies and diet composition. Nentwig (1987) provided a first synthesis dealing with different hunting strategies, size ratios between spiders and prey, and defensive measures of prey organisms. This synthesis highlighted important links between ecological and evolutionary aspects of spider predation by suggesting that convergent evolution, as a consequence of coevolution between spiders and their prey, is more common than generally acknowledged. Nyffeler (1999) was the first to provide a systematic review about spider diets in agricultural fields in Europe and the United States. The author concluded that web-building spiders in agricultural habitats primarily feed upon insect prey and generally have a narrower diet breadth than actively hunting spider species. Birkhofer and Wolters (2012) used a global meta-analysis of actively hunting and web-building spider diets in semi-natural/natural habitats to highlight that diet breadth is positively related to net primary production and that it was generally higher in the tropics. Pekar et al. (2012) demonstrated that the degree of predator specialization in spiders is affected by the phylogenetic relatedness between species. Most recently, Birkhofer et al. (2013) added a review about different approaches to quantify predation by spiders (e.g., molecular gut content analysis; Kuusk and Ekbom, 2010, 2012) and defined four major groups of agrobiont spiders based on field observations of their diet composition.

In the present book chapter, we analyze the effect of land use on quantitative antagonistic networks based on a global metaweb (Figure 13.1b) that is constructed from worldwide literature records of spider diets in 63 web-building species from 39 publications (Figure 13.1a). Publications were searched for in four databases, namely the ISI Web of Knowledge, Scopus, and the databases of the International Society of Arachnology as well as the British Arachnological Society. For this metaweb all prey data from literature were aggregated into 19 prey orders as prey remains are often not identifiable to species level and to make data comparable between different studies (for similar aggregation procedures see Stouffer et al., 2012). To combine different studies, prey numbers were transformed into relative proportions of each prey order in each spider species' diet (see also Birkhofer and Wolters, 2012). To relate networks to land use, each record of a spider species' individual diet was assigned to one of three major ecosystem types (agricultural, semi-natural/natural, or forest) depending on the habitat description in the original study (Table 13.1a–c). Agricultural ecosystems were defined as monocultures or pastures that are directly targeted by a range of agronomic practices (e.g., fertilization, mowing, livestock grazing, or tillage); semi-natural/natural ecosystems combine habitat types that are not agriculturally managed and are not dominated by large trees; forest ecosystems are dominated by large trees. Note that the majority of studies did not provide data on real networks with diverse spider communities, but only included data about single species per habitat type (for exceptions see Nyffeler and Sterling, 1994; Bardwell and Averill, 1997; Pérez-de la Cruz et al., 2007; Diehl et al., 2013). For the analyses of land-use effects three independent metawebs were generated: one for each major ecosystem type. Figure 13.2 illustrates these individual predator–prey networks between web-building spider species and prey orders in agricultural (Figure 13.2a), semi-natural/natural (Figure 13.2b), or forest (Figure 13.2c)

Table 13.1 Habitat types in the three major ecosystem classes. Agricultural ecosystems are defined as monocultures or pastures that are directly targeted by a range of agronomic practices (e.g., fertilization, mowing, livestock grazing, or tillage); semi-natural/natural ecosystems combine habitat types that are not agriculturally managed and are not dominated by large trees; forest ecosystems are dominated by large trees.

Ecosystem classes	Number of studies	Countries[a]
(a) Agricultural		
wheat	5	DE, GB, CH
unspecified	5	PH, CH
cotton	4	US
pastures	3	DE, NE, US
apple	2	US
cocoa	2	MX
soybean	2	US
meadows	1	CH
maize	1	DE
potato	1	DE
(b) Semi-natural/natural		
unmanaged grassland	5	PL, US
ruderal sites	5	JP, PA
scrubland	5	ES, NZ, NG
riparian	3	JP, CH
unspecified	3	CH, DK, PA
heath	2	DE
cave	1	GB
old garden	1	CH
hedges	1	DE
semi-desert	1	NE
(c) Forest		
deciduous	15	BR, JP, PL, US
rain	2	AU
coniferous	1	US

[a] The two-letter code for countries is defined by the International Organization for Standardization.

ecosystems. Network structure of each predator–prey network was characterized by calculating eight network metrics out of six major categories for each metaweb (Box 13.1; for categories see Miranda *et al.*, 2013).

13.3   Predator–Prey Networks in Different Ecosystems

13.3.1   Prey Composition

The relative contributions of beetles (Coleoptera), flies (Diptera), true bugs (Hemiptera), hymenopterans (Hymenoptera), and butterflies (Lepidoptera) to spider diets differ between agricultural, semi-natural/natural, and forest ecosystems. Diets of web-building spiders in

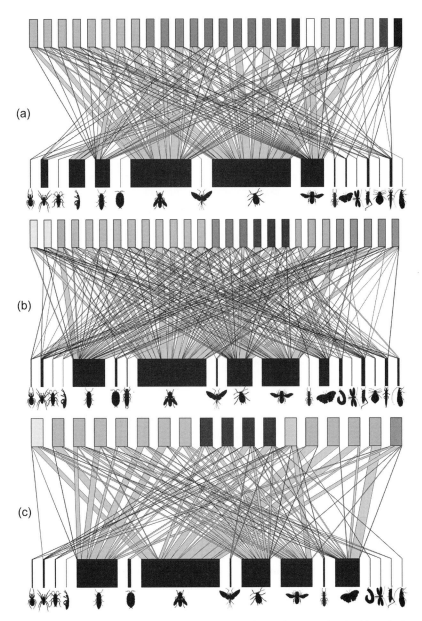

Figure 13.2 Metawebs for web-building spider species and prey orders in (a) agricultural (N = 26 spider species), (b) semi-natural/natural (N = 27), and (c) forest (N = 18) ecosystems. For details on prey orders and color codes for spider species please refer to Figure 13.1; for details on habitat types in each ecosystem class please refer to Table 13.1. (A black and white version of this figure will appear in some formats. For the color version, please refer to the plate section.)

agricultural habitats are characterized by a low contribution of Coleoptera, Hymenoptera, and Lepidoptera prey and high contributions of Hemiptera prey (Figure 13.3).

Spider diets in semi-natural/natural ecosystems, in contrast, are characterized by high contributions of Hymenoptera prey (Figure 13.3). The diet of spiders in forest

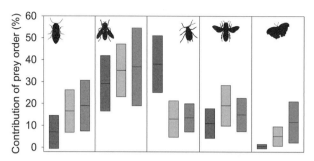

Figure 13.3 The average relative contribution (±95% CI) of prey orders to the diet of web-building spiders in different ecosystems (agricultural, red; semi-natural/natural, green; forest, blue) for the most discriminating prey orders (Coleoptera, Diptera, Hemiptera, Hymenoptera, and Lepidoptera) according to similarity percentage analysis (individual contribution >10%). (A black and white version of this figure will appear in some formats. For the color version, please refer to the plate section.)

ecosystems is characterized by a high percentage of Lepidoptera and a low percentage of Hemiptera prey (Figure 13.3). Diptera prey are generally important in all three ecosystems. The dominance of Hemiptera prey in diets of spiders in agricultural ecosystems (≈40% of all prey items) is most likely a consequence of the high abundance of Hemiptera pests in agricultural systems, with several economically important pests belonging to Hemiptera: Aphididae, Auchenorrhyncha, or Heteroptera (van Emden and Harrington, 2007; Alford, 2011; Birkhofer *et al.*, 2015). The high abundances of spiders in agricultural fields (up to 600 ind m^{-2}; Nyffeler and Sunderland, 2003), high prey consumption rates (up to 200 kg of prey ha^{-1} a^{-1}; Nyffeler, 1999) and the high contribution of Hemiptera prey in the agricultural metaweb highlight the importance of web-building spiders in conservation biological control (Marc *et al.*, 1999). In contrast, the relatively low percentage of Coleoptera and Lepidoptera prey in agricultural habitats is surprising, but may be an artifact of the crop types present in the data (Table 13.1). The order Coleoptera includes many common pests of oilseed rape or vegetable crops, and Lepidoptera are among the major pests in maize and vegetables (McKinlay, 1992; Alford, 2003). These crops are absent or under-represented in the data (Table 13.1a).

Many of the semi-natural/natural ecosystems represented in this study are habitats with limited levels of anthropogenic disturbance (e.g., natural grasslands or heathland) and have a high value for the conservation of plant and insect diversity (e.g., Pyle *et al.*, 1981). The high percentage of Hymenoptera prey in spider diets in these ecosystems with richer floral resources may reflect the activity and abundance of pollinators in these systems compared to resource poor agricultural sites (e.g., cereal fields). The forest data are dominated by prey records from deciduous forests and Coleoptera and Lepidoptera (including economically important pests; Müller *et al.*, 2008; Netherer and Schopf, 2010) are abundant insect orders in these systems.

In terms of the context dependency of trophic links it is important to highlight that a few prey orders were exclusively observed in subsets of ecosystem types. Myriapoda (centipedes and millipedes) and Crustacea (woodlice) prey were only recorded in the diet of spiders in semi-natural/natural or forest ecosystems and assumingly are more abundant in these ecosystems compared to agricultural fields. Trichoptera, Thysanoptera (including economically important pest species; Lewis, 1997), and Dermaptera prey on the contrary were only observed in spider diets of semi-natural/natural or agricultural ecosystems, but not in forests. Only the metaweb for semi-natural/natural ecosystems featured all 19 prey orders, suggesting that agricultural management or the presence of canopy cover in forests may reduce the range of available prey orders (see also Laliberté and Tylianakis, 2010 for parasitoid–host networks). Different levels of heterogeneity in habitat types within the ecosystem classes may affect our estimates of prey composition and network indices, but we argue that our results still represent important differences that hold in the presence of potential heterogeneity.

The observed differences in prey composition of web-building spiders between three major terrestrial ecosystem types were partly expected, as the availability of prey in these ecosystems may to some extent affect diet composition. The question remains, to what extent the observed differences in relative prey composition are exclusively caused by different abundances of prey orders in each ecosystem type. It is important to acknowledge that prey composition partly reflects prey availability (density-dependent predation; Birkhofer et al., 2013). However, this relationship is not as simple as assumed. Prey can be over- or under-represented in spider webs compared to its local abundance (Diehl et al., 2013) as webs are not just "filters" that sample prey in a strictly density-dependent manner (Nentwig, 1983a). Web-building spiders actively select web sites and relocate to prey-rich areas (Harwood et al., 2001) and web architecture may discriminate against certain prey types (e.g., the sticky silk of the Araneidae versus the non-sticky silk of the Linyphiidae) or can even be adapted to local conditions (Sandoval, 1994). Defensive measures or nutrient composition further affect prey utilization in spiders (Jensen et al., 2011) and biotic interactions such as competition, cannibalism, and predation lead to the selection of sub-optimal web sites (Birkhofer et al., 2007). To further test if the prey composition of species is context dependent, we created individual networks for each species that had records from more than one major ecosystem type (Figure 13.4; note that each network comes from a single study).

The same spider species may either have a comparable prey composition with the same dominant prey order in different ecosystems (Figure 13.4a–d), or different dominant prey orders in two ecosystems (Figure 13.4e–g). Aphids, Diptera, or Hymenoptera can, for example, be the dominant prey of a web-building spider species in grasslands and arable fields (Figure 13.4a–d), whereas for other species dominant prey orders differ between these ecosystem classes (Figure 13.4e, g). This pattern suggests that complex relationships between predator and prey taxa, local prey availability, habitat characteristics, and stochastic effects (e.g., wind gusts) drive prey composition in web-building spiders. Nevertheless, this study also demonstrates that web-building spiders show some consistent patterns in prey utilization in different ecosystems and that this predator group contributes to the suppression of economically important pests. However, prey orders that include beneficial arthropods are also part of the diet (e.g., Hymenoptera as pollinators of crops), particularly in non-agricultural ecosystems that may act as important source habitats for these organisms (Smith et al., 2014).

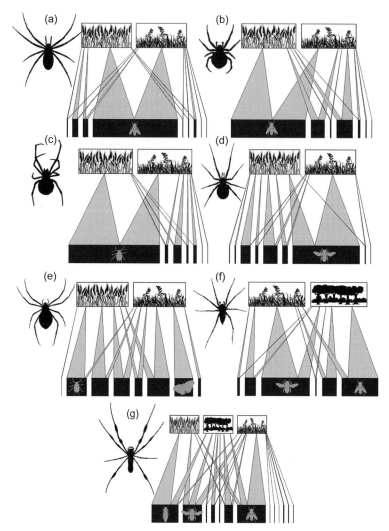

Figure 13.4 Comparison of predator–prey networks for web-building spider species with individual records from more than one ecosystem type. The symbols in the upper level of the interaction network illustrate if a record was derived from an agricultural (wheat tillers), semi-natural/natural (grassland with flowers), or forest ecosystem (trees). The lower level represents different prey orders (only showing symbols for orders that are dominant in the diet of a spider species in at least one ecosystem type), and the size of the link reflects the relative importance of each prey order in a spider species' diet. Spider species are (a) *Argiope bruenichi*, (b) *Araneus quadratus*, (c) *Phylloneta impressa*, (d) *Achaearanea riparia*, (e) *Argiope aurantia*, (f) *Agelena limbata*, and (g) *Nephila clavipes*.

13.3.2   Network Structure

It has been noted that antagonistic networks in deforested ecosystems (e.g., agricultural fields and grasslands) should resemble each other more than those of forest ecosystems (Tylianakis *et al.*, 2007; Laliberté and Tylianakis, 2010). These patterns are observed in

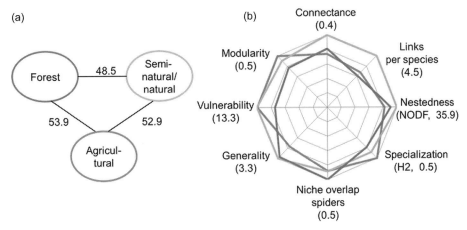

Figure 13.5 Comparison of network structure between ecosystem types. (a) Distance between ecosystem metawebs based on all network metrics (standardized by maximum) and Euclidean distances. Distances between circles are scaled according to Euclidean distance values shown at each link (higher values = higher dissimilarity). (b) Network metrics standardized by the maximum value (100%) for each spider–prey metaweb in different ecosystems (agricultural, red; semi-natural/natural, green; forest, blue). Maximum values are given in parentheses with each network metric. (A black and white version of this figure will appear in some formats. For the color version, please refer to the plate section.)

parasitoid–host networks where the severity of habitat modification and openness in deforested systems results in more uniform network patterns. Similar studies have not been performed in terrestrial predator–prey networks. Based on the constructed metawebs and the resulting network metrics, we compared the similarity of network structure between agricultural, semi-natural/natural, and forest ecosystems. Our results, which are based on a comparison of Euclidean distances that were calculated from standardized network metrics between ecosystem classes (Box 13.1, metrics in categories 1–5) do not support the assumption of a lower resemblance between predator–prey networks in deforested and forested ecosystems (forest vs. semi-natural/natural, or forest vs. agricultural ecosystems) compared to resemblance within deforested ecosystems (agricultural vs. semi-natural/natural ecosystems, Figure 13.5a). In contrast, the predator–prey network that stands out in the present comparison is the one from agricultural systems, with a higher dissimilarity to forest and semi-natural/natural ecosystems (Euclidean distances to both systems >50) when compared to the lower dissimilarity between networks in forest and semi-natural/natural ecosystems (Euclidean distance = 48.5; Figure 13.5a).

When comparing metrics that characterize network structure it becomes apparent that the predator–prey network in agricultural fields is characterized by a relatively low degree of connectivity and generality, and the highest level of specialization and vulnerability, compared to semi-natural/natural or forest ecosystems (Figure 13.5b). These results highlight that web-building spiders in agricultural ecosystems are feeding on a single or few numerically dominant prey orders, with little support for their assumed role as generalist predators in these ecosystems (see also Nyffeler, 1999).

The high level of specialization is most likely not a result of predator preferences, but rather reflects the dominance of single prey orders commonly observed as pests in intensively managed agricultural systems (Tilman, 1999; for parasitoids see Tylianakis et al., 2007). To further understand the drivers of specialization, we compared the networks that were created for species that had records from more than one major ecosystem type. In summary, six out of seven species for which prey records existed from agricultural and semi-natural/natural ecosystems had a higher diet breadth in the semi-natural/natural system (Figure 13.4). Web-building spider species in the metaweb of semi-natural/natural ecosystems (Figure 13.2) show the opposite pattern compared to spiders in agricultural fields with higher connectivity, nestedness, and generality values. Both the higher prey order richness in semi-natural/natural ecosystems (see Section 13.3.1) and the assumingly lower prey order richness in agricultural ecosystems may have contributed to the observed differences in network metrics.

It has been suggested that some network properties are generally less prone to being affected by anthropogenic disturbance (e.g., connectance), whereas others are more sensitive to land-use change (e.g., vulnerability; Tylianakis et al., 2007; Layer et al., 2011; Heleno et al., 2012). Vulnerability, for example, can be positively correlated to parasitism rates, which are often highest in modified agricultural habitats (Tylianakis et al., 2007). Connectance and the number of links per species are the two network metrics that separate agricultural from non-agricultural systems. Vulnerability of prey, however, does not differ between metawebs in agricultural and semi-natural/natural systems. Whether agricultural management alters the modularity or nestedness of networks (Box 13.1) in a systematic manner is unclear (Bohan et al., 2013). Antagonistic networks are rather thought to be characterized by high modularity (Bohan et al., 2013) and our results suggest modularity to be higher in metawebs in agricultural systems compared to forests, and nestedness to be lower in agricultural systems compared to semi-natural/natural ecosystems (Figure 13.5). The generally high nestedness and modularity in all metawebs lies within the expected values for antagonistic networks, which seem to be characterized by a more nested (Kondoh et al., 2010; but see Bascompte et al., 2003) and compartmentalized structure compared to other network types (Lewinsohn et al., 2006; but see Thebault and Fontaine, 2008).

13.3.3   Network Resemblance

It is important to understand how differences in spider and prey composition or biogeographic origins of data in each metaweb contribute to the observed resemblance between metawebs in terms of the six major categories of network metrics (Box 13.1). This is important to reveal sources of systematic bias and to understand drivers of network differences. In Figure 13.6, we use a second stage non-metric multidimensional scaling (NMDS) ordination to illustrate how metawebs of the three ecosystem classes resemble each other based on properties of the predator or prey communities and major categories of network metrics (Box 13.1). Resemblance between the three ecosystem classes was individually calculated for each of six major network topological indices (Box 13.1), spider family composition (square-root transformed percentage of spider species per

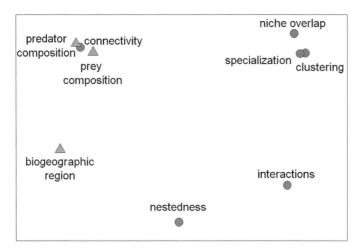

Figure 13.6 Second stage non-metric multidimensional scaling (NMDS) ordination (stress <0.001) based on Spearman correlation coefficients showing how resemblance patterns between the composition of predator families, prey orders, and biogeographic origins in the three metawebs (▲) relate to resemblance patterns based on six major categories of network topological indices (●; categories classified according to Miranda *et al.*, (2013). Two symbols close to each other in ordination space indicate that resemblance patterns between the three metawebs are comparable based on these two parameters (e.g., metawebs that resembled each other in predator composition were also similar in connectivity of networks).

family in each ecosystem class), prey order composition (square-root transformed percentage of prey items per order in each ecosystem class), and biogeographic origin of data (square-root transformed percentage of datasets per biogeographic region in each ecosystem class). The results of these nine resemblance matrices between ecosystem types were plotted in a single NMDS ordination based on how they characterize resemblance between the three ecosystem classes (for interpretation see also legend Figure 13.6).

Metawebs from ecosystems that resemble each other in terms of spider family and prey order composition are, for example, also similar in terms of connectivity (Figure 13.6, symbols for connectivity, predator, and prey composition are clustered close to each other in the upper left corner of the second stage NMDS ordination). This relationship intuitively makes sense, as certain spider families show particularly strong links to certain prey orders (e.g., linyphiids to Collembola) and as certain familiies include more specialized species than others (for specialization in spiders see Pekar *et al.*, 2012). The three ecosystem metawebs further resemble each other based on network metrics that describe niche overlap, specialization, and clustering (cluster of symbols in the upper right corner). Networks with more specialized predators can therefore be expected to have lower niche overlap and clustering values.

Data for spider diets in the studied ecosystem types are unevenly distributed across three major regions, with a higher number of records from agricultural ecosystems in Europe and the Americas and a higher number of records from forest ecosystems in

Australasia (Figure 13.1a). This geographic bias may create patterns in prey composition and network topology that appear to be caused by differences between ecosystem
types. The resemblance in the composition of biogeographic regions from which diet
records are taken did not relate to resemblances between metawebs based on any of the
major categories of network topology (Figure 13.6). It was further not related to
resemblance between metawebs from ecosystems in terms of spider family or prey
order composition. This finding suggests a limited bias due to study origin for the
analyses of prey composition and network topology.

13.4   Conclusions

Our ability to predict how land-use changes will affect biotic interactions is very
limited, which is particularly worrying as the loss of ecological functions may spillover to affect regions beyond those that are directly affected by land-use conversion
(Bohan et al., 2013; Smith et al., 2014). It is important to realize that spillover of
organisms from semi-natural to agricultural (Landis et al., 2000) and from agricultural
to semi-natural habitats (Rand et al., 2006; Blitzer et al., 2012) influences important
regulating ecosystem services (e.g., biological control or pollination). The predicted
global increase in agricultural land and decreasing forest area (see Section 13.1) will
affect the functional role of web-building spider communities in a region considerably.
This study highlights that web-building spiders are rather specialized predators
(Figure 13.5) that mainly feed on hemipteran prey in agricultural ecosystems and
that their impact on beneficial prey taxa is rather low (Figure 13.3). The agricultural
metaweb has the lowest connectivity and agricultural management generally leads to
lower prey order richness. Our study shows that connectivity is closely related to the
family composition of spider communities and the prey order composition
(Figure 13.6), suggesting that low connectivity in the agricultural metaweb is
a consequence of simplified predator and prey communities in anthropogenically
modified ecosystems. In contrast, metawebs of forest and, in particular, semi-natural
/natural ecosystems have higher prey order richness and are characterized by web-
building spiders that rather act as true generalist predators. Web-builders feed more
frequently on hymenopteran prey in semi-natural/natural ecosystems, which may
provide a disservice as this may cause a reduced spillover of pollinators into surrounding agricultural fields. Future abandonment of arable land (e.g., long-term set-aside) or
conversion to flower fields (e.g., ecological focus areas) may therefore not only lead to
higher pollinator numbers, but at the same time may result in a trade-off by promoting
the disservice of higher predation rates on pollinators. This study therefore directly
supports the important role of network studies to develop future management strategies for the provision of multiple ecosystem services (Hines et al., 2015).

Our comparison of network metrics of web-building spiders in metawebs between
different ecosystems (Figure 13.2), but also between the same species in different
ecosystems (Figure 13.4), shows the potential of environmental associations to alter
a species' functional role in terrestrial food webs. Finlay-Doney and Walter (2012)

recently emphasized the importance to consider environmental associations of generalist predator species prior to declaring species as "useful" in biological control strategies. It is not evident from our study, what the main drivers of prey preferences of web-building spider species are in different ecosystems. Our results, however, suggest that a complex combination of prey availability, structural habitat properties, and stochastic drivers interact to affect prey composition in web-building spider species. Future research should focus on identifying and establishing local conditions that promote web-building spider species and communities, which are suitable to act as antagonists of economically important pests. This aspect may be particularly important for future land-use type conversions, as these system changes offer a chance to develop novel systems with specific structural characteristics and disturbance regimes.

Analyzing the differences between antagonistic networks in different terrestrial ecosystems in this study has proven to be an important first step that contributes to a better understanding of the future consequences of land-use change for the delivery of ecosystem services to human societies.

Acknowledgments

We are grateful to two anonymous referees that helped improve a previous version of the manuscript. This review was supported by the project grants ÖkoService (BMBF, 03V0217), ICON (DFG, WO 670/14–2), and by the Swedish Research Council for Environment, Agricultural Sciences, and Spatial Planning (FORMAS).

References

Alford, D. V. (2011). *Plant Pests*. London: Harper Collins Publishers.

Alford, D. V., Nilsson, C., and Ulber, B. (2003). Insect pests of oilseed rape crops. In *Biocontrol of Oilseed Rape Pests*, ed. D. V. Alford, Oxford: Blackwell Publishing, pp. 9–41.

Almeida-Neto, M. and Ulrich, W. (2011). A straightforward computational approach for measuring nestedness using quantitative matrices. *Environmental Modelling and Software*, **26**, 173–178.

Almeida-Neto, M., Guimaraes, P., Guimaraes, Jr., P. R., Loyola, R. D., and Ulrich, W. (2008). A consistent metric for nestedness analysis in ecological systems: reconciling concept and measurement. *Oikos*, **117**, 1227–1239.

Angelsen, A. (2010). Policies for reduced deforestation and their impact on agricultural production. *Proceedings of the National Academy of Sciences of the United States of America*, **107**, 19639–19644.

Attwood, S. J., Maron, M., House, A. P. N., and Zammit, C. (2008). Do arthropod assemblages display globally consistent responses to intensified agricultural land use and management? *Global Ecology and Biogeography*, **17**, 585–599.

Barbosa, P. (1998). Agroecosystems and conservation biological control. In *Conservation Biological Control*, ed. P. Barbosa, San Diego: Academic Press, pp. 39–54.

Bardwell, C. J. and Averill, A. L. (1997). Spiders and their prey in Massachusetts cranberry bogs. *Journal of Arachnology*, **25**, 31–41.

Bascompte, J., Jordano, P., Melian, C. J., and Olesen, J. M. (2003). The nested assembly of plant–animal mutualistic networks. *Proceedings of the National Academy of Sciences of the United States of America*, **100**, 9383–9387.

Bilsing, S. W. (1920). Quantitative studies in the food of spiders. *Ohio Journal of Science*, **20**, 215–260.

Birkhofer, K. and Wolters, V. (2012). The global relationship between climate, net primary production and the diet of spiders. *Global Ecology and Biogeography*, **21**, 100–108.

Birkhofer, K., Scheu, S., and Wise, D. H. (2007). Small-scale spatial pattern of web-building spiders (Araneae) in *Alfalfa*: relationship to disturbance from cutting, prey availability, and intraguild interactions. *Environmental Entomology*, **36**, 801–810.

Birkhofer, K., Entling, M., and Lubin, Y. (2013). Agroecology: trait composition, spatial relationships, trophic interactions. In *Spider Research in the 21st Century: Trends and Perspectives*, ed. D. Penney, Manchester: Siri Scientific Press, pp. 200–229.

Birkhofer, K., Arvidsson, F., Ehlers, D., *et al.* (2015) Landscape complexity and organic farming independently affect the biological control of hemipteran pests and yields in spring barley. *Landscape Ecology*, **31**, 567–579. DOI: 10.1007/s10980-015–0263-8.

Blitzer, E. J., Dormann, C. F., Holzschuh, A., *et al.* (2012). Spillover of functionally important organisms between managed and natural habitats. *Agriculture Ecosystems & Environment*, **146**, 34–43.

Bluthgen, N., Menzel, F., and Bluthgen, N. (2006). Measuring specialization in species interaction networks. *BMC Ecology*, **6**, 9.

Bohan, D. A. and Woodward, G. (2013). Editorial commentary: the potential for network approaches to improve knowledge, understanding, and prediction of the structure and functioning of agricultural systems. *Advances in Ecological Research*, **49**, xiii–xviii.

Bohan, D. A., Raybould, A., Mulder, C., *et al.* (2013). Networking agroecology: integrating the diversity of agroecosystem interactions. *Advances in Ecological Research*, **49**, 1–67.

Dąbrowska-Prot, E., Łuczak, J., and Tarwid, K. (1968). Prey and predator density and their reactions in the process of mosquitoes reduction by spiders in field experiments. *Ekologia Polska*, **16**, 773–819.

DeFries, R. S., Rudel, T., Uriarte, M., and Hansen, M. (2010). Deforestation driven by urban population growth and agricultural trade in the twenty-first century. *Nature Geoscience*, **3**, 178–181.

Diehl, E., Mader, V. L., Wolters, V., and Birkhofer, K. (2013). Management intensity and vegetation complexity affect web-building spiders and their prey. *Oecologia*, **173**, 579–589.

Dormann, C. F. and Strauss, R. (2014). A method for detecting modules in quantitative bipartite networks. *Methods in Ecology and Evolution*, **5**, 90–98.

Dormann, C. F., Gruber, B., and Fründ, J. (2008). Introducing the bipartite package: analysing ecological networks. *R News*, **8**, 8–11.

Dunne, J. A. (2006). The network structure of food webs. In *Ecological Networks: Linking Structure to Dynamics in Food Webs*, ed. M. Pascual and J. A. Dunne, Santa Fe, NM: Santa Fe Institute Studies in the Sciences of Complexity, pp. 27–86.

Dunne, J. A., Williams, R. J., and Martinez, N. D. (2002). Food-web structure and network theory: the role of connectance and size. *Proceedings of the National Academy of Sciences of the United States of America*, **99**, 12917–12922.

Estes, J. A., Terborgh, J., Brashares, J. S., *et al.* (2011). Trophic downgrading of planet Earth. *Science*, **333**, 301–306.

Finlay-Doney, M. and Walter, G. H. (2012). The conceptual and practical implications of interpreting diet breadth mechanistically in generalist predatory insects. *Biological Journal of the Linnean Society*, **107**, 737–763.

Gray, C., Baird, D. J., Baumgartner, S., *et al.* (2014). Ecological networks: the missing links in biomonitoring science. *Journal of Applied Ecology*, **51**, 1444–1449.

Guseinov E. F. (2005). Natural prey of the jumping spider *Salticus tricinctus* (Araneae, Salticidae). *Bulletin of the British Arachnological Society*, **13**, 130–132.

Guseinov, E. F., Cerveira, A. M., and Jackson, R. R. (2004). The predatory strategy, natural diet, and life cycle of *Cyrba algerina*, an araneophagic jumping spider (Salticidae: Spartaeinae) from Azerbaijan. *New Zealand Journal of Zoology*, **31**, 291–303.

Harwood, J. D., Sunderland, K. D., and Symondson, W. O. C. (2001). Living where the food is: web location by linyphiid spiders in relation to prey availability in winter wheat. *Journal of Applied Ecology*, **38**, 88–99.

Havlík, P., Schneider, U. A., Schmid, E., *et al.* (2011). Global land-use implications of first and second generation biofuel targets. *Energy Policy*, **39**, 5690–5702.

Heleno, R., Devoto, M., and Pocock, M. (2012). Connectance of species interaction networks and conservation value: is it any good to be well connected? *Ecological Indicators*, **14**, 7–10.

Hines, J., van der Putten, W. H., de Deyn, G., *et al.* (2015). Towards an integration of biodiversity–ecosystem functioning and food web theory to advance the understanding of connections between multiple ecosystem functions and service provisioning. *Advances in Ecological Research*, **253**, 161–199.

Huseynov, E. F. o. (2005). Natural prey of the jumping spider *Menemerus taeniatus* (Araneae: Salticidae). *European Journal of Entomology*, **102**, 797–799.

Huseynov, E. F. o. (2006a). The prey of the lynx spider *Oxyopes globifer* (Araneae, Oxyopidae) associated with a semidesert dwarf shrub in Azerbaijan. *Journal of Arachnology*, **34**, 422–426.

Huseynov, E. F. o. (2006b). Natural prey of the jumping spider *Heliophanus dunini* (Araneae: Salticidae) associated with Eryngium plants. *Bulletin of the British Arachnological Society*, **13**, 293–296.

Huseynov, E. F. o. (2007a). Natural prey of the crab spider *Thomisus onustus* (Araneae: Thomisidae), an extremely powerful predator of insects. *Journal of Natural History*, **41**, 2341–2349.

Huseynov, E. F. o. (2007b). Natural prey of the lynx spider *Oxyopes lineatus* (Araneae: Oxyopidae). *Entomologica Fennica*, **18**, 144–148.

Huseynov, E. F. o. (2007c). Natural prey of the crab spider *Runcinia grammica* (Araneae: Thomisidae) on Eryngium plants. *Bulletin of the British Arachnological Society*, **14**, 93–96.

Huseynov, E. F. o. (2008). Natural prey of the jumping spider *Philaeus chrysops* (Araneae: Salticidae) in different types of microhabitat. *Bulletin of the British Arachnological Society*, **14**, 262–268.

Huseynov, E. F., Cross, F. R., and Jackson, R. R. (2005). Natural diet and prey-choice behaviour of *Aelurillus muganicus* (Araneae: Salticidae), a myrmecophagic jumping spider from Azerbaijan. *Journal of Zoology*, **267**, 159–165.

Ings, T. C., Bascompte, M. J. M., Blüthgen, N., *et al.* (2009). Ecological networks: beyond food webs. *Journal of Animal Ecology*, **78**, 253–269.

Ives, A. R., Cardinale, B. J., and Snyder, W. E. (2005). A synthesis of subdisciplines: predator–prey interactions, and biodiversity and ecosystem functioning. *Ecology Letters*, **8**, 102–116.

Jensen, K., Mayntz, D., Toft, S., Raubenheimer, D., and Simpson, S. J. (2011). Prey nutrient composition has different effects on *Pardosa* wolf spiders with dissimilar life histories. *Oecologia*, **165**, 577–583.

Kankaanpää, S. and Carter, T. R. (2004). *An Overview of Forest Policies Affecting Land Use in Europe*. The Finnish Environment 706, Helsinki: Finnish Environment Institute.

Kiritani, K., Kawahara, S., Sasaba, T., and Nakasuji, F. (1972). Quantitative evaluation of predation by spiders on the green rice leafhopper *Nephotettix cincticeps* by a sight count method. *Researches on Population Ecology*, **13**, 187–200.

Kondoh, M., Kato, S., and Sakato, Y. (2010). Food webs are built up with nested subwebs. *Ecology*, **91**, 3123–3130.

Kuusk, A.-K. and Ekbom, B. (2010). Lycosid spiders and alternative food: feeding behavior and implications for biological control. *Biological Control*, **55**, 20–26.

Kuusk, A.-K. and Ekbom, B. (2012). Feeding habits of lycosid spiders in field habitats. *Journal of Pest Science*, **85**, 253–260.

Laliberté, E. and Tylianakis, J. M. (2010). Deforestation homogenizes tropical parasitoid–host networks. *Ecology*, **91**, 1740–1747.

Lambin, E. F. and Meyfroidt, P. (2011). Global land use change, economic globalization, and the looming land scarcity. *Proceedings of the National Academy of Sciences of the United States of America*, **108**, 3465–3472.

Landis, D. A., Wratten, S. D., and Gurr, G. M. (2000). Habitat management to conserve natural enemies of arthropod pests in agriculture. *Annual Review of Entomology*, **45**, 175–201.

Layer, K., Hildrew, A. G., Jenkins, G., *et al.* (2011). Long-term dynamics of a well-characterised food web: four decades of acidification and recovery in the Broadstone stream model system. *Advances in Ecological Research*, **44**, 69–117.

Lewinsohn, T. M., Prado, P. I., Jordano, P., Bascompte, J., and Olesen, J. M. (2006). Structure in plant–animal interaction assemblages. *Oikos*, **113**, 174–184.

Lewis, T. (1997). Pest thrips in perspective. In *Thrips as Crop Pests*, ed. T. Lewis, Wallingford: CABI, pp. 1–13.

Marc, P., Canard, A., and Ysnel, F. (1999). Spiders (Araneae) useful for pest limitation and bioindication. *Agriculture Ecosystems & Environment*, **74**, 229–273.

McKinlay, R. G. (1992). *Vegetable Crop Pests*. Boca Raton: CRC Press.

Miranda, M., Parrini, F., and Dalerum, F. (2013). A categorization of recent network approaches to analyse trophic interactions. *Methods in Ecology and Evolution*, **4**, 897–905.

Müller, J., Bussler, H., Gossner, M. M., Rettelbach, T., and Duelli, P. (2008). The European spruce bark beetle *Ips typographus* in a national park: from pest to keystone species. *Biodiversity and Conservation*, **17**, 2979–3001.

Nentwig, W. (1982). Analyses of the prey of cribellate spiders (Araneae: Filistatidae, Dictynidae, Eresidae). *Entomologische Mitteilungen aus dem zoologischen Museum Hamburg*, **7**, 233–244.

Nentwig, W. (1983a). The non-filter function of orb webs in spiders. *Oecologia*, **58**, 418–420.

Nentwig, W. (1983b). The prey of web-building spiders compared with feeding experiments (Araneae, Araneidae, Linyphiidae, Pholcidae, Agelenidae). *Oecologia*, **56**, 132–139.

Nentwig, W. (1985). Prey analysis of 4 species of tropical orb-weaving spiders (Araneae, Araneidae) and a comparison with Araneids of the temperate zone. *Oecologia*, **66**, 580–594.

Nentwig, W. (1987). The prey of spiders. In *Ecophysiology of Spiders*, ed. W. Nentwig, Berlin: Springer Verlag, pp. 249–263.

Netherer, S. and Schopf, A. (2010). Potential effects of climate change on insect herbivores in European forests: general aspects and the pine processionary moth as specific example. *Forest Ecology and Management*, **259**, 831–838.

Nyffeler, M. (1999). Prey selection of spiders in the field. *Journal of Arachnology*, **27**, 317–324.

Nyffeler, M. and Benz, G. (1978). Prey selection by web spiders *Argiope bruennichi* (Scop.), *Araneus quadratus* (Cl.), and *Agelena labyrinthica* (Cl.) on fallow land near Zurich, Switzerland. *Revue Suisse De Zoologie*, **85**, 747–757.

Nyffeler, M. and Benz, G. (1979). Overlap of the niches concerning space and prey of crab spiders (Araneae, Thomisidae) and wolf spiders (Araneae, Lycosidae) in cultivated meadows. *Revue Suisse De Zoologie*, **86**, 855–865.

Nyffeler, M. and Benz, G. (1981a). Field studies on the feeding ecology of spiders: observations in the region of Zurich (Switzerland). *Anzeiger für Schädlingskunde Pflanzenschutz Umweltschutz*, **54**, 33–39.

Nyffeler, M. and Benz, G. (1981b). Some observations on the feeding ecology of the wolf-spider *Pardosa lugubris* (walck). *Deutsche Entomologische Zeitschrift*, **28**, 297–300.

Nyffeler, M. and Benz, G. (1988a). Prey and predatory importance of micryphantid spiders in winter-wheat fields and hay meadows. *Journal of Applied Entomology*, **105**, 190–197.

Nyffeler, M. and Benz, G. (1988b). Feeding ecology and predatory importance of wolf spiders (*Pardosa* spp.) (Araneae, Lycosidae) in winter-wheat fields. *Journal of Applied Entomology*, **106**, 123–134.

Nyffeler, M. and Benz, G. (1988c). Prey analysis of the spider *Achaearanea riparia* (Blackw.) (Araneae, Theridiidae), a generalist predator in winter-wheat fields. *Journal of Applied Entomology*, **106**, 425–431.

Nyffeler, M. and Sterling, W. L. (1994). Comparison of the feeding niche of polyphagous insectivores (Araneae) in a Texas cotton plantation: estimates of niche breadth and overlap. *Environmental Entomology*, **23**, 1294–1303.

Nyffeler, M. and Sunderland, K. D. (2003). Composition, abundance and pest control potential of spider communities in agroecosystems: a comparison of European and US studies. *Agriculture Ecosystems and Environment*, **95**, 579–612.

Nyffeler, M., Dean, D. A., and Sterling, W. L. (1986). Feeding-habits of the spiders *Cyclosa turbinata* (Walckenaer) (Araneae, Araneidae) and *Lycosa rabida* Walckenaer (Araneae, Lycosidae). *Southwestern Entomologist*, **11**, 195–201.

Nyffeler, M., Dean, D. A., and Sterling, W. L. (1987). Predation by green lynx spider, *Peucetia viridans* (Araneae, Oxyopidae), inhabiting cotton and woolly croton plants in east Texas. *Environmental Entomology*, **16**, 355–359.

Nyffeler, M., Dean, D. A., and Sterling, W. L. (1988). Prey records of the web-building spiders *Dictyna segregata* (Dictynidae), *Theridion australe* (Theridiidae), *Tidarren haemorrhoidale* (Theridiidae), and *Frontinella pyramitela* (Linyphiidae) in a cotton agroecosystem. *Southwestern Naturalist*, **33**, 215–218.

Nyffeler, M., Dean, D. A., and Sterling, W. L. (1992). Diets, feeding specialization, and predatory role of 2 lynx spiders, *Oxyopes salticus* and *Peucetia viridans* (Araneae, Oxyopidae), in a Texas cotton agroecosystem. *Environmental Entomology*, **21**, 1457–1465.

Olesen, J. E. and Bindi, M. (2002). Consequences of climate change for European agricultural productivity, land use and policy. *European Journal of Agronomy*, **16**, 239–262.

Pekar, S., Coddington, J. A., and Blackledge, T. A. (2012). Evolution of stenophagy in spiders (Araneae): evidence based on the comparative analysis of spider diets. *Evolution*, **66**, 776–806.

Perez-De la Cruz, M., Sanchez-Soto, S., Ortiz-Garcia, C. F., Zapata-Mata, R., and De la Cruz-Perez, A. (2007). Diversity of insects captured by weaver spiders (Arachnida: Araneae) in the cocoa agroecosystem in Tabasco, Mexico. *Neotropical Entomology*, **36**, 90–101.

Poisot, T., Canard, E., Mouillot, D., Mouquet, N., and Gravel, D. (2012) The dissimilarity of species interaction networks. *Ecology Letters*, **15**, 1353–1361.

Pyle, R., Bentzien, M., and Opler, P. (1981). Insect conservation. *Annual Review of Entomology*, **26**, 233–258.

R Development Core Team (2008). *R: A Language and Environment for Statistical Computing*. Vienna: R Foundation for Statistical Computing.

Rabbinge, R. and van Diepen, C. A. (2000). Changes in agriculture and land use in Europe. *European Journal of Agronomy*, **13**, 85–99.

Rand, T. A., Tylianakis, J. M., and Tscharntke, T. (2006). Spillover edge effects: the dispersal of agriculturally subsidized insect natural enemies into adjacent natural habitats. *Ecology Letters*, **9**, 603–614.

Rounsevell, M. D. A., Reginster, I., Araújo M. B., *et al.* (2006). A coherent set of future land use change scenarios for Europe. *Agriculture Ecosystems and Environment*, **114**, 57–68.

Sala, O. E., Chaplin, F. S., Armesto, J. J., *et al.* (2000). Biodiversity: global biodiversity scenarios for the year 2100. *Science*, **287**, 1770–1774.

Sandoval, C. P. (1994). Plasticity in web design in the spider *Parawixia bistriata*: a response to variable prey type. *Functional Ecology*, **8**, 701–707.

Schmitz, C., van Meijl, H., Kyle, P., *et al.* (2014). Land-use change trajectories up to 2050: insights from a global agro-economic model comparison. *Agricultural Economics*, **45**, 69–84.

Smith, H. G., Birkhofer, K., Clough, Y., *et al.* (2014) Beyond dispersal: the role of animal movement in modern agricultural landscapes. In *Animal Movement Across Scales*, ed. L. A. Hansson and S. Åkesson, Oxford: Oxford University Press.

Smith, P., Gregory, P. J., Van Vuuren, D., *et al.* (2010). Competition for land. *Philosophical Transactions of the Royal Society B: Biological Sciences*, **365**, 2941–2957.

Spittlehouse, D. L. and Stewart, R. B. (2004). Adaptation to climate change in forest management. *Journal of Ecosystems and Management*, **4**, 1–11.

Stouffer, D. B., Sales-Pardo, M., Sirer, M. I., and Bascompte, J. (2012). Evolutionary conservation of species' roles in food webs. *Science*, **335**, 1489–1492.

Sunderland, K. D., Powell, W., and Symondson, W. O. C. (2005). Populations and communities. In *Insects as Natural Enemies: A Practical Perspective*, ed. M. A. Jervis, Dordrecht: Springer, pp. 299–434.

Thebault, E. and Fontaine, C. (2008). Does asymmetric specialization differ between mutualistic and trophic networks? *Oikos*, **117**, 555–563.

Thompson, R. M., Brose, U., Dunne, J. A., *et al.* (2012). Food webs: reconciling the structure and function of biodiversity. *Trends in Ecology and Evolution*, **27**, 689–697.

Tilman, D. (1999). Global environmental impacts of agricultural expansion: the need for sustainable and efficient practices. *Proceedings of the National Academy of Sciences of the United States of America*, **96**, 5995–6000.

Tilman, D., Fargione, J., Wolff, B., *et al.* (2001). Forecasting agriculturally driven global environmental change. *Science*, **292**, 281–284.

Tixier, P., Peyrard, N., Auberlot, J.-N., *et al.* (2013). Modelling interaction networks for enhanced ecosystem services in agroecosystems. *Advances in Ecological Research*, **49**, 437–480.

Traugott, M., Kamenova, S., Ruess, L., Seeber, J., and Plantegenest, M. (2013). Empirically characterising trophic networks: what emerging DNA-based methods, stable isotope and fatty acid analyses can offer. *Advances in Ecological Research*, **49**, 177–224.

Tylianakis, J. M., Tscharntke, T., and Lewis, O. T. (2007). Habitat modification alters the structure of tropical host–parasitoid food webs. *Nature*, **445**, 202–205.

Tylianakis, J. M., Laliberté, E., Nielsen, A., and Bascompte, J. (2010). Conservation of species interaction networks. *Biological Conservation*, **143**, 2270–2279.

Uetz, G. W. and Hartsock, S. P. (1987). Prey selection in an orb-weaving spider *Micrathena gracilis* (Araneae, Araneidae). *Psyche*, **94**, 103–116.

Uetz, G. W., Johnson, A. D., and Schemske, D. W. (1978). Web placement, web structure, and prey capture in orb-weaving spiders. *Bulletin of the British Arachnological Society*, **4**, 141–148.

van Emden, H. F., and Harrington, R. (2007). *Aphids as Crop Pests*. Wallingford: CABI.

van der Putten, W. H., de Ruiter, P. C., Bezemer, T. M., *et al.* (2004). Trophic interactions in a changing world. *Basic and Applied Ecology*, **5**, 487–494.

Wise, D. H. (1993). *Spiders in Ecological Webs*. Cambridge, Cambridge University Press.

Wise, D. H. and Barata, J. L. (1983). Prey of 2 syntopic spiders with different web structures. *Journal of Arachnology*, **11**, 271–281.

14 Ecological Networks in Managed Ecosystems: Connecting Structure to Services

Christian Mulder, Valentina Sechi, Guy Woodward, and David Andrew Bohan

14.1 Introduction

Ecological networks represent a cornerstone of ecology: they describe and evaluate the links between form and function in multispecies systems, such as food-web structure and dynamics, and they connect different scales and levels of biological organization (Moore and de Ruiter, 2012; Wall *et al.*, 2015). These properties of being able to elucidate both the structure within complex systems and their scaling indicate that ecological networks and network theory could be widely applied to practical problems, including management decision-making processes such as the design of nature reserves and the preservation of ecosystem services. While the study of networks – initially food-web compartments, then community assemblages, and more recently mutualistic networks – is now firmly embedded in ecology (Levins, 1974; Cohen, 1978; Hunt *et al.*, 1987; Beare *et al.*, 1992; Solé and Montoya, 2001; Berlow *et al.*, 2004; Moore *et al.*, 2004; Cohen and Carpenter, 2005; Thébault and Fontaine, 2010; Moore and de Ruiter, 2012; Pocock *et al.*, 2012; Neutel and Thorne, 2014), the application of such approaches to managed ecosystems has lagged far behind. There are many explanations for this disconnection between agro-ecology and ecology, not least the pervasive view that because they are human managed and disturbed agro-systems are fundamentally "unnatural" and different from natural ecosystems: most ecologists prefer to study so-called natural ecosystems, even though most of these have in fact been heavily influenced by mankind for centuries either directly by local activity or indirectly by long-distance pollution.

Network approaches have rarely been applied to agriculture and forestry, which is perhaps surprising given that much of the early, integrated management research (e.g., from the seminal works by Von Carlowitz, 1713, and Von Liebig, 1840, onwards) and the study of networks that stimulated major advances in ecological theory was grounded in attempts to improve agricultural and timber production (Wardle, 2002; Schröter *et al.*, 2003; Coleman *et al.*, 2004; Moore and de Ruiter, 2012, and the references therein). The last two decades have seen a hiatus in advances in agro-ecology in this area, while new network theory and empirical studies have elucidated the roles of body size in ecosystems and the study of plant–pollinator networks and other mutualistic webs have redefined our understanding of general ecology. Recently, the isolationism of both fields has been challenged, with a notable increase in intellectual exchange and adoption of ecological theory in agro-ecology, much of which has been driven by the emergence of the ecosystem approach to resource use, and a growing focus on the sustainable provisioning of ecosystem goods and services.

The previously limited and apparently slow uptake of ecological research into agro-ecology can be ascribed to a number of both real and perceived influences, among which a simple lack of interaction and understanding between disciplines (Bohan *et al.*, 2013; Mulder *et al.*, 2015). First, agriculture has frequently been long regarded by policy- and decision-makers almost exclusively as a system of production driven by human manage-ment, with little role for biodiversity or its related processes. Second, the definition of risk standards for potentially harmful compounds has also enabled stakeholders to mitigate human-induced effects, a Pyrrhic victory attributable to both chemists and ecologists. Third, despite the derivation of quality criteria, managed ecosystems became progressively more marginalized in terms of the research effort devoted to their study in "pure" ecology relative to the supposedly wilder areas that have traditionally provided the playground for basic research. This combination of factors accounts for much of the disconnect between the regulatory definitions of risk and harm that were targeted at reducing the impact of anthropogenic effects on biodiversity, on the one hand, and the majority of ecologists who believed that perturbation due to management rendered agro-ecosystems uninteresting (Martin *et al.*, 2012).

Many environmental predictors are being measured increasingly to identify their importance in altering ecosystem services and human well-being, an investigation made possible by the compilation of monitoring datasets. A range of environmental indicators are being used to measure and, in some cases, modify management practices associated with the delivery of ecosystem services, though these approaches are still simplistic and narrow in scope, with the number and type of services being restricted to the few that are easiest to measure (Perrings *et al.*, 2011; Mulder *et al.*, 2015). In a world where we need to produce food while protecting the environment in the face of global drivers of change, agriculture can no longer afford this poor level of connection to (and limited interaction with) ecology, and that by adopting network-based approaches agro-ecology can both benefit from ecology and inform ecologists.

In particular, there is a need to use network approaches, by capitalizing on advances made in general ecology and elsewhere, to unite the human, biotic, and the abiotic factors that drive agricultural production on one hand, with the networks that exist in all the habitats that comprise complex, agro-ecological landscapes on the other: this requires a more realistic integration of the terrestrial and aquatic, below and above ground, and managed and natural habitats that form the interconnected parts of the patchwork landscape in which we live and work. Using the example of the environ-mental frameworks now widely adopted in freshwater biology (Woodward *et al.*, 2012; Gray *et al.*, 2014), this could produce a new eco-stoichiometrical understanding of ecological processes (Mulder, 2010; Mulder *et al.*, 2013; Sechi *et al.*, 2015), which could be used to inform a predictive agro-ecology that could aid the more effective and sustainable delivery of ecosystem functions and services.

14.2   A New Conceptual Framework for Land Management

Agriculture represents the longest running and largest scale replicated field experiment yet conducted by our species. Understanding how artificial selection and land-use practices associated with agriculture and forestry have molded much of the Earth's surface will undoubtedly help us to design more effective management of these systems to maximize the return of goods and services of value to human societies (Mulder *et al.*, 2015), beyond the more immediate and obvious ones related simply to the supply of crop yields and timber production. Recent network research applied to agriculture and forestry has demonstrated that many of these managed systems are surprisingly supporting species- and trait-rich communities that have key functions that are, either directly or indirectly, associated with the main crop or the cultivated tree (e.g., Struebig *et al.*, 2011; Bohan *et al.*, 2013). As a matter of fact, even intensively cultivated arable fields in European temperate regions possess ecological networks that can contain hundreds of species and interactions among the macrobiota alone (Mulder *et al.*, 2012). Given that most process rates saturate at just a few species (Loreau *et al.*, 2002; Scherber *et al.*, 2010; Duncan *et al.*, 2015), this should be more than sufficient biodiversity to be able to support reasonable levels of ecosystem functioning.

A synthesis of community and ecosystem ecology has emerged in supposedly undisturbed ecosystems (Lavorel *et al.*, 1997, 2013; Hooper *et al.*, 2005; Loreau, 2009; Kattge *et al.*, 2011). Traits have often been used as a proxy measure that can help to connect community structure to (usually inferred) ecosystem functioning, as the full role of species is often difficult or impossible to measure directly, especially in their synecological context that takes into account how they interact with one another in the wider ecological network. Some species traits have been linked very closely with both network structure and dynamics: body size and elemental composition are among the most familiar, and these variables can capture most of a given network's key functional attributes in just a few dimensions. Comparing managed with natural (which, in reality, are mostly semi-natural at best) ecosystems, or terrestrial with aquatic ecosystems, clear macroecological patterns in traits emerge.

For instance, human activities result locally in overfertilization accompanied with increased nutrient leaching and globally rising anthropogenic air pollutants (Wolters *et al.*, 2000), as addition of (mineral) fertilizers and atmospheric deposition of N (and to a lesser extent, of P and K) influence the soil C:N:P:K ratios. Changes in these elemental ratios are therefore dependent not only on the active management practices of agroecosystems by farmers but are also dependent on the worldwide input of N, P, and K by atmospheric deposition, regardless of whether the ecosystem is managed or not. To visualize these inputs in relative terms, the global increase in the deposition of N is now comparable to the naturally fixed N in the biosphere (Peñuelas *et al.*, 2013); the deposition of P resulted in a worldwide net loss from terrestrial ecosystems to the oceans (Mahowald *et al.*, 2008); and K, mostly as aerosol, combines direct and indirect anthropogenic effects, from coal sources and biofuel combustion and from forest fires and biomass burning (Hsu *et al.*, 2009). These anthropogenic changes influence most organisms and affect both the managed and the natural ecosystems.

Managed and natural ecosystems have different structural and functional character-istics, and consequently their robustness to drivers of change such as nutrient availability differ. These functional characteristics can be evaluated as critical thresholds, below which ecosystem vulnerability increases and resilience decreases. A threshold elemental ratio (TER) depends on a consumer's individual nutrient-to-carbon ratio, its maximum growth efficiency GE^{max} for each elemental factor (here, N, P, or K), and its maximum growth efficiency for C (Urabe *et al.*, 2010). Obviously, a TER strongly reflects the nutrient availability (N, P, or K) of the resource/prey, and hence the food quality (including the soil or litter quality, such as the amount of fresh organic matter). Given the extent to which we are deliberately manipulating agro-ecosystems, it can be shown (focusing on P and C) that stoichiometrical effects modulate populations and commu-nities as:

$$TER_{C:P} = \left(\frac{GE_P^{max}}{GE_C^{max}}\right) \times \left(\frac{C}{P}\right) \tag{14.1}$$

where TER is the nutrient-to-carbon ratio of a resource below which the consumer's growth rate will be limited by the resource's elemental content and the maximal growth efficiency has been computed for phosphorus, GE_P^{max} and for carbon, GE_C^{max} (Sterner and Elser, 2002). For Allen and Gillooly (2009), TER has to be a significant allometric component. The extent to which a nutrient-to-carbon supply is a robust component can be summarized in the stoichiometrical signal depicted by the departure from the universal $(-)\frac{3}{4}$ scaling in the model $\log_{10}(n) = \alpha + \beta \times \log_{10}(m) + \varepsilon$, where n is the abundance of one species and m its body-mass average, with α, β, and ε being respec-tively the elevation (intercept) of the linear regression, the slope of the linear regression, and the departure from the mass–abundance model. If we define this departure ε as a function of elemental ratios, we can rewrite the allometric model as:

$$\begin{aligned}
\log_{10}(n) &= \alpha - \beta \times \log_{10}(m) + \varepsilon \\
&= \alpha - 3/4 \times \log_{10}(m) + \left(\frac{C}{P}\right) \\
&= \alpha - 3/4 \times \log_{10}(m) + TER_{C:P} \times \left(\frac{GE_C^{max}}{GE_P^{max}}\right)
\end{aligned} \tag{14.2}$$

Then the cross-product between the *quality* of any resource (i.e., soil fertility for the basal species occupying the lowest trophic level, i.e., bacteria, fungi, and plants; food elemental quality for the grazing consumers at the second trophic level; and prey for predatory species occupying the higher trophic levels) and GE^{max} (gross percentage of consumed food converted to mass) would be predictors for the abundance of soil biota and, ultimately, for the same supporting ecosystem services (e.g., crop biomass production). The availability of phosphorus in the soil is strictly correlated with soil acidity. Phosphate reacts quickly with calcium and magnesium in alkaline soils and with aluminum and iron in acidic soils: in both cases P-ions tend to form less soluble compounds and invertebrates

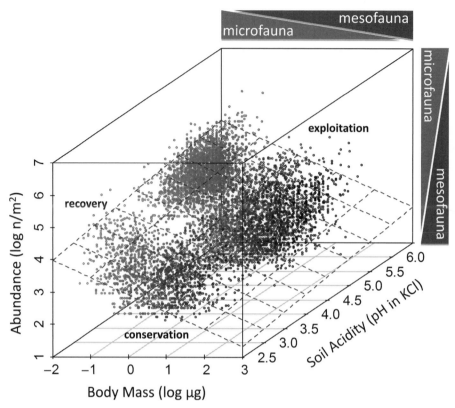

Figure 14.1 3D-scatter of the occurrence of soil microfauna (nematodes) and mesofauna (mites, collembolans, and enchytraeids) in Dutch ecosystems either under management (liming, grazing, and fertilization) or in natural conditions (Scots Pine forests). The empirical trend derived from 7134 populations of soil invertebrates belonging to 135 edaphic communities (soil biota and food-web data publicly downloadable from Cohen and Mulder, 2014), here as stretched 3D plane, corroborates previous results in managed pastures and abandoned grasslands (Mulder and Elser, 2009). A multiple-lines surface (here as blue grid) contains parallel mass–abundance regression lines (each of them representing one soil food web of a single location with its specific soil pH) and shows how allometric scaling tends to become steeper in strongly acidic soils and shallower in slightly acidic, more neutral soils. (A black and white version of this figure will appear in some formats. For the color version, please refer to the plate section.)

physiologically need as much P as possible in their diet. Hence, in managed soils with higher P availability, the total biomass of the larger invertebrates is greater relative to the biomass of the smaller invertebrates than in managed soils with lower P availability (Mulder and Elser, 2009; Cohen and Mulder, 2014), resulting in trait-mediated shifts in the mass–abundance relationships (Figure 14.1). From an eco-stoichiometrical perspective, larger sized invertebrates – having on average less P in their tissues (Sterner and Elser, 2002; Allen and Gillooly, 2009) – are rather mobile, mostly omnivores, and can easily link different networks together.

Several recent papers have raised new questions and provided novel approaches for network research in agriculture, as well as revealing the potential for agro-ecological network research to benefit ecology enormously in a reciprocal feedback – mirroring the close links that once existed in the early developments of both disciplines (QUINTESSENCE Consortium, 2016). From an empirical standpoint, the rapid rate of growth of applied studies would suggest that it is only a matter of time before managed ecosystems return from their current position on the sidelines into the far more pivotal position they previously occupied in ecological research. Although the integration of these fields via the use of network approaches is clearly still embryonic, major empirical and theoretical advances have already been made. In particular, the recent explicit mapping of goods and services onto ecological and socio-economic networks and the incorporation of multiple interaction types into a single network (e.g., Pocock *et al.*, 2012; Mulder *et al.*, 2015; QUINTESSENCE Consortium, 2016) have been pioneered in studies of managed ecosystems and complex systems. This work has also highlighted that there is a need to better integrate into our thinking and understanding the interaction between ecosystem services, certainly more explicitly than has been done to date. It is also clear that greater replication in agro-ecological network research (and ecology in general) is necessary in order to evaluate how certain we can be about management being any practice at the same time a cause of change (for an ecologist) and a solution (for a stakeholder).

In addition to the need to have more replicated or gradient-based studies of networks, the scale at which such studies are conducted needs to be increased in both time and space. These issues of limited replication and scale are common to ecological network research in general, where they have long formed a bottleneck that has slowed progress in the field, and are now being addressed in both pure and applied disciplines, paving the way for their natural extension into agro-ecosystems, which operate in a particularly fragmented landscape (Hagen *et al.*, 2012). Most ecological studies, in fact, were conducted in individual, unreplicated systems that were often wrongly assumed to be isolated, closed systems. Given new developments that are on the horizon, we can start to turn our attention to the key questions that can start to be addressed using a network-based approach.

14.2.1   To What Extent Do Networks in Agro-Ecosystems Differ from Those in Natural Systems, Such That General Ecological Theory Can Be Applied?

The obvious differences between managed and natural systems are that the former are under extreme artificial selection and are maintained in essentially a non-equilibrial and inherently unstable state by active management and external subsidies of propagules and nutrients. Although it is only the crop itself that is directly selected for, artificial selection also operates on many of the other species connected to it in the network. This mix means that rapid evolution can occur in these systems and feedbacks can be especially powerful. However, besides the implications for dynamic processes, such as energy flows from autotrophs to heterotrophs, arguably the single most important aspect of agro-ecosystems is the convergence of two different "Worlds," the green world above ground and the brown world of the soil. In the original green world perspective (Hairston *et al.*, 1960) it was hypothesized that globally the biomasses of herbivores and plants are

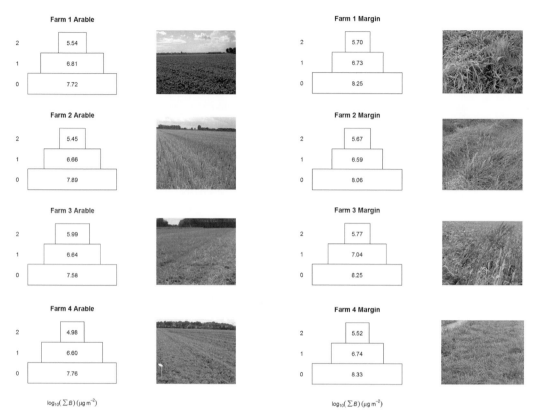

Figure 14.2 Eltonian biomass pyramids for four Dutch farms and their corresponding treatments (arable fields vs. field margins) as computed in R-Cheddar (Hudson *et al.*, 2013). Trophic level 0 (basal microbial resources) includes bacteria and fungi; trophic level 1, the specialized bacterivore, fungivore, substrate feeding and herbivore invertebrates; and trophic level 2, the predating and omnivore invertebrates.

negatively correlated and top–down effects flourish, although at many experimental scales (e.g., the "Sourhope Experiment" and the "Jena Experiment"), bottom–up forces predominate (Fitter *et al.*, 2005 and Rzanny *et al.*, 2013, respectively). In the brown world (Allison, 2006; Kaspari and Yanoviak, 2009), the biomasses of consumers and microbes are positively correlated and bottom–up effects dominate (Figure 14.2). These patterns imply that the green world is much more complicated than the donor-like brown world, as in the green world top–down and bottom–up effects are a matter of scale: at a large scale, top–down predominates (De Visser *et al.*, 2011), at a small scale, the differences with the brown world are less. To a certain extent, top–down and bottom–up effects complement one another in the feedback regulation of producers by consumers and in the energy and nutrient flow from autotrophs to heterotrophs (Wardle, 2002; Moore *et al.*, 2004; Scherber *et al.*, 2010; Strong and Frank, 2010; Moore and de Ruiter, 2012). Agro-ecosystems are suitable environments to investigate at the same time bottom–up and top–down forces.

14.2.2   Can Network Engineering, Via the Introduction of Specifically Sized Organisms Counteract the Negative Effects of Land-Use Intensification?

Invertebrates of different sizes have different effects on soil processes, and body size (or mass) is a major driver of both network structure and dynamics. Several attributes, such as dispersal rate, number of offspring, maximum lifespan, and territory, vary predictably with body size (many of these arise from an autocorrelation with metabolic rate), and new evidence is emerging from research performed in adjacent natural and artificial systems. Food webs have been used to assess the unintended consequences of the relocation of large invertebrates, such as burrowing earthworms (which also act as keystone ecosystem engineers) that have clear economic value for sustainable management. The disruption of supporting ecosystem services due to the loss of non-target, large insect species, can also be evaluated by network approaches. It may be argued that increasing unmanaged areas, such as field edges and margins in a fragmented agricultural landscape – i.e., "land-sparing" approaches to landscape management – could offer a wider selection of plants for large mutualistic pollinators and resources to large predatory species. Indeed, the greater body sizes of individuals at higher trophic levels (here, the intermediate and top species), and the mechanical damage of the fractal-like surfaces of arable fields due to ploughing and mowing, have been used to test and support general ecological theories such as the size–grain hypothesis (Kaspari and Weiser, 2007). For example, "longer legs" or "armed legs" allow predatory arthropods (gamasid mites, spiders, etc.) to walk easily over the litter in the field margins, whereas the same traits seem to increase the difficulties in litter penetration for the predatory arthropods occurring in the arable fields (Figure 14.2).

14.2.3   What Effects Do Agrochemicals Have on Agricultural and Adjacent Networks?

Cascading effects in soil and water biota can be ascribed to a mix of direct (toxicological) and subtler chronic, sub-lethal, or trait-mediated interactions, as well as indirect effects mediated via the network of interconnected species (Posthuma et al., 2014). Abrupt, extensive environmental and landscape changes (such as land conversion) affect ecological processes and lead, therefore, to alterations in food-web structure. For instance, aquatic hyphomycete fungi that decompose allochthonous leaf litter in streams and rivers are essential for cellulolytic decomposition of a fundamental basal resource (Hieber and Gessner, 2002; Alemanno et al., 2007). Decomposition underpins secondary production and covaries with the stoichiometry of senescent leaves in all ecosystems (e.g., Hladyz et al., 2011; Leitch et al., 2014), but as it is mainly fueled by fungal activity it remains sensitive to the use of fungicides elsewhere (i.e., in each of the basins crossed by the streams). In Europe, fungicides are widely used to improve crop productivity. Leached fungicides from the soil of agro-ecosystems will not only limit non-target aquatic fungi, but they will also impact freshwater invertebrates at the next trophic level in the network and, therefore, the entire rate of decomposition. The widespread presence of pesticide and nutrient leachates and their impact of agro-ecosystem networks, often in interaction with other sources of pollution in the form of chronic nutrient addition in adjacent agricultural areas and chemical release in distant urban areas, demands comprehensive frameworks for monitoring changes in network structure (McMahon *et al.*, 2012; Kapo *et al.*, 2014).

14.2.4 How Can We Manipulate Network Structure and Functioning to Maximize the Sustainable Provision of Ecosystem Goods and Services We Need?

We need a new and clear conceptual framework to provide robust parameters to assess carrying capacity, to forecast ecosystem functioning, and finally to quantify ecosystem services. Ecosystem goods, like land products, are easy to quantify directly, but the economical evaluation of the more esoteric ecosystem services is often more difficult. Evaluating the economic costs and benefits of ecosystem services at large scales is even more challenging, often due to disagreements between stakeholders on the value of a particular landscape (e.g., pesticides and fertilizers are often assessed with food-chain approaches). Scaling up our empirical observations can be a solution because the spatial structure of the environment is: (a) relevant for the complex interactions between freshwater and terrestrial systems, and the interchange of chemical stressors and biological entities within and across their respective networks; (b) crucial to integrate physical and chemical variables into landscape ecology and macroecology; (c) a key determinant for predatory, pollinating, and invasive species; and (d) necessary for a comprehensive assessment of the sustainability of multiple management practices in a given area. This idea of layers of interacting networks across agricultural landscapes is being pioneered in network ecology. It has clear implications for both pure and applied ecology (Figure 14.3), especially those related to the conservation and the management of biodiversity hotspots in fragmented habitats, as it includes explicitly multiple scales and ecosystems from field-to-field patches through to aquatic–terrestrial linkages and interdependencies.

In conclusion, we can see huge potential for network approaches to revolutionize our understanding of, and ultimately our ability to manage effectively, the land that supplies us with both the food we need and the biodiversity we value. Despite this huge potential, considerable challenges remain, especially because this is an unavoidably complex and multifaceted area of research. If it is to realize its full potential, we will clearly need to move beyond current narrow modes of thinking, operation, and funding models to deliver the integrated large-scale and multidisciplinary cross-sectorial funding that is needed. We are still some way off from that goal, but there is no doubt that the "silo-thinking" that has dogged this field and slowed its advance must now be abandoned and replaced with a far more holistic and realistic view of the world, and how best to balance the competing demands we place upon it.

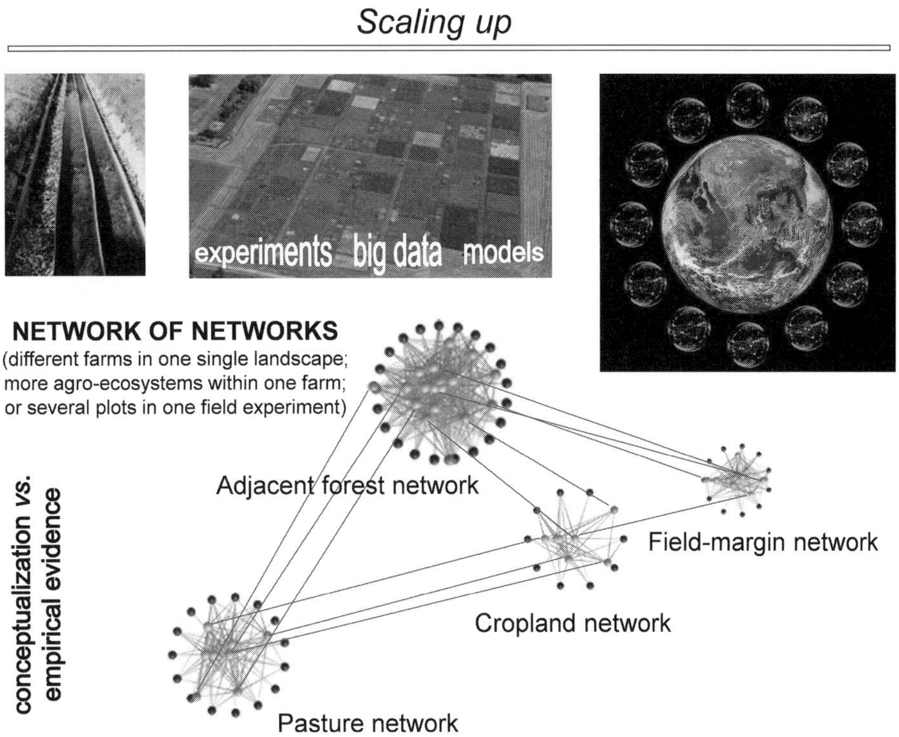

Scaling up

Figure 14.3 To unravel testable biodiversity–ecosystem functioning relationships, the synergy between data and models across the blue, the brown, and the green worlds must become much higher. Ecological networks enable us to focus on emerging critical thresholds, such as effects of exceeding nutrients and pollutants, which can then be validated in both experimental sites and mesocosms (Stewart *et al.*, 2013). Independent ecological networks can be related together at different spatial scales (here as ecosystems within one farm or as plots in an experimental site) by linking all species that interact in some way (Macfadyen *et al.*, 2011; Bohan *et al.*, 2013). (Photo credits: Winfried Voigt and Murray S. A. Thompson.)

References

Alemanno, S., Mancinelli, G., and Basset, A. (2007). Effects of invertebrate patch use behaviour and detritus quality on reed leaf decomposition in aquatic systems: a modelling approach. *Ecological Modelling*, **205**, 492–506.

Allen, A. P. and Gillooly, J. F. (2009). Towards an integration of ecological stoichiometry and the metabolic theory of ecology to better understand nutrient cycling. *Ecology Letters*, **12**, 369–384.

Allison, S. D. (2006). Brown ground: a soil carbon analogue for the green world hypothesis? *American Naturalist*, **167**, 619–627.

Beare, M. H., Parmelee, R. W., Hendrix, P. F., *et al.* (1992). Microbial and faunal interactions and effects on litter nitrogen and decomposition in agroecosystems. *Ecological Monographs*, **62**, 569–591.

Berlow, E. L., Neutel, A.-M., Cohen, J. E., *et al.* (2004). Interaction strengths in food webs: issues and opportunities. *Journal of Animal Ecology*, **73**, 585–598.

Bohan, D. A., Raybould, A., Mulder, C., *et al.* (2013). Networking agroecology: integrating the diversity of agroecosystem interactions. *Advances in Ecological Research*, **49**, 1–67.

Cohen, J. E. (1978). *Food Webs and Niche Space*. Princeton, NJ: Princeton University Press.

Cohen, J. E. and Carpenter, S. R. (2005). Species' average body mass and numerical abundance in a community food web: statistical questions in estimating the relationship. In *Dynamic Food Webs: Multispecies Assemblages, Ecosystem Development, and Environmental Change*, ed. P. C. de Ruiter, V. Wolters, and J. C. Moore, San Diego: Academic Press, pp. 137–156.

Cohen, J. E. and Mulder, C. (2014). Soil invertebrates, chemistry, weather, human management, and edaphic food webs at 135 sites in the Netherlands: SIZEWEB. *Ecology*, **95**, 578.

Coleman, D. C., Crossley, Jr., D. A., and Hendrix, P. F. (2004). *Fundamentals of Soil Ecology*, 2nd edn. San Diego: Academic Press.

De Visser, S. N., Freymann, B. P., and Olff, H. (2011). The Serengeti food web: empirical quantification and analysis of topological changes under increasing human impact. *Journal of Animal Ecology*, **80**, 484–494.

Duncan, C., Thompson, J. R., and Pettorelli, N. (2015). The quest for a mechanistic understanding of biodiversity–ecosystem services relationships. *Proceedings of the Royal Society B: Biological Sciences*, **282**, 20151348.

Fitter, A. H., Gilligan, C. A., Hollingworth, K., *et al.* (2005). Biodiversity and ecosystem function in soil. *Functional Ecology*, **19**, 369–377.

Gray, C., Baird, D. J., Baumgartner, S., *et al.* (2014). Ecological networks: the missing links in biomonitoring science. *Journal of Applied Ecology*, **51**, 1444–1449.

Hagen, M., Kissling, W. D., Rasmussen, C., *et al.* (2012). Biodiversity, species interactions and ecological networks in a fragmented world. *Advances in Ecological Research*, **46**, 89–210.

Hairston, N. G., Smith, F. E., and Slobodkin, L. B. (1960). Community structure, population control, and competition. *American Naturalist*, **94**, 421–425.

Hieber, M. and Gessner, M. O. (2002). Contribution of stream detrivores, fungi, and bacteria to leaf breakdown based on biomass estimates. *Ecology*, **83**, 1026–1038.

Hladyz, S., Ábjörnsson, K., Chauvet, E., *et al.* (2011). Stream ecosystem functioning in an agricultural landscape: the importance of terrestrial–aquatic linkages. *Advances in Ecological Research*, **44**, 211–276.

Hooper, D. U., Chapin, III, F. S., Ewel, J. J., *et al.* (2005). Effects of biodiversity on ecosystem functioning: a consensus of current knowledge and needs for future research. *Ecological Monographs*, **75**, 3–35.

Hsu, S. C., Liu, S. C., Huang, Y.-T., *et al.* (2009). Long-range southeastward transport of Asian biosmoke pollution: signature detected by aerosol potassium in Northern Taiwan. *Journal of Geophysical Research: Atmospheres*, **114**, D14301.

Hudson, L. N., Emerson, R., Jenkins, G. B., *et al.* (2013). Cheddar: analysis and visualisation of ecological communities in R. *Methods in Ecology and Evolution*, **4**, 99–104.

Hunt, H. W., Coleman, D. C., Ingham, E. R., *et al.* (1987). The detrital food web in a shortgrass prairie. *Biology and Fertility of Soils*, **3**, 57–68.

Kapo, K. E., Holmes, C. M., Dyer, S. D., De Zwart, D., and Posthuma, L. (2014). Developing a foundation for eco-epidemiological assessment of aquatic ecological status over large geographic regions utilizing existing data resources and models. *Environmental Toxicology and Chemistry*, **33**, 1665–1677.

Kaspari, M. and Weiser, M. (2007). The size–grain hypothesis: do macroarthropods see a fractal world? *Ecological Entomology*, **32**, 279–282.

Kaspari, M. and Yanoviak, S. P. (2009). Biogeochemistry and the structure of tropical brown food webs. *Ecology*, **90**, 3342–3351.

Kattge, J., Díaz, S., Lavorel, S., *et al.* (2011). TRY: a global database of plant traits. *Global Change Biology*, **17**, 2905–2935.

Lavorel, S., McIntyre, S., Landsberg, J., and Forbes, T. D. A. (1997). Plant functional classifications: from general groups to specific groups based on response to disturbance. *Trends in Ecology and Evolution*, **12**, 474–478.

Lavorel, S., Storkey, J., Bardgett, R. D., *et al.* (2013). A novel framework for linking functional diversity of plants and other trophic levels for the quantification of ecosystem services. *Journal of Vegetation Science*, **22**, 942–948.

Leitch, A. R., Leitch, I. J., Trimmer, M., Guignard, M. S., and Woodward, G. (2014). Impact of genomic diversity in river ecosystems. *Trends in Plant Science*, **19**, 361–366.

Levins, R. (1974). The qualitative analysis of partially specified systems. *Annals of the New York Academy of Sciences*, **231**, 123–138.

Loreau, M. (2009). Linking biodiversity and ecosystems: towards a unifying ecological theory. *Philosophical Transactions of the Royal Society B: Biological Sciences*, **365**, 49–60.

Loreau, M., Downing, A. L., Emmerson, M. C., *et al.* (2002). A new look at the relationship between diversity and stability. In *Biodiversity and Ecosystem Functioning. Synthesis and Perspectives*, ed. M. Loreau, S. Naeem, and P. Inchausti, Oxford: Oxford University Press, pp. 79–91.

Macfadyen, S., Gibson, R. H., Symondson, W. O. C., and Memmott, J. (2011). Landscape structure influences modularity patterns in farm food webs: consequences for pest control. *Ecological Applications*, **21**, 516–524.

Mahowald, N., Jickells, T. D., Baker, A. R., *et al.* (2008). Global distribution of atmospheric phosphorus sources, concentrations and deposition rates, and anthropogenic impacts. *Global Biogeochemical Cycles*, **22**, GB4026.

Martin, L. J., Blossey, B., and Ellis, E. (2012). Mapping where ecologists work: biases in the global distribution of terrestrial ecological observations. *Frontiers in Ecology and the Environment*, **10**, 195–201.

McMahon, T. A., Halstead, N. T., Johnson, S., *et al.* (2012). Fungicide-induced declines of freshwater biodiversity modify ecosystem functions and services. *Ecology Letters*, **15**, 714–722.

Moore, J. C. and de Ruiter, P. C. (2012). *Energetic Food Webs: An Analysis of Real and Model Ecosystems*. Oxford: Oxford University Press.

Moore, J. C., Berlow, E. L., Coleman, D. C., *et al.* (2004). Detritus, trophic dynamics and biodiversity. *Ecology Letters*, **7**, 584–600.

Mulder, C. (2010). Soil fertility controls the size–specific distribution of eukaryotes. *Annals of the New York Academy of Sciences*, **1195**, E74–81.

Mulder, C. and Elser, J. J. (2009). Soil acidity, ecological stoichiometry and allometric scaling in grassland food webs. *Global Change Biology*, **15**, 2730–2738.

Mulder, C., Boit, A., Mori, S., *et al.* (2012). Distributional (in)congruence of biodiversity–ecosystem functioning. *Advances in Ecological Research*, **46**, 1–88.

Mulder, C., Ahrestani, F. S., Bahn, M., *et al.* (2013). Connecting the green and brown worlds: elemental factors and trait-driven predictability of ecological networks. *Advances in Ecological Research*, **49**, 67–173.

Mulder, C., Bennett, E. M., Bohan, D. A., *et al.* (2015). Ten years later: revisiting priorities for science and society a decade after the Millennium Ecosystem Assessment. *Advances in Ecological Research*, **53**, 1–53.

Neutel, A.-M. and Thorne, M. A. S. (2014). Interaction strengths in balanced carbon cycles and the absence of a relation between ecosystem complexity and stability. *Ecology Letters*, **17**, 651–661.

Peñuelas, J., Poulter, B., Sardans, J., *et al.* (2013). Human-induced nitrogen–phosphorus imbalances alter natural and managed ecosystems across the globe. *Nature Communications*, **4**, 2934.

Perrings, C., Naeem, S., Ahrestani, F. S., *et al.* (2011). Ecosystem services, targets, and indicators for the conservation and sustainable use of biodiversity. *Frontiers in Ecology and the Environment*, **9**, 512–520.

Pocock, M. J. O., Evans, D. M., and Memmott, J. (2012). The robustness and restoration of a network of ecological networks. *Science*, **335**, 973–977.

Posthuma, L., Bjørn, A., Zijp, M. C., *et al.* (2014). Beyond safe operating space: finding chemical footprinting feasible. *Environmental Science and Technology*, **38**, 6057–6059.

QUINTESSENCE Consortium (2016). Networking our way to better ecosystem service provision. *Trends in Ecology and Evolution*, **31**. 10.1016/j.tree.2015.12.003.

Rzanny, M., Kuu, A., and Voigt, W. (2013). Bottom–up and top–down forces structuring consumer communities in an experimental grassland. *Oikos*, **122**, 967–976.

Scherber, C., Eisenhauer, N., Weisser, W. W., *et al.* (2010). Bottom–up effects of plant diversity on multitrophic interactions in a biodiversity experiment. *Nature*, **468**, 553–556.

Schröter, D., Wolters, V., and de Ruiter, P. C. (2003). C and N mineralisation in the decomposer food webs of a European forest transect. *Oikos*, **102**, 294–308.

Sechi, V., Brussaard, L., De Goede, R. G. M., Rutgers, M., and Mulder, C. (2015). Choice of resolution by functional trait or taxonomy affects allometric scaling in soil food webs. *American Naturalist*, **185**, 142–149.

Solé, R. V. and Montoya, J. M. (2001). Complexity and fragility in ecological networks. *Proceedings of the Royal Society B: Biological Sciences*, **268**, 2039–2045.

Sterner, R. W. and Elser, J. J. (2002). *Ecological Stoichiometry*. Princeton: Princeton University Press.

Stewart, R. I. A., Dossena, M., Bohan, D. A., *et al.* (2013). Mesocosm experiments as a tool for ecological climate-change research. *Advances in Ecological Research*, **48**, 71–181.

Strong, D. R. and Frank, K. T. (2010). Human involvement in food webs. *Annual Review of Environment and Resources*, **35**, 1–23.

Struebig, M. J., Kingston, T., Petit, E. J., *et al.* (2011). Parallel declines in species and genetic diversity in tropical forest fragments. *Ecology Letters*, **14**, 582–590.

Thébault, E. and Fontaine, C. (2010). Stability of ecological communities and the architecture of mutualistic and trophic networks. *Science*, **329**, 853–856.

Urabe, J., Naeem, S., Raubenheimer, D., and Elser, J. J. (2010). The evolution of biological stoichiometry under global change. *Oikos*, **119**, 737–740.

Von Carlowitz, H. C. (1713). *Sylvicultura oeconomica, oder Haußwirthliche Nachricht und Naturmäßige Anweisung zur wilden Baum-Zucht*. Leipzig: Johann Friedrich Braun.

Von Liebig, J. (1840). *Die Organische Chemie in ihrer Anwendung auf Agricultur und Physiologie*. Braunschweig: Vieweg.

Wall, D. H., Nielsen, U. N., and Six, J. (2015). Soil biodiversity and human health. *Nature*, **528**, 69–76.

Wardle, D. A. (2002). *Communities and Ecosystems: Linking the Aboveground and Belowground Components*. Princeton: Princeton University Press.

Wolters, V., Silver, W. L., Bignell, D. E., *et al.* (2000). Effects of global changes on above- and belowground biodiversity in terrestrial ecosystems: implications for ecosystem functioning. *BioScience*, **50**, 1089–1098.

Woodward, G., Gessner, M. O., Giller, P. S., *et al.* (2012). Continental-scale effects of nutrient pollution on stream ecosystem functioning. *Science*, **336**, 1438–1440.

15 Trait-Based and Process-Oriented Modeling in Ecological Network Dynamics

Marco Scotti, Martin Hartvig, Kirk O. Winemiller, Yuanheng Li, Frank Jauker, Ferenc Jordán, and Carsten F. Dormann

15.1   Introduction

Trophic interactions are one of the most important aspects shaping ecological communities, and the food-web paradigm has played a major role in the development of ecology as a science. Early food-web models attempted to simulate the flow of energy and biomass within local communities (Odum, 1956) or describe the structure of feeding relationships (Pimm, 1982). Species or "trophospecies" (i.e., groups of species that supposedly share the same sets of predators and prey; see Yodzis and Winemiller, 1999) have served as the building blocks of both types of trophic networks (usually referred to as flow webs and topological webs, respectively). Topological food webs are static "snapshots" lacking magnitudes (i.e., estimates of the rate of energy flow or the strength of trophic links) and therefore have limited utility for examination of ecological dynamics. The strength of trophic interactions can be estimated in different ways (e.g., observing magnitudes of biomass/energy transfers, modeling of consumer feeding preferences, functional responses or metabolic constraints, varying interaction coefficients in Lotka–Volterra multispecies competition models, and quantifying responses from manipulative field experiments; see Berlow *et al.*, 2004, 2009). Food-web models have been used to predict risks of secondary extinction (Allesina and Bodini, 2004), examine consequences of biodiversity loss for ecosystem stability (McCann, 2000), and study direct and indirect effects of predators on prey populations (Bondavalli and Ulanowicz, 1999).

Food-web models that lump individuals into species and trophospecies lose much valuable information concerning influential attributes associated with age, body size, foraging history, location, and reproductive tactics. Such lumping may inflate the number of trophic interactions associated with species or trophospecies and thus fails to take into account how variation in feeding preferences between individuals affects food-web structure and dynamics (Bolnick *et al.*, 2007). Trait-based approaches have gained momentum in community and evolutionary ecology, and most ecologists now recognize the relevance of variation at intraspecific as well as interspecific levels (Bolnick *et al.*, 2011). This recognition calls for exploration of new approaches to food-web ecology. Intraspecific trait variation can alter the number and strength of interspecific interactions, especially when food-web interactions are a function of body size

(Otto *et al.*, 2007; Berlow *et al.*, 2009). Hump-shaped deviations from log-linear global models in allometry and temperature dependence have been found both for taxonomically narrow groups of consumer–resource pairs and at the intraspecific level (Rall *et al.*, 2012), but it should be noted that inter- and intraspecific scaling relationships often differ. When the probability that a given predator taxon feeds on a given prey taxon is quantified by a non-linear function that considers the body size of both, the overall attack rate is obtained through the weighted average attack rate across all predator/prey phenotype combinations (i.e., all combinations of predator/prey body mass have to be considered at the individual level; see Bolnick *et al.*, 2011). The same principle applies to any species-specific demographic parameter that varies according to metabolic scaling laws (i.e., given the non-linear nature of metabolic scaling laws, the mean metabolic rate of species is not predicted by the mean species body size; see Jensen's inequality – Ruel and Ayres, 1999).

In this chapter we explore the potential for individual-based modeling (IBM; see Grimm and Railsback, 2013) to incorporate functional traits into food-web ecology. Individual-based models (in the rest of the chapter we will equate "individual-based modeling" with "agent-based modeling" – ABM; see Box 15.1) simulate intraspecific variation explicitly. In IBMs, the macroscopic features and the adaptive dynamical behavior of complex systems result from rule-based description of individual-level traits or properties (e.g., individual-based models have demonstrated that fish may school using local information only, in the absence of higher order and external stimuli; see Kunz and Hemelrijk, 2003). Currently, IBMs have been applied to address hypotheses at population, community, and landscape levels, and some even have attempted to model the interactions between these levels in the organizational hierarchy (e.g., consequences of landscape dispersal on social structure of local populations, or effect of local predator pressure on landscape dispersal; see DeAngelis and Gross, 1992; Giacomini *et al.*, 2013; Scotti *et al.*, 2013).

Box 15.1 Glossary

Ordinary differential equations (ODEs): equations involving functions of independent variables and their derivatives. The deterministic dynamics of populations in continuous time are traditionally described using coupled, first-order ordinary differential equations. While this approach is accurate for large systems, it is often inadequate for small systems where key species may be present in small numbers or key interactions occur at a low rate.

Stochastic simulation algorithms (SSAs): computational methods applied to model time evolution of discrete processes in spatially homogeneous systems; they represent an alternative to continuous, deterministic approaches and give rise to discrete-state models, in the form of a Markov chain.

Markov chain: a collection of random variables that undergo transitions from one state to another; transitions to a new state only depend on the current state of the system and not on the sequence of events that preceded it.

Markov model: a stochastic model in which transitions depend upon the most recent state and not on other previous states.

Markov property: if the probability that a system in any particular state $X(t)$ at time t depends only on the state of the system $X(t-1)$ at time $t-1$ and not on previous states, the system is said to satisfy the Markov property.

Master equation: the fundamental equation that specifies how the probability of the system being in a given state changes with time; it represents the continuous time version of a Markov chain.

Gillespie's stochastic simulation algorithm: a procedure (computer program) that generates time-evolution trajectories (i.e., possible solutions) of finite populations in continuous time, using stochastic equations; it is a variety of a Monte-Carlo method and has become the standard for stochastic models in biology.

Sensitivity analysis (SA): is used to systematically examine the importance of parameters. Sensitivity analysis explores how sensitive model outputs are to changes in parameter values. High sensitivity to a parameter indicates that the process linked to that parameter controls model outputs and system behavior more than other processes.

Uncertainty analysis (UA): analyzes how uncertainty in parameter values affects the reliability of model results. Uncertainty analysis provides a way to assess the confidence of model results.

Robustness analysis (RA): explores the robustness of results and conclusions of a model to changes in its structure. While SA quantifies model responses to changes in parameter values, RA focuses on responses to changes in model structure. Most robust systems display low sensitivity to changes in model structure.

Inverse methods: are used to deduce values of unknown parameters from a set of observations and a model of the system. Usually, numerical experiments are carried out to describe the behavior of state variables (e.g., standing stocks or population sizes) from known parameter values and initial conditions. The inverse method is the reverse of that procedure: observations of the state variables are used to infer the parameters. Inverse methods show strong similarities with regression analysis, but are characterized by a low ratio of observations to unknowns. For example, the inverse formalism can be applied in food-web analysis to estimate carbon flow rates between species starting from time series of species standing stocks, energetic and physiological constraints.

Agent-based models (ABMs): stochastic models that have a finite number of agents or individuals. Usually they define the activity of these unique and autonomous entities in terms of computer programs. Agents (e.g., molecules, organisms, and humans) locally interact with each other and the environment by pursuing a certain goal. Each agent is unique and differs from the others in characteristics such as size, location, resource reserves, and history. Local interaction refers to the fact that each agent does not interact with all other agents but only with its neighbors (in geographic space or network structure). Agents are autonomous: they act independently of each other and pursue their own goals. Important elements of ABMs are emergence, adaptive behavior, and sensing. The algorithmic approach adopted for constructing ABMs allows many details but often leads to models that are too complex for numerical analysis.

Individual-based models (IBMs): are composed of a finite number of individuals (i.e., discrete variables) and usually require a smaller number of attributes than ABMs. IBMs and ABMs are often equivalent and refer to the ways computational models are defined in biology and computer science, respectively. Rules for interactions between individuals and with the local environment are formulated in terms of probabilities, usually as a Markov process, and these models are stochastic.

Emergence: is the most basic concept of agent-based modeling. It refers to global model outcomes that emerge from the behavior of local agents.

Adaptive behavior: is related to the way agents make decisions and change their state in response to changes involving their properties, the direct neighbors, and the local environment.

Sensing: since the ability of agents to respond and adapt to the environment and other agents depends on what information they have, it is essential to know the kind of information available and how this is obtained.

Ecology is a field with a long tradition of bottom–up, individual-based modeling (see Table 15.1 for a set of examples developed using different software platforms). Bottom–up here means that global properties of the studied system are derived from individual-level, mechanistic rules (i.e., modeling details to predict emergent phenomena at higher levels of organization). The first IBMs were developed to simulate forest growth (Botkin et al., 1972; Shugart, 1984) and fish cohort dynamics (DeAngelis et al., 1980). Grimm et al. (2006) suggested that clarity and reproducibility are the reasons behind the predictive capabilities of these models (i.e., they are characterized by an extensive use of mathematics, which encouraged testing, further model development, and applications). Among the most successful early IBMs were some that addressed fish population dynamics. For example, Rose et al. (1999) developed an IBM to examine effects of alternative prey and compensatory responses on walleye and yellow perch population dynamics. Their model simulated daily growth, mortality, and spawning of individuals during three ecologically distinct periods; the model predictions concerning patterns of abundance, growth, and survival were consistent with empirical data. Huse et al. (2004) simulated the movement, foraging, growth, and mortality of cod and capelin in the Barents Sea. Their model was built and calibrated using extensive empirical data on relationships between size, age, behavior, and spatial distributions, which resulted in realistic descriptions of daily movements at a fine environmental scale.

Other IBM-based studies have focused on trophic interactions within communities. For example, anthropogenic impacts in the Florida Everglades change how drought affects small-fish populations, density-dependent foraging by piscivorous fish and birds, and population dynamics of the latter two groups (DeAngelis et al., 1997). van Nes et al.

Table 15.1 Examples of IBM applications and software platforms used for their development in ecology.

Short description	Platform	Reference
Gap model (i.e., simulating forest dynamics in a gap area) that describes ecological patterns and processes over a long period of time (i.e., forest succession) in a mixed-species forest.	JABOWA	Botkin _et al._, 1972
Model simulating a flock of birds (schooling-like behavior). It contributed to the development of some of the first behavioral animations.	Boids	Reynolds, 1987
The model adopts a probabilistic process algebra approach to represent the processes executed by ants; it shows that an age structure emerges when the activities of ants are allocated to various tasks.	WSCCS	Tofts, 1993
Study of ecological epidemiology (i.e., spread of rabies among red foxes in central Europe) that illustrates how local, rare events may have profound impacts on large-scale global patterns.	C++	Jeltsch _et al._, 1997
The model includes size-based opportunistic predation. It studies the role of size distribution of fish communities on the functioning of marine food webs and analyzes the ecosystem effects of fishing.	OSMOSE	Shin and Cury, 2001
Population-viability analysis of salmonids in fragmented streams. The model is age structured and density dependent, and it incorporates both demographic and environmental stochasticity.	FORTRAN	Morita and Yokota, 2002
IBM of stream salmonids used to simulate habitat selection and test foraging theory. This is a good example of "pattern-oriented modeling" as defined by Grimm _et al._ (2005).	inSTREAM	Railsback and Harvey, 2002
The model analyzes multispecies fish communities in the Frisian Lakes (the Netherlands). Monte-Carlo sensitivity analysis was applied to group parameters with similar effects on the model results.	Piscator	van Nes _et al._, 2002
Individual-based coyote population model that explicitly incorporates behavioral features (i.e., dominance and territoriality). This simple model is insensitive to individual parameter estimates.	Swarm	Pitt _et al._, 2003
Model that simulates the spatio-temporal dynamics of Central European natural beech forests. Multiple patterns observed in remnants of natural beech forests were reproduced by this model.	BEFORE	Rademacher _et al._, 2004
Population dynamics models of the copepod _Eurytemora affinis_. The model was parameterized with experimental data and developed by using a generic, open-source software platform.	_Mobidyc_	Souissi _et al._, 2005
IBM that assesses the effect of environmental changes for five overwintering birds in a site of southern UK. Conservation issues under climate change scenarios (e.g., sea-level rise) are identified.	MORPH	dit Durell _et al._, 2006

Table 15.1 (cont.)

Short description	Platform	Reference
Model of the population dynamics of savanna woody species. It explains tree–grass coexistence in semi-arid savannas as a function of precipitation-driven cyclical succession at the (local) patch scale.	SATCHMO	Meyer *et al.*, 2007
IBM that considers predation among size-structured populations in a fish community assembly. It includes allometric constraints, movement, life-history, and interactions among individuals.	MATLAB	Giacomini *et al.*, 2009
IBM that simulates processes and interactions in a hierarchical ecological system that links population, community, and metacommunity dynamics. Emergence of global patterns is investigated.	BlenX	Scotti *et al.*, 2013
Model that shows how the microbial-derived extracellular enzymes influence carbon dynamics in the soil depending on spatial arrangement of resources, the input of detritus, and trophic structure.	NetLogo	Moore *et al.*, 2014

Short descriptions of relevant IBM applications are provided together with details on different software platforms used for simulations and literature sources. This is not an exhaustive review of software platforms available for IBMs; this list reveals a range of tools and IBMs.

(2002) created an IBM to quantify the effects of a multispecies fishery on community dynamics. To evaluate the consequences of population-size structure for fish community assembly and food-web dynamics along environmental gradients, IBMs were simulated considering explicit relationships between trophic interactions and functional traits (Giacomini *et al.*, 2009, 2013). An IBM created to simulate the dynamics of the Prince William Sound food web used a relatively small number of basic demographic parameters (i.e., birth and death rates) integrated with a network of predator–prey interactions for which relative feeding preferences were available (Livi *et al.*, 2011; Gjata *et al.*, 2012). The authors changed, in turns, the population size of each species (i.e., by halving or doubling the abundance) and measured the impacts on other species in the food web. Relative impacts were computed by comparing the new simulated population sizes with the reference scenario (i.e., the scenario representing empirical data without any simulated change). Gjata *et al.* (2012) found no significant differences between the intensity of simulated effects that spread through short and long pathways, and the most important trophic groups were at intermediate trophic levels (Livi *et al.*, 2011). Melián *et al.* (2011) used information from a well-studied food web (more than 25 000 individual prey and predators sampled) to develop an individual-based model that takes into account intraspecific variability (i.e., population genetic variation), microevolutionary processes (by considering sexual reproduction and speciation), and trophic interactions. They showed that high intraspecific variance in individual feeding habits has strong effects on macroscopic properties of food webs. They also observed that high intraspecific variance may determine high rates of change in genetically based phenotypic traits that can lead to convergence of ecological and evolutionary trajectories.

Mechanistic understanding of functional traits and feeding behavior can be made explicit through IBMs. IBMs may describe size-structured populations and allow simulating intraspecific variations in each life stage. Hartvig and Andersen (2013) showed how trophic flows that vary among life stages affect the coexistence of small communities in which species experience ontogenetic trophic-niche shifts. Non-feeding interactions (e.g., habitat modification, predator interference and facilitation) can alter community assembly and often display a wide range of intraspecific and between-system variation (McGill *et al.*, 2006). Some attempts have been made to integrate traditional food-web models with non-feeding interactions. Kéfi *et al.* (2012) suggested that non-feeding interactions should be classified based on how they affect (1) feeding parameters (e.g., refuge from predators, predator interference), (2) non-feeding parameters (e.g., 3D structures provided by kelps and coral reefs), and (3) flows across system boundaries (e.g., herbivore-induced emission of volatile metabolites that attract predators to herbivore-damaged plants). IBMs could provide an effective framework to link changes in population diet diversity to trait variation (Giacomini *et al.*, 2009, 2013; Melián *et al.*, 2011). IBMs can be used to simulate biotic and abiotic factors that do not regulate trophic interactions; for example, Moore *et al.* (2014) demonstrated with simulations that the influence of extracellular enzymes on carbon dynamics in soil depends on the spatial arrangement of resources and is regulated by input rate and input intervals of detritus. IBMs also can help experimental ecologists by alleviating consequences of sample-size limitation in empirical studies. Meyer *et al.* (2009) proposed that empirical studies of above- and belowground controls of trophic structures would benefit from the statistical power of individual-based simulations for evaluating the reliability of biomass estimates and the probability of experimental failures due to missing values.

Here we consider strategies for implementing IBMs in food-web ecology. We demonstrate how this modeling approach can be suitable to simulate the influence of functional trait variability on ecological dynamics. We emphasize that this modeling approach has the potential to greatly expand the scope of traditional food-web analyses, and we present several possible applications. Finally, we discuss future challenges and limitations that can impair applications of IBMs in food-web ecology.

15.2   Dimensions of Variability

Individual-based models simulate the behavior of simple entities that interact to form complex systems (e.g., they model individuals of a population within a given habitat or different species within a local community; see Livi *et al.*, 2011). System-level patterns can be viewed as determined by concurrent and elementary processes that regulate the behavior of single individuals. For example, outcomes from IBMs of food-web dynamics have been compared with empirical datasets that (1) summarize individual diets and feeding rates at the community-level, or (2) report macro-ecological patterns of species diversity in food webs (Giacomini *et al.*, 2009; Melián *et al.*, 2011). IBMs consist of an alternative approach if compared to research studies that focus on global rules to describe food-web dynamics (e.g., deterministic models that rely on ordinary differential equations – ODEs; see Gross *et al.*, 2009; Stouffer and Bascompte, 2011).

Individual-based models are composed of individuals, autonomous (and discrete) entities that interact locally with neighbors and perceive and respond to environmental conditions. Individuals are often defined as computer algorithms and simulated adopting a stochastic approach (e.g., Gillespie's stochastic simulation algorithm). The propensity of interactions and the occurrence of specific events are defined in terms of probabilities. As a result of their interactions, individuals may change their state and alter their behavior. High-level properties emerge from the complex interplay between single individuals, and this feature makes IBMs attractive for community ecology (i.e., an intriguing goal is to understand which individual-level properties drive the emergence of community-level patterns; see Box 15.2). Although the idea of coding individual-level rules could be seen as an elegant and unbiased approach for the study of food webs

---

Box 15.2 Global Properties That Could Emerge from Individual-Based Models

Abundance: population size (e.g., total sum of individuals pertaining to a given species); also the temporal or spatial explicit version of this measure could be evaluated (e.g., population size during jellyfish bloom or coyote abundance in a given landscape patch).

Diversity: can be computed using population (and sub-population) sizes or trophic flows; multispecies assemblies and life stages of a single species refer to populations and sub-populations, respectively. Diversity (of sizes and flows) can be quantified with Shannon's index of diversity and average mutual information.

Food-web connectance: the fraction of all possible trophic links that are realized in a network.

Link density: the average number of trophic links per species (or individuals).

Fractions of basal, intermediate, and top species in food webs: are used to define the relative importance of (1) species/individuals that are consumed but have no prey (basal; e.g., primary producers), (2) species/individuals that are both predators and prey (intermediate), and (3) predators that have no predators (top; i.e., apical predators).

Fraction of omnivorous and omnivory: the first index quantifies the relative number of species (or individuals) that feed on different trophic levels while the second characterizes the variety in the trophic levels of the food of a consumer (i.e., omnivores are species/individuals that receive energy from food chains of different lengths).

Fraction of cannibalistic taxa: the relative number of cannibalistic species (or individuals) in a food web (or in a population, in the case of different life stages).

Fraction of herbivores: percentage of herbivores plus detritivores (taxa/individuals that feed on autotrophs or detritus).

> **Effective trophic level:** represents the weighted average distance that energy must take to get from an energy source (e.g., sunlight entering the system) to a given taxon (or individual).
>
> **Mean chain length:** mean food-chain length, averaged over all species (or individuals).
>
> **Cycling:** can be quantified by the fraction of matter/energy that is recycled with respect to the total amount that circulates in the ecosystem (e.g., Finn cycling index) or by characterizing the number of cycles and their structure (i.e., length of cycles and taxon/individuals involved).
>
> **Metacommunity size and heterogeneity:** can be measured in spatially explicit IBMs (where two or more metapopulations are simulated in different landscape patches) and are computed as the sum and the standard deviation of individuals that belong to different metapopulations, respectively.

(i.e., it does not require a formulation of global principles regulating system dynamics), it also poses some challenges. Essential for the development of IBMs is the identification of key individual-level properties that are expected to be the main drivers of whole-system dynamics, which is far from trivial. The way IBMs are constructed depends strongly on the choice of the parameters for simulating a given set of dynamics at given levels of the associated organizational hierarchy. Two possible strategies can be adopted for achieving the required level of "individuality." IBMs can be implemented either starting from a restricted and well-characterized set of individual-level properties (bottom–up model construction), or using deterministic models as a scaffold by increasing the relative importance of individuality step by step (top–down model construction). Hybrid models that include both discrete and continuous variables (i.e., deterministic and individual-based approaches coexist) have also been presented (Gamarra and Solé, 2002). We suggest that the integration of individual-based rules in a deterministic framework should start from those components for which discrete variables are expected to be more relevant. It is inevitable that hybrid models are mostly specific for the system under investigation, thus representing a trade-off between the flexibility needed to incorporate variability at the individual level and the generalizability across systems. Population dynamics of a rare or threatened species might be modeled using an IBM that simulates individual reproduction, survival, growth, and feeding, while abundant species could be modeled using a more traditional aggregate and deterministic approach (see Scotti *et al.*, 2013 for a theoretical approach). Rare species (i.e., ones characterized by small population size) are likely to display high variance in comparison to average population behavior (i.e., the variability conferred by single individuals can be wider than in the case of large populations; see Carnicer *et al.*, 2012). Thus, considering the individual-level dynamics of rare species takes into account deviations from average population behavior, leading to a more realistic representation of what occurs in nature. Hybrid models could be developed by including discrete variables that have ubiquitous effects, such as stochastic environmental factors that affect all individuals, independently of whether or not these belong to rare species or large populations (Morita and Yokota, 2002).

To build bottom–up IBMs, modelers start from a small subset of rules to develop architecture capable of predicting outcomes at both the individual level and higher levels of organization. Features such as predation and other interspecific interactions, spatial dynamics, and genetic variation may be added to create a more complex model. This incremental IBM construction is called "composability" and refers to the chance of programming individual-level dynamics in terms of distinct elementary modules (e.g., a module for simulating the consequences of social dynamics on spatial movement can be added and run in parallel to a module that determines probability of feeding interactions as a function of predator and prey densities; see Jordán *et al.*, 2011). Different interaction types (e.g., feeding preferences and propensity to establish social interactions with conspecific individuals) can be expressed in terms of probability to occur in a given time frame, thus allowing consideration of several concurrent (i.e., parallel) processes under a single integrative framework.

The parallel simulation of distinct events that involve different hierarchical levels can help to extend food-web studies with non-trophic processes (Kéfi *et al.*, 2012 suggested a possible scheme to frame non-trophic interactions in the context of food-web analysis). IBMs and their modular structure represent an ideal tool to model elements (e.g., abiotic factors) and consider multiple interaction types and processes (e.g., facilitation, mutualism, and genetics) that influence key traits in community ecology. Given the broad impact that non-trophic relationships and processes can have at the level of food webs (Kéfi *et al.*, 2012) we highlight the need for their integration in community-level research studies. For example, specific death rates can affect equilibrium feeding rates (Neutel *et al.*, 2002) and, in some systems, the strength of trophic interactions may be inferred from body-size relationships (Berlow *et al.*, 2009). The frequency of social interactions and alarm calls is intensified in the presence of predators, and changes in patterns of social behavior can trigger differences in food distribution and feeding rates (Barton *et al.*, 1996). Coral reef ecosystems provide refuge (i.e., facilitation) to fish and motile invertebrates (Pratchett *et al.*, 2009), and desert shrubs buffer environmental stress thus facilitating the establishment and persistence of more heterogeneous plant communities (Pugnaire *et al.*, 1996). At the landscape scale, migration can reduce predation (Hebblewhite and Merrill, 2007) and habitat fragmentation can release top predator pressure on mesopredators, leading to local extinction of species at lower trophic levels (Crooks and Soulé, 1999). Modeling the multiplicity of processes and factors that act in parallel and regulate ecosystem (and landscape) dynamics can result in an explosion of complexity. Modelers are challenged with the task of characterizing system dynamics in the most realistic way, but this cannot come at the expense of model "readability" (i.e., if the model is too complex for being able to mechanistically understand main drivers of its outcomes, which means that the modeler has failed in the phase of model construction).

A central issue in individual-based modeling is the need to identify key mechanisms and parameters that are expected to have the strongest effects on system dynamics. This can be done either by deciding *a priori* the most significant parameters (e.g., the inclusion of body size in a model that describes predator–prey interactions) or by reducing a large initial set of parameters through sensitivity analysis (i.e., by detecting which parameters determine the largest changes in model output when varied). Sensitivity analysis is carried out by performing simulation experiments and might help to clarify how the dynamics of a system emerges from the characteristics of individuals. Sensitivity analysis can also be applied to investigate the importance of species and the role of particular interactions in a community context. In the Prince William Sound ecosystem, the halibut has been shown to be the fish species generating the largest community response when disturbed (Livi *et al.*, 2011). Scotti *et al.* (2012) quantified the dynamical effects of apical predators on other species and compared the simulated impacts with the strength of interactions in the weighted food web. They observed a positive correlation between average simulated impacts and feeding prefer-ences (i.e., the simulated impacts that display largest magnitude travel through the strongest links in the weighted food web), but highest simulated variability was observed for the weakest links of the weighted food web. From a methodological point of view, sensitivity analysis can be performed with a variety of techniques. Parry *et al.* (2013) implemented a Bayesian method for rapid and thorough sensitivity analysis, with the goal of identifying model parameters that require more rigorous parameterization. van Nes *et al.* (2002) proposed an approach to find clusters of parameters that have roughly the same effects on the model results. In an IBM constructed for analyzing the effect of turbulence on feeding of larval fish, Monte-Carlo simulations using Latin hypercube sampling methods were adopted for sensitivity analysis (Megrey and Hinckley, 2001).

Despite the fact that very large numbers of traits and processes could be considered when constructing IBMs, the number needed to explain all interspecific interactions in ecological networks may be small (i.e., less than ten; see Eklöf *et al.*, 2013). Therefore it is critically important to identify the most significant traits required for the description of ecological interactions. To this end, techniques for feature selection and feature extrac-tion could be applied. These techniques are often used when a wide array of features (e.g., traits) exists and only a few empirical data are available (Devijver and Kittler, 1982; Saeys *et al.*, 2007). The main goal is to extract the minimum and most informative subset of features needed to characterize system behavior (e.g., by combining the bill gape of birds with fruit size, species attributes can be linked to large-scale frugivore network structure; see Eklöf *et al.*, 2013). Ideally, such a process should be driven entirely by data (e.g., following the pattern-oriented modeling paradigm using approx-imate Bayesian computation: Grimm *et al.*, 2005; Hartig *et al.*, 2011; see also inverse methods in Box 15.1). An alternative option for dimensionality reduction is to focus on specific ecological network modules (in graph theory the more strict terminology for modules is communities). This is because the nodes of a network can often be grouped into subsets that display a high density of internal connections (i.e., nodes belonging to a module show more connections within the module than with nodes of other modules). If the structural grouping of nodes (e.g., species in a food web) reflects some ecological properties (e.g., module composition matches benthic and pelagic sub-communities; see Allesina *et al.*, 2005), then construction of IBMs that are module specific could be justified. Dynamical simulation of network modules is reasonable when traits that reflect a well-defined biological process are significantly over-represented in such structural modules (e.g., if the goal is simulating salt-marsh dynamics, then the solution could be to focus on the network module that is significantly enriched with species showing adaptation to high-salinity concentration). The focus on structural modules might facilitate inclusion of groups of species that are clearly associated to an ecological phenomenon (e.g., resistance to high salinity) as well as other species that are densely connected to the first group (these latter species may not necessarily have an intuitive connection to the process under investigation; e.g., species that lack salinity tolerance). We suggest that strategies for reducing the number of traits might also have an indirect positive effect on the complexity of the parameter space (i.e., fewer traits might also require fewer parameters). In addition to quantifying the relevance of parameters through sensitivity analysis (i.e., to detect which are the main parameters controlling model outputs), uncertainty analysis and robustness analysis could be applied (Box 15.1). Uncertainty analysis assesses the reliability of model predictions as a function of uncertainties in model input and design, while robustness analysis focuses on the sensitivity of model output to changes in the model design. Model validation and testing can represent a further step to evaluate the relevance of traits and processes included in IBMs (i.e., to investigate whether model design is appropriate and the parameters are the most representative to capture system-level dynamics). Model validation could be oriented toward system-level properties (Box 15.2) that emerge from individual-level rules (e.g., to analyze discrepancies between population-level heterogeneity of feeding interactions in empirical and simulated food webs; see Melián *et al.*, 2011). Although the reproducibility of system-level patterns is a necessary but not sufficient condition to validate IBMs (i.e., also models that rely on unrealistic equations and parameters may lead to the emergence of system-level patterns that are consistent with empirical data), we suggest this might represent a preliminary step for their testing. Once coherent system-level patterns are obtained, modelers should explore the mechanistic functioning of IBMs in more detail by focusing on specific interactions or modules (e.g., to analyze population-level dynamics associated with cascading effect or apparent competition; see Jordán *et al.*, 2011).

Inclusion of the most representative traits (i.e., key traits that are expected to be the most relevant for simulating the emergence of system-level properties that match empirical data), and as a consequence the use of related parameters, can be limited by data availability. Constraints associated with data availability are a common obstacle in ecological modeling and do not represent a peculiarity of IBMs. However, since IBMs often rely on a highly parameterized description of individual-level dynamics, they are particularly susceptible to the risk of data deficiency. Scotti *et al.* (2013) developed a theoretical model of a hierarchical ecological system to understand how processes and interactions that occur at the level of population, community, and metacommunity may affect the emergence of global patterns (i.e., metapopulation size and spatial heterogeneity). Although they started from empirical datasets, data sources were very heterogeneous and combined details concerning different taxonomic levels: most of the social relationships describing intrapopulation dynamics were obtained from mammals and social insects, food-web interactions were mainly gathered from aquatic systems, and landscape dynamics were inferred from spatial movements of various species (mostly carabid beetles). High-quality databases could also promote better understanding of model outcomes; Melián *et al.* (2011) confronted their neutral eco-evolutionary dynamics model with (1) data on prey consumption per individual predator and (2) prey–predator diversity in several environmental situations. They used a large dataset describing prey and predators at multiple locations in the estuary of the Guadalquivir River in Spain. In summary, the scarcity of datasets characterizing the same traits for multiple species and the fact that most of these data, if available, are not collected for the primary purpose of modeling represent two serious drawbacks that impair the growth of IBMs in community ecology.

Early IBM applications were criticized for the lack of clear descriptions that limited the chances of any rigorous evaluation (Lorek and Sonnenschein, 1999). The introduction of the ODD (overview, design concepts, and details) protocol contributed to standardize the description of IBMs, enhanced rigorous model formulation, and led to a more understandable and complete explanation of methods and results (Grimm *et al.*, 2010). The ODD protocol attenuated criticism on irreproducible outcomes and further stimulated the application of IBMs in ecology. Three main issues have been identified in the ODD protocol: (1) redundancy; (2) long descriptions are required even for simple models; (3) units of object-oriented implementations are illustrated in different sections (i.e., properties and methods have to be presented separately). Despite such limitations, it is clear that this kind of standard protocol is highly relevant in the context of community ecology. Indeed, a multiplicity of processes is likely to be included in IBMs simulating community dynamics, and this complexity calls for unambiguous model descriptions. The ODD protocol represents a guide to how to build an IBM and summarizes the most relevant decisions taken in the phase of model construction (e.g., list of the entities simulated, declaration of the variables used, and explanation of the processes and interactions modeled). In Box 15.3 we use the ODD protocol to describe a simple IBM that investigates the effect of hunger level of a predator on the functional response. In this example the most relevant variables that regulate the predator feeding

---

Box 15.3 The ODD Protocol: An Example

We introduce a simple model for studying the effect of hunger level of a predator on the functional response. The description of the model follows the ODD protocol format (Grimm *et al.*, 2006, 2010). Unused concepts are omitted in the description.

Overview

1. Purpose

The model investigates the effect of initial hunger level of a predator on the functional response (i.e., intake rate of a predator as a function of prey density and hunger). It mimics feeding experiments performed in the laboratory with one predator and various numbers of prey. The simulated data are fitted to functional responses and the parameters of functional responses for different levels of hunger of the predator are compared.

2. Entities, State Variables, and Scales

Agents: various numbers of prey and a single predator are modeled as agents. Prey state variables include individual identity, spatial coordinates, and body mass. The predator is characterized by spatial coordinates, body mass, and a set of state variables related to hunting, food processing, and digestion. Such variables are gut fill (i.e., milligrams of food that is in the gut) and still-handling (i.e., how many time steps are still needed for processing the prey; this variable is set to a value when the predator catches a prey); their value determines the behavior of the predator, defining whether it is in resting, hunting, or chewing stage. Body mass is a super trait and the biologically relevant variables that describe the predator feeding activity are parameterized according to allometric rules (e.g., handling time, time needed to kill and chew a prey item; digestion rate, amount of food a predator can digest per time; gut capacity, maximum milligrams of food). Also predator and prey velocities are regulated via allometric relationships and depend on constant scaling exponents.

Spatial units: a grid represents the experimental arena. The predator and prey are placed in a square grid of 100 cm × 100 cm, with cell resolution of 1 cm × 1 cm.

Time units: one time step represents one second and each simulation lasts for one hour (i.e., 3600 discrete time steps).

3. Process Overview and Scheduling

The simulations mimic the predation cycle. Feeding is a process executed by the predator and follows a fixed order: (a) predator digestion (this is the only action that occurs at each step of the simulation); (b) prey handling, if the predator is in chewing stage; (c) search for hidden prey, if the predator is in hunting stage (i.e., to check the presence of a prey in the same cell of the predator); (d) hunting, if the predator is in hunting stage (i.e., when the predator attacks the prey there is a certain probability that the prey can escape); (e) prey killing (and update of killed prey; i.e., the number of prey in the model is fixed and when a prey is eaten another prey is included to replace the consumed one) if hunting succeeds. The predator can also be in resting stage; in this case the actions (b)–(e) (i.e., prey handling, search for hidden prey, hunting, and prey killing) are not activated. Moreover, simulations include the random move of the prey and the random move of the predator, if the gut fullness of this latter is less than a hunting threshold (i.e., when in hunting stage). The time is modeled as discrete time steps (one step corresponds to a second). State variables (e.g., spatial distribution of prey and predator) are updated at each step of the simulations.

Design Concepts

Basic principles: the model simulates the number of prey consumed by the predator. It investigates the persistence of prey and predator in response to changes in the satiety level of the predator. Such changes affect predator behavior (i.e., resting vs. hunting mode). Allometric scaling is used to model the feeding behavior of the predator (e.g., probability of successful attack and handling time) as well as predator and prey velocity.

Emergence: the simulated data are used to fit different functional responses that quantify the intake rate of the predator as a function of prey density and hunger level of the predator.

Adaptation: the predator makes decisions on whether resting or hunting according to its gut fullness.

Objectives: in feeding experiments carried out in the laboratory, predators are always starved beforehand. However, in natural populations each predator may have different hunger levels. The main goal of the model is to test how different initial hunger levels affect functional response parameters and population dynamics of a predator–prey system.

Sensing: the predator and prey are able to detect each other when they are in the same cell (i.e., in each cell of size 1 cm × 1 cm). Both predator and prey are bounded in the arena of size 100 cm × 100 cm.

Interaction: the predator can feed on prey when they are in the same cell (i.e., successful hunting). Successful hunting involves prey killing, and is followed by prey handling and prey digestion. The eaten prey is replaced by another prey to which a random set of spatial coordinates is assigned.

Stochasticity: the predator and prey do random walks. Random numbers are used to initialize the spatial coordinates of all individuals (to mimic well-mixed initial conditions) and control the movement direction of agents when random walk applies; the distance covered is determined by velocity and depends on allometric scaling. Time is a discrete variable and Bernoulli random numbers are drawn to decide handling time for processing prey.

Observation: at the end of each simulation, the total intake of the predator is recorded (i.e., all prey eaten after 3600 time steps are recorded). This serves to determine the shape of functional responses in the presence of different satiety levels of the predator. Functional responses are used in Rosenzweig–MacArthur models.

Details

1. **Initialization:** at the beginning of each simulation, the grid, the number of prey, and the state variables (e.g., still-handling) are defined. Four scenarios with different levels of initial gut fill (quantified by dimensionless numbers in the interval [0, 1]) are

simulated (i.e., initial gut fill equals to 0, 0.4, 0.8, or randomly chosen). The body mass of predator and prey are 100 mg and 1 mg. For each parameter set, the initial number of prey follows a geometric sequence with a common ratio of 2. Initial values of hunting threshold and still-handling are 0.6 and 0, respectively.

2. **Input data:** the model does not use input data to represent time-varying processes.

3. **Submodels:** the model is implemented in C++. Allometric scaling is adopted for computing predator and prey velocity, gut size, digestion rate, handling time, and probability of successful attack. Velocity, gut size, and digestion rate (X) follow the same equation format, being regulated by body mass (M), a scaling exponent (a_x), and the coefficient x_0 (15.1). Handling time (T_h) (15.2) depends on: the body mass of predator and prey (M_p and M_n), the scaling exponents for predator ($a_{t,p}$) and prey ($a_{t,n}$), and the coefficient t_0. Probability of successful attack (S_a) (15.3) depends on predator–prey body mass ratio (M_p/M_n) and two coefficients (s_1, s_2).

$$X = x_0 M^{a_x} \tag{15.1}$$

$$T_h = t_0 M_p^{a_{t,p}} M_n^{a_{t,n}} \tag{15.2}$$

$$S_a = s_1 \frac{M_p}{M_n} e^{1 - s_2 \frac{M_p}{M_n}} \tag{15.3}$$

Parameter values in these allometric equations are based on empirical data from different sources (see Table 15.2 for references and units).

Table 15.2 Model parameters.

Parameter	Identifier	Unit	Reference
Velocity	V	cm s^{-1}	Peters, 1993
Gut size	G	mg	Ibarrola *et al.*, 2012
Digestion rate	D	mg s^{-1}	Ibarrola *et al.*, 2012
Handling time	T_h	s	(data from) Rall *et al.*, 2012
Probability of successful attack	S_a	–	(data from) Gergs and Ratte, 2009

Note: identifiers, units, and literature sources are summarized for each parameter. In some cases, data from literature have been re-analyzed for model construction (e.g., handling time).

activity are expressed as a function of body mass, showing how the individual-based approach represents an ideal platform for studying trait-based processes. In Box 15.4 we discuss the results obtained by simulating the simple model. Options for gradually refining the structure of the model and issues related to data availability and model parameterization are highlighted.

Box 15.4 Analysis of a Trait-Based IBM: An Example

We show the results of the IBM described in Box 15.3. The model investigates the effect of hunger level of a predator on the functional response. The merits, limits, and future developments of such a model are summarized.

Simulation experiments were carried out by performing five runs for each gut-fill scenario, except for randomly chosen initial gut fill (in this latter case the simulation was repeated 50 times; the random scenario is supposed to be representative of a natural population).

The estimated parameters of functional response change with different initial gut fill (Figure 15.1). The Hill exponent determines the shape of the functional response

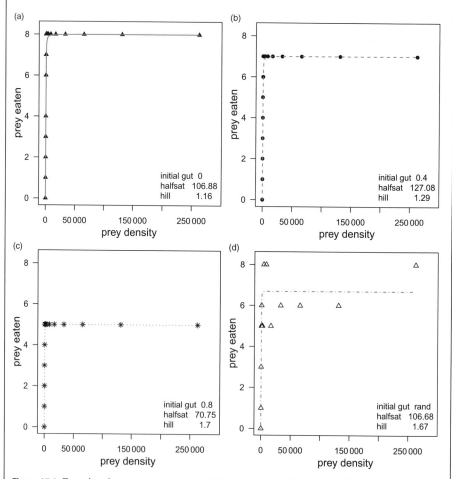

Figure 15.1 Functional response curves for different values of initial gut fill. Curves refer to simulations carried out with the following values of initial gut fill (see the bottom right, initial gut): (a) 0, (b) 0.4, (c) 0.8, and (d) randomly chosen. Functional response parameters are shown at the bottom right of each panel: half saturation density (halfsat) and Hill exponent (hill).

and increases with initial gut fill. In case of randomly chosen initial gut fill the Hill exponent is in the range of the first three scenarios. We then used the functional responses in Rosenzweig–MacArthur models (Williams and Martinez, 2004; Brose, 2008) and analytically solved them. The results show that different levels of initial gut fill influence the dynamics of a predator–prey system (Figure 15.2).

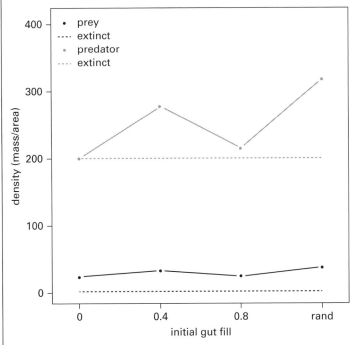

Figure 15.2 Population densities in equilibrium and extinction boundaries. Population densities of prey (black) and predator (gray) are expressed as biomass per area and calculated using functional response parameters in scenarios with different initial gut fullness of the predator: 0, 0.4, 0.8, and randomly chosen. Functional response curves and parameters are shown in Figure 15.1.

In this IBM various dynamical aspects (e.g., handling time and probability of successful attack) are trait based and depend on body mass (see Equations 15.1 to 15.3 in Box 15.3). The allometric rates are based on real empirical data and the IBM described here allows saving a substantial amount of time instead of performing many time-consuming feeding experiments in the laboratory. Results of the IBM can suggest the most interesting scenarios that deserve to be explored with empirical experiments. New components can be progressively added in the IBM to further characterize the behavior of agents without having to rewrite equations (i.e., composability; see Jordán et al., 2011). The IBM of this example could be extended to add other biological processes such as day and night differences, clustering, or hiding behaviors of prey. The main issues are related to model parameterization. It is often complicated to have the complete set of parameters from the system under

investigation. In many cases, the equations used in IBMs include parameters estimated from literature data (as for some of the parameters in the IBM discussed here).
Additional efforts should be dedicated to develop tools for parameter estimation,
sensitivity analysis, and dimensionality reduction in IBMs. We suggest that the
creation of standardized databases on traits and functional traits might also alleviate
concerns related to uncertainties of parameters in IBMs.

15.3   Uses of Individual-Based Models and Potential Applications

Stochastic IBMs may help achieve better understanding for a broad range of ecological
processes and management problems. In this section we discuss possible applications
that involve: (1) strategies for biodiversity conservation of rare species; (2) fishing
management of multispecies assemblages; (3) eco-evolutionary dynamics and plausible
mechanisms for speciation; (4) trait-mediated indirect effects; and (5) the emergence of
synchronized community modules with comparable turnover rates.

   The individual-level representation of species interactions could be useful for biodiversity conservation, especially when related to rarity. IBMs are a helpful tool for
simulating the dynamics of rare and critically endangered species for which a few
dozen of individuals still survive worldwide (e.g., the northern hairy-nosed wombat
in Australia and Javan rhinoceros in Indonesia). In these cases, even the sex ratio can be
decisive for population persistence, and IBMs are thus natural tools for studying rare
species. However, the concept of rarity can also be extended to describe larger metapopulations that live in highly fragmented landscapes (e.g., the coyote is exposed to risks of
local extinction from some landscape patches in Southern California; see Crooks and
Soulé, 1999) or be relevant for a restricted time interval (i.e., certain species can be rare
during particular seasons as in the case of the epiphytic diatom, *Isthmia nervosa*, which
is seasonally rare in winter; see Ruesink, 1998). When population size is small, such as
during a transitory phase or within a restricted spatial scale, demographic variance is
pronounced and can greatly affect population dynamics and species coexistence
(Bolnick *et al.*, 2011).

   In ecology, initial applications of IBMs were aimed at projecting population dynamics
(DeAngelis *et al.*, 1980; Rose *et al.*, 1999), but there is growing interest toward the use of
IBMs to simulate patterns and processes in ecological communities (Livi *et al.*, 2011).
Changes in the functional niche of individuals can be effectively modeled using IBMs,
and this can be particularly relevant with regard to size variation within population
cohorts and life stages (van Nes *et al.*, 2002). Giacomini *et al.* (2009, 2013) analyzed
predation in size-structured fish populations; they considered allometric constraints to
energetic requirement, spatial dynamics, life history, and trophic interactions among
individuals. Functional traits had different relative importance according to environmental conditions, but body growth rate was a key factor influencing community
stability and coexistence (Winemiller, 2017). With IBMs, diversity and variability of
several species traits can be studied simultaneously thus providing an advantage compared to deterministic models of differential equations (see also the "composability"

property that allows one to gradually increase model complexity; Jordán *et al.*, 2011). Thus, IBMs represent an appealing option to traditional aggregated population models. They can be useful for simulating multispecies fish communities, possibly providing support to decision-makers in fisheries management.

Trait variation alters ecological processes (1) through direct effects or (2) because of indirect mechanisms (i.e., heritable intraspecific variation that permits eco-evolutionary dynamics; see Bolnick *et al.*, 2011). The opportunity of including both intraspecific trait variation and population genetics makes IBM an excellent tool for investigating feedback between ecological processes and trait evolution (Melián *et al.*, 2011). We suggest that IBMs could help to clarify the contribution of genetic variance to eco-evolutionary dynamics (i.e., heritability) by distinguishing its role from non-evolutionary processes (e.g., assessing the effects of climate change, seasonality, and stochasticity of climatic conditions). Moreover, IBMs could be used to (1) quantify the strength of trait evolution or (2) analyze the consequences of changes in genetic variation on individual phenotypes and ecological dynamics (e.g., for determining whether the migration of new individuals to a given ecosystem increases genetic variance, thus promoting faster adaptation to environmental change). IBMs are ideal for adding the genetic component to studies of trait variability. Intriguing applications have been developed for testing neutral biodiversity theory (with genetic speciation for individual-based food webs; see Melián *et al.*, 2011) and explicitly including genetic correlations as a constraint to the pathway of multi-trait evolution (Jones *et al.*, 2003). The main concerns are related to the computational feasibility of simulations and empirical data availability for model validation.

Besides heritable trait variation, IBMs can be used for identifying other types of trait-mediated indirect effects (e.g., consequences of age classes and life-history strategies; see Giacomini *et al.*, 2013). In the simulation of a multispecies fish community, ecosystem productivity regulated the feeding activities of fish in a bottom–up manner (Giacomini *et al.*, 2013). Greater productivity increased basal food resources and favored the persistence of species that reached large sizes, including most top predators. In another study, the strength of direct and indirect effects was quantified with individual-based stochastic simulations of the Prince William Sound food web (Gjata *et al.*, 2012). The authors observed a significant prevalence of bottom–up (i.e., prey–predator) mechanisms without finding any difference between indirect effects traveling through short and long paths. Aided by the individual-based stochastic framework, they were also able to investigate variance associated with food-web modules (e.g., mutual consumption, predator–prey, prey–predator, and trophic cascade), with no significant differences detected. Understanding module dynamics may be a bridge between defining local rules and understanding emergent global patterns. The ability to implement stochastic simulations using an individual-based approach is a major advantage of IBMs. Stochasticity was essential for (1) capturing the unexpected importance of weak links in weighted food webs (Scotti *et al.*, 2012), (2) connecting the demographic variation among individuals with population and metacommunity dynamics (Melián *et al.*, 2011), and (3) detecting the spatial heterogeneity of a metapopulation due to different migration rates (Scotti *et al.*, 2013).

Given the relevance of stochastic processes in metapopulation (and metacommunity) dynamics, we emphasize the possible role of IBMs to simulate individual movement through landscape patches. IBMs could be used to capture the emergence of synchronized modules with comparable turnover rates. This feature deserves attention because sub-population dynamics that are asynchronous tend to reduce the risk of local extinction (Hanski *et al.*, 1995). Moreover, environmental drivers can synchronize food-web interactions: using time-series data on Central European lakes, Straile (2002) found that North Atlantic oscillations have advanced the timing of the interaction between algae and herbivores by approximately two weeks. This was due to the faster growth of herbivores in warmer water. The use of IBMs could help to characterize relationships linking environmental and demographic stochasticity (e.g., to understand whether interannual differences in climatic conditions and frequency of extreme events translate into increased variability in the demography of herbivores). At the landscape scale IBMs can also investigate mechanisms of plant species coexistence. Meyer *et al.* (2007) found that shrub cover of the previous year, precipitation, and their interaction explain tree–grass coexistence in semi-arid savannas.

15.4   Challenges and Future Developments

An early criticism of IBMs was related to their poorly documented description (Lorek and Sonnenschein, 1999). This led to difficulties in replicating the results obtained by IBMs and generated skepticism on such a modeling approach. To address this issue, Grimm *et al.* (2006, 2010) introduced the ODD protocol. The ODD protocol supports unambiguous model description and represents an essential tool for promoting the standardization of IBMs. In addition to stressing the need for clarity in descriptions of model formulation, we summarize other aspects that should be taken into account to improve model reliability. First, a common criticism of IBMs for simulating population dynamics is related to their high degree of system specificity (Grimm, 1999). It has been pointed out that there are no straightforward ways to extrapolate general ecological properties from population-level IBMs (i.e., lack of generality). Although IBMs often produce patterns that seem ecologically plausible for a given population within a particular environmental setting, they generally have little application for the same species in a different context, or other species within the same habitat. Second, IBMs produce detailed descriptions of population dynamics but they require very large amounts of empirical data on traits and processes for model construction and validation; consequently, a great deal of time and effort are needed to find and fine tune available data to construct the model. A lack of data hinders model construction, and this problem is particularly acute when considering community (and metacommunity) dynamics (Scotti *et al.*, 2013). Third, the complexity of IBMs poses serious concerns on the interpretation of their outcomes; it is often challenging to quantify uncertainty and error propagation, analyze results, and validate models (Black and McKane, 2012). Lack of generality as well as difficulties in extrapolating system-level properties could be mitigated through mathematical analysis of the master equation (Black and McKane,

2012; see also Box 15.1). Techniques for analysis of the master equation are used to determine macroscopic behavior of the system, and approximation methods can be applied when the master equation cannot be solved analytically (Newman *et al.*, 2004). In general, we emphasize the need for studying IBMs analytically, rather than simply relying on their definition in terms of computer algorithms. Emergence of food-web properties from processes operating at lower levels is a viable alternative to mathematical analysis, especially when the modeling effort is supported by a large dataset (Melián *et al.*, 2011). Modelers could use IBMs for testing general questions in theoretical ecology, bearing in mind that nature will behave differently than their model. Hartvig and Andersen (2013) discussed the role of body size for coexistence between two species with structured populations (i.e., species that undergo ontogenetic trophic niche shifts). In this case, coexistence is regulated by a trade-off between competition (when the size at maturation is similar for individuals of both species) and predator–prey interactions (including cannibalism). The use of a stochastic model illustrated how niche variation promotes coexistence among species competing within a one-dimensional niche space (Yamauchi and Miki, 2009). This effect was restricted to particular situations when the environment was constant, while it generally fostered diversity in the case of niche-independent (i.e., environmental) fluctuations. Scotti *et al.* (2013) found that intermediate rates of spatial dispersal yielded largest metapopulation sizes, a more heterogeneous metapopulation distribution, and more cohesive social structures among individuals of the same species.

Uncertainty and error propagation can be embraced and identified in a rigorous way by combining sensitivity analysis with "composability" (Jordán *et al.*, 2011). If IBMs are built with a modular structure (i.e., by incrementally adding new details to individual-level description), the sensitivity of model outcomes to new components can be tested systematically. With sensitivity analysis, the relevance of each new parameter included in the model can be classified according to its influence on simulations. Knowledge about the relative importance of each parameter on model dynamics could be used to highlight essential data, possibly suggesting the need for ad hoc sampling efforts.

Incomplete datasets are a constant issue for modelers. However, some examples of high-quality online repositories exist and their extensive use shows their potential to stimulate research studies: (1) FishBase is a global database of fish species (it includes information on taxonomy, spatial distribution, biometrics and morphology, behavior and habitats, ecology and population dynamics, as well as reproductive, metabolic, and genetic data; Froese and Pauly, 2000, see also www.fishbase.org); (2) the Knowledge Network for Biocomplexity allows for storing and sharing diverse environmental data (https://knb.ecoinformatics.org); and (3) the Ocean Biogeographic Information System database is an open-access repository for marine biogeographic data (www.iobis.org). We suggest that ecologists should strengthen and standardize the use of public repositories for archiving ecological data. Major efforts should be made to incorporate data reporting on individual-level variability (e.g., genetics, life history, and morphology). Public online repositories for ecological data might have indirect positive impact on individual-based modeling: (1) automatizing input/output data flows could shorten the time needed to build the models; (2) the concerns related to missing data could be buffered through the use of information from other comparable sites (e.g., if no feeding preferences of a fish are known in the study area, they could be approximated through the feeding habit of the same species in similar environmental conditions); and (3) such data may serve as prior information for parameters in the IBM, narrowing down the range of values over which to carry out sensitivity analysis.

We see the creation and management of public datasets as a key future challenge for raising the success of IBMs. Reliable and extensive databases represent an upstream requirement to the phase of model construction also if traditional datasets on traits and spatial distributions cannot solve the issues related to model design (i.e., which processes to include in IBMs). The relevance of different processes could be assessed according to the impact that their inclusion in IBMs has on the emergence of high-level properties (e.g., consequences of migration on species coexistence and diversity in landscape patches). By focusing on the construction step, we have discussed two alternative strategies. IBMs could be developed either by (1) including some elements of individuality in existing deterministic models (i.e., hybrid models) or (2) modeling the system dynamics from scratch, using elementary rules and increasing model complexity in a progressive way (i.e., "composability"). Hybrid modeling could shorten simulation time, representing a significant improvement in terms of computational costs. However, this comes at certain costs, such as adopting proper methodologies to separate the main system into different subsystems (e.g., dynamics of large populations could be simulated in a deterministic fashion, while IBMs could be used to simulate the dynamics of rare species), dynamically switching between different mathematical formalisms (i.e., deterministic vs. stochastic simulations), and checking for the consistency of system parameters between different levels of abstraction. The exclusive use of elementary rules emphasizes the strict bottom–up construction philosophy; its efficacy is impaired by the need for identifying the most significant basic rules (and processes) required for capturing the essence of system dynamics. Algorithms for feature selection could assist modelers in this task. Feature selection could also be complemented by methodologies for determining the optimal level of aggregation (i.e., to find which processes and components should be simulated); this objective could be achieved either through sensitivity analysis of different model configurations or focusing on structural cohesive models of ecological networks (e.g., food webs). Starting from the network structure, modelers could define which species should be lumped together (e.g., on the basis of their trophic similarities) or select sub-parts of the ecological community that are significantly enriched with species of interest (e.g., network modules where species sensitive to the consequences of climate change are over-represented, if the objective of the modeler is simulating the potential impact of global warming).

Besides the challenges related to data availability and methodological aspects (i.e., strategies for building IBMs and methods for entity aggregation), we envision a third layer for IBM development and applications. In the future, more IBMs will be developed to simulate community dynamics (either in the context of food webs or, more generally, of ecological networks of multiple interactions). IBMs represent a powerful tool for

(1) considering the effects of (stochastic) trait variations (Giacomini *et al.*, 2009, 2013); (2) including non-trophic interactions in food-web modeling (see Kéfi *et al.*, 2012 for possible strategies); (3) integrating multiple and concurrent interaction types (i.e., the behavior of individual entities can be coded in terms of probabilities and this allows us to integrate food-web models with other interactions such as facilitation and mutualism; see Jordán *et al.*, 2011); and (4) describing ecosystems at various hierarchical levels (i.e., from population ecology to metacommunities; see Scotti *et al.*, 2013). IBMs could also be used to investigate whether species that belong to the same food-web module (i.e., a sub-part of the network in which most interactions occur within the module rather than with species of other modules) display synchronous dynamics. This could raise interesting questions, such as exploring the role of interspecific variability of feeding habits. For example, it has been shown that generalist species often have populations composed of a wide array of specialized individuals, and this feature is ignored by static food-web representations (i.e., the degree of each species in binary food webs lumps together the contribution of all individuals; see Bolnick *et al.*, 2011) as well as dynamic food-web models based on aggregate variables.

Future developments are partially constrained by the availability of high-quality data. As high-performance computational resources and standardized protocols for model construction supported the rise of IBMs in ecology during the last 20 years (Grimm, 1999; Grimm *et al.*, 2006, 2010), greater availability of data should allow the next generation of models to simulate processes across multiple levels of the ecological hierarchy, including food webs. IBMs would benefit from a new type of database storing details on ecosystem processes in addition to traditional data on traits, abundances, and distributions. Another challenge for the future is related to the validation of the output of IBMs. These models often contain many interacting variables and relationships, thus facing risks of error propagation. Uncertainties associated with the choice of the most significant traits, model structure, and parameter selection can be partially alleviated by standard protocols (Grimm *et al.*, 2006, 2010), but we suggest that IBMs should be tested by considering their ability to reproduce general patterns (Box 15.2). Once this preliminary condition is satisfied, modelers should focus on the mechanistic functioning of specific processes (e.g., trophic cascade and apparent competition involving some of the simulated species), possibly being driven by insights from sensitivity and robustness analyses (Box 15.1). IBMs are usually constructed to simulate specific systems, but ecologists should not abandon the search for general patterns of theoretical population and community ecology (Grimm, 1999).

References

Allesina, S. and Bodini, A. (2004). Who dominates whom in the ecosystem? Energy flow bottlenecks and cascading extinctions. *Journal of Theoretical Biology*, **230**, 351–358.

Allesina, S., Bodini, A., and Bondavalli, C. (2005). Ecological subsystems via graph theory: the role of strongly connected components. *Oikos*, **110**, 164–176.

Barton, R. A., Byrne, R. W., and Whiten, A. (1996). Ecology, feeding competition and social structure in baboons. *Behavioral Ecology and Sociobiology*, **38**, 321–329.

Berlow, E. L., Neutel, A. M., Cohen, J. E., *et al.* (2004). Interaction strengths in food webs: issues and opportunities. *Journal of Animal Ecology*, **73**, 585–598.

Berlow, E. L., Dunne, J. A., Martinez, N. D., *et al.* (2009). Simple prediction of interaction strengths in complex food webs. *Proceedings of the National Academy of Sciences of the United States of America*, **106**, 187–191.

Black, A. J. and McKane, A. J. (2012). Stochastic formulation of ecological models and their applications. *Trends in Ecology and Evolution*, **27**, 337–345.

Bolnick, D. I., Svanbäck, R., Araújo, M. S., and Persson, L. (2007). Comparative support for the niche variation hypothesis that more generalized populations also are more heterogeneous. *Proceedings of the National Academy of Sciences of the United States of America*, **104**, 10075–10079.

Bolnick, D. I., Amarasekare, P., and Araújo, M. S. (2011). Why intraspecific trait variation matters in community ecology. *Trends in Ecology and Evolution*, **26**, 183–192.

Bondavalli, C. and Ulanowicz, R. E. (1999). Unexpected effects of predators upon their prey: the case of the American alligator. *Ecosystems*, **2**, 49–63.

Botkin, D. B., Janak, J. F., and Wallis, J. R. (1972). Some ecological consequences of a computer model of forest growth. *Journal of Ecology*, **60**, 849–872.

Brose, U. (2008). Complex food webs prevent competitive exclusion among producer species. *Proceedings of the Royal Society B: Biological Sciences*, **275**, 2507–2514.

Carnicer, J., Brotons, L., Stefanescu, C., and Penuelas, J. (2012). Biogeography of species richness gradients: linking adaptive traits, demography and diversification. *Biological Reviews*, **87**, 457–479.

Crooks, K. R. and Soulé, M. E. (1999). Mesopredator release and avifaunal extinctions in a fragmented system. *Nature*, **400**, 563–566.

DeAngelis, D. L. and Gross, L. J. (1992). *Individual-Based Models and Approaches in Ecology: Populations, Communities and Ecosystems*. New York: Chapman and Hall.

DeAngelis, D. L., Cox, D. K., and Coutant, C. C. (1980). Cannibalism and size dispersal in young-of-the-year largemouth bass: experiment and model. *Ecological Modelling*, **8**, 133–148.

DeAngelis, D. L., Loftus, W. F., Trexler, J. C., and Ulanowicz, R. E. (1997). Modeling fish dynamics and effects of stress in a hydrologically pulsed ecosystem. *Journal of Aquatic Ecosystem Stress and Recovery*, **6**, 1–13.

Devijver, P. A. and Kittler, J. (1982). *Pattern Recognition: A Statistical Approach (Vol. 761)*. London: Prentice-Hall.

dit Durell, S. E., Stillman, R. A., Caldow, R. W., *et al.* (2006). Modelling the effect of environmental change on shorebirds: a case study on Poole Harbour, UK. *Biological Conservation*, **131**, 459–473.

Eklöf, A., Jacob, U., Kopp, J., *et al.* (2013). The dimensionality of ecological networks. *Ecology Letters*, **16**, 577–583.

Froese, R. and Pauly, D. (eds.) (2000). *FishBase 2000: Concepts, Design and Data Sources (No. 1594)*. WorldFish.

Gamarra, J. G. and Solé, R. V. (2002). Complex discrete dynamics from simple continuous population models. *Bulletin of Mathematical Biology*, **64**, 611–620.

Gergs, A. and Ratte, H. T. (2009). Predicting functional response and size selectivity of juvenile *Notonecta maculata* foraging on *Daphnia magna*. *Ecological Modelling*, **220**, 3331–3341.

Giacomini, H. C., De Marco, Jr., P., and Petrere, Jr., M. (2009). Exploring community assembly through an individual based model for trophic interactions. *Ecological Modelling*, **220**, 23–39.

Giacomini, H. C., DeAngelis, D. L., Trexler, J. C., and Petrere, Jr., M. (2013). Trait contributions to fish community assembly emerge from trophic interactions in an individual-based model. *Ecological Modelling*, **251**, 32–43.

Gjata, N., Scotti, M., and Jordán, F. (2012). The strength of simulated indirect interaction modules in a real food web. *Ecological Complexity*, **11**, 160–164.

Grimm, V. (1999). Ten years of individual-based modelling in ecology: what have we learned and what could we learn in the future? *Ecological Modelling*, **115**, 129–148.

Grimm, V. and Railsback, S. F. (2013). *Individual-Based Modeling and Ecology*. Princeton University Press.

Grimm, V., Revilla, E., Berger, U., *et al.* (2005). Pattern-oriented modeling of agent-based complex systems: lessons from ecology. *Science*, **310**, 987–991.

Grimm, V., Berger, U., Bastiansen, F., *et al.* (2006). A standard protocol for describing individual-based and agent-based models. *Ecological Modelling*, **198**, 115–126.

Grimm, V., Berger, U., DeAngelis, D. L., *et al.* (2010). The ODD protocol: a review and first update. *Ecological Modelling*, **221**, 2760–2768.

Gross, T., Rudolf, L., Levin, S. A., and Dieckmann, U. (2009). Generalized models reveal stabilizing factors in food webs. *Science*, **325**, 747–750.

Hanski, I., Pakkala, T., Kuussaari, M., and Lei, G. (1995). Metapopulation persistence of an endangered butterfly in a fragmented landscape. *Oikos*, **72**, 21–28.

Hartig, F., Calabrese, J. M., Reineking, B., Wiegand, T., and Huth, A. (2011). Statistical inference for stochastic simulation models: theory and application. *Ecology Letters*, **14**, 816–827.

Hartvig, M. and Andersen, K. H. (2013). Coexistence of structured populations with size-based prey selection. *Theoretical Population Biology*, **89**, 24–33.

Hebblewhite, M. and Merrill, E. H. (2007). Multiscale wolf predation risk for elk: does migration reduce risk? *Oecologia*, **152**, 377–387.

Huse, G., Johansen, G. O., Bogstad, B., and Gjøsæter, H. (2004). Studying spatial and trophic interactions between capelin and cod using individual-based modelling. *ICES Journal of Marine Science*, **61**, 1201–1213.

Ibarrola, I., Arambalza, U., Navarro, J. M., Urrutia, M. B., and Navarro, E. (2012). Allometric relationships in feeding and digestion in the Chilean mytilids *Mytilus chilensis* (Hupé), *Choromytilus chorus* (Molina) and *Aulacomya ater* (Molina): a comparative study. *Journal of Experimental Marine Biology and Ecology*, **426–427**, 18–27.

Jeltsch, F., Müller, M. S., Grimm, V., Wissel, C., and Brandl, R. (1997). Pattern formation triggered by rare events: lessons from the spread of rabies. *Proceedings of the Royal Society B: Biological Sciences*, **264**, 495–503.

Jones, A. G., Arnold, S. J., and Bürger, R. (2003). Stability of the G-matrix in a population experiencing pleiotropic mutation, stabilizing selection, and genetic drift. *Evolution*, **57**, 1747–1760.

Jordán, F., Scotti, M., and Priami, C. (2011). Process algebra-based computational tools in ecological modelling. *Ecological Complexity*, **8**, 357–363.

Kéfi, S., Berlow, E. L., Wieters, E. A., *et al.* (2012). More than a meal. . . integrating non-feeding interactions into food webs. *Ecology Letters*, **15**, 291–300.

Kunz, H. and Hemelrijk, C. K. (2003). Artificial fish schools: collective effects of school size, body size, and body form. *Artificial Life*, **9**, 237–253.

Livi, C. M., Jordán, F., Lecca, P., and Okey, T. A. (2011). Identifying key species in ecosystems with stochastic sensitivity analysis. *Ecological Modelling*, **222**, 2542–2551.

Lorek, H. and Sonnenschein, M. (1999). Modelling and simulation software to support individual-based ecological modelling. *Ecological Modelling*, **115**, 199–216.

McCann, K. S. (2000). The diversity–stability debate. *Nature*, **405**, 228–233.

McGill, B. J., Enquist, B. J., Weiher, E., and Westoby, M. (2006). Rebuilding community ecology from functional traits. *Trends in Ecology and Evolution*, **21**, 178–185.

Megrey, B. A. and Hinckley, S. (2001). Effect of turbulence on feeding of larval fishes: a sensitivity analysis using an individual-based model. *ICES Journal of Marine Science*, **58**, 1015–1029.

Melián, C. J., Vilas, C., Baldó, F., *et al.* (2011). Eco-evolutionary dynamics of individual-based food webs. *Advances in Ecological Research*, **45**, 226.

Meyer, K. M., Wiegand, K., Ward, D., and Moustakas, A. (2007). The rhythm of savanna patch dynamics. *Journal of Ecology*, **95**, 1306–1315.

Meyer, K. M., Mooij, W. M., Vos, M., Hol, W. H. G., and van der Putten, W. H. (2009). The power of simulating experiments. *Ecological Modelling*, **220**, 2594–2597.

Moore, J. C., Boone, R. B., Koyama, A., and Holfelder, K. (2014). Enzymatic and detrital influences on the structure, function, and dynamics of spatially-explicit model ecosystems. *Biogeochemistry*, **117**, 205–227.

Morita, K. and Yokota, A. (2002). Population viability of stream-resident salmonids after habitat fragmentation: a case study with white-spotted charr *Salvelinus leucomaenis* by an individual based model. *Ecological Modelling*, **155**, 85–94.

Neutel, A. M., Heesterbeek, J. A., and de Ruiter, P. C. (2002). Stability in real food webs: weak links in long loops. *Science*, **296**, 1120–1123.

Newman, T. J., Ferdy, J. B., and Quince, C. (2004). Extinction times and moment closure in the stochastic logistic process. *Theoretical Population Biology*, **65**, 115–126.

Odum, H. T. (1956). Efficiencies, size of organisms, and community structure. *Ecology*, **37**, 592–597.

Otto, S. B., Rall, B. C., and Brose, U. (2007). Allometric degree distributions facilitate food-web stability. *Nature*, **450**, 1226–1229.

Parry, H. R., Topping, C. J., Kennedy, M. C., Boatman, N. D., and Murray, A. W. (2013). A Bayesian sensitivity analysis applied to an agent-based model of bird population response to landscape change. *Environmental Modelling and Software*, **45**, 104–115.

Peters, R. H. (1993). *The Ecological Implications of Body Size*. Cambridge University Press.

Pimm, S. L. (1982). *Food Webs*. London: Chapman and Hall.

Pitt, W. C., Box, P. W., and Knowlton, F. F. (2003). An individual-based model of canid populations: modelling territoriality and social structure. *Ecological Modelling*, **166**, 109–121.

Pratchett, M. S., Wilson, S. K., Graham, N. A. J., *et al.* (2009). Coral bleaching and consequences for motile reef organisms: past, present and uncertain future effects. In *Coral Bleaching*, eds. M. J. H. van Oppen and J. M. Lough, Berlin Heidelberg: Springer, pp. 139–158.

Pugnaire, F. I., Haase, P., and Puigdefábregas, J. (1996). Facilitation between higher plant species in a semiarid environment. *Ecology*, **77**, 1420–1426.

Rademacher, C., Neuert, C., Grundmann, V., Wissel, C., and Grimm, V. (2004). Reconstructing spatiotemporal dynamics of Central European natural beech forests: the rule-based forest model BEFORE. *Forest Ecology and Management*, **194**, 349–368.

Railsback, S. F. and Harvey, B. C. (2002). Analysis of habitat-selection rules using an individual-based model. *Ecology*, **83**, 1817–1830.

Rall, B. C., Brose, U., Hartvig, M., *et al.* (2012). Universal temperature and body-mass scaling of feeding rates. *Philosophical Transactions of the Royal Society B: Biological Sciences*, **367**, 2923–2934.

Reynolds, C. W. (1987). Flocks, herds and schools: a distributed behavioral model. *Computer Graphics*, **21**, 25–34.

Rose, K. A., Rutherford, E. S., McDermot, D. S., Forney, J. L., and Mills, E. L. (1999). Individual-based model of yellow perch and walleye populations in Oneida Lake. *Ecological Monographs*, **69**, 127–154.

Ruel, J. J. and Ayres, M. P. (1999). Jensen's inequality predicts effects of environmental variation. *Trends in Ecology and Evolution*, **14**, 361–366.

Ruesink, J. L. (1998). Variation in per capita interaction strength: thresholds due to nonlinear dynamics and nonequilibrium conditions. *Proceedings of the National Academy of Sciences of the United States of America*, **95**, 6843–6847.

Saeys, Y., Inza, I., and Larrañaga, P. (2007). A review of feature selection techniques in bioinformatics. *Bioinformatics*, **23**, 2507–2517.

Scotti, M., Gjata, N., Livi, C. M., and Jordán, F. (2012). Dynamical effects of weak trophic interactions in a stochastic food web simulation. *Community Ecology*, **13**, 230–237.

Scotti, M., Ciocchetta, F., and Jordán, F. (2013). Social and landscape effects on food webs: a multi-level network simulation model. *Journal of Complex Networks*, **1**, 160–182.

Shin, Y. J. and Cury, P. (2001). Exploring fish community dynamics through size-dependent trophic interactions using a spatialized individual-based model. *Aquatic Living Resources*, **14**, 65–80.

Shugart, H. H. (1984). *A Theory of Forest Dynamics: The Ecological Implications of Forest Succession Models*. New York: Springer-Verlag.

Souissi, S., Seuront, L., Schmitt, F. G., and Ginot, V. (2005). Describing space-time patterns in aquatic ecology using IBMs and scaling and multi-scaling approaches. *Nonlinear Analysis: Real World Applications*, **6**, 705–730.

Stouffer, D. B. and Bascompte, J. (2011). Compartmentalization increases food-web persistence. *Proceedings of the National Academy of Sciences of the United States of America*, **108**, 3648–3652.

Straile, D. (2002). North Atlantic Oscillation synchronizes food-web interactions in central European lakes. *Proceedings of the Royal Society B: Biological Sciences*, **269**, 391–395.

Tofts, C. (1993). Algorithms for task allocation in ants. (A study of temporal polyethism: theory.) *Bulletin of Mathematical Biology*, **55**, 891–918.

van Nes, E. H., Lammens, E. H., and Scheffer, M. (2002). PISCATOR, an individual-based model to analyze the dynamics of lake fish communities. *Ecological Modelling*, **152**, 261–278.

Williams, R. J. and Martinez, N. D. (2004). Stabilization of chaotic and non-permanent food-web dynamics. *European Physical Journal B-Condensed Matter and Complex Systems*, **38**, 297–303.

Winemiller, K. O. (2017). Food web dynamics when divergent life history strategies respond to environmental variation differently: a fisheries ecology perspective. In *Adaptive Food Webs: Stability and Transitions of Real and Model Ecosystems*, ed. P. C. de Ruiter, V. Wolters, K. S. McCann, and J. C. Moore. Cambridge: Cambridge University Press, pp. 305–323.

Yamauchi, A. and Miki, T. (2009). Intraspecific niche flexibility facilitates species coexistence in a competitive community with a fluctuating environment. *Oikos*, **118**, 55–66.

Yodzis, P. and Winemiller, K. O. (1999). In search of operational trophospecies in a tropical aquatic food web. *Oikos*, **87**, 327–340.

16 Empirical Methods of Identifying and Quantifying Trophic Interactions for Constructing Soil Food-Web Models

Amber Heijboer, Liliane Ruess, Michael Traugott, Alexandre Jousset, and Peter C. de Ruiter

16.1 Introduction

Food-web models, which depict the trophic relationships between organisms within a community, form a powerful and versatile approach to study the relationships between community structure and ecosystem functioning. Although food-web models have recently been applied to a wide range of ecological studies (Memmott, 2009; Sanders et al., 2014), such approaches can be greatly improved by introducing high-resolution trophic information from empirical studies and experiments that realistically describe topological structure and energy flows (de Ruiter et al., 2005). Over the last decades major technological advances have been made in empirically characterizing trophic networks by describing, in detail, the connectedness and flows in food webs. Existing empirical techniques, such as stable isotope probing (SIP) (Layman et al., 2012), have been refined and new approaches have been created by combining methods, e.g., combining Raman spectroscopy or fatty acid analysis with SIP (Ruess et al., 2005a; Li et al., 2013). These empirical methods can provide insight into different aspects of food webs and together form an extensive toolbox to investigate trophic interactions. It is crucial to recognize the potential and limitations of a range of empirical approaches in order to choose the right method in the design of empirically based food-web studies.

Empirically based food webs are generally classified according to the type of input information that is required. In the following lines we will provide an overview of four types of food-web model: connectedness webs, semi-quantitative webs, energy-flow webs, and functional webs. Paine (1980) introduced three of those webs, which are widely accepted and applied in food-web studies across ecosystems. We propose to add a fourth type of empirically based food web, the semi-quantitative web. All of these food webs have the same basic structure, but the conceptual webs differ in the type of trophic information they describe and represent (Figure 16.1). *Connectedness webs* (Figure 16.1a) define the basic structure of a food web by describing the food-web connections *per se*. The food web consists of species connected by arrows, visualizing the direction of matter and energy flows. Due to the complexity of interactions, taxa are often lumped into feeding guilds, whose members have a similar trophic level and diet, and comparable function in the food web. *Semi-quantitative webs* (Figure 16.1b) differ

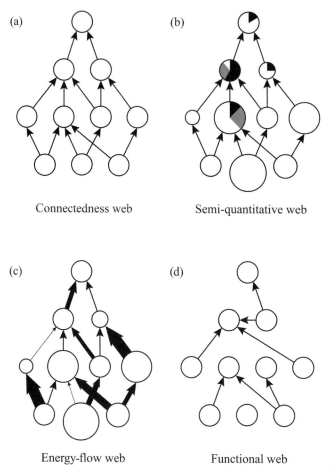

Figure 16.1 Four different types of empirically based food web, three of them (a, c, and d) as defined by Paine (1980). (a) Connectedness web visualizing qualitative feeding relationships; feeding guilds of species connected by arrows, visualizing the direction of matter and energy flows. (b) Semi-quantitative web visualizing the abundances/biomasses of species (groups) and the frequency of feeding interactions. (c) Energy-flow web visualizing the biomasses of species (groups), connected by vectors visualizing the amount of material and energy flow. (d) Functional web visualizing the effect of species manipulation on the population size of other species in the food web, highlighting the functional role of species including non-trophic effects (Figure adapted from Paine, 1980 and Selakovic et al., 2014).

from connectedness webs in that they contain quantitative information on the abundances or biomass of the food resources and the consumers. Additionally, they provide a semi-quantitative measure to feeding relationships such as the frequency of feeding interactions between (groups of) species, which can provide a good proxy for trophic interaction strength (Baker et al., 2014). *Energy-flow webs* (Figure 16.1c) aim to assess nutrient flows quantitatively. They thus contain quantitative biomass information, and the feeding relationships are fully quantified by vectors summarizing both the direction and the amount of material and energy flows. Finally, *functional webs* (Figure 16.1d) are characterized by experimental manipulation (e.g., species removal, resource exclosure or amendment) to assign the functional role of species. In this chapter, we examine how empirical trophic information can be used to construct such empirically based food webs.

There is an increasing demand for empirical data and experiments to improve empirically based food webs, e.g., in the fields of ecosystem engineering (Sanders *et al.*, 2014) and applied ecology (Memmott, 2009). Especially in the field of soil food-web ecology, trophic interactions are still poorly understood, since it is difficult to define the trophic roles of belowground organisms. Ideally, a food web would be described at high taxonomic resolution, representing the present species and their interspecific trophic interactions. However, species-level approaches are often difficult or impractical since the assignment of trophic interactions to specific soil organisms, and especially soil microbes, is hard due to their small size, difficult extraction from their habitat, and huge taxonomic and functional diversity (Eggers and Jones, 2000). These facts particularly hamper the determination of microbial-faunal food-web interactions. To circumvent this limitation, ecologists aggregate groups of species into feeding guilds based on diet and life-history characteristics. This approach has been especially applied to soils, the class of ecosystems on which this chapter will focus. New techniques introduced in this chapter provide better information on biological diversity and functionality within the food web, and in this way improve significantly the level of detail and realism in empirically based food webs.

Molecular and biochemical techniques that have been developed over the last decades have opened new windows in (soil) food-web ecology to study which food sources sustain specific soil heterotrophs and to assess small-scale activity and trophic links in food webs. These empirical techniques offer a great opportunity to further unravel feeding guilds and to create predictive models that are able to provide answers on specific soil organisms and their roles in trophic networks. However, a clear overview of what type of information empirical methods can offer for soil food-web modeling is missing. This chapter provides an up-to-date overview of the molecular and biochemical methods applicable in trophic soil studies. We start from a theoretical perspective of soil food webs to present the rationales and requirements of empirically based food-web modeling. Thereafter, an overview is provided on existing and upcoming empirical techniques and the research fields where they can be applied. The chapter concludes with recommendations on how to use the outcomes from empirical techniques to improve and fully exploit the use of empirically based soil food-web models as well as suggestions for future research priorities. Our goal is to provide an overview of state-of-the-art empirical approaches that can be used to create and improve food-web models described in other chapters of this book, with a special focus on food-web structure and flows.

16.2   From Empirical Data to Food-Web Models

Transferring the outcomes of empirical studies into useful data for soil food-web modeling remains one of the main challenges when combining empirical and theoretical food-web research. This paragraph provides an overview of the types of data resulting from molecular and biochemical trophic studies that are essential to create soil food-web models. The information is categorized by the different types of food webs as introduced above: connectedness web, semi-quantitative web, energy-flow web, and functional web.

16.2.1   Connectedness Webs

Connectedness webs are the basic form of food webs, often visualized by the use of a simple box-and-arrow diagram (Figure 16.1a). The boxes and arrows represent respectively (groups of) species and the direction of their trophic interactions. This type of information can give important insights with respect to the stability and complexity of a food-web model. What does an empirically based connectedness web require? The construction of a connectedness web from empirical data requires at least information on the presence of organisms (binary) and the direction of feeding interactions: "who eats whom?" arranged per feeding guild, but preferably on a more highly resolved taxonomic level. By constructing a connectedness web, valuable information on horizontal (i.e., within a trophic level) and vertical (i.e., between trophic levels) food-web diversity is gained.

16.2.2   Semi-Quantitative Webs

Where connectedness webs only provide information on which species and feeding relationships are present, semi-quantitative webs also give information on the frequency of the trophic interactions (f_j) and the population sizes of present taxa on both the prey and consumer side (B_j) (Figure 16.1b). Frequency of interaction is based on the proportion of predators that contain remains of a specific prey, and can be calculated as:

$$f_j = \frac{F_{tot}}{N} \tag{16.1}$$

In this equation, F_{tot} refers to the total number of interactions detected and N is the total number of consumers for which the average number of interactions is calculated. This type of semi-quantification is not often used in soil ecology, but has shown great potential in aquatic food webs and host–parasitoid webs. In the past, semi-quantitative measurements of detection frequency have been criticized as a rough way of quantification, since they give no information on the biomass of prey eaten by a predator (Hyslop, 1980). Yet it gives a robust and interpretable description of the dietary composition of organisms, as reviewed by Baker et al. (2014). Determining the frequency of interactions makes it possible to gain quantifiable data from presence/absence data of the dietary composition that can be relatively easily gained compared to empirical methods that are necessary to quantitatively describe energy and carbon flows.

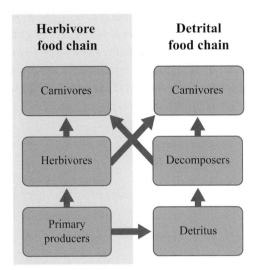

Figure 16.2 The two major soil food chains: (left) the herbivore food chain (primary-production based); (right) the detrital food chain (decomposition based).

16.2.3   Energy-Flow Webs

While the links in connectedness and semi-quantitative webs provide a measure of food-web complexity and structure, the arrows and boxes in energy-flow webs are weighted in terms of population sizes and the rate with which material is transferred from the resources to the consumers (Figure 16.1c). To construct energy-flow webs, information is therefore required on the biomass of resources and soil organisms, as well as the rates of flows of matter among the resources and species within the food web (Moore and de Ruiter, 2012).

Energy flux rates can be calculated by means of the detrital food-web model (DFWM), originally proposed by O'Neill (1969) and subsequently applied to various food webs from native (Hunt *et al.*, 1987; Berg *et al.*, 2001; Schröter *et al.*, 2003) and agricultural soils (de Ruiter *et al.*, 1993). The steady-state assumption underlies the DFWM model, i.e., that the production of a population balances the rate of loss through natural death and predation. The model was originally applied to decomposition-based (detrital) food-web models, relying on dead organic matter or detritus as a source of energy. However, the model can also be applied well to primary production-based (herbivory) food chains in soil, where carbon and nutrients are originating from living plant biomass. Figure 16.2 combines those two important food chains in soil, showing a great potential of linking them together via soil food-web modeling.

Using the steady-state assumption, the feeding rates can be calculated as:

$$F_j = \frac{d_j B_j + M_j}{e_j} \tag{16.2}$$

In this equation, F_j refers to the feeding rate of group j ($kg_{Carbon}ha^{-1}yr^{-1}$), d_j to its specific death rate (yr^{-1}), B_j to the average annual population size ($kg_{Carbon}ha^{-1}$), M_j to the death rate due to predation ($kg_{Carbon}ha^{-1}yr^{-1}$), e_j to the energy conversion efficiency. For predators feeding on more than one prey type, the feeding rate per prey type (F_{ij}) is calculated assuming that the predator feeds on a prey type according to the relative abundance of this prey type and on prey preferences:

$$F_{ij} = \frac{w_{ij}B_i}{\sum_{k=1}^{n} w_{kj}B_k} * F_j \qquad (16.3)$$

In this equation, w_{ij} refers to the preference of predator j for prey i over its other prey types and n is the number of trophic groups. k is the numerator of the summation over all (n) trophic groups. The model calculates the feeding rates in a top–down sequence. It starts with the top predators, for which only natural death is assumed, i.e. $M_j = 0$. Hence in this step all necessary parameter values are available. Then the model proceeds working backwards to the lowest trophic levels. The M_j values then become available through the calculations in the former steps.

What does this DFWM approach require in terms of input data? In addition to the obtained values for population biomasses, only the preferences w_{ij} and the energy conversion efficiencies e_j are required. The energy conversion efficiency is known for most species of soil organisms, but the species diets and preferences are still largely unknown on soil-species level. New techniques (introduced below), combined with controlled laboratory experiments, will provide information on the relative preferences w_{ij} of predator j for prey i up to a higher taxonomic resolution. Such detailed information will make it possible to construct energy-flow web models with an increased level of detail and realism.

16.2.4   Functional Webs

Where energy-flow webs focus on the biomasses of feeding guilds and their interconnecting rates of energy transfer, functional webs originally describe the influence of the species manipulation (not only feeding interactions) on the population sizes of the remaining species (MacArthur, 1972; Paine, 1992) (Figure 16.1d). The relative impact of one species on another within functional webs is described by the interaction strength of a species relationship, comprising both trophic and non-trophic interactions. Where the removal of a weakly interacting species will not have large consequences for food-web structure, the removal of a strongly interacting species can have severe consequences. There are, however, multiple definitions of interaction strengths, and these vary between empiricists and theoreticians (Berlow et al., 2004; Moore and de Ruiter, 2012). In this chapter, we will restrict our discussion to one of the theoretical approaches of interaction strengths by showing how the use of only trophic information can provide interaction strengths, by making use of a community (Jacobian) matrix as was first formulated by May (1974).

The strength of the trophic links is described by the interaction strength. In the theoretical approach, interaction strengths are defined as the "per capita" (or "per biomass") effects of the populations upon one another (May, 1972; Pimm, 1982). These values of interaction strengths are used as the entries of a Jacobian matrix representation of the food webs, which can be used to analyze the stability of the food webs. This procedure is based on the Lotka–Volterra approach, in which the dynamics of the trophic groups are described as:

$$\frac{dX_i}{dt} = -d_i X_i - \sum_{j=1}^{n} c_{ij} X_i X_j + \sum_{h=1}^{n} e_i c_{hi} X_h X_i \tag{16.4}$$

In this equation, X represents the population sizes of the trophic groups, c_{ij} the coefficient of interaction between group i and group j, and d and e have a similar meaning as in Eq. (16.2). Sometimes, modifications of this equation are used, for example the equation describing the dynamics of soil organic matter (Moore et $al.$, 1993). From these Lotka–Volterra equations, one can derive interaction strengths as the partial derivatives of the differential equations near equilibrium:

$$\alpha_{ij} = \left(\frac{\delta \dot{X}_i}{\delta X_j}\right)^* \tag{16.5}$$

Here α_{ij} denotes the interaction strength imposed by group j on group i, $\dot{X}_t = \frac{dX_i}{dt}$, and * denotes at equilibrium. Hence if we obtain empirically based values for α_{ij} we can build a Jacobian community matrix, analyze stability, and evaluate the importance of food-web components to food-web stability. What is required to obtain empirically based values for the interaction strengths α_{ij}?

If we take the partial derivatives of Eq. (16.2), we obtain the following formulae for interaction strength:

$$\alpha_{ij} = -c_{ij} X_i^* \tag{16.6}$$

for the per $capita$ effect of predator j on prey i and

$$\alpha_{ji} = e_j c_{ij} X_j^* \tag{16.7}$$

for the per $capita$ effect of prey i on predator j.

Values of c_{ij} are difficult to obtain, given the measure of this parameter expressed as "per amount per time." However, we can rewrite the formula for interaction strengths by using the following substitution, i.e., replace the terms $c_{ij} X_i^* X_j^*$ in the Lotka–Volterra equation by the feeding rates, F_{ij}, as estimated by the DFWM, and assuming that the equilibrium population sizes, X_i^*, X_j^*, are represented by the observed population sizes, B_i and B_j. Then we obtain:

$$\alpha_{ij} = -c_{ij}X_i^* = -\frac{F_{ij}}{B_j} \tag{16.8}$$

for the *per capita* effect of predator j on prey i

$$\alpha_{ji} = e_{ij}c_{ij}X_j^* = \frac{e_jF_{ij}}{B_i} \tag{16.9}$$

for the *per capita* effect of prey i on predator j.

From these reformulations of interaction strengths, we see that they can be directly derived from observed biomasses, calculated feeding rates, and known energy-conversion efficiencies. Then we are able to evaluate the stability of the Jacobian matrix and hence of the food web. This, however, does not provide an empirically based functional web. To fill in this gap, Neutel *et al.* (2002) proposed to look at food-web stability in terms of lengths and weights of trophic interaction loops. A trophic interaction loop describes a pathway of interactions (i.e., not feeding rates) from a species through the web back to the same species without visiting the species more than once; hence a loop is a closed chain of trophic links. An example of such a loop is the soil microbial loop (e.g., Bonkowski, 2004), where carbon is allocated from plant roots to rhizosphere bacteria, which are linked to micro-faunal predators, mainly protozoa. Grazing of protozoa on soil bacteria causes nutrients to be released, which are taken up by plants. These processes are running mainly between bacteria and protozoa with no higher trophic food-web levels involved, thereby forming a trophic interaction loop. Such loops may vary in length; the *loop length* being the number of trophic groups visited, and in weight, the *loop weight* being the geometric mean of the interaction strengths in the loop, defined as the per capita effects of the Jacobian matrices. The maximum of all loop weights is an indicator of food-web stability. Looking at the weights of trophic loops has a twofold meaning. First of all, it allows us to better understand the patterns in interaction strengths underlying stability. Second, it identifies food-web components that are key to food-web stability, which is close to the functional webs derived from manipulation experiments (sensu Paine 1980, 1992), while they can be calculated by using the obtained values for interaction strength.

To summarize, all required parameters for creating empirically based soil food-web models are compiled in Table 16.1. Not all of those parameters are necessary to measure and qualify in empirical experiments, since some of the parameters can be derived as explained above. The following paragraphs will therefore focus on those parameters (f_j, F_j, B_j, and w_{ij}) that cannot be derived from formulae (such as parameters Mj, c_{ij}, and α_{ij}) or are already known on species level (such as parameters e_j and d_j), but need real-world quantification in order to create empirically based soil food-web models.

16.3   Incorporating Empirical Information into Food-Web Models

The empirical design of food-web research largely depends on the questions addressed and, correspondingly, the type of food-web model one wishes to construct. We therefore

Table 16.1 Overview of parameters used to create a connectedness, semi-quantitative, energy-flow, or functional web. Each food-web model is described by the given parameters for that specific type of model, including all of the above.

	Name	Description	Unit
Connectedness web		Presence of organisms	[a]
		Feeding relationships: "who eats whom"	[a]
Semi-quantitative web	f_j	Frequency of interaction	[a]
	B_j	Population size	$kg_{Carbon}ha^{-1}$
Energy-flow web	F_j	Feeding rates	$kg_{Carbon}ha^{-1}yr^{-1}$
	d_j	Specific death rate	yr^{-1}
	M_j	Death rate due to predation[b]	$kg_{Carbon}ha^{-1}yr^{-1}$
	e_j	Energy conversion efficiency	[c]
	w_{ij}	Prey preferences	[a]
Functional web	c_{ij}	Coefficient of interaction[d]	$kg_{Carbon}^{-1}yr^{-1}$

Note: units are based on their frequent use in detrital-based food-web models. Not all parameters need to be measured empirically, since some parameters (e.g., coefficient of interactions and interaction strengths) can be derived from equations.
[a] Dimensionless; [b] Derived from Eqs. (16.1) and (16.2) (assuming $M_j = 0$ for top predators); [c] Dimensionless if units for prey and consumer population sizes are the same; [d] Derived from Eqs. (16.7) and (16.8).

provide a comprehensive overview of the available empirical techniques in function of the type of information they can provide and how they can be combined with different theoretical models.

16.3.1 Connectedness Webs

Soil food webs usually consist of diverse communities of species arranged along the subterranean herbivore and detrital food chains (Scheu, 2002). Disentangling the feeding interactions in these complex communities is not easy as the opaque habitat makes it impossible to observe who is feeding on whom directly. Furthermore, the small sizes of the interacting species, liquid feeding, and extra-oral digestion complicate the use of a morphology-based assessment of trophic interactions. Soil food-web research has thus greatly benefited from the development of techniques that overcome these hurdles, such as molecular biological methods and biomarker approaches.

16.3.1.1 DNA-Based Techniques on Dietary Samples
DNA-based approaches can identify the DNA of food remains at high specificity and sensitivity, thereby opening up new possibilities to examine feeding relationships and how organisms are trophically connected in natural communities (Pompanon *et al.*,

2012; Traugott *et al.*, 2013). In brief, food DNA is extracted from the dietary sample, specific short fragments of it are amplified using the polymerase chain reaction (PCR) technique, and the resulting PCR products are identified either by sequencing or by their length, which is indicative for a specific taxon (i.e., diagnostic PCR see below). Molecular techniques can be used for analyzing almost all of the trophic links expressed within soil food webs, including trophic interactions between mesofauna (e.g., Heidemann *et al.*, 2014a) and macrofauna (Juen and Traugott, 2007; Lundgren and Fergen, 2014) – the only requirement is that amplifiable food DNA is present in the dietary sample. Moreover, these techniques are not restricted to analyzing the consumption of fresh food, but it is also possible to detect the DNA of scavenged prey (Juen and Traugott, 2005) and decaying plant material that has been consumed (Wallinger *et al.*, 2013). It is important to point out that this approach does not take into account what has been metabolized by a consumer but merely detects what has been consumed.

A variety of sample types have been used for the molecular study of feeding interaction. The simplest way is to identify food remains, either directly taken from the consumer (e.g., masticated prey from wasps; Kasper *et al.*, 2004) or collected in the environment, is by examining the DNA present within them. In most cases, however, DNA of consumed food present within either feces, the gut content in the consumer, and regurgitates is examined. Feces are usually employed to study vertebrate food choice while in invertebrates typically whole-body DNA extracts are used to retrieve food DNA from the gut content (King *et al.*, 2008). As such, gut content analysis is lethal to the consumer. In situations where these post-mortem approaches are not appropriate (i.e., in rare or protected species or when multiple feeding events of individual consumers are of interest) fecal pellets (Boyer *et al.*, 2011) and regurgitates (Waldner and Traugott, 2012) provide a means to obtain dietary samples from invertebrates non-invasively. Fecal pellets of invertebrate decomposers can also be used to assign them to their producer at the species level using PCR techniques (Seeber *et al.*, 2010), extending the possibilities for molecular profiling of soil faunal communities (Andújar *et al.*, 2015).

Aside from analyzing dietary samples for trophic information, molecular methods can also be extremely valuable to identify the consumer via diagnostic PCR or DNA barcoding (Wirta *et al.*, 2014), providing food webs that are taxonomically highly resolved on both the consumer/host and food/parasitoid sides. The molecular methods used for analyzing trophic interactions can be classified into two basic approaches of (1) diagnostic PCR and (2) sequence-based identification (Traugott *et al.*, 2013). In the former, taxon-specific primers are employed to amplify short fragments of food DNA followed by electrophoretic separation and visualization. The amplification of these specific fragments diagnoses the presence of the targeted DNA in the sample. The level of taxonomic identification can be set according to the needs of the study, i.e., earthworm prey can be identified either generally on a family level using earthworm-group specific primers (e.g., Harper *et al.*, 2005) or down to species and even lineage level (King *et al.*, 2010). Primer pairs for different prey taxa can be mixed together in multiplex PCRs allowing us to test dietary samples for several targets in parallel, strongly increasing the efficiency of the analysis (Harper *et al.*, 2005; Sint *et al.*, 2012). In sequence-based food identification, primers are employed that target short DNA fragments of a wide range of food taxa (Pompanon *et al.*, 2012). The resulting PCR products are subjected to high-throughput sequencing using next-generation sequencing (NGS) techniques. After quality checking and sorting these sequences, they can be assigned tentative identities using either public sequence databases or specific reference sequences (Pompanon *et al.*, 2012).

While diagnostic PCR is ideally suited for rapid and low-cost screening of a large numbers of samples, this approach will only allow one to detect the a-priori selected taxa targeted by the primers. Although several multiplex PCR assays can be used in parallel to detect several tens of taxa within the consumer's diet, this approach becomes inefficient when a broad range of dietary items needs to be examined. In such cases, sequence-based food detection via NGS is advantageous as it allows us to explore the diet spectrum of generalist consumers or to obtain dietary information on a population level using a pool of individual dietary samples (Deagle *et al.*, 2009). Sequenced-based diet identification, however, can be hampered by potentially poor coverage of sample sequences in public databases. Other problems include excessive co-amplification of consumer DNA, which requires the application of blocking primers (Vestheim and Jarman, 2008), the lack of suitably conserved regions for primer bindings sites to allow for amplification of fragments suitable for barcoding a broad range of target taxa (Deagle *et al.*, 2014), and the comparably high costs for testing large numbers of individual samples. The decision of which approach to use largely depends on the nature of the research project and the questions addressed: diagnostic PCR is typically employed for assessing the detection frequency via individual-based dietary analysis using larger numbers of samples, which would currently be too costly to be processed by NGS. Next-generation sequencing-based food identification, on the other hand, is most efficient for pooled dietary/consumer samples, allowing us to obtain an in-depth picture of the diet of a specific consumer on a population level. Moreover, it is important to consider that the detection of food DNA in dietary samples does not necessarily confirm that the specific food taxon was digested and metabolized into the consumer's tissue. For example, nematodes can have a short bacterial residence time in the intestine, which means that not all prokaryote cells are digested (Ghafouri and McGhee, 2007). We advise the reader to consult the latest reviews, such as King *et al.* (2008), Pompanon *et al.* (2012), Symondson (2012), Traugott *et al.* (2013), and Clare (2014), for more detailed information.

16.3.1.2 Lipid Analysis

Feeding interactions are generally drawn from primary producers via herbivores to carnivores, suggesting that population development at any given trophic level is limited by populations in the trophic level below. Such bottom–up control is widespread in soil food webs, as decomposers lack influence on the amount of organic matter, e.g., litter, feces, or necromass, available as basal resource. In soil, the bacterial decomposition pathway is predominantly resource controlled, while the fungal pathway faces greater top–down effects mainly mediated by micro-arthropods (Scheu *et al.*, 2005). A useful way to assess the carbon flux in food webs is the analysis of lipids, namely phospholipid fatty acids (PLFAs) and neutral lipid fatty acids (NLFA). This *in situ* method allows one to assign animal diets and carbon transfer in cryptic systems such as soil food webs (Ruess and Chamberlain, 2010; Traugott *et al.*, 2013), providing information on feeding relationships.

In soil, total fatty acid analyses have successfully been used as a qualitative measure for carbon assimilation in primary and secondary decomposers. To date, most such feeding habits studies have focused on micro- and mesofauna, i.e., nematodes (Chen *et al.*, 2001; Ruess *et al.*, 2002, 2004) and Collembola (Ruess *et al.*, 2004, 2005b; Chamberlain *et al.*, 2005). Only recently have higher trophic levels of the soil food web been considered, such as by taking into account lipid patterns in centipedes and spiders (Haubert *et al.*, 2009; Pollierer *et al.*, 2010; Ferlian *et al.*, 2012). An underlying assumption related to the use of lipids as trophic biomarkers is a concept referred to as "dietary routing," which denotes the transfer of fatty acids from the diet into consumer tissue without modification. This process is well known in vertebrates and used, for example, in food chemistry to assign the origin of dairy products (Molkentin and Giesemann, 2007). Moreover, it was applied frequently in the herbivore food chain of marine ecosystems to monitor predator–prey interactions and carbon flow between phytoplankton and zooplankton (e.g., Müller-Navarra *et al.*, 2000; Stübing *et al.*, 2003; Pond *et al.*, 2006). The basic principle of this approach is that organisms at the base of the food web are capable of synthesizing specific fatty acids, which do not occur in the metabolism of organisms at higher trophic levels, and therefore can be used as biomarkers. Two general types of marker fatty acids have to be distinguished: (1) absolute markers the consumer cannot synthesize, and only appear in the lipid profile when it has fed on the respective diet, and (2) relative markers that are components of consumer metabolism but are additionally highly accumulated from the diet (Ruess *et al.*, 2005b). Ruess and Chamberlain (2010) provide a useful review on method application, advantages, and drawbacks in fatty acids as a tool in soil food-web analysis.

Table 16.2 provides an overview of fatty acid biomarkers for, respectively, the herbivore and detrital food chain (Figure 16.2) based on current knowledge. However, future research is likely to reveal additonal biomarkers. In plant and algal tissues marker fatty acids occur in both the phospholipid and neutral lipid fractions (Ruess *et al.*, 2007; Buse *et al.*, 2013), yet for soil microbes and fungi these are predominantly found in the phospholipids of membranes (White *et al.*, 1996; Zelles, 1999). One exception is the

Table 16.2 Fatty acid biomarkers useful for determination of carbon flows in herbivore and detrital food chains, respectively.

Herbivore food chain	Detrital food chain
Plants	**Bacteria**
18:1ω9	iso/anteiso – gram-positive
18:3ω3,6,9	cyclopropyl – gram-negative
18:3ω6,9,12	
Algae	**Fungi**
16:2ω6,9	16:1ω5 – arbuscular mycorrhiza
16:3ω3,6,9	18:2ω6,9 – ectomycorrhiza and saprotrophs

arbuscular mycorrhiza fungi (AMF), as the marker 16:1ω5 is common in phospholipids of bacteria and fungi, whereas in the neutral lipids it is exclusive to AMF (Ngosong *et al.*, 2012). There is one notable difference in fatty acid transfer, i.e., the direction of feeding interactions: in the classical trophic cascade of the herbivore food chain it is unidirectional, whereas marker fatty acids derived from litter or debris of primary producers also fuel the detrital food chain. An additional blur between both food chains arises from cross-feeding at higher trophic levels. Generally, tritrophic transport occurs as shown for bacteria-based (bacteria–nematodes–Collembola) and fungal-based (fungi–nematodes–Collembola) food chains (Ruess *et al.*, 2004; Chamberlain *et al.*, 2005) up to top predators such as centipedes (Pollierer *et al.*, 2010). Thus a marker fatty acid can indicate feeding on a specific diet or predation on prey also feeding on this diet. On one hand, this can hamper assignment of a binary link, but on the other hand it allows one to follow the feeding relationships across multiple trophic levels of the food web.

One additional fact makes the application of fatty acids in soil food-web studies particularly attractive: after feeding and ingestion of the diet, marker fatty acids are predominantly routed into the neutral lipids of consumers (Ruess *et al.*, 2004; Haubert *et al.*, 2006). As only some actinobacteria possess neutral lipids to a significant extent (Alvarez and Steinbüchel, 2002), the detection of microbial marker fatty acids in consumer storage fat enables distinguishing between viable microbes, in the gut or on the body surface, and the microbial tissue assimilated by the animal grazer. That goes beyond the detection of a bacterial DNA via gut content analysis, as it assigns bacterial carbon allocated in consumer biomass. This is a great advantage in decomposer systems, where bacteria form a basal resource, as it offers the possibility to link microbial and faunal food webs.

16.3.2   Semi-Quantitative Webs

Molecular prey detection, as discussed earlier in this chapter, usually provides an absence/presence matrix for the food DNA detected within a sample. With the use of these data, one can establish a detection frequency of specific food taxa (f_j) as a proxy for the strength of trophic interactions, which is necessary to construct a semi-quantitative food web. With the use of detection frequency of feeding relationships, it will allow us to assess the most important diet resources for the consumer assuming that frequently consumed foods are more important for sustaining the consumer than rarely consumed ones (King *et al.*, 2008; Heidemann *et al.*, 2014b). Moreover, it is important to consider that the analysis of a dietary sample provides a snap-shot picture of the recently consumed food. Therefore the quality and robustness of the trophic data generated is positively correlated with the number of dietary samples analyzed. Factors such as food and consumer identity (Greenstone *et al.*, 2007; Waldner *et al.*, 2013; Wallinger *et al.*, 2013) can affect post-feeding food DNA detection intervals and need to be considered when analyzing and interpreting molecularly derived trophic data (for reviews see King *et al.*, 2008; Pompanon *et al.*, 2012; Symondson, 2012; Traugott *et al.*, 2013; Greenstone *et al.*, 2014).

Although quantitative PCR (qPCR) allows one to estimate the number of food DNA molecules present within a sample (Zhang *et al.*, 2007), it is of little help for quantifying the number of prey items consumed or estimating the meal size from gut content samples. This is because a small and a big meal digested for a short and a long time, respectively, can easily provide a similar number of food DNA molecules (King *et al.*, 2008). As it is usually unknown in a field-collected consumer when a feeding event occurred before it was caught, the number of food DNA molecules cannot be used to estimate meal size or number of prey consumed. However, in feces, which are an end product of digestion, qPCR can provide a semi-quantitative estimate of diet composition (Deagle and Tollit, 2007).

16.3.3   Energy-Flow and Functional Webs

Biomarker and molecular-based techniques are not only useful to qualify feeding interactions, but also to quantify those interactions in terms of energy flow. The following paragraphs give an overview of state-of-the-art techniques that are currently used to gather energy-flow data for soil ecosystems, complemented with future perspectives on cutting-edge techniques to study energy food-web models empirically. Table 16.1 shows that the construction of functional webs does not require additional empirical measurements compared to energy-flow webs, since the necessary parameters can be derived from equations as introduced above in Section 16.2. Techniques discussed in this subsection will therefore provide data for both the construction of energy-flow and functional webs.

16.3.3.1   Stable Isotope Probing

Stable isotope probing (SIP) is one of the empirical approaches that has been upcoming in food-web ecology over the past decade. It combines the use of molecular techniques with the detection of stable isotopes (e.g., ^{13}C and ^{15}N), making it possible to trace flows of matter in food webs on the smallest scale in all trophic levels of the food web. The main idea of SIP is that organisms feeding on a specific stable isotope-enriched substrate can be traced by probing the fate of these stable isotopes into cellular biomarkers of active consumers. The big advantage of SIP is the possibility to observe the link between identity and (metabolic) functioning *in situ*, which can give important information about both the structure and flows in food webs.

The term "stable isotope probing" was used for the first time by Radajewski *et al.* (2000), describing the tracing of a ^{13}C-enriched carbon source into microbial DNA. However, the labeling of metabolically active organisms can be followed by tracing labeled biomarkers such as DNA, RNA, and fatty acids (i.e., DNA–, RNA–, and FA–SIP). Lipids were among the first compounds to be measured after labeling due to the ease in their GC analysis. First approaches using stable isotope-labeled substrates were done by the use of microbial lipid analysis; Boschker *et al.* (1998) traced the fate of ^{13}C-enriched acetate and methane incorporated into PLFAs to link specific environmental processes to the identity of the microbial groups involved. In the context of food-web studies, PLFA and NLFA extraction in combination with stable isotope probing can give important information about the rate (F_j) and fate (w_{ij}) of feeding interactions, and is highly complementary to other culture-independent methods (Maxfield and Evershed, 2011). Moreover, as the ^{13}C value of a specific fatty acid is dependent on the carbon pool it is derived from, i.e., de novo synthesis or dietary routing, a ^{13}C label introduced into a food web can be used to examine the route of the ^{13}C-pulse through the web by FA–SIP (Ruess and Chamberlain, 2010).

The DNA-based stable isotope probing (DNA–SIP) technique separates stable isotope labeled "heavy" DNA from unlabeled "light" DNA by density gradient centrifugation, after which the DNA can be identified both on a high taxonomic resolution. The possibility to derive high-quality taxonomic information regarding the labeled community is the main advantage of DNA–SIP over FA–SIP. On the other hand, DNA–SIP is less sensitive because low DNA synthesis rates limit the enrichment. Therefore a higher level of stable isotope enrichment compared to FA–SIP is required in order to get sufficient labeling. The low sensitivity of DNA–SIP applies to a lesser extent for the method of RNA-based stable isotope probing (RNA–SIP). RNA–SIP was reported for the first time by Manefield *et al.* (2002) who studied the degradation of ^{13}C labeled phenol in an industrial bioreactor. RNA synthesis rates are higher compared to DNA, which strengthens the sensitivity of RNA–SIP. Another advantage of RNA is the large amount of information it can give on both the phylogenetic (rRNA) and the functional (mRNA) gene diversity of the labeled organism. Therefore a combination of RNA–SIP (i.e., identity and functional gene diversity) with FA–SIP (i.e., carbon flux) is ideal to link the identity and function of key biota in the food web.

Over the last decade, SIP was mainly used to trace defined microbial groups responsible for the primary degradation of specific substrates (Radajewski *et al.*, 2000). However, labeling occurs also through secondary assimilation of labeled substrates. This so-called "cross-feeding" already occurred during the DNA–SIP experiments of Radajewski *et al.* (2000). Cross-feeding was often seen as an issue, since organisms that are not involved in the primary assimilation of a substrate become labeled. However, the phenomenon of cross-feeding has great potential to increase our insight into trophic interactions among multiple trophic levels in the same habitat (Friedrich, 2011). Experiments designed in time series provide an excellent opportunity to capture the dynamic nature of carbon flow through soil food webs by studying cross-feeding patterns (Drigo *et al.*, 2010).

Since its first introduction, SIP has been combined with different molecular techniques, and new applications continue to be introduced (Abraham, 2014). Huang *et al.* (2007) studied the incorporation of ^{13}C in microbial cells by combining stable-isotope Raman microscopy with fluorescence *in situ* hybridization (FISH), called Raman–FISH. This method can gain increased insight into the incorporation of carbon in organisms on the cellular level, increasing the detailed understanding of energy flows in food webs. Also the combination of SIP with Raman spectroscopy has led to increased insight in the carbon flows of food webs on the individual microbial level, by achieving a better understanding of energy flows and metabolic pathways in the context of complex food webs (Li *et al.*, 2013). One of the upcoming applications of SIP is to determine both identity and function by targeting organisms with the use of so-called Chip–SIP. This approach uses phylogenetic microarrays and secondary ion mass spectronomy (NanoSIMS) as a high-sensitivity and high-throughput method to test genomics-generated hypotheses about biogeochemical function in any natural environment (Mayali *et al.*, 2012). Although the use of Chip–SIP is still in its developing phase, it is definitely a promising tool to study the functioning of energy flows within food webs.

16.3.3.2   Quantitative Fatty Acid Signature Analysis

A prospect for the future in soil food-web analysis is the application of quantitative fatty acid signature analysis (QFASA), which makes it possible to assign feeding rates (F_j) to predator diets. QFASA was recently developed as a tool to estimate predator diets in marine mammals such as grey seals, polar bears, and seabirds (Iverson *et al.*, 2004; Thiemann *et al.*, 2008; Williams and Buck, 2010). At present, its wide-scale application is hampered by the lack in information on specific (lipid) pathways and current meta-bolism (life cycle, starvation) for most soil animal grazers and predators. Nevertheless, for Collembola, metabolism and pattern of lipids are well known in regard to food quality, environmental factors, and biotic constraints such as life cycle and starvation (Holmstrup *et al.*, 2002; Haubert *et al.*, 2004, 2008; van Dooremalen and Ellers, 2010), which offers a starting point for establishment of QFASA in soil food webs. Other promising areas for further investigations using fatty acids are comparison of resources and fecal profiles to assign both consumption on, as well as propagation of, dietary organisms (Buse *et al.*, 2014) or to disentangle trophic from mutualistic processes in plant–microbe–fauna interactions (Ngosong *et al.*, 2014).

16.3.3.3   Nematode Community Analysis

The empirical method of nematode faunal analysis uses nematode assemblages to assign conditions to the soil micro-food web such as major decomposition pathways, nutrient status, or disturbance (Ferris *et al.*, 2001). Exploring the soil nematode community structure provides an empirical method within the framework of feeding guilds and can be used to determine the interaction of the component species within a guild, but also to ask questions regarding the interactions between the various guilds that compose the larger community. Thus nematode communities can serve as a model for general processes in the soil food web and as a tool to link microbial and faunal food webs. The latter is particularly important as food-web models still lack sufficient quantitative empirical data on carbon and energy fluxes between microbes and fauna.

The soil micro-food web consists of basal organic resources derived from photo-autotrophs (e.g., plant litter and root exudates), the microflora (bacteria and fungi), and the micro- and mesofauna that feed upon the microflora or on each other (Wardle *et al.*, 1998). Within this web, nematodes (Figure 16.3) are the most abundant and diverse multicellular organisms with millions of individuals and up to 200 species per square meter (Yeates, 2010). Moreover, nematodes have established functional groups at each trophic level and feed on bacteria, fungi, algae, or roots as well as on other microfauna (Yeates *et al.*, 1993). Due to these diverse biological interactions, nematodes hold a central position in both bottom–up and top–down controlled food webs (Ferris, 2010a; Yeates, 2010). In particular those nematodes that graze on bacteria and fungi

Table 16.3 Nematode faunal analysis and the respective community indices according to Bongers (1990), Freckman (1988), and Ferris *et al.* (2001) used to assign food-web conditions.

Food-web condition	Nematode faunal analysis
Bottom–up effect of resources	Density and biomass of trophic groups
Decomposition pathways and energy flux	Channel index, fungal-to-bacterial feeder ratio
Enrichment and structure	Enrichment index, structure index
Disturbance or maturity	Maturity index
Natural or managed conditions	Plant parasite index

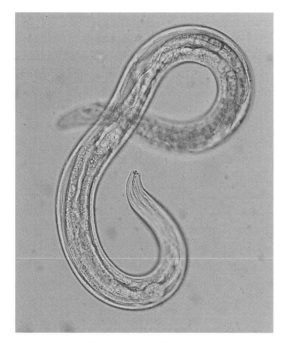

Figure 16.3 *Acrobeloides buetschlii*, an opportunistic nematode species and common bacterial feeder in the soil. (Picture by Veronika Bartel.)

play important roles in influencing soil microbial biomass, activity, and mineralization processes (Bardgett *et al.*, 1999; Griffiths *et al.*, 1999). Thus although nematodes represent only a small amount of biomass in the soil, their key position in the micro-food web impacts on ecosystem-level processes such as energy flow and nutrient cycling (e.g., Yeates *et al.*, 2009; Ferris, 2010a; Neher, 2010).

Nematode abundance (B_j), diversity, and effect on soil processes make nematode assemblages useful indicators of food-web conditions. By addressing the changes in horizontal as well as vertical diversity the nematode faunal analysis concept allows determining structure and function of the food web (Table 16.3). Nematode community indices based on life-history traits or trophic groups are applied to assign soil decomposition pathways (channel index) as well as enrichment and structure (enrichment and structure index) of food webs in grassland, arable, and forest soil (Ferris *et al.*, 2001; Ruess, 2003; Ruess and Ferris, 2004). In particular, the channel index is a useful tool to determine the major fluxes of carbon and energy through the soil food web in terms of feeding rates (F_j). Plant effects, i.e., aboveground impact, are expressed in the plant parasite index (PPI; Bongers, 1990), whereas the maturity index (MI; Bongers, 1990) is a measure for disturbance and successional stage. Moreover, the establishment of functional groups in both primary production-based (herbivore) as well as decomposition-based (detrital) food chains can allow for a linkage between these two fundamental pathways (see Figure 16.2).

Only recently, metabolic footprints of nematodes were introduced as metrics for the magnitude of services provided by feeding guilds in the soil food web (Ferris, 2010b). This approach takes advantage of the standardized morphometric characteristics used in nematode taxonomic description. This comprehensive database facilitates assessment of body volume and weight, which can be converted to carbon metabolism by prescribed coefficients. Until now, soil food-web models generally apply abundance data as a proxy, partly combined with values on respiration or functional response. However, this does not take into account the relationship between prey and predator body sizes, which was reported to systematically differ across habitats and consumer types (Brose *et al.*, 2006). Body-size relationship is an important factor for interaction strength patterns in food webs and thus affects resilience and stability (Jonsson, 2014). Including metabolic footprints into soil food-web analysis, e.g., by making the energy conversion efficiency (e_j) body-size dependent, provides an opportunity for a more detailed interpretation of energy-flow webs and improves the accuracy of quantitative models. Overall, nematode faunal analysis provides a useful tool for assessing the importance of the different energy channels (i.e., bacterial, fungal, plant), as well as food chains (i.e., herbivore, detrital), in soil food webs and thus for determining food-web functioning and energy flux.

16.3.3.4   Controlled Laboratory Experiments

Since soil food webs are often difficult to tease apart because of the large number of intertwined and potentially confounding effects, controlled laboratory experiments offer an ideal complement to field experiments. Controlled small-scale experiments offer the possibility to follow in detail single interactions under defined conditions (von Berg *et al.*, 2012). The results can be then straightforwardly connected to genetic and physiological data on the studied organisms (Brose *et al.*, 2008; Brose, 2010; Neidig *et al.*, 2011). In connection with empirical and theoretical predictions (Baiser *et al.*, 2010), laboratory experiments can serve as a basis for parameterizing more complex food webs (Brose *et al.*, 2008; Brose, 2010). Novel approaches of molecular marking of food items also hold great promise for experiments in such controlled environments, including the quantification of food consumption in functional-response experiments. For example, Mora *et al.* (2014) showed that silica particles containing encapsualted DNA can be used to label food items with the label being detectable for several days post consumption. Moreover, the label gets transferred across trophic levels and includes the possibility to quantify prey uptake using real-time PCR.

Laboratory feeding choice experiments are commonly used to study feeding preferences of soil organisms (w_{ij}). Even generalist predators show distinct feeding preferences. Selective feeding depends on several parameters, such as body mass ratio, and total and relative abundance (Kalinkat *et al.*, 2011). Further, prey properties, such as the presence of defensive structures, and active selection processes by the predators can influence which prey will be consumed first (Jousset, 2012). Prey selection has profound impacts on the stability and evolutionary dynamics of the whole community. In controlled systems, it is possible to mix preys at different relative and total abundances and thus accurately determine under which conditions predators will eat which prey. Microcosms can be set up to mimic several environment types such as plant-root systems (Jousset *et al.*, 2009), lakes (Jürgens and Simek, 2000), or litter (Vuvic-Pestic *et al.*, 2010). Prey abundance in the diet can be tracked with a vast array of available methods. For instance, prey staining and imaging allow us to count remaining, unconsumed prey (Jousset *et al.*, 2009) or even vacuole content for protozoa (Jezbera *et al.*, 2005). Experimental work can subsequently be combined with DNA-based methods and field data, for example for determination of active predation versus scavenging of dead prey (Heidemann *et al.*, 2011).

16.4   Discussion and Conclusions

Soil food-web ecologists have just started to exploit (molecular and biochemical) empirical tools to study subterranean feeding networks. Winemiller and Layman (2005) concluded that empirical food-web research is lagging behind theoretical research. The overview provided in this chapter illustrates many of the advancements that have been made since then in empirically testing soil food-web models. However, a big challenge remains in bringing theoretical and empirical food-web scientists together to take full advantage of the range of possibilities that empirical methods offer for food-web modeling. Table 16.4 provides an overview of the methods discussed and the type of information each can provide to empirically based soil food-web models.

The choice of specific empirical methods will largely depend on the type of questions asked in empirical studies, as well as the properties of food webs that one wishes to obtain. It is therefore especially important to note that the properties of food webs vary depending on the techniques used to reconstruct a food-web model (Wirta *et al.*, 2014). It is essential to have a good consideration of multiple empirical techniques, as displayed in the conceptual diagram of Figure 16.4.

Figure 16.4 provides an overview of the types of empirical methods that are suitable to study specific trophic levels, or the soil food web as a whole. Most methods presented allow for the detection of trophic connections in most of the trophic levels of the soil food web. Exceptions are PLFA–SIP analyses (focusing on the microbial part of the soil food web), but combined with different types of lipid analyses, lipids could be traced further into the soil food web. Nematode community analyses focus mainly on the

Table 16.4 Overview of the empirical methods discussed. For each method it is specified what type of necessary data the method can provide in terms of empirically based soil food-web modeling.

Methods	Type of data
DNA-based techniques on dietary samples	Presence/absence of specific organisms
	Frequency of interaction (f_j)
qPCR on fecal samples	Frequency of interaction, potentially diet composition (f_j)
Lipid analyses	Frequency of interaction (f_j)
Fatty acid analyses on fecal pellets	Frequency of interaction (f_j), assimilation efficiency
Quantitative fatty acid signatures (QFASA)	Feeding rates (F_j)
Stable isotope probing (SIP)	Feeding rates (F_j), prey preferences (w_{ij})
FA–SIP	*Carbon flux*
DNA–SIP	*Identity*
RNA–SIP	*Phylogenetic (rRNA) and functional (mRNA) gene diversity*
Nematode community analyses	Feeding rates (F_j)
	Population size (B_j)
Food-choice experiments	Prey (substrate) preferences (w_{ij})

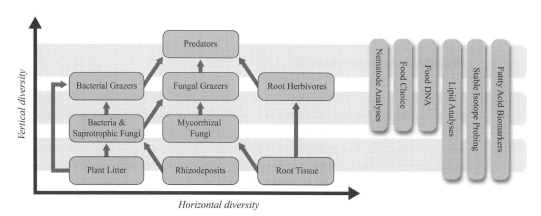

Figure 16.4 Conceptual diagram of a soil food web, showing the main feeding guilds and major pathways of carbon and energy. Horizontal diversity refers to diversity within trophic levels and vertical diversity refers to diversity between trophic levels. The right side of the diagram displays the proposed empirical methods over the range of trophic levels they can be applied to.

higher trophic levels of the soil food web, acting as "connectors" between the unexploited microbial part of the soil food web and the faunal food part of the food web that has been described in much more detail. Identifying food remains with the help of DNA-based techniques offers a high specificity and sensitivity, and opens up entire new possibilities to examine trophic interactions. Especially in combination with controlled feeding experiments, this method is of great value to determine exact feeding interactions, as well as feeding preferences; a combination of results that is of high value for establishing empirically based soil food-web models.

In the history of soil food-web modeling, there has been a strong divergence between soil food web models that were based upon primary producers (herbivory based) and models that relied on dead organic matter (detrital based) (see Figure 16.2). Those two types of food webs have been studied in separate areas of research due to large differences in empirical approaches. We expect that emerging methods, as described in this chapter, will yield large advances in bringing production-based and detrital-based food chains closer together and even link the two fields of research. Not only are the new arising empirical methods able to link different types of food webs, the snap-shot dietary information provided by, for example, DNA-based methods is also ideally suited for assessing the temporal dynamics in soil food webs, a topic that remains largely unexplored. Existing soil food-web models could also be further improved by including host–parasitoid relationships. The molecular techniques discussed also offer an effective way to study endoparasitism by detecting, for instance, the DNA of parasites and parasitoids within the host sample (Agustí *et al.*, 2005; Gariepy *et al.*, 2008; Traugott *et al.*, 2013; Hrček and Godfray, 2015). Pooling empirical techniques into combined detrital- and herbivory-based food webs with host–parasitoid food webs, has therefore a great potential to better understand the detailed interactions within soil food webs, as well as the functioning of soil food webs as a whole. Only recently, combining stable isotope analysis of bulk tissues as well as fatty acids gave new insight into the allocation and transfer of plant-derived carbon through a food web in an arable soil. The study of Pausch *et al.* (2015) revealed that saprotrophic fungi, not bacteria, are most active in these processes, challenging previous views on the dominance of bacteria in root carbon dynamics in arable soil.

Recent advances in empirical methods will open up new possibilities to study important areas of food-web model research, e.g., the link between microbial diversity and the functioning of soil food webs or the link between nematode diversity and their impact on soil food-web structure. Emerging empirical techniques, as described in this chapter, can bring a much higher resolution into food-web models that will certainly revolutionize our view of soil food webs. The high specificity at which trophic links can be identified raises the characterization of trophic niches for soil invertebrates to a completely new level, allowing a critical evaluation of the commonly used grouping of specific species into feeding guilds. Although empirically based soil food-web models that make use of feeding guilds have proven their value and utility (e.g., Hunt *et al.*, 1987; de Ruiter *et al.*, 1993; Berg *et al.*, 2001; Schröter *et al.*, 2003) an increased level of detail will be of great value for predictive models that focus on spatial and temporal patterns, as well as models that highlight the importance of specific parts of the soil food web, such as the soil microbial community, by bringing in greater phylogenetic resolution.

References

Abraham, W.-R. (2014). Applications and impacts of stable isotope probing for analysis of microbial interactions. *Journal of Appled Microbiology and Biotechnology*, **98**, 4817–4828.

Agustí, N., Bourguet, D., Spataro, T., *et al.* (2005). Detection, identification and geographical distribution of European corn borer larval parasitoids using molecular markers. *Molecular Ecology*, **14**, 3267–3274.

Alvarez, H. M. and Steinbüchel, A. (2002). Triacylglycerols in prokaryotic microorganisms. *Journal of Applied Microbiology and Biotechnology*, **60**, 367–376.

Andújar, C., Arribas, P., Ruzicka, F., *et al.* (2015). Phylogenetic community ecology of soil biodiversity using mitochondrial metagenomics. *Molecular Ecology*, **24**(14), 3603–3617. doi:10.1111/mec.13195.

Baiser, B., Russell, G. J., and Lockwood, J. L. (2010). Connectance determines invasion success via trophic interactions in model food webs. *Oikos*, **119**, 1970–1976.

Baker, R., Buckland, A., and Sheaves, M. (2014). Fish gut content analysis: robust measures of diet composition. *Fish and Fisheries*, **15**, 170–177.

Bardgett, R. D., Cook, R., Yeates, G. W., and Denton, C. S. (1999). The influence of nematodes on below-ground processes in grassland ecosystems. *Plant Soil*, **212**, 23–33.

Berg, M., de Ruiter, P. C., Didden, W. A. M., *et al.* (2001). Community food web, decomposition and nitrogen mineralisation in a stratified Scots pine forest soil. *Oikos* **94**, 130–142.

Berlow, E. L., Neutel, A.-M., Cohen, J. E., *et al.* (2004). Interaction strengths in food webs: issues and opportunities. *Journal of Animal Ecology*, **73**, 585–598.

Bongers, T. (1990). The maturity index: an ecological measure of environmental disturbance based on nematode species composition. *Oecologica*, **83**, 14–19.

Bonkowski, M. (2004). Protozoa and plant growth: the microbial loop in soil revisited. *New Phytologist*, **162**, 617–631.

Boschker, H. T. S., Nold, S. C., Wellsbury, P., *et al.* (1998). Direct linking of microbial populations to specific biogeochemical processes by 13C-labelling of biomarkers. *Nature*, **392**, 801–805.

Boyer, S., Yeates, G. W., Wratten, S. D., Holyoake, A., and Cruickshank, R. H. (2011). Molecular and morphological analyses of faeces to investigate the diet of earthworm predators: example of a carnivorous land snail endemic to New Zealand. *Pedobiologia (Jena)*, **54**, 153–158.

Brose, U. (2010). Body-mass constraints on foraging behaviour determine population and food-web dynamics. *Functional Ecology*, **24**, 28–34.

Brose, U., Jonsson, T., Berlow, E. L., *et al.* (2006). Consumer-resource body-size relationships in natural food webs. *Ecology*, **87**, 2411–2417.

Brose, U., Ehnes, R. B., Rall, B. C., *et al.* (2008). Foraging theory predicts predator–prey energy fluxes. *Journal of Animal Ecology*, **77**, 1072–1078.

Buse, T., Ruess, L., and Filser, J. (2013). New trophic biomarkers for Collembola reared on algal diets. *Pedobiologia (Jena)*, **56**, 153–159.

Buse, T., Ruess, L., and Filser, J. (2014). Collembola gut passage shapes microbial communities in faecal pellets but not viability of dietary algal cells. *Chemoecology*, **24**, 79–84.

Chamberlain, P. M., Bull, I. D., Black, H. I. J., Ineson, P., and Evershed, R. P. (2005). Fatty acid composition and change in Collembola fed differing diets: identification of trophic biomarkers. *Soil Biology and Biochemistry*, **37**, 1608–1624.

Chen, J., Ferris, H., Scow, K. M., and Graham, K. J. (2001). Fatty acid composition and dynamics of selected fungal-feeding nematodes and fungi. *Comparative Biochemistry and Physiology Part B: Biochemistry and Molecular Biology*, **130**, 135–144.

Clare, E. L. (2014). Molecular detection of trophic interactions: emerging trends, distinct advantages, significant considerations and conservation applications. *Evolutionary Applications*, **7**(9), 1144–1157. doi:10.1111/eva.12225.

Deagle, B. E. and Tollit, D. J. (2007). Quantitative analysis of prey DNA in pinniped faeces: potential to estimate diet composition? *Conservation Genetics*, **8**, 743–747.

Deagle, B. E., Kirkwood, R., Jarman, S. N. (2009). Analysis of Australian fur seal diet by pyrosequencing prey DNA in faeces. *Molecular Ecology*, **18**, 2022–2038.

Deagle, B. E., Jarman, S. N., Coissac, E., Pompanon, F., and Taberlet, P. (2014). DNA metabarcoding and the cytochrome c oxidase subunit I marker. *Biology Letters*, **10**, dx.doi.org/10.1098/rsbl.2014.0562.

de Ruiter, P. C., Moore, J. C., Zwart, K. B., *et al.* (1993). Simulation of nitrogen mineralization in the below-ground food webs of two winter wheat fields. *Journal of Applied Ecology*, **30**, 95–106.

de Ruiter, P. C., Wolters, V., and Moore, J. C. (2005). *Dynamics Food Webs: Multispecies Assemblages, Ecosystem Development and Environmental Change*. San Diego, CA: Academic Press.

Drigo, B., Pijl, A. S., Duyts, H., *et al.* (2010). Shifting carbon flow from roots into associated microbial communities in response to elevated atmospheric CO_2. *Proceeding of the National Academy of Sciences of the USA*, **107**, 10938–10942.

Eggers, T. and Jones, T. H. (2000). You are what you eat… or are you? *Trends in Ecology and Evolution*, **15**, 265–266.

Ferlian, O., Scheu, S., and Pollierer, M. M. (2012). Trophic interactions in centipedes (Chilopoda, Myriapoda) as indicated by fatty acid patterns: variations with life stage, forest age and season. *Soil Biology and Biochemistry*, **52**, 33–42.

Ferris, H. (2010a). Contribution of nematodes to the structure and function of the soil food web. *Journal of Nematology*, **42**, 63–67.

Ferris, H. (2010b). Form and function: metabolic footprints of nematodes in the soil food web. *European Journal of Soil Biology*, **46**, 97–104.

Ferris, H., Bongers, T., and De Goede, R. G. M. (2001). A framework for soil food web diagnostics: extension of the nematode faunal analysis concept. *Applied Soil Ecology*, **18**, 13–29.

Freckman, D. W. (1988). Bacterivorous nematodes and organic-matter decomposition. *Agriculture Ecosystems and Environment*, **24**, 195–217.

Friedrich, M. W. (2011). Trophic interactions in microbial communities and food webs traced by stable isotope probing of nucleic acids. In *Stable Isotope Probing and Related Technologies*, ed. J. C. Murrell and A. S. Whiteley, American Society for Microbiology Press, pp. 203–232.

Gariepy, T., Kuhlmann, U., Gillott, C., and Erlandson, M. (2008). A large-scale comparison of conventional and molecular methods for the evaluation of host–parasitoid associations in non-target risk-assessment studies. *Journal of Applied Ecology*, **45**, 708–715.

Ghafouri, S. and McGhee, J. (2007). Bacterial residence time in the intestine of *Caenorhabditis elegans*. *Nematology*, **9**, 87–91.

Greenstone, M. H., Rowley, D. L., Weber, D. C., Payton, M. E., and Hawthorne, D. J. (2007). Feeding mode and prey detectability half-lives in molecular gut-content analysis: an example with two predators of the Colorado potato beetle. *Bulletin of Entomological Research*, **97**, 201–209.

Greenstone, M. H., Payton, M. E., Weber, D. C., and Simmons, A. M. (2014). The detectability half-life in arthropod predator–prey research: what it is, why we need it, how to measure it, and how to use it. *Molecular Ecology*, **23**, 3799–3813.

Griffiths, B. S., Ritz, K., Ebblewhite, N., and Dobson, G. (1999). Soil microbial community structure: effects of substrate loading rates. *Soil Biology and Biochemistry*, **31**, 145–153.

Harper, G. L., King, R. A., Dodd, C. S., *et al.* (2005). Rapid screening of invertebrate predators for multiple prey DNA targets. *Molecular Ecology*, **14**, 819–827.

Haubert, D., Häggblom, M. M., Scheu, S., and Ruess, L. (2004). Effects of fungal food quality and starvation on the fatty acid composition of *Protaphorura fimata* (Collembola). *Comparative Biochemistry and Physiology Part B: Biochemistry and Molecular Biology*, **138**, 41–52.

Haubert, D., Häggblom, M. M., Langel, R., Scheu, S., and Ruess, L. (2006). Trophic shift of stable isotopes and fatty acids in Collembola on bacterial diets. *Soil Biology and Biochemistry*, **38**, 2004–2007.

Haubert, D., Häggblom, M. M., Scheu, S., and Ruess, L. (2008). Effects of temperature and life stage on the fatty acid composition of Collembola. *European Journal of Soil Biology*, **44**, 213–219.

Haubert, D., Birkhofer, K., Fließbach, A., *et al.* (2009). Trophic structure and major trophic links in conventional versus organic farming systems as indicated by carbon stable isotope ratios of fatty acids. *Oikos*, **118**, 1579–1589.

Heidemann, K., Scheu, S., Ruess, L., and Maraun, M. (2011). Molecular detection of nematode predation and scavenging in oribatid mites: laboratory and field experiments. *Soil Biology and Biochemistry*, **43**, 2229–2236.

Heidemann, K., Hennies, A., Schakowske, J., *et al.* (2014a). Free-living nematodes as prey for higher trophic levels of forest soil food webs. *Oikos*, **123**, 1199–1211.

Heidemann, K., Ruess, L., Scheu, S., and Maraun, M. (2014b). Nematode consumption by mite communities varies in different forest microhabitats as indicated by molecular gut content analysis. *Experimental and Applied Acarology*, **64**, 49–60.

Holmstrup, M., Hedlund, K., and Boriss, H. (2002). Drought acclimation and lipid composition in *Folsomia candida*: implications for cold shock, heat shock and acute desiccation stress. *Journal of Insect Physiology*, **48**, 961–970.

Hrček. J. and Godfray. H. C. J. (2015). What do molecular methods bring to host–parasitoid food webs? *Trends in Parasitology*, **31**, 30–35.

Huang, W. E., Stoecker, K., Griffiths, R., *et al.* (2007). Raman–FISH: combining stable-isotope Raman spectroscopy and fluorescence in situ hybridization for the single cell analysis of identity and function. *Environmental Microbiology*, **9**, 1878–1889.

Hunt, H. W., Coleman, D. C., Ingham, E. R., *et al.* (1987). The detrital food web in a shortgrass prairie. *Biology and Fertility of Soils*, **3**, 57–68.

Hyslop, E. J. (1980). Stomach contents analysis: a review of methods and their application. *Journal of Fish Biology*, **17**, 411–429.

Iverson, S. J., Field, C., Bowen, W. D., and Blanchard, W. (2004). Quantitative fatty acid signature analysis: a new method of estimating predator diets. *Ecological Monographs*, **74**, 211–235.

Jezbera, J., Hornák, K., and Simek, K. (2005). Food selection by bacterivorous protists: insight from the analysis of the food vacuole content by means of fluorescence in situ hybridization. *FEMS Microbiol Ecology*, **52**, 351–363.

Jonsson, T. (2014). Trophic links and the relationship between predator and prey body sizes in food webs. *Community Ecology*, **15**, 54–64.

Jousset, A. (2012). Ecological and evolutive implications of bacterial defences against predators. *Environmental Microbiology*, **14**, 1830–1843.

Jousset. A., Rochat, L., Péchy-Tarr, M., *et al.* (2009). Predators promote defence of rhizosphere bacterial populations by selective feeding on non-toxic cheaters. *ISME Journal*, **3**, 666–674.

Juen, A. and Traugott, M. (2005). Detecting predation and scavenging by DNA gut-content analysis: a case study using a soil insect predator–prey system. *Oecologia*, **142**, 344–352.

Juen, A. and Traugott, M. (2007). Revealing species-specific trophic links in soil food webs: molecular identification of scarab predators. *Molecular Ecology*, **16**, 1545–1557.

Jürgens, K. and Simek, K. (2000). Functional response and particle size selection of *Halteria* cf. *grandinella*, a common freshwater oligotrichous ciliate. *Aquatic Microbial Ecology*, **22**, 57–68.

Kalinkat, G., Rall, B. C., Vucic-Pestic, O., and Brose, U. (2011). The allometry of prey preferences. *PLoS One*, **6**, e25937.

Kasper, M. L., Reeson, A. F., Cooper, S. J. B, Perry, K. D., and Austin, A. D. (2004). Assessment of prey overlap between a native (*Polistes humilis*) and an introduced (*Vespula germanica*) social wasp using morphology and phylogenetic analyses of 16S rDNA. *Molecular Ecology*, **13**, 2037–2048.

King, R. A., Read, D. S., Traugott, M., and Symondson, W. O. C. (2008). Molecular analysis of predation: a review of best practice for DNA-based approaches. *Molecular Ecology*, **17**, 947–963.

King, R. A., Vaughan, I. P., Bell, J. R., Bohan, D. A., and Symondson, W. O. C. (2010). Prey choice by carabid beetles feeding on an earthworm community analysed using species- and lineage-specific PCR primers. *Molecular Ecology*, **19**, 1721–1732.

Layman, C. A., Araujo, M. S., Boucek, R., *et al.* (2012). Applying stable isotopes to examine food-web structure: an overview of analytical tools. *Biological Reviews*, **87**, 545–562.

Li, M., Huang, W. E., Gibson, C. M., Fowler, P. W., and Jousset, A. (2013). Stable isotope probing and Raman spectroscopy for monitoring carbon flow in a food chain and revealing metabolic pathway. *Analytical Chemistry*, **85**, 1642–1649.

Lundgren, J. G. and Fergen, J. K. (2014). Predator community structure and trophic linkage strength to a focal prey. *Molecular Ecology*, **23**, 3790–3798.

MacArthur, R. H. (1972). Strong, or weak, interations? *Transactions of the Connecticut Academy of Arts and Sciences*, **44**, 177–188.

Manefield, M., Whiteley, A. S., Griffiths, R. I., and Bailey, M. J. (2002). RNA stable isotope probing, a novel means of linking microbial community function to phylogeny. *Rapid Communication in Mass Spectrometry*, **16**, 2179–2183.

Maxfield, P. J. and Evershed, R. P. (2011). Phospholipid fatty acid stable isotope probing techniques in microbial ecology. In *Stable Isotope Probing and Related Technologies*, ed. J. C. Murrell and A. S. Whiteley, American Society of Microbiology Press, pp. 37–71.

May, R. M. (1972). Will a large complex system be stable? *Nature*, **238**, 413–414.

May, R. M. (1974). *Stability and Complexity in Model Ecosystems*. Princeton, NJ: Princeton University Press.

Mayali, X., Weber, P. K., Brodie, E. L., *et al.* (2012). High-throughput isotopic analysis of RNA microarrays to quantify microbial resource use. *ISME Journal*, **6**, 1210–1221.

Memmott, J. (2009). Food webs: a ladder for picking strawberries or a practical tool for practical problems? *Philosophical Transactions of the Royal Society B: Biological Sciences*, **364**, 1693–1699.

Molkentin, J. and Giesemann, A. (2007). Differentiation of organically and conventionally produced milk by stable isotope and fatty acid analysis. *Analytical Bioanalytical Chemistry*, **388**, 297–305.

Moore, J. C. and de Ruiter, P. C. (2012). *Energetic Food Webs*. Oxford, UK: Oxford University Press.

Moore, J. C., de Ruiter, P. C., and Hunt, H. W. (1993). Soil invertebrate/micro-invertebrate interactions: disproportionate effects of species on food web structure and function. *Veterinary Parasitology*, **48**, 247–260.

Mora, C. A., Paunescu, D., Grass, R. N., and Stark, W. J. (2014). Silica particles with encapsulated DNA as trophic tracers. *Molecular Ecology Resources*, **15**(2), 231–241, doi:10.1111/1755–0998.12299.

Müller-Navarra, D. C., Brett, M. T., Liston, A. M., and Goldman, C. R. (2000). A highly unsaturated fatty acid predicts carbon transfer between primary producers and consumers. *Nature*, **403**, 74–77.

Neher, D. A. (2010). Ecology of plant and free-living nematodes in natural and agricultural soil. *Annual Review of Phytopathology*, **48**, 371–394.

Neidig, N., Paul, R. J., Scheu, S., and Jousset, A. (2011). Secondary metabolites of *Pseudomonas fluorescens* CHA0 drive complex non-trophic interactions with bacterivorous nematodes. *Microbial Ecology*, **61**, 853–859.

Neutel, A.-M., Heesterbeek, J. A. P., and de Ruiter, P. C. (2002). Stability in real food webs: weak links in long loops. *Science*, **296**, 1120–1123.

Ngosong, C., Gabriel, E., and Ruess, L. (2012). Use of the signature fatty acid 16:1ω5 as a tool to determine the distribution of arbuscular mycorrhizal fungi in soil. *Journal of Lipids*, **2012**. doi:10.1155/2012/236807.

Ngosong, C., Gabriel, E., and Ruess, L. (2014). Collembola grazing on arbuscular mycorrhiza fungi modulates nutrient allocation in plants. *Pedobiologia (Jena)*, **57**, 171–179.

O'Neill, R. V. (1969). Indirect estimation of energy fluxes in animal food webs. *Journal of Theoretical Biology*, **22**, 284–290.

Paine, R. T. (1980). Food webs: linkage, interation strength and community infrastructure. *Journal of Animal Ecology*, **49**, 667–685.

Paine, R. T. (1992). Food-web analysis through field measurement of per capita interaction strength. *Nature*, **355**, 73–75.

Pausch, J., Kramer, S., Scharroba, A., *et al.* (2015). Small but active: pool size does not matter for carbon incorporation in belowground food webs. *Functional Ecology*, **30** (3), 479–489.

Pimm, S. L. (1982). *Food Webs*. London, UK: Chapman and Hall.

Pollierer, M. M., Scheu, S., and Haubert, D. (2010). Taking it to the next level: trophic transfer of marker fatty acids from basal resource to predators. *Soil Biology and Biochemistry*, **42**, 919–925.

Pompanon, F., Deagle, B. E., Symondson, W. O. C., *et al.* (2012). Who is eating what: diet assessment using next generation sequencing. *Molecular Ecology*, **21**, 1931–1950.

Pond, D. W., Leakey, R. J. G., and Fallick, A. E. (2006). Monitoring microbial predator–prey interactions: an experimental study using fatty acid biomarker and compound-specific stable isotope techniques. *Journal of Plankton Research*, **28**, 419–427.

Radajewski, S., Ineson, P., Parekh, N. R., and Murrell, J. C. (2000). Stable-isotope probing as a tool in microbial ecology. *Nature*, **403**, 646–649.

Ruess, L. (2003). Nematode soil faunal analysis of decomposition pathways in different ecosystems. *Nematology*, **5**, 179–181.

Ruess, L. and Chamberlain, P. M. (2010). The fat that matters: soil food web analysis using fatty acids and their carbon stable isotope signature. *Soil Biology and Biochemistry*, **42**, 1898–1910.

Ruess, L. and Ferris, H. (2004). Decomposition pathways and successional changes. *Nematology Monographs and Perspectives*, **2**, 547–556.

Ruess, L., Häggblom, M. M., Garciá Zapata, E. J., and Dighton, J. (2002). Fatty acids of fungi and nematodes: possible biomarkers in the soil food chain? *Soil Biology and Biochemistry*, **34**, 745–756.

Ruess, L., Häggblom, M. M., Langel, R., and Scheu, S. (2004). Nitrogen isotope ratios and fatty acid composition as indicators of animal diets in belowground systems. *Oecologia*, **139**, 336–346.

Ruess, L., Tiunov, A., Haubert, D., *et al.* (2005a). Carbon stable isotope fractionation and trophic transfer of fatty acids in fungal based soil food chains. *Soil Biology Biochemistry*, **37**, 945–953.

Ruess, L., Schütz, K., Haubert, D., *et al.* (2005b). Application of lipid analysis to understand trophic interactions in soil. *Ecology*, **86**, 2075–2082.

Ruess, L., Schütz, K., Migge-Kleian, S., *et al.* (2007). Lipid composition of Collembola and their food resources in deciduous forest stands: implications for feeding strategies. *Soil Biology and Biochemistry*, **39**, 1990–2000.

Sanders, D., Jones, C. G., Thébault, E., *et al.* (2014). Integrating ecosystem engineering and food webs. *Oikos*, **123**, 513–524.

Scheu, S. (2002). The soil food web: structure and perspectives. *European Journal of Soil Biology*, **38**, 11–20.

Scheu, S., Ruess, L., and Bonkowski, M. (2005). Interactions between microorganisms and soil micro- and mesofauna. In *Microorganisms in Soil: Roles in Genesis and Functions*, ed. F. Buscot and A. Varma, Berlin: Springer-Verlag, pp. 253–277.

Schröter, D., Wolters, V., and de Ruiter, P. C. (2003). C and N mineralisation in the decomposer food webs of a European forest transect. *Oikos*, **102**, 294–308.

Seeber, J., Rief, A., Seeber, G. U. H., Meyer, E., and Traugott, M. (2010). Molecular identification of detritivorous soil invertebrates from their faecal pellets. *Soil Biology Biochemistry*, **42**, 1263–1267.

Selakovic, S., de Ruiter, P. C., and Heesterbeek, H. (2014). Infectious disease agents mediate interaction in food webs and ecosystems. *Proceedings of the Royal Society B: Biological Sciences*, **281**(1777), 20132709.

Sint, D., Raso, L., and Traugott, M. (2012). Advances in multiplex PCR: balancing primer efficiencies and improving detection success. *Methods in Ecology and Evolution*, **3**, 898–905.

Stübing, D., Hagen, W., and Schmidt, K. (2003). On the use of lipid biomarkers in marine food web analyses: an experimental case study on the Antarctic krill, *Euphausia superba*. *Limnology and Oceanography*, **48**, 1685–1700.

Symondson, W. O. C. (2012). The molecular revolution: using polymerase chain reaction based methods to explore the role of predators in terrestrial food webs. In *Biodiversity and Insect Pests: Key Issues for Sustainable Management*, ed. G. M. Gurr, S. D. Wratten, W. E. Snyder, and D. M. Y. Read, New York: Wiley and Sons, pp. 166–184.

Thiemann, G. W., Iverson, S. J., and Stirling, I. (2008). Polar bear diets and arctic marine food webs: insights from fatty acid analysis. *Ecological Monographs*, **78**, 591–613.

Traugott, M., Kamenova, S., and Ruess, L. (2013). Empirically characterising trophic networks: what emerging DNA-based methods, stable isotope and fatty acid analyses can offer. *Advances in Ecological Research*, **49**, 177–224, doi:10.1016/B978-0–12-420002–9.00003–2.

van Dooremalen, C. and Ellers, J. (2010). A moderate change in temperature induces changes in fatty acid composition of storage and membrane lipids in a soil arthropod. *Journal Insect Physiology*, **56**, 178–184.

Vestheim, H. and Jarman, S. N. (2008). Blocking primers to enhance PCR amplification of rare sequences in mixed samples: a case study on prey DNA in Antarctic krill stomachs. *Frontiers in Zoology*, **5**, 12, DOI: 10.1186/1742–9994-5–12.

von Berg, K., Traugott, M., and Scheu, S. (2012). Scavenging and active predation in generalist predators: a mesocosm study employing DNA-based gut content analysis. *Pedobiologia (Jena)*, **55**, 1–5.

Vuvic-Pestic, O. K., Birkhofer, K., Rall, B. C., Scheu, S., and Brose, U. (2010). Habitat structure and prey aggregation determine the functional response in a soil predator–prey interaction. *Pedobiologica*, **53**, 307–312.

Waldner, T. and Traugott, M. (2012). DNA-based analysis of regurgitates: a noninvasive approach to examine the diet of invertebrate consumers. *Molecular Ecology Resures*, **12**, 669–675.

Waldner, T., Sint, D., Juen, A., and Traugott, M. (2013). The effect of predator identity on post-feeding prey DNA detection success in soil-dwelling macro-invertebrates. *Soil Biology Biochemistry*, **63**, 116–123.

Wallinger, C., Staudacher, K., Schallhart, N., *et al.* (2013). The effect of plant identity and the level of plant decay on molecular gut content analysis in a herbivorous soil insect. *Molecular Ecology Resures*, **13**, 75–83.

Wardle, D. A., Verhoef, H. A., and Clarholm, M. (1998). Trophic relationships in the soil microfood-web: predicting the responses to a changing global environment. *Global Change Biology*, **4**, 713–727.

White, D. C., Stair, J. O., and Ringelberg, D. B. (1996). Quantitative comparisons of in situ microbial biodiversity by signature biomarker analysis. *Journal of Industrial Microbiology*, **17**, 185–196.

Williams, C. T. and Buck, C. L. (2010). Using fatty acids as dietary tracers in seabird trophic ecology: theory, application and limitations. *Journal of Ornithology*, **151**, 531–543.

Winemiller, K. O. and Layman, C. A. (2005). Food web science: moving on the path from abstraction to prediction. In *Dynamic Food Webs*, ed. P. C. de Ruiter, J. C. Moore, and V. Wolters, San Diego: Academic Press, pp. 10–23.

Wirta, H. K., Hebert, P. D. N., Kaartinen, R., *et al.* (2014). Complementary molecular information changes our perception of food web structure. *Proceedings of the National Academy of Sciences of the United States of America*, **111**, 1885–1890.

Yeates, G. W. (2010). Nematodes in ecological webs. In *Encyclopedia of Life Sciences (ELS)*, New York: John Wiley and Sons. doi:10.1002/9780470015902.a0021913.

Yeates, G. W., Bongers, T., De Goede, R. G., Freckman, D. W., and Georgieva, S. S. (1993). Feeding habits in soil nematode families and genera: an outline for soil ecologists. *Journal of Nematology*, **25**, 315–331.

Yeates, G. W., Ferris, H., Moens, T., and van der Putten, W. H. (2009). The role of nematodes in ecosystems. In *Nematodes as Environmental Bioindicators*, ed. M. Wilson and T. Kakouli-Duarte, Wallingford, UK: CABI, pp. 1–44.

Zelles, L. (1999). Fatty acid patterns of phospholipids and lipopolysaccharides in the characterisation of microbial communities in soil: a review. *Biology and Fertility of Soils*, **29**, 111–129.

Zhang, G.-F., Lü, Z.-C., Wan, F.-H., and Lövei, G. L. (2007). Real-time PCR quantification of *Bemisia tabaci* (Homoptera: Aleyrodidae) B-biotype remains in predator guts. *Molecular Ecology Notes*, **7**, 947–954.

Part III

Food Webs and Environmental Sustainability

17 Integrating Species Interaction Networks and Biogeography

José M. Montoya and Núria Galiana

17.1 Introduction

Species interaction network studies and biogeography have evolved independently from each other, although some remarkable exceptions exist. The prevailing wisdom is that biotic interactions rule in local-scale networks while large spatial scales are the province of climate. In this chapter we suggest that this and other cross-disciplinary boundaries are artificial, and that much progress can be made through the adoption of both a biogeographical perspective in networks and a network perspective in biogeography. We present fundamental ecological questions in which integration is needed to find answers, and highlight recent integrative efforts in both biogeography and network research. In particular, we focus on two topics. First, the importance of multispecies distribution in a changing world. And, second, the existence of large-scale gradients on community structure, by which local interactions may play a secondary role for network structure and dynamics. Moving forward, we suggest that the integration of network and biogeography research represents one of the most promising yet challenging avenues for food-web studies.

17.1.1 The Search for Universalities in Species Interaction Networks

The last decade has witnessed a revolution in the study of large species interaction networks, including food webs and mutualistic networks of free-living species – as those describing plants and their pollinators. Numerous theoretical and empirical studies have identified universal patterns and mechanisms in the way species interact across different habitat types, which in turn affect community dynamics (Dunne, 2006; Montoya *et al.*, 2006; Bascompte, 2009; Ings *et al.*, 2009). These interaction patterns are not only key to understanding biodiversity organization within communities, but also to predicting ecosystem stability and resistance to different components of environmental change (Montoya *et al.*, 2006, 2009; Lurgi *et al.*, 2012a), and important ecosystem functions, such as primary production, biogeochemical cycles, pest control, or pollination (Montoya *et al.*, 2003; Tylianakis *et al.*, 2007; Reiss *et al.*, 2009; Thompson *et al.*, 2012).

Although some authors warned that the idiosyncrasies of individual species and their dynamics might prevent the existence of such regularities (Polis, 1991; Cohen *et al.*, 1993; Paine *et al.*, 1998), most studies suggest the existence of a few universal structural patterns in food webs and mutualistic networks (Warren, 1990; Martínez, 1991; Montoya *et al.*, 2006; Bascompte, 2009; Ings *et al.*, 2009). These universalities include: that most species have a few links to other species while only a few species are generalists (Solé and Montoya, 2001; Dunne *et al.*, 2002; Jordano *et al.*, 2003); the existence of compartments or modules in food webs, where species within the same compartment interact more frequently among them than with species outside them (Krause *et al.*, 2003; Melián and Bascompte, 2004), the presence of nested subsets of species in mutualistic networks, where generalists tend to interact with specialists (Bascompte *et al.*, 2003), or the predominance of weak interaction strengths between consumers and their resources (Berlow *et al.*, 2004; Wootton and Emmerson, 2005).

Most suggested processes responsible for observed structural patterns and dynamics in ecological networks operate on a local scale, as those related with consumer–resource dynamics, habitat occupancy, or foraging behavior. The influence of processes that operate at larger spatial extents (e.g., regional and geographical scales), such as climate or productivity, is hardly considered as a determinant of local network structure and dynamics. Notable exceptions exist. The relationships of some food-web properties, such as connectance and food-chain length, with variables that change geographically, such as species richness, primary productivity, or environmental variability, have received some attention (e.g., Briand and Cohen, 1987; Warren, 1990; Post *et al.*, 2000). In our view, these examples are the exception to the general rule that neglects the effects of processes operating at large spatial scales on the structure and dynamics of local communities.

17.1.2   Biogeography: Looking for Sources of Geographical Variability

In marked contrast to the search for universalities in networks research, biogeographers (and macroecologists) look for the pattern and potential processes of variation of community and species characteristics across large geographical extents. Examples include the latitudinal diversity gradient – i.e., species richness decreases with latitude (Hawkins *et al.*, 2003; Currie *et al.*, 2004) – or the latitudinal gradient in niche breadth – i.e., niches or number of resources per consumer are narrower in the tropics (MacArthur, 1972; Stevens, 1989; but see Vazquez and Stevens, 2004).

While few biogeographers recognize the role of biotic interactions, as we present below using several examples, the majority of biogeographical studies tends to neglect the role and importance of biotic interactions (Araújo *et al.*, 2008; Wisz *et al.*, 2013). First, these are not considered as important drivers of species distributions, with environmental variables (mostly climate) assumed to play the leading role. Second, in marked contrast to the simultaneous consideration of multiple trophic levels and their interactions in network studies, biogeographers usually focus on "one guild at a time," and accumulate evidence to support the observed pattern across different taxa.

However, it is clear that the importance of biotic interactions varies across geographical areas. It is well known that a wide range of interactions is more important at low latitude systems, including higher herbivory and insect predation in the tropics, and the predominance of tropical mutualisms such as cleaning symbioses in marine systems and ant–plant interactions in terrestrial ones (Schemske *et al.*, 2009). Moreover, numerous studies regarding probably the oldest and most intensively studied biogeographical pattern (Hawkins, 2001), the latitudinal diversity gradient, hypothesize that more intense and stable biotic interactions in the tropics may explain the pattern, as we discuss below (Schemske *et al.*, 2009).

Despite some historical and recent integrative efforts, the burgeoning areas of ecological networks and biogeography rarely have been merged. A noticeable but marginal exception is the integration of the classical theory of island biogeography and food webs, initiated by Bob Holt (1993, 1996, 2010), and later formalized by Gravel *et al.* (2011) (see Box 17.1). Moreover, the interplay between local and regional processes is firmly established in the study of local diversity (Ricklefs, 1987; Huston, 1999; Holyoak *et al.*, 2005). Regional processes operating at large spatial scales, such as species dispersal from a regional propagule supply, are important determinants of local diversity and dynamics in both terrestrial and marine systems (Cornell and Harrison, 2013). We now need to determine if regional processes are also necessary for understanding the interaction patterns and dynamics of local communities.

In this chapter we discuss why the gap between biogeography and ecological networks exists and how it is manifested. We argue that the perceived boundaries between both disciplines are largely artificial. We outline how to both bring a community network perspective into biogeographical studies, and adopt a gradient-based, biogeographical perspective into the study of the structure and dynamics of species interaction networks. In particular, we show that multispecies interactions are important for the explanation of the latitudinal diversity gradient and that forecasting species distributions in a changing world requires considering multispecies interactions across large spatial extents. Similarly, we discuss the existence of large-scale gradients on network structure that requires assessing the relative role of spatial and climatic drivers versus local biotic interactions.

17.2   Different Processes Operating at Different Spatial Scales?

A number of conceptual and methodological differences explain the absence of links between biogeography and species interaction network research (Table 17.1; Figure 17.2). The most evident is the spatial scale under consideration. Networks are commonly defined at the local or landscape scale, with processes affecting local species interactions as the main determinants of their structure, dynamics, and functioning. In contrast, biogeography focuses on large-scale (regional to global) patterns, and the processes at play usually relate to environmental conditions, past or present. This is manifested via the main visualization tools used: graphs connecting species in networks versus maps depicting variability in biogeography (Figure 17.3).

Major differences exist when it comes to the modeling frameworks used. Network studies focus on interactive population dynamics models, where species coexistence and

Box 17.1 Island Biogeography Meets Food Webs

MacArthur and Wilson (1967) proposed in their theory of island biogeography (TIB) the diversity-dependent dynamic balance between immigration and extinction as a determinant of island species richness. In their model, the species immigration rate for an island decreases as the number of species on the island increases, and species extinction rate increases with the number of species, implying that diversity will reach an equilibrium. Even though MacArthur and Wilson already stated that the "interference" between species might shape the extinction curve, they did not address explicitly the effect of considering multispecies interactions.

Gravel *et al.* (2011) extended the classic island biogeography theory to account for trophic interactions assuming bottom–up trophic dependencies: a species needs the presence of at least one of its prey items to colonize a patch, and it becomes extinct if its last prey becomes extinct. Given that colonization by higher trophic levels cannot occur until the lower ones have become established, the trophic theory of island biogeography (TTIB) predicts a slower accumulation of species affecting the classic equilibrium point (Figure 17.1). Thus TTIB shows that considering a trophic constraint on species immigration and extinction would affect species richness equilibrium in a local community. This effect is intensified as consumers' diet specialization increases due to the limitation of finding the required prey to colonize.

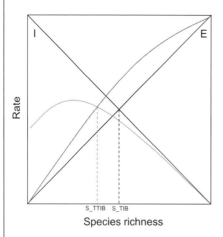

Figure 17.1 Trophic theory of island biogeography adapted from Gravel *et al.* (2011). The classic TIB is depicted in black. Equilibrium species richness (S_TIB) is reached when immigration rate is equal to extinction rate (intersection between I and E; i.e., black dotted line). The TTIB is depicted in colors (orange and blue for immigration and extinction rates, respectively). Equilibrium species richness (S_TTIB) is represented by the red dotted line. (A black and white version of this figure will appear in some formats. For the color version, please refer to the plate section.)

Table 17.1 Major conceptual and methodological differences between research on ecological networks and biogeography.

	Ecological networks	Biogeography
Mechanisms	Biotic interactions	Environment (climate)
Models	Interactive population dynamics	Species occupancy and distribution
Taxonomic spread	Multiple guilds (i.e., trophic groups)	Isolated guilds
Emerging issues	Universal patterns	Variability across environmental gradients

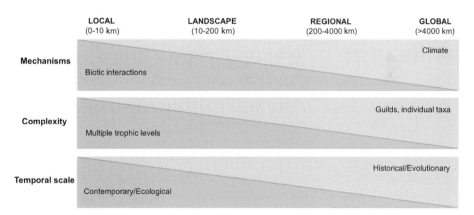

Figure 17.2 Variation of the relative importance of the mechanisms, level of complexity, and temporal scale across the spatial scale under consideration. Local network studies typically focus on biotic interactions across multiple trophic levels over short-term temporal scales. In contrast, global/regional biogeographical studies focus on climate as the main driver within guilds and over evolutionary time scales.

density are regulated by species interactions, including a wide range of top–down and bottom–up dynamics. In contrast, biogeographical models focus on the abiotic determinants – usually climate – of species presence or absence. From the multidimensional niche of a species, most biogeographers only consider the abiotic dimensions while network researchers focus on its biotic component.

The tenet is that large spatial scales are the province of climate – contemporary or past – while biotic interactions rule at local scales (Pearson and Dawson, 2003; Thuiller *et al.*, 2004). Accordingly, Johnson (1980) suggested an integrative framework for the ordering of selection processes operating across spatial scales. He identified first-order selection as the selection of physical or geographical range of species, followed by the selection of the home range of an individual or group, and finally a third- and fourth-order selection determining the feeding interactions with that individual.

However, this tenet has never been tested with datasets containing detailed information on spatially explicit climatic variables and biotic interactions. In the few cases where the relative role of climate, dispersal, and biotic interactions were considered

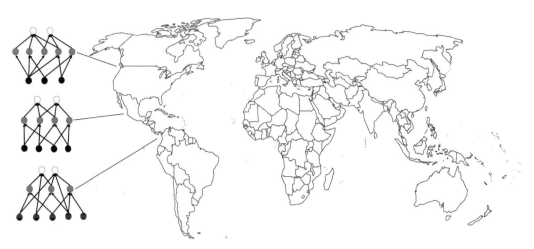

Figure 17.3 Mapping networks. Visualization techniques greatly differ between network studies, using graphs connecting species through biotic interactions (left-hand-side graphs, contrived networks), and biogeography, using maps to illustrate the analyzed pattern. To determine the existence of biogeographical patterns in network structure and dynamics, we need to analyze several networks across large-scale gradients.

together, species interactions were actually estimated from species co-occurrence data, not from direct observations. Boulangeat *et al.* (2012), for instance, used a spatially nested modeling framework to understand the distribution and abundance of plant species in the French Alps, showing that species presence/absence was determined by climatic factors and dispersal, while species abundances were mostly determined by biotic interactions, including competition and facilitation. In other words, abiotic relationships filter the species able to occupy a given environment, determining plant species composition, whereas biotic interactions manifested their importance at local scales affecting species densities. This promising approach would benefit from including observed species interactions not inferred simply from species co-occurrence data. However, such datasets for multispecies systems, and for multiple predator–prey interactions in particular, are scarce, if available at all.

Accumulating evidence suggests that the boundaries between climate and biotic interactions as determinants of regional and local dynamics, respectively, are largely artificial. Within local communities, climate has a strong influence on species interactions, resulting in complex community responses. Theoretical and experimental work has shown that increasing environmental temperature modifies population growth rate, carrying capacities, and metabolic and ingestion rates (Vasseur and McCann, 2005; Rall *et al.*, 2010). Also, warming alters universal properties of food webs, shifting body mass–abundance scaling relationships (Dossena *et al.*, 2012), and altering degree distributions by increasing diet specialization (Lurgi *et al.*, 2012b). Similarly, climatic changes can alter the prevalence of top–down versus bottom–up control in aquatic ecosystems (Shurin *et al.*, 2012) and can affect profoundly several ecosystem processes, including productivity, ecosystem respiration, and decomposition rates (Yvon-Durocher *et al.*,

2010, 2011; Dossena *et al.*, 2012). All taken together these suggest that modifications of large-scale processes, such as climate, have profound consequences in local communities, mostly mediated by species interactions.

Less clear is the relevance of biotic interactions at large spatial extents. In a recent review, Wisz *et al.* (2013) showed compelling evidence on the effects of biotic interactions determining the current (and past) distribution of species across multiple taxa and habitats, including competition, facilitation, herbivory, and predation. Similarly, the new ranges of species that result from climate change are highly dependent on their interactions with other species (Alexander *et al.*, 2015).

Let us reconsider the latitudinal diversity gradient. Most hypotheses suggest that abiotic factors are responsible for finding species richness peaking in equatorial regions and declining toward the poles. Among these factors, the water-energy tandem, and historical climatic stability have received strong support across taxa and continents (Hawkins *et al.*, 2003; Currie *et al.*, 2004). The role of biotic interactions hangs in any discussion about the gradient. Yet its relative importance against abiotic factors is difficult to test because the quality and resolution of biotic interaction data across large spatial scales is very low in comparison to the quality and resolution of most abiotic drivers. However, its conceptual and theoretical basis is firmly established. Dobzhanski (1950) proposed that the benign, constant climate in the tropics led to a greater importance of biotic interactions, resulting in tropical species more specialized and tropical communities harboring greater species diversity. In contrast, the severe and variable climate in temperate regions resulted in the evolution of a few generalized species. Schemske and collaborators (Schemske, 2002; Schemske *et al.*, 2009) formalized this, and they suggested that more intense biotic interactions in the tropics promoted coevolution resulting in faster adaptation and speciation. New species introduce new resources and interactions, hence expanding the number of niches and creating a positive feedback of diversity. Similarly, resource specialization has been suggested as one explanation for the observed latitudinal gradient in species richness: specialization reduces interspecific competition and facilitates species coexistence by partitioning niche space (MacArthur, 1972; Janzen, 1973; Stevens, 1989; but see Vazquez and Stevens, 2004).

Another paradigmatic example where biotic interactions are generally neglected is species distribution modeling (SDMs). Classical SDMs either completely ignore biotic interactions, or at most include the assumption that biotic interactions are equally strong and important across the entire species range, and, more importantly, that the strength of interactions remains constant over space and time. Both assumptions are oversimplifications of reality, and in the case of constant interactions in the face of global change, may result in misleading conclusions, because species interactions are profoundly modified by different components of global change (Tylianakis *et al.*, 2008; Montoya and Raffaelli, 2010).

Recent extensions of SDMs only include the unidirectional influence of one or a few species (see Kissling *et al.*, 2012 for a review). Kissling *et al.* (2010), for example, showed that the prediction of future bird distributions in Kenya as a consequence of climate change was sensitive to the inclusion of bird–woody plant interactions.

Predicted species losses of birds due to climate change were significantly stronger (and even reversed) when lagged response times of woody plants and their association with birds were included in the model. Even when SDMs ignoring biotic interactions successfully predict species presence/absence, they typically fail to explain and predict species abundances across gradients (Sagarin *et al.*, 2006; Boulangeat *et al.*, 2012).

These serious shortcomings of SDMs have stimulated several biogeographers to develop novel approaches to model biotic interactions in multispecies networks at large spatial scales. The challenge is big but progress can be made. Kissling *et al.* (2012) suggest ways to do so by using multispecies co-occurrence datasets across large-scale environmental gradients to infer potential interaction matrices, coupled with comprehensive spatio-temporal data on biotic interactions in large networks and using network theory and natural history information to reduce model complexity. This method can be coupled to that developed by Morales-Castilla *et al.* (2015), which infers biotic interaction networks within regional species pools considering proxies that include functional traits, phylogenies, and geography. Such progress requires intense and creative collaboration among biogeographers and community ecologists.

The temporal scale under consideration also differs between network and biogeographical approaches, somehow echoing processes operating at different spatial scales. Although in both cases present-day species composition, abundances, and distributions are the focus, the hypothesized determinants typically operate over short-term, i.e., ecological, time scales in network studies, while they operate over long-term, i.e., evolutionary and geological, time scales in biogeographical research. There are only a few network studies that consider an evolutionary dimension, allowing for speciation dynamics (McKane and Drossel, 2005; Bell, 2007) or coevolution (Guimaraes *et al.*, 2011), although there is a recent interest in assessing the interplay between ecological and evolutionary dynamics, i.e., eco-evolutionary dynamics, in species interaction networks (Melián *et al.*, 2011; Moya-Laraño *et al.*, 2012). In addition, a recent study showed the influence of historical climate change on the modularity and nestedness of pollination networks (Dalsgaard *et al.*, 2013), suggesting that climatic stability over evolutionary time scales was at least as important as current climate to understand present-day networks.

In contrast, geological and past evolutionary changes are widely used for explaining biogeographical patterns. Time for speciation, for example, is one of the central determinants of the latitudinal diversity gradient for a broad variety of plants and animals, by which past climatic stability in the tropics as opposed to glaciation cycles in northern hemisphere temperate areas could explain the pattern (Hawkins *et al.*, 2007; Romdal *et al.*, 2013).

17.3   Biogeographical Changes in Network Structure

The influence of spatial processes on local networks has received considerable attention. Examples include the theoretical integration of island biogeography and food webs (Holt, 1996, 2010; Holt *et al.*, 2002, Gravel *et al.*, 2011; Massol *et al.*, 2011), the coupling of local webs in space through foraging behavior and movement of species across trophic positions (McCann *et al.*, 2005; Rooney *et al.*, 2006, 2008), or the effects of dispersal of consumers and resources in metacommunities on the dynamics and persistence of relatively simple food webs, so-called modules (Leibold *et al.*, 2004; Holyoak *et al.*, 2005; Amarasekare, 2008).

Despite these advances on the effects of space in networks, if we want to consider large-scale processes in networks, we need approaches that analyze how network characteristics change across a given large-scale gradient. This approach is in marked contrast with the search for universalities typical of research in networks. It requires moving from finding universalities to finding the sources of variability.

Kitching pioneered the study of biogeographical gradients of food-web structure. Using the communities associated to phytotelmata (i.e., water bodies in terrestrial plants) he showed that several food-web properties varied across a latitudinal gradient. In particular, increasing latitude decreased mean food-chain length and predator generalism (Kitching, 2000) – a measure of diet specialization. However, Kitching's studies had small sample sizes and relatively simple food webs. Interestingly though, both food-chain length and diet specialization have received attention in more detailed and comprehensive studies along environmental gradients.

Several hypotheses try to explain observed variation in food-chain length across habitats (see Post, 2002 for a review). Some hypothesized processes vary along geographical gradients. That is the case of resource availability: the more productive the system, the longer the food chains or the trophic position of the top predator. However, resource availability has limited predictive power, limiting food-chain length only in systems with very low resource availability. Ecosystem size (area or volume) appears as the best predictor of food-chain length (Post *et al.*, 2000), although the components of ecosystem size (e.g., habitat availability and heterogeneity, species diversity) that explain the pattern are not yet clear. Thus variation in food-chain length seems to be determined by a combination of local factors and large-scale processes.

Diet specialization is at the core of research in both ecological networks and biogeography. The general perception among ecologists is that biotic specialization increases toward the tropics. More generally, environmental constancy (or stability) leads to higher specialism. Community ecology and biogeography seem to agree historically in this respect. MacArthur (1955) stated that the greater stability and lower seasonality in the tropics lead populations at low latitudes to be more stable than populations at higher latitudes, and, in turn, greater population stability should allow for narrower (and more specialized) diet niches. It has been shown that both past and contemporary climate stability influence biotic specialization (Schleuning *et al.*, 2012). The former reflecting the available time for species coevolution due to temporal stability of local communities (Jansson and Dynesius, 2002) and contemporary climate determining the relative abundances and densities of resource species (by means of species diversity), which, in turn, regulates consumer species searching times (Kreft and Jetz, 2007; MacArthur and Pianka, 1966). Longer search times constrain the specialization of consumer species (Albrecht *et al.*, 2010) and therefore warm climates (i.e., tropical regions with higher diversity of resource species) are supposed to lead to higher specialization of biotic interactions.

Most studies conducted so far across geographical gradients focused on a very limited subset of the interactions within the overall species interaction network. However, patterns of specialization may change when the wider network of biotic interactions is considered. For instance, reciprocal specializations that occur when a consumer specializes on a particular resource and vice versa – illustrated most famously by the Malagasy orchid and the moth Darwin predicted would pollinate it (Darwin, 1862) – are extremely rare when the whole interaction network is considered (Joppa *et al.*, 2009).

A few recent studies have characterized latitudinal patterns in the structure of mutualistic networks, showing non-conclusive results. While Dalsgaard *et al.* (2011) and Trøjelsgaard and Olesen (2013) found an increase in network specialization toward the tropics, Schleuning *et al.* (2012) found the opposite, i.e., less network specialization toward the tropics, and Ollerton and Crammer (2002) found no latitudinal trends. Recently, Morris *et al.* (2014) have published a study of the latitudinal variability in the structure of antagonistic consumer–resource networks where they did not find any pattern of specialization across a latitudinal gradient after correcting for the number of connections and the taxonomic diversity of the community.

Further analyses are needed to determine the existence of biogeographical patterns in network structure, contemplating both present-day and historical determinants. The spatial scale of the network studied is crucial here, as recently showed for host–parasitoid food webs where a latitudinal gradient of network specialization was only observed for the regional web (i.e., the network resulting from aggregating several local communities), not for the local communities. This suggests that the interesting answers may lie at the mesoscale, where the interplay of local dynamics, spatial processes, and processes operating at large spatial scales occur.

17.4   Concluding Remarks

Networks and biogeography have been two of the most fruitful research areas in natural sciences over the last decade. Unfortunately, they have developed in isolation from each other, although remarkable exceptions exist that connect both research areas. The integration of biogeography and networks would provide answers to many open questions at the core of both disciplines. We have shown several examples here including (1) the importance of biotic interactions for the latitudinal diversity gradient; (2) the accurate forecast of future species distributions and community composition at large spatial scales using multispecies interactions in species distribution models; (3) the importance of large-scale processes for local community dynamics and network structure; and (4) the existence of geographical gradients in network structure and dynamics. These questions are particularly relevant in the face of modifications of both large-scale and local-scale processes, such as climate change and biodiversity loss, respectively. This integration will require removing some well-established but highly artificial barriers, namely the association between each discipline and their spatial and temporal scales unequivocally linked to specific processes operating at that scale. In practical terms, it needs the development of integrative theoretical frameworks and methodologies, and a new generation of datasets that allow for assessing the local and regional determinants of multispecies interaction networks.

Acknowledgments

This work is supported by the French Laboratory of Excellence Project "TULIP" (ANR-10-LABX-41; ANR-11-IDEX-0002–02) and by a Region Midi-Pyrenees project (CNRS 121090).

References

Albrecht, M., Riesen, M., and Schmid, B. (2010). Plant–pollinator network assembly along the chronosequence of a glacier foreland. *Oikos*, **119**(10), 1610–1624.

Alexander, J. M., Diez, J. M., and Levine, J. M. (2015). Novel competitors shape species' response to climate change. *Nature*, **525**, 515–518.

Amarasekare, P. (2008). Spatial dynamics of foodwebs. *Annual Review of Ecology and Systematics*, **39**, 479–500.

Araújo, M. B., Nogués-Bravo, D., Diniz-Filho, J. A. F., *et al.* (2008). Quaternary climate changes explain diversity among reptiles and amphibians. *Ecography*, **31**, 8–15.

Bascompte, J. (2009). Mutualistic networks. *Frontiers in Ecology and the Environment*, **8**, 429–436.

Bascompte, J., Jordano, P., Melián, C. J., and Olesen, J. M. (2003). The nested assembly of plant–animal mutualistic networks. *Proceedings of the National Academy of Sciences of the United States of America*, **100**, 9383–9387.

Bell, G. (2007). The evolution of trophic structure. *Heredity*, **99**(5), 494–505.

Berlow, E. L., Neutel, A.-M., Cohen, J. E., *et al.* (2004). Interaction strengths in food webs: issues and opportunities. *Journal of Animal Ecology*, **73**, 585–598.

Boulangeat, I., Gravel, D., and Thuiller, W. (2012). Accounting for dispersal and biotic interactions to disentangle the drivers of species distributions and their abundances. *Ecology Letters*, **15**(6), 584–593.

Briand, F. and Cohen, J. E. (1987). Environmental correlates of food chain length. *Science*, **238**(4829), 956–960.

Cohen, J. E., Pimm, S. L., Yodzis, P., and Saldaña, J. (1993). Body sizes of animal predators and animal prey in food webs. *Journal of Animal Ecology*, **62**, 67–78.

Cornell, H. V. and Harrison, S. P. (2013). Regional effects as important determinants of local diversity in both marine and terrestrial systems. *Oikos*, **122**, 288–297.

Currie, D. J., Mittelbach, G. G., Cornell, H. V., *et al.* (2004). Predictions and tests of climate-based hypotheses of broad-scale variation in taxonomic richness. *Ecology Letters*, **7**(12), 1121–1134.

Dalsgaard, B., Magård, E., Fjeldså, J., *et al.* (2011). Specialization in plant–hummingbird networks is associated with species richness, contemporary precipitation and quaternary climate-change velocity. *PLoS ONE*, **6**(10), e25891.

Dalsgaard, B., Trøjelsgaard, K., Martín González, A. M., *et al.* (2013). Historical climate-change influences modularity and nestedness of pollination networks. *Ecography*, **36**(12), 1331–1340.

Darwin, C. R. (1862). *On the Various Contrivances by Which British and Foreign Orchids Are Fertilised by Insects, and on the Good Effects of Intercrossing.* London, UK: John Murray.

Dobzhansky, T. (1950). Evolution in the tropics. *American Scientist*, **38**(2), 209–221.

Dossena, M., Yvon-Durocher, G., Grey, J., *et al.* (2012). Warming alters community size structure and ecosystem functioning. *Proceedings of the Royal Society B: Biological Sciences*, **279**, 3011–3019.

Dunne, J. A. (2006). The network structure of food webs. In *Ecological Networks: Linking Structure to Dynamics in Food Webs*, ed. M. Pascual and J. A. Dunne, Oxford, UK: Oxford University Press, pp. 27–86.

Dunne, J. A., Williams, R. J., and Martinez, N. D. (2002). Food-web structure and network theory: the role of connectance and size. *Proceedings of the National Academy of Sciences of the United States of America*, **99**, 12917–12922.

Gravel, D., Massol, F., Canard, E., Mouillot, D., and Mouquet, N. (2011). Trophic theory of island biogeography. *Ecology Letters*, **14**(10), 1010–1016.

Guimaraes, Jr., P. R., Jordano, P., and Thompson, J. N. (2011). Evolution and coevolution in mutualistic networks. *Ecology Letters*, **14**(9), 877–885.

Hawkins, B. A. (2001). Ecology's oldest pattern? *Trends in Ecology and Evolution*, **16**, 470.

Hawkins, B. A., Porter, E. E., and Diniz-Filho, J. A. F. (2003). Productivity and history as predictors of the latitudinal diversity gradient of terrestrial birds. *Ecology*, **84**, 1608–1623.

Hawkins, B. A., Diniz-Filho, J. A. F., Jaramillo, C. A., and Soeller, S. A. (2007). Climate, niche conservatism, and the global bird diversity gradient. *American Naturalist*, **170**(S2), S16–S27.

Holt, A. R., Warren, P. H., and Gaston, K. J. (2002). The importance of biotic interactions in abundance–occupancy relationships. *Journal of Animal Ecology*, **71**(5), 846–854.

Holt, R. D. (1993). Ecology at the mesoscale: the influence of regional processes on local communities. In *Species Diversity in Ecological Communities*, ed. R. Ricklefs and D. Schluter, Chicago, IL: University of Chicago Press, pp. 77–88.

Holt, R. D. (1996). Food webs in space: an island biogeographic perspective. In *Food Webs: Contemporary Perspectives*, ed. G. A. Polis and K. Winemiller, London: Chapman and Hall, pp. 313–323.

Holt, R. D. (2010). Towards a trophic island biogeography: reflections on the interface of island biogeography and food web ecology. In *The Theory of Island Biogeography Revisited*, ed. J. B. Losos and R. E. Ricklefs, Princeton, NJ: Princeton University Press, pp. 143–185.

Holyoak, M., Leibold, M. A., and Holt, R. D. (2005). *Metacommunities: Spatial Dynamics and Ecological Communities*. Chicago, IL: University of Chicago Press.

Huston, M. A. (1999). Local processes and regional patterns: appropriate scales for understanding variation in the diversity of plants and animals. *Oikos*, **86**, 393–401.

Ings, T. C., Montoya, J. M., Bascompte, J., *et al.* (2009). Review: ecological networks – beyond food webs. *Journal of Animal Ecology*, **78**(1), 253–269.

Jansson, R. and Dynesius, M. (2002). The fate of clades in a world of recurrent climatic change: Milankovitch oscillations and evolution. *Annual Review of Ecology and Systematics*, **33**, 741–777.

Janzen, D. H. (1973). Sweep samples of tropical foliage insects: effects of seasons, vegetation types, elevation, time of day, and insularity. *Ecology*, **54**, 687–708.

Johnson, D. H. (1980). The comparison of usage and availability measurements for evaluating resource preference. *Ecology*, **61**(1), 65–71.

Joppa, L. N., Bascompte, J., Montoya, J. M., *et al.* (2009). Reciprocal specialization in ecological networks. *Ecology Letters*, **12**, 961–969.

Jordano, P., Bascompte, J., and Olesen, J. M. (2003). Invariant properties in coevolutionary networks of plant–animal interactions. *Ecology Letters*, **6**, 69–81.

Kissling, W. D., Field, R., Korntheuer, H., Heyder, U., and Böhning-Gaese, K. (2010). Woody plants and the prediction of climate-change impacts on bird diversity. *Philosophical Transactions of the Royal Society B: Biological Sciences*, **365**(1549), 2035–2045.

Kissling, W. D., Dormann, C. F., Groeneveld, J., *et al.* (2012). Towards novel approaches to modelling biotic interactions in multispecies assemblages at large spatial extents. *Journal of Biogeography*, **39**(12), 2163–2178.

Kitching, R. L. (2000). *Food Webs and Container Habitats: the Natural History and Ecology of Phytotelmata*. Cambridge, UK: Cambridge University Press.

Krause, A. E., Frank, K. A., Mason, D. M., Ulanowicz, R. E., and Taylor, W. W. (2003). Compartments revealed in food web structure, *Nature*, **426**, 482–485.

Kreft, H. and Jetz, W. (2007). Global patterns and determinants of vascular plant diversity. *Proceedings of the National Academy of Sciences of the United States of America*, **104**(14), 5925–5930.

Leibold, M. A., Holyoak, M., Mouquet, N., *et al.* (2004). The metacommunity concept: a framework for multi-scale community ecology. *Ecology Letters*, **7**(7), 601–613.

Lurgi, M., Lopez, B. C., and Montoya, J. M. (2012a). Climate change impacts on body size and food web structure on mountain ecosystems. *Philosophical Transactions of the Royal Society B: Biological Sciences*, **367**, 3050–3057.

Lurgi, M., Lopez, B. C., and Montoya, J. M. (2012b). Novel communities from climate change. *Philosophical Transactions of the Royal Society B: Biological Sciences*, **367**, 2913–2922.

MacArthur, R. H. (1955). Fluctuations of animal populations, and a measure of community stability. *Ecology*, **36**, 533–536.

MacArthur, R. H. (1972). *Geographical Ecology: Patterns in the Distribution of Species*. New York, NY: Harper and Row.

MacArthur, R. H. and Pianka, E. R. (1966). On optimal use of a patchy environment. *American Naturalist*, **100**, 603–609.

MacArthur, R. H. and Wilson, E. O. (1967). *The Theory of Island Biogeography*. Princeton, NJ: Princeton University Press.

Martínez, N. D. (1991). Artifacts or attributes? Effects of resolution on the Little Rock Lake food web. *Ecological Monographs*, **64**(4), 367–392.

Massol, F., Gravel, D., Mouquet, N., *et al.* (2011). Linking community and ecosystem dynamics through spatial ecology. *Ecology Letters*, **14**(3), 313–323.

McCann, K. S., Rasmussen, J. B., and Umbanhowar, J. (2005). The dynamics of spatially coupled food webs. *Ecology Letters*, **8**, 513–523.

McKane, A. J. and Drossel, B. (2005). Modelling evolving food webs. In *Dynamic Food Webs: Multispecies Assemblages, Ecosystem Development, and Environmental Change*, ed. P. C. de Ruiter, V. Wolters, and J. C. Moore, Oxford, UK: Academic Press, pp. 74–88.

Melián, C. J. and Bascompte, J. (2004). Food web cohesion. *Ecology*, **85**, 352–358.

Melián, C. J., Vilas, C., Baldó, F., González-Ortegón, E., Drake, P., and Williams, R. J. (2011). Eco-evolutionary dynamics of individual-based food webs. *Advances in Ecological Research*, **45**, 225–268.

Montoya, J. M. and Raffaelli, D. (2010). Climate change, biotic interactions and ecosystem services. *Philosophical Transactions of the Royal Society B: Biological Sciences*, **365**(1549), 2013–2018.

Montoya, J. M., Rodriguez, M. Á., and Hawkins, B. A. (2003). Food web complexity and higher-level ecosystem services. *Ecology Letters*, **6**, 587–593.

Montoya, J. M., Pimm, S. L., and Solé, R V. (2006). Ecological networks and their fragility. *Nature*, **442**, 259–264.

Montoya, J. M., Woodward, G., Emmerson, M. C., and Solé, R. V. (2009). Press perturbations and indirect effects in real food webs. *Ecology*, **90**, 2426–2433.

Morales-Castilla, I., Matias, M. G., Gravel, D., and Araújo, M. B. (2015). Inferring biotic interactions from proxies. *Trends in Ecology and Evolution*, **30**, 347–356.

Morris, R. J., Gripenberg, S., Lewis, O. T., and Roslin, T. (2014). Antagonistic interaction networks are structured independently of latitude and host guild. *Ecology Letters*, **17**(3), 340–349.

Moya-Laraño, J., Verdeny-Vilalta, O., Rowntree, J., *et al.* (2012). Chapter 1: Climate change and eco-evolutionary dynamics in food webs. *Advances in Ecological Research*, **47**, 1–80.

Ollerton, J. and Cranmer, L. (2002). Latitudinal trends in plant-pollinator interactions: are tropical plants more specialised? *Oikos*, **98**(2), 340–350.

Paine, R. T., Tegner, M. J., and Johnson, E. A. (1998). Compounded perturbations yield ecological surprises. *Ecosystems*, **1**, 535–545.

Pearson, R. G. and Dawson, T. P. (2003). Predicting the impacts of climate change on the distribution of species: are bioclimate envelope models useful? *Global Ecology and Biogeography*, **12**(5), 361–371.

Polis, G. A. (1991). Complex trophic interactions in deserts: an empirical critique of food web theory. *American Naturalist*, **138**, 123–155.

Post, D. M. (2002). The long and short of food-chain length. *Trends in Ecology and Evolution*, **17**(6), 269–277.

Post, D. M., Pace, M. L., and Hairston, Jr., N. G. (2000). Ecosystem size determines food-chain length in lakes. *Nature*, **405**, 1047–1049.

Rall, B. C., Vucic-Pestic, O., Ehnes, R. B., Emmerson, M., and Brose, U. (2010). Temperature, predator–prey interaction strength and population stability. *Global Change Biology*, **16**(8), 2145–2157.

Reiss, J., Bridle, J. R., Montoya, J. M., and Woodward, G. (2009). Emerging horizons in biodiversity and ecosystem functioning research. *Trends in Ecology and Evolution*, **24**(9), 505–514.

Ricklefs, R. E. (1987). Community diversity: relative roles of local and regional processes. *Science*, **235**, 167–171.

Romdal, T. S., Araújo, M. B., and Rahbek, C. (2013). Life on a tropical planet: niche conservatism and the global diversity gradient. *Global Ecology and Biogeography*, **22** (3), 344–350.

Rooney, N., McCann, K. S., Gellner, G., and Moore, J. (2006). Structural asymmetry and the stability of diverse food webs. *Nature*, **442**, 265–269.

Rooney, N., McCann, K. S., and Moore, J. C. (2008). A landscape theory for food web architecture. *Ecology Letters*, **11**(8), 867–881.

Sagarin, R. D., Gaines, S. D., and Gaylord, B. (2006). Moving beyond assumptions to understand abundance distributions across the ranges of species. *Trends in Ecology and Evolution*, **21**(9), 524–530.

Schemske, D. W. (2002). Ecological and evolutionary perspectives on the origins of tropical diversity. In *Foundations of Tropical Forest Biology*, ed. R. L. Chazdon and T. C. Whitmore, Chicago, IL: University of Chicago Press, pp. 163–173.

Schemske, D. W., Mittelbach, G. G., Cornell, H. V., Sobel, J. M., and Roy, K. (2009). Is there a latitudinal gradient in the importance of biotic interactions? *Annual Review of Ecology and Systematics*, **40**, 245–269.

Schleuning, M., Fründ, J., Klein, A.-M., *et al.* (2012). Specialization of mutualistic interaction networks decreases toward tropical latitudes. *Current Biology*, **22**, 1925–1931.

Shurin, J. B., Clasen, J. L., Greig, H. S., Kratina, P., and Thompson, P. L. (2012). Warming shifts top–down and bottom–up control of pond food web structure and function. *Philosophical Transactions of the Royal Society B: Biological Sciences*, **367**, 3008–3017.

Solé, R. V. and Montoya, J. M. (2001). Complexity and fragility in ecological networks. *Proceedings of the Royal Society B: Biological Sciences*, **268**, 2039–2045.

Stevens, G. C. (1989). The latitudinal gradients in geographical range: how so many species co-exist in the tropics. *American Naturalist*, **133**, 240–256.

Thompson, R. M., Brose, U., Dunne, J. A., *et al.* (2012). Food webs: reconciling the structure and function of biodiversity. *Trends in Ecology and Evolution*, **27**(12), 689–697.

Thuiller, W., Brotons, L., Araújo, M. B., and Lavorel, S. (2004). Effects of restricting environmental range of data to project current and future species distributions. *Ecography*, **27**(2), 165–172.

Trøjelsgaard, K. and Olesen, J. M. (2013). Macroecology of pollination networks. *Global Ecology and Biogeography*, **22**(2), 149–162.

Tylianakis, J. M., Tscharntke, T., and Lewis, O. T. (2007). Habitat modification alters the structure of tropical host–parasitoid food webs. *Nature*, **445**(7124), 202–205.

Tylianakis, J. M., Didham, R. K., Bascompte, J., and Wardle, D. A. (2008). Global change and species interactions in terrestrial ecosystems. *Ecology Letters*, **11**(12), 1351–1363.

Vasseur, D. A. and McCann, K. S. (2005). A mechanistic approach for modeling temperature-dependent consumer–resource dynamics. *American Naturalist*, **166**(2), 184–198.

Vázquez, D. P. and Stevens, R. D. (2004). The latitudinal gradient in niche breadth: concepts and evidence. *American Naturalist*, **164**(1), E1–19.

Warren, P. H. (1990). Variation in food-web structure: the determinants of connectance. *American Naturalist*, **136**(5), 689–700.

Wisz, M. S., Pottier, J., Kissling, W. D., *et al.* (2013). The role of biotic interactions in shaping distributions and realised assemblages of species: implications for species distribution modelling. *Biological Reviews*, **88**(1), 15–30.

Wootton, J. T. and Emmerson, M. (2005). Measurement of interaction strength in nature. *Annual Review of Ecology, Evolution, and Systematics*, **5**, 419–444.

Yvon-Durocher, G., Jones, J., Trimmer, M., Woodward, G., and Montoya, J. M. (2010). Warming alters the metabolic balance of ecosystems. *Philosophical Transactions of the Royal Society B: Biological Sciences*, **365**(1549), 2117–2126.

Yvon-Durocher, G., Montoya, J. M., Woodward, G., Jones, J., and Trimmer, M. (2011). Warming increases the proportion of primary production emitted as methane from freshwater mesocosms. *Global Change Biology*, **17**(2), 1225–1234.

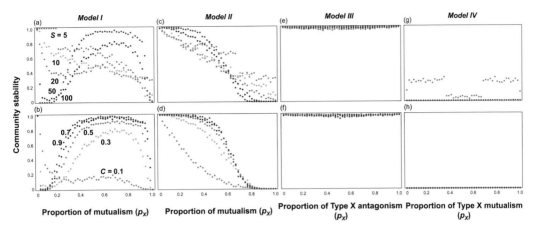

Figure 2.2 Effect of interaction-type mixing on community stability with varying complexity for Model I (a, b), Model II (c, d), Model III (e, f), and Model IV (g, h). For upper panels (a, c, e, g), red ($S = 5$), orange ($S = 10$), green ($S = 20$), blue ($S = 50$) and purple ($S = 100$); $C = 0.3$. For lower panels (b, d, f, h), red ($C = 0.1$), orange ($C = 0.3$), green ($C = 0.5$), blue ($C = 0.7$), and purple ($C = 0.9$); $S = 50$. Community stability was measured as the proportion of locally stable community models.

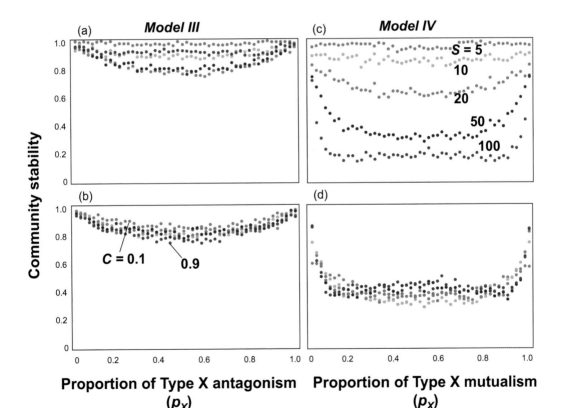

Figure 2.3 Effect of interaction-type mixing on community stability with varying complexity for Model III (a, b; antagonism) with weaker self-regulation and IV (c, d; mutualism) with stronger self-regulation. For upper panels (a, c), red ($S = 5$), orange ($S = 10$), green ($S = 20$), blue ($S = 50$), and purple ($S = 100$); $C = 0.3$; $s_i = 0.0$–0.5. For lower panels (b, d), red ($C = 0.1$), orange ($C = 0.3$), green ($C = 0.5$), blue ($C = 0.7$), and purple ($C = 0.9$); $S = 50$; $s_i = 0.0$–10.0. Community stability was measured as the proportion of locally stable community models.

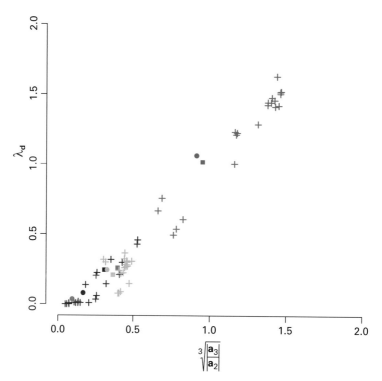

Figure 3.4 Correlation between key feedbacks and stability. Three-link and two-link predator–prey feedbacks $\sqrt[3]{\dfrac{|a_3|}{|a_2|}}$, where a_3 and a_2 are coefficients of the characteristic polynomial of the community matrices representing the sum of the feedbacks of the 3rd and 2nd levels, respectively (Neutel and Thorne, 2014), and system vulnerability λ_d for empirical parameterizations of the interaction strengths of the Antarctic dry and wet tundra systems and 21 soil food webs (where detritus does not affect stability) with plus signs for the 21 soil webs, compared with symmetric (yellow symbols) and asymmetric (green symbols) parameterizations ($N = 71$, $R^2 = 0.97$, $P < 10^{-15}$). Redrawn with permission. © 2014 Neutel and Thorne. *Ecology Letters* published by John Wiley & Sons Ltd and CNRS.

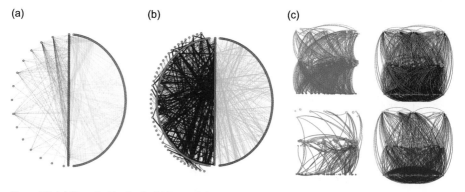

Figure 6.2 (a) Donaña Ecological Network based on data from Melián *et al.* (2009) showing herbivores (left), plants (middle), and pollinators and seed dispersers (right). Antagonistic links are in blue and mutualistic ones in orange. (b) Species interaction network at Norwood Farm, Somerset, UK based on data from Pocock *et al.* (2012). Antagonistic links are in blue and mutualistic ones in orange plants in the middle. (c) Chilean interaction web based on data from Kéfi *et al.* (2015). Trophic network in yellow (top left), positive non-trophic network in green (bottom left), negative non-trophic network in blue (bottom right), and the top-right network shows all trophic (in yellow) and all non-trophic links (positive and negative) in purple; between the 104 species of the community.

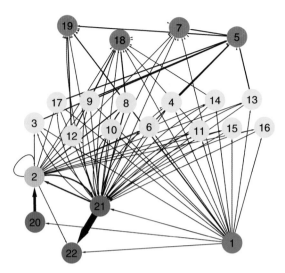

Figure 11.1 The food web of the Lake Scuro ecosystem. Keys are: 1 phytoplankton; 2 microbial loop; 3 *Keratella cochlearis*; 4 *Ascomorpha* spp.; 5 *Synchaeta* sp.; 6 *Polyarthra* sp.; 7 predator rotifers; 8 *Conochilus unicornis-hippocrepis*; 9 other rotifers; 10 *Diaphanosoma brachyurum*; 11 *Daphnia longispina*; 12 other cladocerans; 13 *Eudiaptomus nauplii*; 14 *Eudiaptomus* CI-CII-CIII; 15 *Eudiaptomus* CIV-CV; 16 *Eudiaptomus intermedius* adults; 17 Cyclopoid copepods nauplii; 18 Cyclopoid copepods copepodites; 19 Cyclopoid copepods adults; 20 WDOC; 21 WPOC; 22 BPOC.

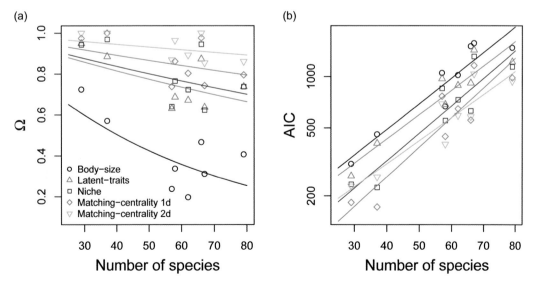

Figure 12.1 Performance of the five statistical models. Panel (a) shows the fraction of correctly fitted trophic links as a function of the number of species in the eight fitted food webs. More complex models in terms of the number of latent traits provide better fits. Panel (b) gives the AIC as a function of the number of species. It indicates that, based on this criterion, the increase in the complexity of the models is justified.

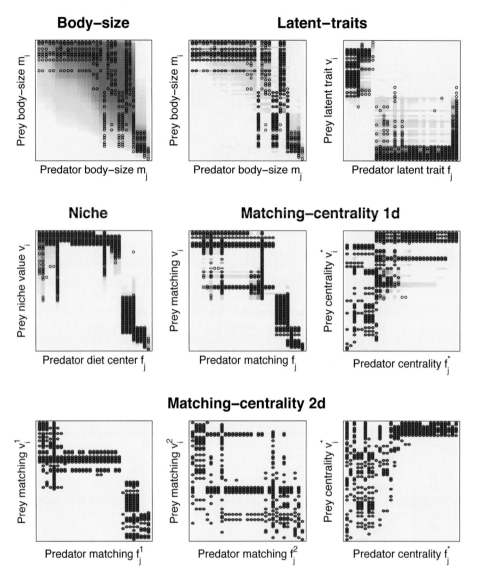

Figure 12.2 Representation of the food web of Tuesday Lake in the latent-traits space. Each panel represents the adjacency matrix, with the dots indicating a trophic interaction between a predator (columns) and a prey (row). The color, from yellow to red, indicates increasing fitted linking probability of the respective models. Species are ordered according to the relevant variable or latent trait (see axis legends).

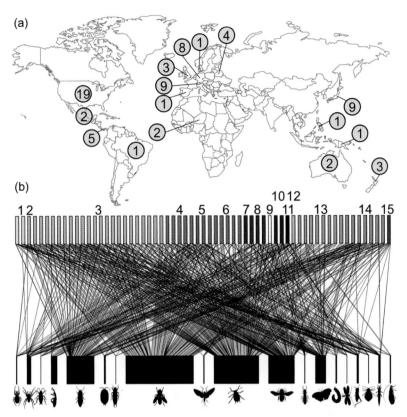

Figure 13.1 (a) Global map with the number of spider species per country for which diets were reported with sufficient level of detail to be included here. All data come from visual searches for web-building spiders and their prey in primary studies. For spider species that occurred more than once in the database only the dataset with most prey items was included in the global metaweb. (b) Global metaweb of web-building spider species (N = 63) and the relative predation on prey orders (N = 19). Spider species are color coded according to family identity and numbered: 1 Agelenidae, 2 Amaurobiidae, 3 Araneidae, 4 Dictynidae, 5 Eresidae, 6 Linyphiidae, 7 Mimetidae, 8 Nephilidae, 9 Pholcidae, 10 Pisauridae, 11 Scytodidae, 12 Sicariidae, 13 Tetragnathidae, 14 Theridiidae, 15 Uloboridae. Prey orders from left to right are: Acari, Araneae, Blattodea, Collembola, Coleoptera, Isopoda, Dermaptera, Diptera, Ephemeroptera, Hemiptera, Hymenoptera, Isoptera, Lepidoptera, Myriapoda, Neuroptera, Orthopteroidea, Psocoptera, Thysanoptera, and Trichoptera.

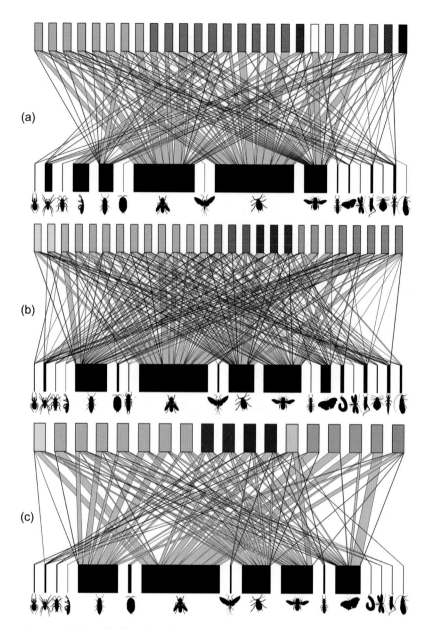

Figure 13.2 Metawebs for web-building spider species and prey orders in (a) agricultural (N = 26 spider species), (b) semi-natural/natural (N = 27), and (c) forest (N = 18) ecosystems. For details on prey orders and color codes for spider species please refer to Figure 13.1; for details on habitat types in each ecosystem class please refer to Table 13.1.

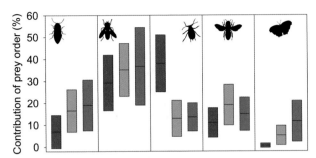

Figure 13.3 The average relative contribution (±95% CI) of prey orders to the diet of web-building spiders in different ecosystems (agricultural, red; semi-natural/natural, green; forest, blue) for the most discriminating prey orders (Coleoptera, Diptera, Hemiptera, Hymenoptera, and Lepidoptera) according to similarity percentage analysis (individual contribution >10%).

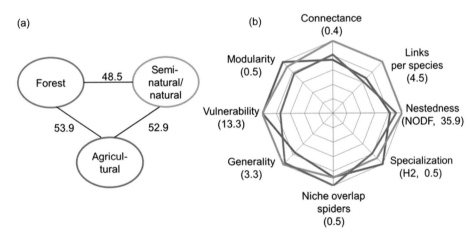

Figure 13.5 Comparison of network structure between ecosystem types. (a) Distance between ecosystem metawebs based on all network metrics (standardized by maximum) and Euclidean distances. Distances between circles are scaled according to Euclidean distance values shown at each link (higher values = higher dissimilarity). (b) Network metrics standardized by the maximum value (100%) for each spider–prey metaweb in different ecosystems (agricultural, red; semi-natural/natural, green; forest, blue). Maximum values are given in parentheses with each network metric.

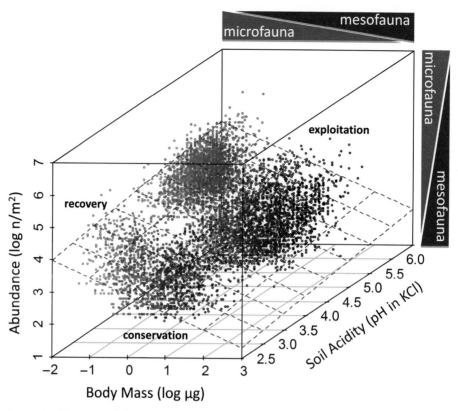

Figure 14.1 3D-scatter of the occurrence of soil microfauna (nematodes) and mesofauna (mites, collembolans, and enchytraeids) in Dutch ecosystems either under management (liming, grazing, and fertilization) or in natural conditions (Scots Pine forests). The empirical trend derived from 7134 populations of soil invertebrates belonging to 135 edaphic communities (soil biota and food-web data publicly downloadable from Cohen and Mulder, 2014), here as stretched 3D plane, corroborates previous results in managed pastures and abandoned grasslands (Mulder and Elser, 2009). A multiple-lines surface (here as blue grid) contains parallel mass–abundance regression lines (each of them representing one soil food web of a single location with its specific soil pH) and shows how allometric scaling tends to become steeper in strongly acidic soils and shallower in slightly acidic, more neutral soils.

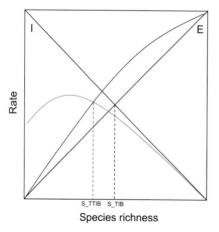

Figure 17.1 Trophic theory of island biogeography adapted from Gravel *et al.* (2011). The classic TIB is depicted in black. Equilibrium species richness (S_TIB) is reached when immigration rate is equal to extinction rate (intersection between I and E; i.e., black dotted line). The TTIB is depicted in colors (orange and blue for immigration and extinction rates, respectively). Equilibrium species richness (S_TTIB) is represented by the red dotted line.

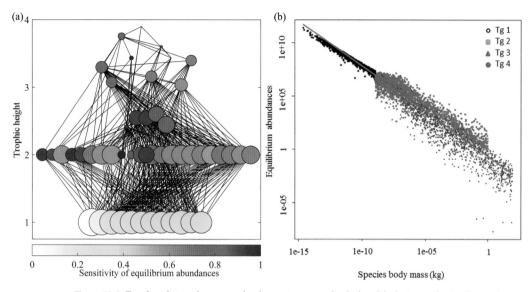

Figure 19.1 Food-web topology, species importance, and relationship between body size and abundance. (a) The trophic structure of one of 100 model communities generated by the sequential assembly algorithm, and the equilibrium abundances and structural importance of its species (as determined by CSA). Each node represents a species and the edges the feeding relationships. Node size corresponds to the relative equilibrium abundance of species, with node color denoting the relative sensitivity of equilibrium abundances (see color bar) to a perturbation of the growth/mortality rate of the species. (b) Equilibrium abundance as a function of species average body mass (m) for all species in the 100 model communities. The red line represents the fixed effect estimate of the linear mixed effect model: $log(N^*_{kj}) = \beta + \omega * log(m_{k,j}) + b_j + e_{kj}$, where N_{kj}^* is the equilibrium abundance of species k in food web j, β is the fixed effect intercept, $m_{k,j}$ is the average body mass of species k in food web j, b_j is the random effect on the intercept of each food web j, and e_{kj} is the residual abundance of species k in food web j. The slope of the fixed effect line (ω) equals –0.85 with a standard deviation of 0.0029 and is within the range of what has been observed empirically (Brown *et al.*, 2004; Reuman *et al.*, 2008). (AIC indicated that random effect on slopes was not needed.)

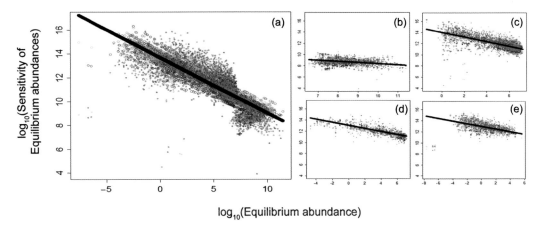

Figure 19.2 Species structural importance is inversely related to abundance in ecological communities. (a) Sensitivity of the distribution of equilibrium densities, to a perturbation of the growth/mortality rate of a species, as a function of equilibrium abundance of the species, across all communities. (b)–(e) The same relationship as in (a) but within trophic groups 1 (b), 2 (c), 3 (d) and 4 (e), respectively. The bold black line in each subplot represents the global relationship (fixed effects only) of mixed-effect models treating communities as a random effect on model intercepts and slopes (for all individual random effects, see Table S19.5 in the Appendix, and for estimates of model parameters see Table 19.1). Globally, sensitivity is significantly and inversely related to species abundance within all trophic groups (b)–(e) as well as within whole communities (a) (P <0.0001 for all models).

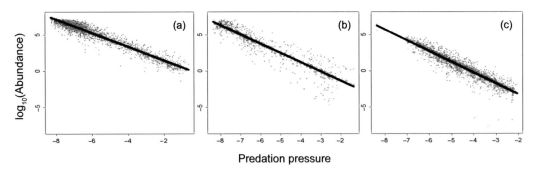

Figure 19.3 Consumer species with low abundance tend to be strong interactors. The relationship between predation pressure and equilibrium abundances for (a) trophic group 2, (b) trophic group 3, and (c) trophic group 4. The bold black line in each subplot represents the global relationship across all 100 ecological communities in the mixed-effect model, i.e., with random effects depending on community identity (for all individual random effects see Table S19.6 in the Appendix, and for estimates of model parameters see Table S19.5 in the Appendix). All slopes are significantly different from zero, P <0.001.

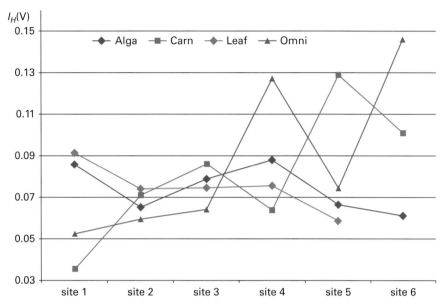

Figure 20.2 The relative importance of four particular trophic groups during the simulations, along the food-web gradient. In site 1, algae and leaf particles have larger effects on community dynamics than all other nodes. In site 6, omnivores and carnivores are the most influential group. This suggests a transition from bottom–up to top–down control. The x axis shows space (sampling sites along the river) and the y axis shows the $I_H(V)$ simulated importance values.

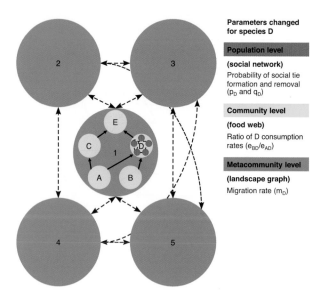

Figure 21.1 Structure of the hierarchical model. Five metapopulations occupy and migrate through five patches in a complete landscape graph. In each patch, individuals of the five species interact in a food web and conspecific individuals form and remove social interactions. The food web and social network (of species D only) are illustrated here for the central patch but they apply to the whole landscape graph.

18 Food-Web Dynamics When Divergent Life-History Strategies Respond to Environmental Variation Differently: A Fisheries Ecology Perspective

Kirk O. Winemiller

18.1 Introduction

The fundamental goal of food-web ecology is to understand essential ecosystem components, their features, and interactions such that predictions about dynamics can be made with reasonable precision and accuracy. The degree to which ecologists can be successful in this endeavor is critical for management of fisheries, invasive species, agricultural pests, infectious diseases, and a host of other ecological challenges. A frequently recognized, yet rarely addressed, challenge in food-web modeling is the diversity of responses to environmental variation by organisms with different combinations of functional traits, such as those that define life-history strategies. Spatial and temporal variation of influential environmental variables differentially affects recruitment and population dynamics of species having divergent life-history strategies. This, in turn, affects their interactions with resources and each other. This chapter briefly examines the relationship between life-history theory and food-web ecology, and then discusses how this relationship affects ecological applications using fisheries management as an example. Models that focus solely on networks of consumer–resource interactions without accounting for interspecific differences in responses to environmental variation cannot simulate real-world food-web dynamics. Individual-based models can simulate interactions among organisms with divergent traits as well as responses to environmental factors, and recently developed individual-based models have shown good potential for predicting food-web dynamics. Identification of manageable sets of influential functional traits is critical for success. Validation of individual-based models requires extensive empirical data at multiple hierarchy levels, ranging from individual to population, and spatio-temporal scales. This chapter briefly explores evidence for abiotic environmental mediation of population and trophic dynamics, and the potential for individual-based modeling approaches to simulate the dynamics of food webs supporting multispecies fisheries.

18.2   Food-Web Paradigms

Food-web ecology has progressed from crude descriptions toward modeling dynamics in complex systems. The basic food-web paradigm originated with descriptions of predator–prey networks (Camerano, 1880; Summerhayes and Elton, 1923), followed by estimates of productivity and energy assimilation within networks of functional groups (Lindeman, 1942). Although these depictions of food webs appear crude by today's standards, these works ushered in new perspectives in ecology that shifted lines of inquiry away from autecology with a focus on organismal responses to environmental gradients and patterns of species abundance and distribution, toward an emphasis on species interactions and local population dynamics. Food-web research has subsequently followed multiple tracks, including theoretical models seeking to understand relationships among constructs such as complexity, stability, and invasibility, and simulation models attempting to simulate population and trophic and dynamics, or nutrient, biomass, and energy flow.

Theoretical ecologists have examined the relationship between complexity and diversity using continuous-time, linear models that depict each interacting population as a single parameter representing abundance, N_i. The classic study of complexity and stability by May (1973) has been followed by a succession of models examining alternative stability criteria based on assumptions of equilibrium dynamics and using alternative food-web structures (Haydon, 1994; Allesina and Pascual, 2008; Alcántara and Rey, 2012; McCann, 2012; Tang et al., 2014). Among theoretical models testing stability criteria, only a few have divided populations into life stages (van Kooten et al., 2005; Rudolf and Lafferty, 2011) or spatial subunits (Hastings, 1996). Theoretical models created to examine the complexity–stability relationship generally have not simulated abiotic environmental influences on population dynamics and species interactions, and for this reason they have limited capabilities to predict the dynamics of real-world systems. Moreover, the search for equilibrium dynamics seems misguided given evidence that most real-world food webs exhibit non-equilibrium or transient dynamics (Hastings et al., 1993; McCann, 2012).

There exists a long history of network modeling to simulate flow of biomass and energy among food-web compartments, but attempts to simulate population dynamics within real-world food webs are quite recent. A major challenge for the latter is that population dynamics are influenced by more than just predator–prey interactions. Additional factors regulating populations include (1) habitat availability and quality; (2) ecosystem productivity; (3) reproduction and recruitment; (4) dispersal; and (5) types of population interactions other than consumer–resource (Figure 18.1). Habitat availability and quality are fundamental to fitness and population abundance, and are the basis for the Grinnellian niche concept and species-distribution models (Soberón, 2007).

Ecosystem productivity determines the nutrients and energy available to organisms, which in turn influence population abundance and community structure (Worm and Duffy, 2003), including the distribution of biomass within and between vertical trophic positions (Post, 2002).

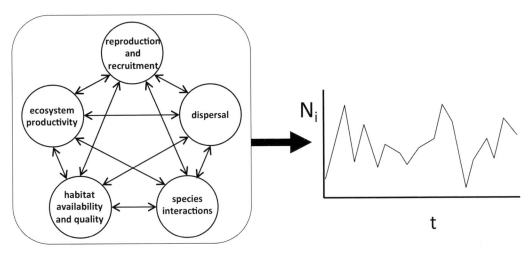

Figure 18.1 Scheme highlighting fundamental ecological factors that interact to influence population dynamics. Most food-web models deal explicitly with species interactions and make assumptions about the other aspects.

Any attempt to simulate population and biomass dynamics in food webs must recognize that multiple factors are influenced by environmental variation. From the classic studies of Andewartha and Birch (1954) to the present, the ecological literature has been dominated by research demonstrating how abiotic environmental drivers influence population dynamics (Turchin, 2003). Temperature and precipitation are particularly influential drivers of population dynamics of terrestrial plants, insects, and vertebrates (Lawton, 1995; Madsen and Shine, 2000; Sinclair, 2003) as well as soil microbial communities (Zeglin *et al.*, 2013). River and stream communities are strongly influenced by hydrology (Power *et al.*, 2008), plankton in lakes are affected by seasonal changes in temperature and insolation (Vasseur and Gaedke, 2007), and oceanographic currents strongly affect marine communities (Woodson *et al.*, 2012). Trophic interactions can be directly influenced by temperature (Siddon *et al.*, 2013), precipitation (Barton and Ives, 2014), hydrology (Winemiller, 1990), and other abiotic environmental factors, as well as dispersal within metacommunities (Holt, 2009; Bellmore *et al.*, 2015).

18.3   Environmental Variation, Life-History Strategies and Food-Web Dynamics

A major challenge in food-web modeling is the diversity of demographic responses to environmental variation by species with different life-history strategies (Polis *et al.*, 1996). Populations with divergent life-history strategies respond to temporal and spatial environmental variation in ways not captured by models that focus solely on networks of consumer–resource interactions. Life-history theory predicts how populations respond to patterns of environmental variation (Pianka, 1970; Grime, 1977; Southwood, 1977; Stearns, 1992), and therefore can guide the construction of models seeking to simulate

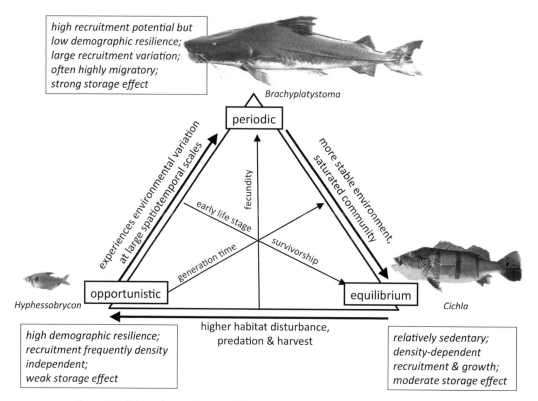

high recruitment potential but low demographic resilience; large recruitment variation; often highly migratory; strong storage effect

Brachyplatystoma

periodic

experiences environmental variation at large spatiotemporal scales

more stable environment, saturated community

fecundity

early life stage

generation time

survivorship

opportunistic

equilibrium

Hyphessobrycon

Cichla

high demographic resilience; recruitment frequently density independent; weak storage effect

higher habitat disturbance, predation & harvest

relatively sedentary; density-dependent recruitment & growth; moderate storage effect

Figure 18.2 Triangular continuum of life-history strategies based on trade-offs among three fundamental demographic parameters (generation time, fecundity, survival) with Amazon freshwater fishes providing examples of three endpoint strategies (opportunistic, periodic, and equilibrium). Different environmental conditions select for each endpoint strategy, and species' positions within the continuum are predicted to influence relative fitness and population dynamics for a given environment. Based on Winemiller (1989, 1992, 2005) and Winemiller and Rose (1992).

real-world food webs. An example is a triangular continuum of life-history strategies (Winemiller, 1989, 1992, 2005; Winemiller and Rose, 1992) that predicts fitness differences associated with contrasting patterns of environmental variation. The model contrasts three endpoint strategies defined by different combinations of traits constrained by physical, physiological, and demographic trade-offs (Figure 18.2). Demographic trade-offs are defined by the relationship $r = \Sigma(l_x{*}m_x)/T$, so that no endpoint strategy can maximize fitness (indexed here by r, the intrinsic rate of increase) by simultaneously maximizing survivorship (l_x), fecundity (m_x), and minimizing generation time (T). The most effective means to increase r is to minimize T, and this defines an opportunistic endpoint strategy adaptive in environmental settings that favor colonization or high population resilience in response to high juvenile and adult mortality.

Opportunistic strategists are small organisms that have early maturation and high reproductive effort with frequent reproductive bouts, typically with minimal parental investment in individual offspring.

Alternatively, in environmental settings with strong density-dependent influences on survival, growth, and reproduction (relatively saturated communities within relatively stable habitats), selection favors delayed maturation that allows greater investment in somatic functions allowing for greater competitive and predator defense capabilities, and greater investment in individual offspring (greater offspring size, parental care). In this *equilibrium strategy*, delayed maturation trades off with short generation time, and greater investment in individual offspring (allowing higher survivorship during juvenile life stages) trades off with fecundity.

A third strategy, the *periodic strategy*, maximizes fecundity at the expense of investment per offspring (clutch size–offspring size trade-off). Periodic strategists tend to reproduce in discrete bouts, often annually, that coincide with environmental periodicity that has intervals favorable for early life stage survival without parental care. Periodic strategists tend to have high temporal and spatial variation in recruitment success that can result in dominant cohorts within populations of long-lived species, such as most large fishes (Winemiller and Rose, 1992) and trees (Higgins *et al.*, 2000). The periodic strategy also tends to be associated with migration that exploits large-scale patterns of spatial and temporal environmental variation.

Figure 18.2 provides a schematic of this triangular life-history model, showing examples of three endpoint strategies among diverse fishes of the Amazon River in South America. Opportunistic strategists, *Hyphessobrycon* species mature at small sizes, breed frequently by scattering eggs onto aquatic vegetation and provide no parental care, and typically suffer high mortality from predation at all life stages, with a life span of one to two years. These small fishes inhabit shallow habitats in streams and along margins of rivers and lakes and have high population resiliency in response to changing environmental conditions associated with hydrologic dynamics. As an example of the equilibrium strategy, the tucunaré (*Cichla* species) are predatory fishes of rivers and lakes. Tucunarés have relatively low fecundity but aggressively protect their offspring for extended periods, and juvenile survivorship is comparatively high. The piraíba (*Brachyplatystoma filamentosum*) grows to over 2.5 m and 200 kg and provides an extreme example of the periodic strategy. This apex predator migrates long distances and is an important fishery resource. During annual flood pulses, this giant catfish spawns batches of hundreds of thousands of tiny eggs, most of which obviously fail to survive for more than a few days even under optimal conditions. Like most periodic strategists, only a very tiny fraction of an annual cohort needs to survive during favorable but infrequent periods (episodic recruitment) in order to maintain viable populations of this long-lived species. In diverse biological communities, species with divergent life-history strategies perceive and respond to environmental variation at different spatiotemporal scales, which poses a difficult challenge for food-web modeling.

A generally under-appreciated aspect of life-history strategies that greatly affects population dynamics is the storage effect (Chesson and Warner, 1981; Chesson, 2000b). Organisms often benefit from favorable periods and are able to store these benefits to enhance survival or competitive ability during unfavorable periods of resource scarcity or crowding and intensified competition and predation. Variation in environmental favorableness can occur either in time (e.g., seasonality) or space (landscape heterogeneity). Large-scale environmental variation can induce resource pulses that affect early life stage survival (Koenig and Liebhold, 2005) causing episodic recruitment (Cheal *et al.*, 2007; Menge *et al.*, 2011) as commonly observed in populations of periodic strategists. The storage effect can increase species coexistence under competition scenarios if species respond differently to patterns and degrees of environmental variation for relevant factors in a manner that reduces the negative impact of unfavorable periods. As a result, individual and population gains during favorable periods are stored to buffer against losses during unfavorable periods (or within bad patches on landscapes).

The storage effect can increase species coexistence when (1) demographic decoupling of population dynamics of coexisting species arises from partially uncorrelated responses to environmental variation; (2) the strength of competition covaries with environmental conditions; and (3) certain life-history traits, such as seed banks or long-lived adults, limit the impact of competition under unfavorable environmental conditions (Angert *et al.*, 2009). All communities probably have each of these characteristics, but they seem especially apparent in freshwater and marine communities that support important fisheries. Most large fishes have a strongly periodic life-history strategy that embodies the storage effect in long-lived adults that undergo repeated attempts at reproduction (often annual spawning bouts) that result in strong recruitment during favorable periods, and little or no reproductive success during times and in places where conditions for early life stage survival are unfavorable. Thus the periodic strategy represents a bet-hedging strategy. For these species, recruitment dynamics are expected to be weakly or only rarely density dependent (Rose *et al.*, 2001; Winemiller, 2005). Migration is another mechanism that can produce the storage effect when organisms are able to leave areas with degraded conditions or depleted resources and enter more favorable areas. Long-lived, live-bearing species, including marine mammals and many large sharks and rays (equilibrium strategists), often manifest the storage effect in this manner by migrating seasonally to nursery/calving areas to give birth. Organisms with less longevity would manifest a weaker storage effect; however, even for small species, there sometimes are physiological mechanisms for energy and nutrient storage that transfer gains during periods of resource abundance to survival during periods of resource scarcity. Plants achieve the storage effect via seed dormancy that allows fitness gains during favorable conditions and persistence under unfavorable conditions. Microbes, insects, and even some aquatic animals (e.g., brine shrimp, fairy shrimp, and annual killifishes) have evolved this propagule-banking capability.

Given the dominance of periodic strategists in marine fish and forest vegetation communities, the food webs of these communities should be fundamentally influenced by the storage effect. In marine ecosystems, large-scale patterns of environmental variation strongly affect fish-stock dynamics, because episodic recruitment may sustain breeding populations of long-lived species for many years or even decades. Bakun and Broad (2003) proposed that efforts to model fish-stock dynamics in the ocean should move away from conventional trophodynamics and instead emphasize what they call *loopholes* among factors controlling early life stage survival. Loopholes involve heterogeneity in ocean productivity, predator density, and disruptive perturbations associated with gyres, upwellings, and other kinds of currents driven by climate patterns and dynamics interacting with geographic features. Loopholes can result in skewed patterns of reproductive success of the sort that lead to pulses of recruitment that yield stocks dominated by strong year classes (i.e., a strong storage effect in long-lived fishes with a periodic strategy). Citing several examples, they contend that large-scale climatic influences can overwhelm local factors in determining the dynamics of regionally inter-connected marine fish stocks. Food-web models that assume density dependence and equilibrium dynamics are unsuitable tools for marine fisheries management. Instead, fish-eries management should focus on opportunities for greater harvest during periods when environmental conditions support high recruitment and stock abundance, with harvest reduced when such conditions are lacking in order to conserve reproductive potential.

The influence of life-history strategies on trophic dynamics can be detected in commu-nity biomass distribution. In wasp-waist food webs, a mid trophic-level species exerts top–down control on its food and bottom–up control on its predators. Community biomass is dominated by one or few species at middle trophic levels, with lower and higher levels comprising numerous species. Most examples of wasp-waist food webs involve opportu-nistic strategists in marine pelagic systems subject to strong interannual variation in climate and oceanographic conditions (e.g., domination by anchovies, sardines, and krill responsive to the El Niño Southern Oscillation), but also may apply to terrestrial systems with herbivorous periodic strategists that undergo outbreaks and migrations in response to large-scale variation in rainfall and temperature affecting vegetation (e.g., locust plagues). Mass migrations spatially redistribute top–down effects (foraging footprints) in local food webs (Atkinson *et al.*, 2014). At the same time, these aggregations of mid trophic-level species provide major resource pulses for their predators (bottom–up effects), with major ramifications for local food-web dynamics.

Irrespective of the environmental drivers involved, it is clear that pulsing of resources and consumer populations has profound implications for food-web dynamics. Recurrent resource pulses can facilitate coexistence via the storage effect, or, conversely, pulses can interfere with coexistence mechanisms that may be effective in stable environments (Holt, 2008). Holt (2008) proposed that short-term responses to resource pulses are qualitatively different from longer term responses, and that pulses can induce transitions between alternative stable states. This leads to another fundamental challenge for modeling food-web dynamics – differential rates of response to abiotic environmental factors and species interactions as functions of life-history strategies. Carpenter and Turner (2001) offered the following modeling generalization: "slow processes are treated as parameters, whereas fast processes may be solved at equilibrium." They noted that turnover times in ecosystem components vary by at least 12 orders of magnitude, from nutrient cycling at one end of the spectrum to weathering of rocks at the other. Population generation time varies by more than two orders of magnitude among fishes alone, and varies by at least six orders of magnitude across all types of living organisms. Carpenter and Turner (2001) discussed how disparities in turnover times of ecosystem components results in time lags, which in the extreme case produce what they called *ecological legacies*, enduring features that can influence the dynamics of ecological components. Time lags and ecological legacies within real-world food webs pose additional challenges for conventional food-web models based on continuous-time, linear equations (see McCann, 2012, for a discussion of modeling time lags).

18.4   Integrating Life-History Traits into Food-Web Models

How then can we capture the pervasive influence of environmental variation and life-history strategies in models simulating food-web dynamics? Over the past 25 years, several investigators have included size and age structure in models simulating population dynamics and species interactions (Yodzis and Innes, 1992; de Roos *et al.*, 2003a; Hartvig *et al.*, 2011; Rudolf and Lafferty, 2011). Recently, Zhou *et al.* (2013) created continuous-time versus discrete-time versions of predator–prey models with or without stage-structured populations in which predation was continuous and reproduction was seasonal. A discrete-time model with seasonal reproduction, whether stage structured or non-stage structured, produced locally stable equilibria, but a continuous-time model with an instantaneous approximation of seasonal reproduction produced a limit cycle. The continuous-time model seemed unable to match life-history attributes, and, importantly, seasonal reproduction may have a stabilizing effect on predator–prey interactions. Because life-history parameters in the stage-structured models affected asymptotic dynamics, they further concluded that discrete-time models have the best potential for modeling real-world systems.

The influence of life history on trophic dynamics is further demonstrated by a study that compared food-web models with and without stage-structured populations (Fujiwara, 2016). These models allowed simulated populations to invade and become extinct over time. In every case, randomly formed food webs were unstable, with invasion and extinction rates declining over time. Rules preventing trophic interactions among similar-sized organisms greatly increased the persistence of consumer populations in the unstructured food web, but had relatively little influence on the structured food web. Increasing the carrying capacity of primary producers reduced consumer extinction in the unstructured food web, but increased it in the structured food web. Mean trophic level was higher in the structured food web. These kinds of outcomes would not be possible from aggregate models that, for example, simulate populations as singular units interacting according to coefficients in numerical response functions.

The next challenge is to integrate environmental variation with life-history strategies and trophic interactions in a food-web model (Yang and Rudolf, 2010; Varughese, 2011). Such models are rare for obvious reasons – formalization of conceptual models with different domains (i.e., parameters, scales, turnover times, etc.) into a quantitative model or set of linked quantitative models is challenging, and acquisition of empirical data needed to build and test models at relevant scales is extremely challenging. Considering various other kinds of network interactions (besides consumer–resource) also can influence population dynamics (Olff *et al.*, 2009), the task appears even more daunting. Given that it would be virtually impossible to modify traditional continuous-time, linear models to simulate effects from life-history strategies, storage effects, and non-consumptive aspects of species interactions, perhaps it is time to explore a fundamentally different approach for modeling food-web dynamics.

Individual-based (or agent-based) modeling approaches have been used extensively to model complex ecological systems (Grimm *et al.*, 2005), including influences of environmental drivers and life-history strategies on population dynamics (Van Winkle *et al.*, 1993), and recent research suggests that this approach can be extended to model food-web dynamics. Individual-based models simulate traits and processes affecting growth, survival, and interspecific interactions of individual organisms or life-stage cohorts. Such approaches may be feasible for simulating food-web dynamics of systems for which there are sufficient empirical data, including long-term data for climatic and abiotic environmental factors and population abundance as well as knowledge of behavioral and physiological mechanisms. Schmitz and Booth (1997) created a spatially explicit, individual-based model to investigate the influence of organism behavior on the dynamics of a tri-trophic food chain. Their model allowed consumers to select food resources based on both perception and intentional behavior under high and low degrees of trophic efficiency. Model food webs were persistent only when consumers foraged adaptively, especially when trophic efficiencies were set at about 10%, the level reported for many real-world systems. More recently, Melián *et al.* (2014) created an individual trait-based, stochastic model and showed that intrapopulation variation in the strength of prey selection by different predator phenotypes had a strong effect on food-web dynamics. Strongly connected predators preferentially consumed common prey, and weakly connected predators selected rare prey. They called for compilation of multispecies datasets resolved at the individual level that can be used to discriminate among alternative food-web models.

Hartvig *et al.* (2011) present an individual-based model that combined features of traditional food-web models lacking size structure with trait-based modeling that incorporated size-related allometry and physiological functions to simulate individual growth and ontogenetic niche shifts. Their model parameters were scaled according to body size and size at maturation, and thus were species independent. Their model could be solved analytically by assuming that the community-size spectrum follows a power law, and they were able to analyze outputs from a series of food-web simulations at community, species, individual, and trait levels. Siddon *et al.* (2013) developed an individual-based model to investigate how interannual variation in water temperature, prey composition, and prey quality affected growth of juvenile walleye pollock in the Gulf of Alaska. Their model included a mechanistic feeding component dependent on larval development and behavior, local prey densities and size, and physical oceanographic conditions. Spatial variation in water temperature and prey composition created hot spots conducive to enhanced growth and survival, providing support for the match–mismatch hypothesis for episodic recruitment in marine fish.

Giacomini *et al.* (2013) created an individual-based model that simulated fish community assembly, trophic interactions, and different life-history strategies involving multiple physiological trade-offs that constrained performance. To produce a continuum of life-history strategies, their model integrated a biphasic growth model with size-structured foraging models. The model also used a type III functional response to simulate a gradient of prey accessibility. Fishes were represented explicitly as individuals, but resources were modeled as biomass compartments as in traditional state variable models. Carrying capacity varied seasonally, following a sine function with a period of one year. Prey accessibility was modeled according to a constraint on maximum attack rates of predators' functional responses. Each fish had an age, developmental stage, and weight that was divided among irreversible mass, reversible mass, and gonad mass (Figure 18.3). Following the biphasic growth model, an adult fish

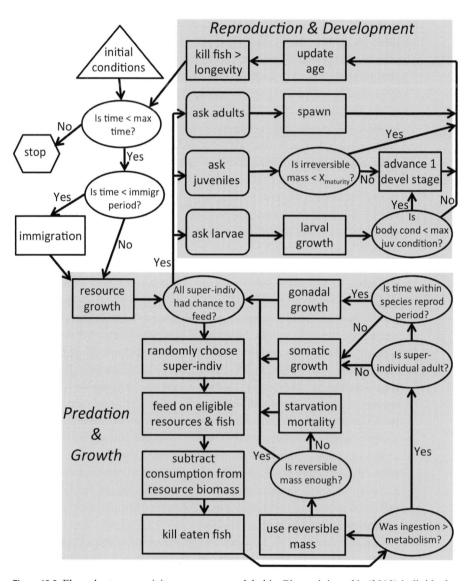

Figure 18.3 Flow chart summarizing processes modeled in Giacomini *et al.*'s (2013) individual-based model of fish community assembly. Their model simulated immigration, feeding interactions, growth, survival, development, and reproduction. (Rectangles = processes, ovals = decision criteria; redrawn from Figure 1 in Giacomini *et al.*, 2013.)

from year to year allocated increasing proportions of available biomass and energy to gonad production.

This determined the degree to which a fish was a fractional versus a batch spawner. A timing parameter determined the reproductive period within a year. Fecundity was an emergent property that depended on the realized amount of gonad production divided by egg size. Simulations were performed to evaluate the relative influences of life history and feeding traits along gradients of productivity and prey accessibility. Among 750 assembled communities, 5387 species persisted to the end of the model simulations. As would be predicted by life-history theory, there was large variation in the influence of life-history traits on species sorting along productivity and prey availability gradients. Community composition and stability were most strongly influenced by body growth rate, followed by egg size and maximum body size. Relatively large, fast growing, and fecund species tended to dominate assemblages when productivity or prey accessibility were high. Overall, they found that the distribution of species traits and the consequences for community dynamics are strictly dependent on how the benefits and costs of these traits are balanced across different conditions. Although they did not explicitly model investment in parental care, their results generally matched predictions of the triangular life-history model (see Figure 18.2). At very low resource availability and predation risk, there was a trend toward opportunistic strategies (smaller size, earlier maturation, variable egg size), and as resource availability and predation increased, they observed a trend toward the periodic strategy of smaller egg size and greater fecundity.

Giacomini et al. (2013) present an individual-based model that demonstrates well the manner in which the balance of trait costs/benefits is determined by environmental conditions, and how this yields fitness differences. Among the countless combinations of life-history trait values that could be conceived by a modeler, only restricted sets of traits (i.e., strategies) actually are observed in nature. Of course, organismal traits influence not only individual fitness, but also community and food-web dynamics. By defining trait trade-offs and responses to selection gradients, life-history theory provides essential guidance for the development of agent-based models of food-web dynamics in variable environments. By defining trait trade-offs and responses to selection gradients, life-history theory provides essential guidance for development of agent-based models of food-web dynamics. The model developed by Giacomini et al. (2013) also simulated a seasonal environment, such as those found in river–floodplain ecosystems (Power et al., 2008; Winemiller et al., 2014). For food-web modeling using an individual-based or agent-based approach, an advantage of this sort of seasonal, or pulsing, system is that abiotic drivers often have strong and fairly well understood effects on nutrient dynamics, reproduction, and species interactions. For example, seasonal flood pulses in tropical floodplains result in dilution of dissolved inorganic nutrients, lower aquatic primary production, lower fish densities, and greater reliance of the aquatic food web on terrestrial sources of primary production (Winemiller, 1990; Winemiller et al., 2014). In addition, fish reproduction and recruitment are strongly tied to the annual hydrological cycle of tropical rivers (Winemiller, 1989; Bailly et al., 2008). An individual-based approach would seem to provide the only feasible means to model variation in abiotic environmental drivers, demographic responses of species with alternative life-history strategies, and trait-dependent species interactions.

18.5   Applications of Food-Web Models: Fisheries Management

A major application of food-web models is fisheries management. According to Travis *et al.* (2014) fisheries management has progressed through two phases and is poised to enter a third. Early fisheries management viewed stocks as isolated entities and relied heavily on models predicting maximum sustainable yield. The concept of maximum sustainable yield is strongly tied to assumptions about density dependence in population regulation, which for many fish stocks is a dubious assumption (Rose *et al.*, 2001; Winemiller, 2005). From the 1980s to the present, there has been greater appreciation for the influence of bottom–up abiotic environmental drivers and variation in large-scale oceanographic processes on productivity, recruitment, and maximum sustainable yield. Food-web models that simulate instantaneous biomass fluxes, such as Ecopath/Ecosim (Christensen and Walters, 2004; Christensen, 2013), have been used extensively to simulate alternative climatic and harvest scenarios.

Travis *et al.* (2014) proposed that fisheries management will embark upon a new phase in which non-linear responses of communities and ecosystems are recognized as common outcomes of networks of direct and indirect species interactions, including feedback loops. They provided case studies showing how environmental stress and anthropogenic influences result in phase shifts in marine communities, and argued that this dynamic should be particularly prevalent when functional diversity is low. To be useful within this new fisheries management paradigm, food-web models would need to incorporate life-history strategies (responding differentially to environmental variation and influencing the storage effect) as well as non-linear species interactions as suggested by the authors. Nonetheless, given that even simple non-linear food-web models yield chaotic dynamics, complex models likely will have limited applications with regard to long-term projections.

Given the pervasive influence of abiotic environmental drivers on ecosystem processes and population dynamics, how could any dynamic food-web model be expected to simulate the diversity of population responses to environmental variation and species interactions? The Ecopath/Ecosim modeling approach in fisheries management can be used to illustrate this challenge. Ecopath employs a mass-balance approach by using linear top–down formulations with the input being fish catches and outputs as rates of production by phytoplankton and other food-web components required to produce those catches. Ecosim is a biomass-flow model used to simulate aquatic and marine food-web dynamics, and has been applied to the management of dozens of fisheries worldwide. In a model of the Western and Central Pacific Ocean pelagic food web (Allain, 2005; Allain *et al.*, 2007), planktonic species were aggregated into functional groups that actually represent diverse life-history strategies with different maximum intrinsic rates of increase (r_{max}), population turnover times, degrees of demographic resilience, responses to environmental variation at different scales of space and time, and strengths

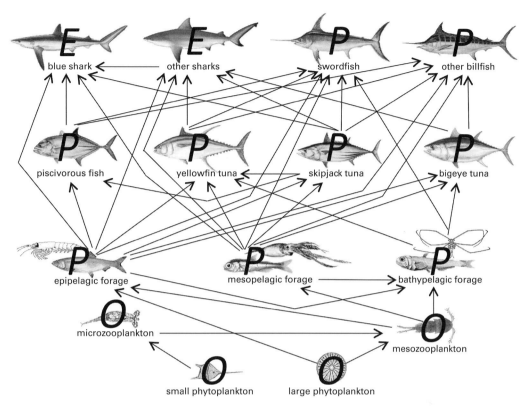

Figure 18.4 Ecosim/Ecopath model of the Western and Central Pacific Ocean pelagic food web from Allain (2005). Species were aggregated into functional groups, and these groups can be identified as having a relatively opportunistic (*O*), periodic (*P*), or equilibrium (*E*) life-history strategies. The manner in which populations respond to predation, human impacts, and abiotic variation at different spatio-temporal scales depends on recruitment dynamics, population turnover, demographic resilience, and the storage effect as conferred by life-history traits. Considerable life-history variation also is contained within each functional group.

of the storage effect (Figure 18.4). Major intrinsic differences among and within food-web compartments should be a major source of bias in aggregate simulation models. Compared to fishes, birds, and mammals, phytoplankton and zooplankton are opportunistic strategists in the extreme, and are expected to have more rapid population turnover and greater demographic resilience in response to predation.

Small pelagic fishes would be considered relative opportunistic strategists in relation to large pelagic fishes that are extreme periodic strategists with broadcast spawning, episodic recruitment in response to oceanographic environmental drivers, and strong storage effect. Sharks and marine birds and mammals are equilibrium strategists with populations that should be responsive to density-dependent influences with a strong storage effect and slower response times to changes in resources.

Parameterization, of course, is a critically important aspect of simulation modeling. Like all network-biomass/energy-flow models, Ecopath/Ecosim requires estimates for standing biomass, rates of production, rates of consumption, and rates for other kinds of inputs and losses. Allain *et al.* (2007) categorized each of their input parameters according to their degree of confidence based on whether data were derived from field studies of the modeled system, field studies of other systems, modeling efforts, or "guesstimates" (educated guesses). Eighteen of 99 parameter estimates were classified as high confidence and 31 were considered guesstimates. The authors therefore urged caution in direct application of their simulation findings to fisheries management. Another major challenge in simulation modeling is validation, and rarely do we have datasets for large complex systems sufficient for evaluation of not only parameter values, but also the equations linking model subcomponents and simulated outcomes.

In recent years, fisheries ecology has seen greater development of so-called, end-to-end models to deal with large-scale environmental factors on ecosystem dynamics and fish stocks. Rose (2012) summarized the description by Travers *et al.* (2007) of an end-to-end model "as one that: (1) aims to represent the entire food web and the associated abiotic environment, (2) requires the integration of physical and biological processes at different scales, (3) implements two-way interaction between ecosystem components, and (4) accounts for the dynamic forcing effect of climate and human impacts at multiple trophic levels." Rose (2012) expanded the description presented by Travers *et al.* (2007) to require inclusion of multiple species or functional groups at each trophic level, including top predators, with physics modeled in a manner responsive to climate inputs and human impacts modeled in a dynamic (state-dependent) manner. He discouraged the repackaging of poorly performing component submodels, and recommended that more work is needed on theoretically oriented models to guide data collection and development of site-specific models for fisheries management.

Dynamic food-web models that simulate human impacts on fisheries (e.g., harvest, climate change, dams) will play an increasingly important role in management and policy. For example, Jacobsen *et al.* (2014) developed a trait-based model with individual predator–prey interactions to compare the effects of balanced harvesting (species are exploited in accordance with their productivity) with traditional selective harvesting aimed at protecting juvenile fish from exploitation. They found that balanced fishing produced a slightly larger total maximum sustainable yield. Because fishing reduced competition, predation, and cannibalism, the total maximum sustainable yield was achieved at fairly high exploitation rates. However, yields from balanced fishing were dominated by small fish species. Individual-based models, such as those created recently by Giacomini *et al.* (2013) and Jacobsen *et al.* (2014), provide a good foundation for the development of dynamic food-web models with foraging dynamics and life-history trade-offs responsive to environmental variation for application to fisheries management.

18.6   Conclusions

Application of a food-web approach should be fundamental for fisheries management and other natural resource challenges, such as control of agricultural pests, invasive species, and diseases. At the same time, ecologists realize that food webs are complex, dynamic, and under the influence of abiotic and biotic environmental drivers that vary across multiple scales of space and time. Models constructed to simulate food-web dynamics therefore face considerable challenges. Such challenges evoke a familiar metaphor – *this is not rocket science – it is far more difficult!* How can a single model capture cause and effect relationships involving stage-based trophic interactions among species with contrasting life-history strategies that respond differentially to temporal and spatial environmental variation at multiple scales? This chapter has briefly outlined some key aspects of this problem, and concludes that process-oriented, trait-based modeling approaches may provide a viable avenue for future research (see also Chapter 15 in this volume). Mathematical models created to test theories about complexity and stability, for example, usually lack population structure, key functional relationships between consumers and their resources, and the influence of environmental variation. There exists, of course, a trade-off between generality, which often requires use of abstract or aggregate parameters, versus specificity, which often requires great detail to simulate a real-world system. This specificity also necessitates great amounts of empirical data and understanding of important cause–effect relationships – information that is available for relatively few study systems.

Food webs have been compared to road maps (Paine, 1988). Both portray structure in the form of relationships among entities. In the case of maps, the entities are locations in 2D or 3D space; in the case of food webs, the entities are organisms, populations, functional groups, and other ecosystem components. Maps and food webs both are tools created for specific purposes. The primary purpose of maps is navigation, and for food webs the primary purpose is predicting ecological dynamics. Whereas various degrees of simplification and aggregation will always be required in food-web modeling, construction of abstract webs to test general theories, such as the relationship between network complexity and stability, likely will not advance our ability to predict the dynamics of real-world systems. Here I have argued the case for trait-based, process-oriented food-web models as a means to simulate food-web dynamics. This endeavor will require research that accumulates large empirical datasets as well as comparative and experimental studies that identify the most influential parameters and their cause–effect relationships. Akin to cartography, such work is laborious, requires attention to detail with regard for effects of scale and resolution, and tends to advance slowly and incrementally. Technological advancements have been responsible for leaps in precision and accuracy of navigation (e.g., from sextant to GPS), leading one to anticipate how future technological innovations will ramp up progress in trophic ecology, life-history theory, computer science, and ecological modeling.

Acknowledgments

I thank Caroline Arantes, Eduardo Cunha, Daniel Fitzgerald, and two anonymous reviewers for very helpful comments that greatly improved this chapter; however, they bear no responsibility for any omissions, inaccuracies, or faults of logic. I also thank the organizers and participants of the 4th Decadal Food Web Symposium for their stimulating ideas and insightful discussions.

References

Alcántara, J. M. and Rey, P. J. (2012). Linking topological structure and dynamics in ecological networks. *American Naturalist*, **180**, 186–199.

Allain, V. (2005). Ecopath model of the pelagic ecosystem of the Western and Central Pacific Ocean. *First Regular Session of the Scientific Committee of the Western and Central Pacific Fisheries Commission*, WCPFC-SC1-EB WP-10, 1–19.

Allain, V., Nicol, S., Essington, T., *et al*. (2007). An Ecopath with Ecosim model of the Western and Central Pacific Ocean warm pool pelagic ecosystem. *Third Regular Session of the Scientific Committee of the Western and Central Pacific Fisheries Commission*, WCPFC-SC3-EB SWG/IP-8, 1–42.

Allesina, S. and Pascual, M. (2008). Network structure, predator–prey modules, and stability in large food webs. *Theoretical Ecology*, **1**, 55–64.

Andrewartha, H. G. and Birch, C. (1954). *The Distribution and Abundance of Animals*. Chicago, IL: University of Chicago Press.

Angert, A. L., Huxman, T. E., Chesson, P., and Venable, D. L. (2009). Functional tradeoffs determine species coexistence via the storage effect. *Proceedings of the National Academy of Sciences of the United States of America*, **106**, 11641–11645.

Atkinson, A., Hill, S. L., Barange, M., *et al*. (2014). Sardine cycles, krill declines, and locust plagues: revisiting "wasp-waist" food webs. *Trends in Ecology and Evolution*, **29**, 309–316.

Bailly, D., Agostinho, A. A., and Suzuki, H. I. (2008). Influence of the flood regime on the reproduction of fish species with different reproductive strategies in the Cuiabá River, Upper Pantanal, Brazil. *River Research and Applications*, **24**, 1218–1229.

Bakun, A. and Broad, K. (2003). Environmental "loopholes" and fish population dynamics: comparative pattern recognition with focus on El Niño effects in the Pacific. *Fisheries Oceanography*, **12**, 458–473.

Barton, B. T. and Ives, A. R. (2014). Species interactions and a chain of indirect effects driven by reduced precipitation. *Ecology*, **95**, 486–494.

Bellmore, J. R., Baxter, C. V., and Connolly, P. J. (2015). Spatial complexity reduces interaction strengths in the meta-food web of a river floodplain mosaic. *Ecology*, **96**, 274–283.

Camerano, L. (1880). Dell'equilibrio dei viventi merce la reciproca distribuzione. *Atti della Scienze di Torino*, **15**, 393–414.

Carpenter, S. R. and Turner, M. G. (2001). Hares and tortoises: interactions of fast and slow variables in ecosystems. *Ecosystems*, **3**, 495–497.

Cheal, A. J., Delean, S., and Thompson, A. A. (2007). Spatial synchrony in coral reef fish populations and the influence of climate. *Ecology*, **88**, 158–169.

Chesson, P. (2000). General theory of competitive coexistence in spatially varying environments. *Theoretical Population Biology*, **58**, 211–237.

Chesson, P. and Warner, R. (1981). Environmental variability promotes coexistence in lottery competitive systems. *American Naturalist*, **117**, 923–943.

Christensen, V. (2013). Ecological networks in fisheries: predicting the future? *Fisheries*, **38**, 76–81.

Christensen, V. and Walters, C. J. (2004). Ecopath with Ecosim: methods, capabilities and limitations. *Ecological Modelling*, **172**(2–4), 109–139.

De Roos, A. M., Persson, L., and McCauley, E. (2003). The influence of size-dependent life history traits on the structure and dynamics of populations and communities. *Ecology Letters*, **6**, 473–487.

Fujiwara, M. (2016). Incorporating demographic diversity into food web models: effects on community structure and dynamics. *Ecological Modelling*, **322**, 10–18.

Giacomini, H. C., DeAngelis, D. L., Trexler, J. C., and Petrere, Jr, M. (2013). Trait contributions to fish community assembly emerge from trophic interactions in an individual-based model. *Ecological Modelling*, **251**, 32–43.

Grime, J. P. (1977). Evidence for existence of three primary strategies in plants and its relevance to ecological and evolutionary theory. *American Naturalist*, **111**, 1169–1194.

Grimm, V., Revilla, E., Berger, U., *et al.* (2005). Pattern-oriented modeling of agent-based complex systems: lessons from ecology. *Science*, **310**, 987–991.

Hartvig, M., Andersen, K. H., and Beyer, J. E. (2011). Food web framework for size-structured populations. *Journal of Theoretical Biology*, **272**, 113–122.

Hastings, A. (1996). What equilibrium behavior of Lotka–Volterra models does not tell us about food webs. In *Food Webs: Integration of Patterns and Dynamics*, ed. G. A. Polis and K. O. Winemiller, New York, NY: Chapman & Hall, pp. 211–217.

Hastings, A., Hom, C., Ellner, S., Turchin, P., and Godfray, H. C. J. (1993). Chaos in ecology: is mother nature a strange attractor? *Annual Reviews of Ecology and Systematics*, **24**, 1–33.

Haydon, D. (1994). Pivotal assumptions determining the relationship between stability and complexity: an analytical synthesis of the stability-complexity debate. *American Naturalist*, **144**, 14–29.

Higgins, S., Pickett, S. T. A., and Bond, W. J. (2000). Predicting extinction risks for plants: environmental stochasticity can save declining populations. *Trends in Ecology and Evolution*, **15**, 516–520.

Holt, R. D. (2008). Theoretical perspectives on resource pulses. *Ecology*, **89**, 671–681.

Holt, R. D. (2009). Towards a trophic island biogeography: reflections on the interface of island biogeography and food web ecology. In *The Theory of Island Biogeography Revisited*, ed. J. B. Losos and R. E. Ricklefs, Princeton, NJ: Princeton University Press, pp. 143–185.

Jacobsen, N. S., Gislason, H., and Andersen, K. H. (2014). The consequences of balanced harvesting of fish communities. *Proceedings of the Royal Society B: Biological Sciences*, **281**(1775), DOI: 10.1098/rspb.2013.2701.

Koenig, W. D. and Liebhold, A. M. (2005). Effects of periodical cicada emergences on abundance and synchrony of avian populations. *Ecology*, **86**, 1873–1882.

Lawton, J. H. (1995). Population dynamic principles. In *Extinction Rates*, ed. J. H. Lawton and R. M. May, Oxford, UK: Oxford University Press, pp. 147–163.

Lindeman, R. L. (1942). The trophic-dynamic aspect of ecology. *Ecology*, **23**, 399–418.

Madsen, T. and Shine, R. (2000). Rain, fish and snakes: climatically driven population dynamics of Arafura filesnakes in tropical Australia. *Oecologia*, **124**, 208–215.

May, R. M. (1973). *Stability and Complexity in Model Ecosystems*. Princeton, NJ: Princeton University Press.

McCann, K. (2012). *Food Webs*. Princeton, NJ: Princeton University Press.

Melián, C. J., Baldó, F., Matthews, B., *et al.* (2014). Individual trait variation and diversity in food webs. *Advances in Ecological Research*, **50**, 207–241.

Menge, B. A., Gouhier, T., Friedenburg, T., *et al.* (2011). Linking long-term, large-scale climatic and environmental variability to patterns of marine invertebrate recruitment: toward explaining "unexplained" variation. *Journal of Experimental Marine Biology and Ecology*, **400**, 236–249.

Olff, H., Alonso, D., Berg, M. P., *et al.* (2009). Parallel ecological networks in ecosystems. *Philosophical Transactions of the Royal Society B: Biological Sciences*, **364**, 1755–1779.

Paine, R. T. (1988). On food webs: road maps of interactions or the grist for theoretical development? *Ecology*, **69**, 1648–1654.

Pianka, E. R. (1970). On r- and K-selection. *American Naturalist*, **104**, 592–597.

Polis, G. A., Holt, R. D., Menge, B. A., and Winemiller, K. O. (1996). Time, space, and life history: influences on food webs. In *Food Webs: Integration of Patterns and Dynamics*, ed. G. A. Polis and K. O. Winemiller, New York, NY: Chapman and Hall, pp. 435–460.

Post, D. M. (2002). The long and short of food-chain length. *Trends in Ecology and Evolution*, **17**, 269–277.

Power, M. E., Parker, M. S., and Dietrich, W. E. (2008). Seasonal reassembly of a river food web: floods, droughts, and impacts of fish. *Ecological Monographs*, **78**, 263–282.

Rose, K. A. (2012). End-to-end models for marine ecosystems: are we on the precipice of a significant advance or just putting lipstick on a pig? *Scientia Marina*, **76**, 195–201.

Rose, K. A., Cowan, J. H., Winemiller, K. O., Myers, R. A., and Hilborn, R. (2001). Compensatory density-dependence in fish populations: importance, controversy, understanding, and prognosis. *Fish and Fisheries*, **2**, 293–327.

Rudolf, V. H. W. and Lafferty, K. D. (2011). Stage structure alters how complexity affects stability of ecological networks. *Ecology Letters*, **14**, 75–79.

Schmitz, O. J. and Booth, G. (1997). Modelling food web complexity: the consequences of individual-based, spatially explicit behavioural ecology on trophic interactions. *Evolutionary Ecology*, **11**, 379–398.

Siddon, E. C., Kristiansen, T., Mueter, F. J., *et al.* (2013). Spatial match-mismatch between juvenile fish and prey provides a mechanism for recruitment variability across contrasting climate conditions in the Eastern Bering Sea. *PLOS One*, **8**(12), e84526.

Sinclair, A. R. E. (2003). Mammal population regulation, keystone processes and ecosystem dynamics. *Philosophical Transactions of the Royal Society B: Biological Sciences*, **358**, 1729–1740.

Soberón, J. (2007). Grinnellian and Eltonian niches and geographic distributions of species. *Ecology Letters*, **10**, 1115–1123.

Southwood, T. R. E. (1977). Habitat, the templet for ecological strategies? *Journal of Animal Ecology*, **46**, 337–365.

Stearns, S. C. (1992). *The Evolution of Life Histories*. Oxford, UK: Oxford University Press.

Summerhayes, V. S. and Elton, C. S. (1923). Contribution to the ecology of Spitsbergen and Bear Island. *Journal of Ecology*, **11**, 214–286.

Tang, S., Pawar, S., and Allesina, S. (2014). Correlation between interaction strengths drives stability in large ecological networks. *Ecology Letters*, **17**, 1094–1100.

Travers, M., Shin, Y.-J., Jennings, S., and Cury, P. (2007). Towards end-to-end models for investigating the effects of climate and fishing in marine ecosystems. *Progressive Oceanography*, **75**, 751–770.

Travis, J., Coleman, F. C., Auster, P. J., *et al.* (2014). Integrating the invisible fabric of nature into fisheries management. *Proceedings of the National Academy of Sciences of the United States of America*, **111**(2), 581–584.

Turchin, P. (2003). *Complex Population Dynamics. A Theoretical/Empirical Synthesis.* Princeton, NJ: Princeton University Press.

van Kooten, T., de Roos, A. M., and Persson, L. (2005). Bistability and an Allee effect as emergent consequences of stage-specific predation. *Journal of Theoretical Biology*, **237**, 67–74.

Van Winkle, W., Rose, K. A., and Chambers, R. C. (1993). Individual-based approach to fish population dynamics: an overview. *Transactions of the American Fisheries Society*, **122**, 397–403.

Varughese, M. M. (2011). A framework for modelling ecological communities and their interactions with the environment. *Ecological Complexity*, **8**, 105–112.

Vasseur, D. and Gaedke, U. (2007). Spectral analysis unmasks synchronous and compensatory dynamics in plankton communities. *Ecology*, **88**, 2058–2071.

Winemiller, K. O. (1989). Patterns of variation in life history among South American fishes in seasonal environments. *Oecologia*, **81**, 225–241.

Winemiller, K. O. (1990). Spatial and temporal variation in tropical fish trophic networks. *Ecological Monographs*, **60**, 331–367.

Winemiller, K. O. (1992). Life history strategies and the effectiveness of sexual selection. *Oikos*, **62**, 318–327.

Winemiller, K. O. (2005). Life history strategies, population regulation, and their implications for fisheries management. *Canadian Journal of Fisheries and Aquatic Sciences*, **62**, 872–885.

Winemiller, K. O. and Rose, K. A. (1992). Patterns of life-history diversification in North American fishes: implications for population regulation. *Canadian Journal of Fisheries and Aquatic Sciences*, **49**, 2196–2218.

Winemiller, K. O., Roelke, D. L., Cotner, J. B., *et al.* (2014). Top–down control of basal resources in a cyclically pulsing ecosystem. *Ecological Monographs*, **84**, 621–635.

Woodson, C. B., McManus, M. A., Tyburczy, J. A., *et al.* (2012). Coastal fronts set recruitment and connectivity patterns across multiple taxa. *Limnology and Oceanography*, **57**, 582–596.

Worm, B. and Duffy, J. E. (2003). Biodiversity, productivity and stability in real food webs. *Trends in Ecology and Evolution*, **18**, 628–632.

Yang, L. H. and Rudolf, V. H. W. (2010). Phenology, ontogeny and the effects of climate change on the timing of species interactions. *Ecology Letters*, **13**, 1–10.

Yodzis, P. and Innes, S. (1992). Body size and consumer-resource dynamics. *American Naturalist*, **139**, 1151–1175.

Zeglin, L. H., Bottomley, P. J., Jumpponen, A., *et al.* (2013). Altered precipitation regime affects the function and composition of soil microbial communities on multiple time scales. *Ecology*, **94**, 2334–2345.

Zhou, C., Fujiwara, M., and Grant, W. E. (2013). Dynamics of a predator–prey interaction with seasonal reproduction and continuous predation. *Ecological Modelling*, **268**, 25–36.

19 Rare but Important: Perturbations to Uncommon Species Can Have a Large Impact on the Structure of Ecological Communities

Tomas Jonsson, Sofia Berg, Torbjörn Säterberg, Céline Hauzy, and Bo Ebenman

19.1   Introduction

Climate change, overexploitation of natural populations, invasive species, and habitat degradation are affecting species and ecosystems all over the world on a scale that is predicted to lead to the sixth mass extinction (Barnosky *et al.*, 2011). Such dramatic perturbations to ecological communities are likely to affect the structure, functioning, and stability of ecosystems (Montoya *et al.*, 2006; Pereira *et al.*, 2010: Isbell *et al.*, 2011; Hooper *et al.*, 2012), invoking the important question if changes to, or even loss of, some species can be expected to have more far-reaching consequences than others. Or put another way: are some species more important than others in upholding the structure and functioning of ecological systems? Despite the urgency of this question, our understanding of the contribution of different species to community structure and functioning is incomplete and contradictory. For example, some models describing patterns in ecosystems suggest that all species are "equivalent" (i.e., differences are neutral with respect to community properties, e.g., MacArthur and Wilson, 1967; Hubbell, 2001), while others suggest that some nodes are disproportionately important for maintaining community structure and functioning (Saavedra *et al.*, 2011) and can be considered keystone species (Paine, 1966, 1969; Cottee-Jones and Whittaker, 2012). Due to the practical and ethical constraints associated with perturbing species in real ecosystems, many researchers have used theoretical approaches to address these questions, for example studying the effect of small as well as large disturbances (Curtsdotter *et al.*, 2011; Berg *et al.*, 2015; Jonsson *et al.*, 2015) to different kinds of species in model communities and monitoring the response. Theoretical studies of species extinctions have among other things documented the obvious importance of highly connected species for the permanence of the remaining food web (Eklöf and Ebenman, 2006), and identified large body size, high trophic position, low vulnerability, or low numerical abundance (Berg *et al.*, 2015) as traits of potential keystone species that, if lost, are likely to cause cascades of secondary extinctions (Borrvall *et al.*, 2000; Ebenman and Jonsson, 2005). Here we also use a theoretical approach to look for species that, when perturbed, will affect a community more than others, and analyze if there are species with particular attributes whose ecological importance currently are under-appreciated. One such attribute might be the abundance of a species.

Although abundance may differ among populations within a species, and often fluctuates over time, the vast majority of the world's species have comparably low abundances, while relatively few species are in fact really common (May, 1975; Gaston, 2010). Still, because a high proportion of the total biomass and productivity often can be attributed to the common species, these species – although few in number of species – should build up the core structure of most ecosystems upon which many other species depend (Ellison *et al.*, 2005; Gaston and Fuller, 2008; Gaston, 2011). Thus it has been argued that a prerequisite for ecological importance is that a species is relatively common, while rare species, due to their low abundance, are likely to be comparably insignificant players (Thompson, 2010). However, strongly interacting consumers that suppress the abundance of their resource(s), and thus limit their own abundance, is a well-known phenomenon for some predator–prey and biological-control systems (Beddington *et al.*, 1978; Morin, 2011). Based on this, it could be hypothesized that perturbations to such rare consumers, via indirect effects, could have far-reaching cascading effects on many other species, with potentially large consequences for community structure and functioning. Whether this translates to any general relationship between abundance and importance of species in multispecies systems is yet unknown, but a few experimental and empirical studies have suggested that rare species, despite their low abundances, actually could contribute significantly to ecosystem structure, functioning, and stability (Lyons and Schwartz, 2001; Lyons *et al.*, 2005; Hol *et al.*, 2010; Estes *et al.*, 2011; Bracken and Low, 2012). Although there are some indications of the potential importance of rare species, the effects of less common species on ecosystem structure and dynamics are poorly understood in general, and further studies, using a variety of approaches for evaluating the importance of rare species, have been called for (Lyons *et al.*, 2005). To this end we present a theoretical analysis of the relationship between the abundance of a species and its influence on structural properties of the system of which it is a part.

19.2 Methods: Food Webs, Metrics, and Analytical Approaches

19.2.1 Sequential Food-Web Assembly

The communities used in this study were generated through a sequential assembly process and have been used elsewhere to address other questions (Säterberg *et al.*, 2013, Berg *et al.*, 2015). Briefly, species were one at a time allowed to attempt to successfully invade an evolving food web (initially consisting of seven basal species), until a locally stable, 50-species community, had been built up. Community dynamics during the assembly process were described by Lotka–Volterra predator–prey interactions generalized to n species:

$$\frac{dN_i}{dt} = N_i\left(r_i + \sum_{j=1}^{n} \alpha_{ij} N_j\right)$$ (19.1)

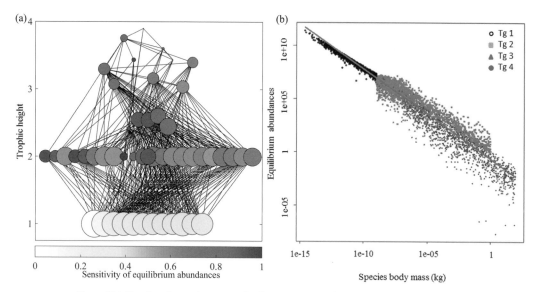

Figure 19.1 Food-web topology, species importance, and relationship between body size and abundance. (a) The trophic structure of one of 100 model communities generated by the sequential assembly algorithm, and the equilibrium abundances and structural importance of its species (as determined by CSA). Each node represents a species and the edges the feeding relationships. Node size corresponds to the relative equilibrium abundance of species, with node color denoting the relative sensitivity of equilibrium abundances (see color bar) to a perturbation of the growth/ mortality rate of the species. (b) Equilibrium abundance as a function of species average body mass (m) for all species in the 100 model communities. The red line represents the fixed effect estimate of the linear mixed effect model: $log(N_{kj}^*) = \beta + \omega * log(m_{k,j}) + b_j + e_{kj}$, where N_{kj}^* is the equilibrium abundance of species k in food web j, β is the fixed effect intercept, $m_{k,j}$ is the average body mass of species k in food web j, b_j is the random effect on the intercept of each food web j, and e_{kj} is the residual abundance of species k in food web j. The slope of the fixed effect line (ω) equals –0.85 with a standard deviation of 0.0029 and is within the range of what has been observed empirically (Brown *et al.*, 2004; Reuman *et al.*, 2008). (AIC indicated that random effect on slopes was not needed.) (A black and white version of this figure will appear in some formats. For the color version, please refer to the plate section.)

Here N_i is the abundance of species i in a community with n species, r_i is a species' intrinsic growth (if i is a basal species) or mortality rate (if i is a consumer species), α_{ij} is the per capita effect of species j on the per capita growth (or mortality) rate of species i. Values of model parameters (i.e., vital rates of species and interaction strengths) were related to body size (drawn at random from predefined intervals for primary producers, herbivores, and carnivore species, respectively) by metabolic theory (Peters, 1983; Brown *et al.*, 2004; McCoy and Gillooly, 2008; Berg *et al.*, 2011; see Säterberg *et al.*, 2013, for details of the algorithm and how parameters were related to body size). Using this approach, the network topology is an outcome of the dynamical assembly process rather than pre-set before adding dynamics. The final model communities (see Figure 19.1a for an example) have a realistic topology (Säterberg *et al.*, 2013; Berg *et al.*, 2015) and the local mass–abundance slopes (i.e., log abundance as a function of log average body mass across all species) are within the ranges of those observed in real communities (Figure 19.1b).

19.2.2   Community Sensitivity Analysis

Community sensitivity analysis (CSA; Berg *et al.*, 2011) was applied on the assembled food webs to study the effect of small permanent changes in the intrinsic growth, or mortality, rate of a species on community structure (the distribution of equilibrium abundances of species). Sensitivity of community structure $(S_{\hat{\mathbf{N}}})$ to a change in the intrinsic growth, or mortality, rate (b) of species k are given by:

$$S_{\hat{\mathbf{N}}}(b_k) = \sum_i |\gamma_{ik}| \tag{19.2}$$

Here, $\hat{\mathbf{N}}$ is the vector of equilibrium abundances, and γ_{ik} is element ik in the inverse interaction matrix $(\mathbf{A}^{-1})$, which specify the effect of species k on species i (i.e., $\gamma_{ik} = \dfrac{\partial \hat{N}_i}{\partial b_k}$, Bender *et al.*, 1984). Using this approach the association between the structural importance of a species, and its abundance, can be analytically assessed (i.e., without the need to actually perform a perturbation to species k and numerically simulate the response of other species). Here a high importance is equivalent to species with a high structural sensitivity value (i.e., species that, when perturbed, affect the abundances of species in a community to a greater extent than other species in the food web).

19.2.3   Statistical Analyses

As trophic position (tp), defined as the distance (average number of links plus one) of a consumer from a primary producer, and species abundances are highly correlated in real communities (Jonsson *et al.*, 2005) as well as in our model systems (see Table S19.1 in the Appendix), it is important to separate the effect of species abundance from the effect of trophic position. Thus we distinguish between four trophic groups (Tg i, $i = 1, \ldots, 4$): Tg 1 includes all primary producers (i.e., species with tp $= 1$), Tg 2 corresponds to all primary consumers (i.e., species with tp $= 2$), Tg 3 corresponds to species with $2 < \text{tp} \leq 3$, and Tg 4 to species with $3 < \text{tp} \leq 4$. Statistical analyses were restricted to trophic groups with at least three species. We assessed the effect of species abundances on the sensitivity of equilibrium abundances (Eq. 19.2), across all species, as well as within trophic groups, using linear mixed-effect models. Random effects, depending on individual webs, were introduced on the slope and intercept. Sensitivity of equilibrium abundances and species abundances were log transformed to improve the normality of residuals. The trophic group was considered as a categorical variable with four levels and the abundance was considered as a continuous variable. The equation for the mixed-effect model was as follows:

$$Y_{i,j,k} = \beta_i + \rho_i X_{i,j,k} + b_{i,j} + p_{i,j} X_{i,j,k} + e_{i,j,k} \tag{19.3}$$

where $Y_{i,j,k}$ is the log sensitivity of abundance in trophic group i, in food web j when species k is disturbed, $X_{i,j,k}$ is the log abundance of species k in trophic group i, in food web j. The intercept and the slope of the fixed effect for trophic group i are given by β_i and ρ_i. For each trophic group the random effects on the intercept and on the slope are given by $b_{i,j}$ and $p_{i,j}$. $e_{i,j,k}$ is the residual variability for species k in trophic group i, in food web j, that is not explained by fixed or random effects.

Furthermore, to shed light on potential mechanisms affecting the structural importance of rare and common species, we analyzed the relationship between the abundance of a species and the strength of its interaction with other species in the community. How strongly a species interacts with other species was measured by two metrics, based on the Lotka–Volterra per capita interaction coefficients (α_{ij}, see Eq. 19.1): (1) *predation pressure* and (2) *predation stress*. Predation pressure (PP_j) measures how strongly individuals of consumer species j interact with their entire set of resource species (Γ_j) and was defined as the sum of their per capita trophic interactions with their prey:

$$PP_j = \sum_{i \in \Gamma_j} \alpha_{ij} \tag{19.4}$$

Predation stress (PS_i) measures how exposed individuals of resource species i are to predation from their entire set of consumer species (Ψ_i) and was defined as the sum of their trophic interactions with their predators:

$$PS_i = \sum_{j \in \Psi_i} \alpha_{ij} \tag{19.5}$$

The effect of predation pressure and stress on species abundance was analyzed within each trophic group using a linear mixed effect model for each group. Random effects, depending on individual webs, were introduced on the intercept. Abundances and predation pressure and stress were log transformed to improve normality of the distribution residuals. Predation pressure and stress were considered continuous variables. The equation for the mixed-effect model was as follows:

$$Y_{i,j,k} = \beta_i + \rho_i X_{i,j,k} + b_{i,j} + e_{i,j,k} \tag{19.6}$$

Here, $Y_{i,j,k}$ is the log abundance of species k in trophic group i, in food web j. $X_{i,j,k}$ is the predation pressure or stress of species k in trophic group i, in food web j. For trophic group i, the fixed effect intercept is β_i, and the (fixed effect) slope is ρ_i. Random effects on the intercept are given by $b_{i,j}$ and $e_{i,j,k}$ is the residual variability for species k in trophic group i, in food web j.

All linear mixed-effect models were fitted using package *nlme* (Pinheiro *et al.*, 2012) in R (R Development Core Team, 2012), by means of restricted maximum likelihood. In order to select parsimonious models, we used likelihood ratio tests and Akaike's information criterion (AIC) to simplify random effects. P-values were calculated according to Pinheiro and Bates (2004) and should be regarded as indications of significance only.

Table 19.1 Species importance as a function of species equilibrium abundance, as indicated by linear mixed-effect model analysis.

Sensitivity of community structure	$\eta_i\,(v_i)$	Parameters	Estimates	s.e.m.	t-value	P
All species	100 (5000)	β	13.681	0.072	189.045	<0.0001
		ρ	−0.465	0.008	−55.795	<0.0001
Trophic group 1	100 (1106)	β	10.419	0.326	31.961	<0.0001
		ρ	−0.203	0.037	−5.471	<0.0001
Trophic group 2	100 (2610)	β	14.031	0.161	87.312	<0.0001
		ρ	−0.422	0.026	−16.417	<0.0001
Trophic group 3	86 (449)	β	13.071	0.091	143.592	<0.0001
		ρ	−0.276	0.017	−15.980	<0.0001
Trophic group 4	98 (798)	β	12.969	0.068	191.184	<0.0001
		ρ	−0.242	0.016	−14.684	<0.0001

Note: Random effects depending on individual webs were introduced on the slopes and intercepts. Parameters ρ and β report the fixed effects estimates of the slope and intercept, respectively. For random effects see Table S19.5. P-values were calculated according to Pinheiro and Bates (2004) and should be regarded as indications of significance only. η_i: Number of food webs with at least three species in trophic group i, v_i: number of species in trophic group i from all food webs with at least three species in this trophic group, s.e.m.: standard error.

19.3   Results

We found a significant and strong relationship, across all webs and species, between the equilibrium abundance of species and their effect on the structure of communities (Figure 19.2a; Table 19.1). Furthermore, we found the same qualitative relationship within a trophic group (i.e., between species with a similar trophic position) as across the entire community; changes in the growth rate of rare species have larger effects on community structure than changes in the growth rate of common species (Figure 19.2; Table 19.1). The relationship is weaker (but still significant) for primary producers (i.e., trophic group 1) than for consumers (i.e., trophic groups 2 to 4). Within trophic groups 1 and 2 all species had the same trophic position (exactly one or two, respectively), but within trophic groups 3 and 4 there existed some variability in trophic position among species (see Table S19.2 in the Appendix). However, using a mixed-effect model where the contribution of trophic position to abundance was controlled for does not change our conclusions (see Table S19.3 in the Appendix).

Finally, we found that the abundance of a consumer is strongly and negatively related to the predation pressure exerted by the species on its resources (in all consumer trophic groups, Figure 19.3, see also Table S19.4 in the Appendix). Predation stress experienced by a species on the other hand, had a very low and non-significant explanatory power for species abundance.

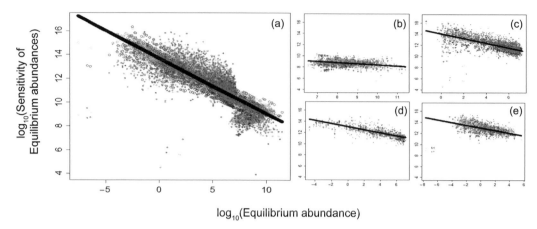

Figure 19.2 Species structural importance is inversely related to abundance in ecological communities. (a) Sensitivity of the distribution of equilibrium densities, to a perturbation of the growth/mortality rate of a species, as a function of equilibrium abundance of the species, across all communities. (b)–(e) The same relationship as in (a) but within trophic groups 1 (b), 2 (c), 3 (d) and 4 (e), respectively. The bold black line in each subplot represents the global relationship (fixed effects only) of mixed-effect models treating communities as a random effect on model intercepts and slopes (for all individual random effects, see Table S19.5 in the Appendix, and for estimates of model parameters see Table 19.1). Globally, sensitivity is significantly and inversely related to species abundance within all trophic groups (b)–(e) as well as within whole communities (a) (P <0.0001 for all models). (A black and white version of this figure will appear in some formats. For the color version, please refer to the plate section.)

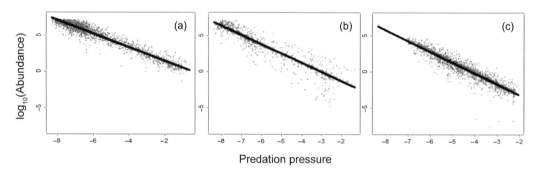

Figure 19.3 Consumer species with low abundance tend to be strong interactors. The relationship between predation pressure and equilibrium abundances for (a) trophic group 2, (b) trophic group 3, and (c) trophic group 4. The bold black line in each subplot represents the global relationship across all 100 ecological communities in the mixed-effect model, i.e., with random effects depending on community identity (for all individual random effects see Table S19.6 in the Appendix, and for estimates of model parameters see Table S19.5 in the Appendix). All slopes are significantly different from zero, P <0.001. (A black and white version of this figure will appear in some formats. For the color version, please refer to the plate section.)

19.4   Discussion

Using a theoretical approach we have shown that perturbations to the growth or mortality rate of uncommon species in ecological communities are predicted to affect the distribution of abundances relatively more than changes to more common species. Thus our results imply that, although a species is rare and does not contribute much to total community biomass or productivity, it may still be important by means of a disproportional effect on community structure. The effect, or importance, of different species were quantified using a metric from community sensitivity analysis (Berg *et al.*, 2011), and this structural sensitivity metric (Eq. 19.2) is very similar to the so-called net effects of Montoya *et al.* (2005, 2009). These theoretically predicted structural sensitivity values of species have been shown to be significantly correlated to empirical measurements of the same metric in a marine intertidal system (O'Gorman *et al.*, 2010), supporting that our analytically calculated sensitivities here carry relevant information on how the distribution of species abundances in a system will respond to perturbations. Furthermore, the importance of rare species for community structure has also been indicated in experimental/field studies. For instance, a recent study of a rocky shore community showed that losses (a large perturbation) of rare species had a disproportionately large impact on the biomass of species at higher trophic levels (Bracken and Low, 2012).

Our finding that the effects of a change in the growth rate of a species on community structure is inversely related to its abundance may seem counter-intuitive. How can this surprising result be understood and explained? We suggest that the explanation is to be sought in how the abundance of a species is related to its interactions with other species in the community. To explore this hypothesis, we analyzed the relationship between the abundance of a species and two interaction metrics – *predation pressure* (Eq. 19.4) and *predation stress* (Eq. 19.5). We found, in our model systems, that the abundance of a consumer tends to be negatively related to the predation pressure exerted by the species on its resources (in all consumer trophic groups; Figure 19.3; Table S19.4 in the Appendix). Predation stress experienced by a species, on the other hand, did not have any significant explanatory power for species abundance.

These findings suggest that (important) rare species interact more strongly, on a per capita basis, with other species (mainly their resources) than do many more abundant species. This is in accordance with the findings of a recent empirical study; in intertidal communities, Wood *et al.* (2010) found the population effect (i.e., the per capita effect times abundance) of a species, on other species, to be independent of its trophic position. Since abundance and trophic position are negatively correlated, this indicates that per capita effects are negatively correlated to abundance, and thus implies that individuals of rare species in these communities, as in our model systems, interact more strongly with other species than do individuals of more abundant species.

Theory predicts that consumer species having strong per capita effects on their prey will also have low equilibrium/average abundances (Case, 2000). Classical empirical examples come from biological control, where control agents exerting very high per capita pressure on their target species can induce low densities of their targets and, as a result, low abundances of the agents themselves (Beddington *et al.*, 1978; Morin, 2011). Because of their strong interactions and controlling effect, even small changes to their abundance can be expected to affect species that are directly linked to these species. However, how far-reaching the repercussions of this effect can be (both in terms of number of species affected and magnitude of the response), and if it implies any general relationship (i.e., beyond the strongest interactors) between the abundance of a species and its structural importance in multispecies systems has been unclear. Here we have shown that a negative correlation between the abundance of a species and its effect on other species can indeed be expected among species that are dynamically linked in an ecological network. In real systems, species can be rare for many reasons, however (e.g., due to anthropogenic stress or abiotic factors). Thus we are not suggesting that every uncommon species is important. Neither do we suggest that common species are ecologically unimportant. For instance, in a recent theoretical study of competitive communities (i.e., communities with a single trophic level), it was found that deletion of abundant species led to a higher risk of cascading extinctions than deletion of rare species (Fowler, 2010). Rather, our study suggests that, if judging by the abundance and contribution to particular ecosystem processes of a species only, the structural importance of many species in ecological communities may be much under-rated. Although particularly striking for some really rare species, it should be true for many species that are not so abundant for dynamical reasons (i.e., due to strong interactions with other species), including many moderately uncommon species.

Our analysis of the ecological importance of rare species is based on generalized Lotka–Volterra models with linear functional responses. Such models have been used in many recent studies to explore the dynamics of food webs and their response to different stressors (Ives and Cardinale, 2004; Chesson and Kuang, 2008; Sahasrabudhe and Motter, 2011; Allesina and Tang, 2012). Although many alternatives for modeling trophic dynamics exist (e.g., using models with non-linear functional responses; Yodzis and Innes, 1992; Berlow *et al.*, 2009) that may more realistically capture the foraging success of individual consumers across large ranges of prey densities, little empirical support exists for the superiority of these alternatives over the simpler model used here, for the population response over more moderate ranges of prey densities. Indeed, Novak (2010) estimated interaction strengths empirically and analyzed the potential non-linearity of observed functional responses, finding that models assuming linear functional responses actually performed better than others. Novak's study contributes to a growing literature suggesting that "trophic interactions are approximately linear in the range of mean prey densities actually observed in nature, especially in multispecies settings" (Novak, 2010; see also Wootton and Emmerson, 2005). Furthermore, some theoretical studies have compared model results based on linear and non-linear functional responses, and found that relationships between structure and stability and general patterns in the response of food webs to species loss were similar for the two scenarios (e.g., Borrvall and Ebenman, 2006; Petchey *et al.*, 2008; Mougi and Kondoh, 2012). Overall, these studies suggest that our results, based on models with linear functional responses, have something to say about the role of rare species in real ecosystems.

In addition to the structural importance of rare species documented here, a recent theoretical study (Berg *et al.*, 2015) has shown that, if lost, some rare species can cause disproportionally many secondary extinctions, and may thus be keystone structures that play an important role in upholding the structure of ecological communities. However, at the same time as the ecological importance of many rare species may be under-appreciated, these species are also the ones that tend to be most prone to extinction (Lande, 1993).

This indicates a potential dilemma: a negative relationship in ecological networks between a species contribution to properties related to network persistence and the vulnerability of the species itself. A recent study of mutualistic networks points in the same direction; species contributing most strongly to network persistence were also those that were most vulnerable to extinction (Saavedra *et al.*, 2011).

19.5   Conclusions

We found that across all species, as well as within trophic groups, community structure tends (on average) to be more sensitive to perturbations to rare species than perturbations to more abundant species. While maybe an unexpected finding at first thought, we present an ecological mechanism for this: individuals of (important) rare species interact more strongly with other species (mainly their resources) than do individuals of more abundant species. Thus our results suggest that there is an inverse relationship between abundance determined by trophic interactions and importance in ecological communities. In other words, common species cannot a priori be assumed to be keystone species in terms of their effect on community structure and dynamics. Instead, among the species whose abundance is mainly affected by biotic interactions, the rare consumers are expected to be the most dynamically and structurally important. An important implication of our study is that rare species should be major targets for conservation efforts, not only for ethical and moral reasons, but in some cases also for ecological reasons.

Acknowledgments

We thank Stefan Sellman, Joel E. Cohen, Otso Ovaskainen, and Donald DeAngelis for comments and discussion. This research was supported by Formas (the Swedish Research Council for Environment, Agricultural Sciences and Spatial Planning).

Appendix

Table S19.1 Relationship between trophic position and abundance of species in individual food webs.

Food web	R	P	Food web	R	P	Food web	R	P
1	−0.847	<0.001	35	−0.845	<0.001	69	−0.836	<0.001
2	−0.818	<0.001	36	−0.824	<0.001	70	−0.877	<0.001
3	−0.769	<0.001	37	−0.743	<0.001	71	−0.827	<0.001
4	−0.928	<0.001	38	−0.854	<0.001	72	−0.85	<0.001
5	−0.88	<0.001	39	−0.827	<0.001	73	−0.868	<0.001
6	−0.948	<0.001	40	−0.761	<0.001	74	−0.803	<0.001
7	−0.835	<0.001	41	−0.771	<0.001	75	−0.876	<0.001
8	−0.901	<0.001	42	−0.868	<0.001	76	−0.816	<0.001

Table S19.1 (cont.)

Food web	R	P	Food web	R	P	Food web	R	P
9	−0.797	<0.001	43	−0.878	<0.001	77	−0.764	<0.001
10	−0.836	<0.001	44	−0.874	<0.001	78	−0.801	<0.001
11	−0.863	<0.001	45	−0.85	<0.001	79	−0.891	<0.001
12	−0.779	<0.001	46	−0.836	<0.001	80	−0.86	<0.001
13	−0.854	<0.001	47	−0.869	<0.001	81	−0.898	<0.001
14	−0.826	<0.001	48	−0.851	<0.001	82	−0.839	<0.001
15	−0.82	<0.001	49	−0.894	<0.001	83	−0.872	<0.001
16	−0.87	<0.001	50	−0.876	<0.001	84	−0.841	<0.001
17	−0.714	<0.001	51	−0.855	<0.001	85	−0.793	<0.001
18	−0.876	<0.001	52	−0.922	<0.001	86	−0.915	<0.001
19	−0.883	<0.001	53	−0.862	<0.001	87	−0.899	<0.001
20	−0.814	<0.001	54	−0.851	<0.001	88	−0.863	<0.001
21	−0.828	<0.001	55	−0.768	<0.001	89	−0.893	<0.001
22	−0.822	<0.001	56	−0.823	<0.001	90	−0.887	<0.001
23	−0.799	<0.001	57	−0.807	<0.001	91	−0.887	<0.001
24	−0.845	<0.001	58	−0.839	<0.001	92	−0.803	<0.001
25	−0.885	<0.001	59	−0.875	<0.001	93	−0.853	<0.001
26	−0.875	<0.001	60	−0.826	<0.001	94	−0.856	<0.001
27	−0.876	<0.001	61	−0.852	<0.001	95	−0.896	<0.001
28	−0.831	<0.001	62	−0.901	<0.001	96	−0.852	<0.001
29	−0.867	<0.001	63	−0.818	<0.001	97	−0.822	<0.001
30	−0.898	<0.001	64	−0.785	<0.001	98	−0.868	<0.001
31	−0.872	<0.001	65	−0.83	<0.001	99	−0.851	<0.001
32	−0.886	<0.001	66	−0.903	<0.001	100	−0.82	<0.001
33	−0.84	<0.001	67	−0.887	<0.001			
34	−0.861	<0.001	68	−0.903	<0.001			

Note: R is the Pearson product moment correlation coefficient, and P the significance for the association between species trophic position and log equilibrium abundance.

Table S19.2 Species equilibrium abundance as a function of trophic position, as indicated by a linear mixed-effect model.

	$\eta_i (v_i)$	Parameters	Estimates	s.e.m.	t-value	P
Trophic group 3	86 (449)	κ	2.833	0.019	149.056	<0.0001
		μ	−0.070	0.004	−17.099	<0.0001
Trophic group 4	98 (798)	κ	3.436	0.011	322.795	<0.0001
		μ	−0.082	0.003	−23.579	<0.0001

Note: Random effects, depending on individual webs, were introduced on the intercepts for both groups and on the slope for trophic group 3. Parameters κ and μ report the fixed effects estimates of the intercept and slope, respectively. For random effects see Table S19.6. η_i: Number of food webs with at least three species in trophic group i; v_i: number of species in trophic group i from all food webs with at least three species in this trophic group; s.e.m.: standard error.

Table S19.3 Species importance as a function of species residual abundance, as indicated by linear mixed-effect model analysis.

	$\eta_i\ (v_i)$	Parameters	Estimates	s.e.m.	t-value	P
Sensitivity of community structure:						
All species	100 (5000)	β	11.328	0.051	224.208	<0.0001
		ρ	−0.334	0.021	−15.818	<0.0001
Trophic group 3	86 (449)	β	12.128	0.061	199.984	<0.0001
		ρ	−0.300	0.021	−14.245	<0.0001
Trophic group 4	98 (798)	β	12.816	0.067	191.068	<0.0001
		ρ	−0.241	0.022	−10.881	<0.0001

Notes: Random effects, depending on individual webs, were introduced on the slopes and on the intercepts. Parameters ρ and β report the fixed effects estimates of the slope and intercept, respectively. P values were calculated according to Pinheiro and Bates (2004) and should be regarded as indications of significance only. η_i: number of food webs with at least three species in trophic group i; v_i: number of species in trophic group i from all food webs with at least three species in this trophic group; s.e.m.: standard error.

Table S19.4 Predation pressure as a function of species equilibrium abundance, as indicated by a linear mixed-effect model.

	$\eta_i\ (v_i)$	Parameters	Estimates	s.e.m.	t-value	P
Trophic group 2	100 (2610)	β	−0.474	0.054	−8.782	<0.0001
		ρ	−0.946	0.007	−129.947	<0.0001
Trophic group 3	86 (449)	β	−2.736	0.131	−20.847	<0.0001
		ρ	−1.106	0.019	−55.638	<0.0001
Trophic group 4	98 (798)	β	−6.140	0.128	−48.152	<0.0001
		ρ	−1.478	0.024	−59.524	<0.0001

Notes: Random effects, depending on individual webs, were introduced on the intercepts. Parameters β and ρ report the fixed effects estimates of the intercept and predation pressure. For random effects see Table S19.6. P values were calculated according to Pinheiro and Bates (2004) and should be regarded as indications of significance only. η_i: number of food webs with at least three species in trophic group i; v_i: number of species in trophic group i from all food webs with at least three species in this trophic group; s.e.m.: standard error.

Table S19.5 Random effects on intercept (β) and slope (ρ) for each food web of the relationship between equilibrium abundances and sensitivity of equilibrium abundance (Figure 19.2b–e) for all species and for each trophic group.

Food web	All species		Tg 1		Tg 2		Tg 3		Tg 4	
	β	ρ	β	ρ	β	ρ	β	ρ	β	ρ
1	−0.012	0.018	2.919	−0.325	0.378	−0.031	0.417	0.03	−0.102	−0.033
2	0.053	−0.023	−1.981	0.239	0.137	−0.055	0.392	−0.111	0.2	−0.036
3	−1.152	0.071	−0.942	0.033	−2.316	0.317	−0.833	0.084	−0.566	0.014
4	1.081	0.004	−3.031	0.411	1.271	−0.063	−0.624	0.076	0.855	−0.105
5	−0.555	0.058	−0.144	0.007	−0.413	0.05	0.842	−0.179	−0.213	0.043

Table S19.5 (cont.)

Food web	All species β	ρ	Tg 1 β	ρ	Tg 2 β	ρ	Tg 3 β	ρ	Tg 4 β	ρ
6	−0.284	0.056	−3.362	0.382	1.388	−0.272	−0.336	0.066	−1.002	0.156
7	0.348	−0.035	3.04	−0.306	1.906	−0.32	−0.185	0.042	−0.232	0.029
8	−0.373	0.083	1.649	−0.132	0.13	−0.018	−0.432	0.065	−0.415	0
9	−0.284	0.011	0.989	−0.159	−1.039	0.145	0.296	−0.026	−0.143	0.052
10	0.389	−0.025	0.402	0.005	1.039	−0.147	0.131	−0.029	0.321	−0.006
11	0.509	−0.033	−2.434	0.313	1.251	−0.149	−0.354	0.029	0.663	−0.138
12	−0.071	0	−0.072	0.002	0.551	−0.122	0.157	−0.046	−0.144	−0.026
13	−0.722	0.027	−2.341	0.235	−1.685	0.232	0.154	0.016	−0.345	0.055
14	−0.598	0.088	1.074	−0.108	−7.031	1.113	−0.166	0.041	0.552	−0.099
15	0.066	−0.011	−2.372	0.267	−0.62	0.106	−0.318	0.048	0.347	−0.077
16	−0.103	0.024	−1.567	0.17	−0.93	0.205	−0.55	0	−2.088	0.444
17	−1.361	0.117	16.218	−1.919	−0.669	0.005	−0.253	−0.044	−0.702	0.076
18	−0.862	0.035	−2.889	0.26	−1.165	0.092	0.1	0.003	1.209	−0.089
19	1.238	−0.049	−2.238	0.318	1.573	−0.151	0.002	−0.014	−0.213	−0.001
20	−0.263	−0.005	−2.306	0.256	0.402	−0.086	0.415	−0.055	0.189	−0.053
21	0.303	−0.035	2.037	−0.227	−0.088	0.008	−0.289	0.028	−0.238	0.071
22	−0.296	0.016	0.67	−0.08	−0.101	−0.031	−0.033	−0.045	−0.157	−0.004
23	0.059	−0.022	0.623	−0.075	0.627	−0.115	0.244	0.001	−0.126	0.019
24	−0.132	−0.012	−1.925	0.198	0.054	−0.053	−0.285	0.061	−0.276	0
25	−0.255	−0.04	−2.242	0.188	0.43	−0.149	0.595	−0.037	−0.284	0.015
26	0.156	−0.026	−1.617	0.196	0.926	−0.185	−0.41	0.032	0.239	−0.056
27	−0.114	0.007	0.015	0.024	−0.081	0.019	0.588	−0.042	0.548	−0.028
28	0.55	−0.015	−0.093	0.07	1.282	−0.178	0.757	−0.138	0.361	−0.011
29	1.219	−0.092	1.924	−0.156	2.243	−0.284	0.132	−0.036	−0.232	−0.003
30	−0.692	0.027	0.606	−0.099	−0.771	0.015	0.031	0.04	0.517	−0.025
31	0.976	−0.064	1.791	−0.134	2.144	−0.27	−0.128	−0.054	0.337	0.006
32	0.276	−0.024	3.184	−0.308	1.424	−0.261	−0.042	0.003	−0.168	0.037
33	0.026	−0.007	−2.064	0.232	0.329	−0.048	0.183	0.026	0.417	−0.065
34	0.26	0.001	−1.416	0.184	−0.413	0.137	0.127	0.004	−0.097	−0.049
35	0.12	−0.054	0.129	−0.047	1.085	−0.222	−0.554	0.074	−0.065	−0.045
36	−0.142	−0.001	0.244	−0.012	0.576	−0.159	0.148	0.067	0.783	−0.082
37	0.146	−0.017	−0.288	0.04	−0.081	0.004	0.106	−0.002	0.022	0.052
38	−0.049	−0.009	−0.648	0.069	−0.504	0.07	−0.091	−0.019	0.014	0
39	0.078	0.017	2.008	−0.22	−0.47	0.146	1.18	−0.066	0.768	−0.054
40	0.676	−0.126	−2.445	0.252	0.982	−0.218	−0.085	−0.073	0.252	−0.094
41	−0.406	0.083	0.574	−0.038	−3.104	0.542	0.184	0.008	−0.376	0.041
42	0.125	−0.029	−1.865	0.186	−0.256	0.065	0.551	−0.041	−0.041	−0.005
43	0.092	−0.034	−0.891	0.097	0.344	−0.099	−0.578	0.074	0.671	−0.098
44	0.242	−0.026	−0.194	0.023	0.654	−0.06	0.339	−0.006	−0.379	0.032
45	0.637	0.01	−0.185	0.09	0.662	0.019	−0.055	0.044	−1.002	0.125
46	−0.111	−0.015	−0.375	−0.018	0.405	−0.099	0.476	−0.054	0.888	−0.063
47	−0.466	0.067	2.111	−0.311	0.873	−0.139	−0.043	−0.034	−0.412	0.032
48	0.612	−0.006	1.536	−0.127	0.721	0.009	0.504	−0.098	−0.454	0.039
49	−0.345	0.066	−0.16	0.024	0.409	−0.068	0.192	0.025	0.992	−0.102
50	−0.278	0.054	−0.949	0.11	−0.291	0.072	−0.707	0.078	0.532	−0.062

Table S19.5 (cont.)

Food web	All species β	ρ	Tg 1 β	ρ	Tg 2 β	ρ	Tg 3 β	ρ	Tg 4 β	ρ
51	1.282	−0.086	4.36	−0.435	1.933	−0.213	−0.301	0.06	0.431	−0.155
52	0.619	−0.08	1.308	−0.176	−0.068	0.031	0.05	0.008	−0.054	−0.002
53	0.041	−0.04	−2.028	0.226	−0.117	−0.012	−0.295	0.008	0.294	−0.071
54	−0.383	0.002	−1.654	0.162	−0.347	−0.008	−0.072	−0.006	−0.176	0.154
55	0.192	−0.023	0.776	−0.094	−0.695	0.142	0.658	−0.009	−0.034	−0.008
56	0.356	0.014	4.682	−0.527	0.098	0.042	−0.067	−0.034	0.458	−0.086
57	−1.066	0.116	−1.006	0.113	−3.335	0.54	−0.193	−0.088	0.532	−0.137
58	−0.642	0.057	−2.111	0.222	−3.989	0.647	0.283	−0.044	0.425	−0.08
59	0.387	−0.083	0.832	−0.138	−0.157	−0.006	−0.235	−0.019	−0.051	0.026
60	−0.231	0	−4.661	0.503	−0.917	0.147	0.548	−0.048	−0.873	0.065
61	−0.142	−0.001	1.225	−0.142	0.862	−0.187	−0.148	−0.016	0.938	−0.14
62	0.206	0.019	2.877	−0.365	1.313	−0.176	0.35	−0.032	−0.014	0.057
63	0.503	−0.019	1.934	−0.174	−0.1	0.074	0.262	−0.023	−0.713	0.012
64	0.086	−0.005	−2.949	0.344	0.338	−0.033	−0.408	−0.03	−1.455	0.223
65	−0.948	0.016	−7.353	0.795	−0.995	0.037	−0.271	−0.056	0.108	−0.006
66	−0.739	0.033	0.271	−0.108	0.672	−0.27	−0.292	0.003	−0.461	0.095
67	0.256	−0.039	1.329	−0.142	0.459	−0.079	0.548	−0.041	0.993	−0.103
68	−0.082	−0.007	1.406	−0.174	0.157	−0.037	−0.343	0.022	−0.333	0.024
69	0.622	−0.023	3.428	−0.31	0.265	0.016	1.244	−0.069	0.148	0.018
70	−0.346	0.027	−2.882	0.308	−0.039	0.025	0.271	−0.033	0.009	−0.021
71	0.395	−0.058	−2.585	0.284	0.817	−0.135	0.268	−0.017	−1.006	0.093
72	0.333	−0.04	0.203	−0.014	1.006	−0.16	−0.253	0.047	−0.478	0.055
73	−1.174	0.04	−7.94	0.879	−1.633	0.136	0.519	−0.072	−0.189	−0.012
74	−0.66	−0.01	−2.228	0.177	−1.157	0.061	−0.12	0.057	0.068	0.011
75	0.448	0	3.132	−0.295	1.484	−0.203	−1.495	0.169	−0.102	−0.009
76	0.143	−0.027	−2.742	0.306	−0.774	0.086	0.03	−0.049	1.167	−0.199
77	−0.556	0.043	2.218	−0.277	−1.339	0.154	−0.078	0.023	0.121	−0.026
78	1.28	−0.078	3.955	−0.394	1.576	−0.109	−0.597	0.11	0.706	−0.114
79	0.479	−0.037	−1.292	0.168	1.261	−0.189	−0.019	−0.024	0.323	−0.128
80	0.614	−0.053	−0.364	0.067	0.365	−0.013	−0.927	0.107	−0.883	0.144
81	0.097	−0.042	−0.246	0.013	0.36	−0.097	−0.519	0.066	0.752	−0.075
82	−0.491	0.043	1.679	−0.241	0.606	−0.147	−0.204	0.024	−0.104	0.033
83	1.027	−0.107	3.585	−0.375	2.43	−0.362	−0.243	0.029	−2.078	0.472
84	−0.958	0.087	0.085	−0.053	−3.562	0.627	0.459	0.097	−0.269	0.071
85	−2.164	0.241	3.071	−0.491	−1.625	0.243	−0.449	0.08	−0.186	−0.065
86	0.33	−0.047	0.873	−0.103	1.316	−0.225	0.143	0.009	−0.111	−0.065
87	−0.393	−0.002	1.406	−0.201	−0.128	−0.039	NA	NA	−0.302	0.007
88	0.172	−0.077	−3.257	0.343	0.907	−0.204	NA	NA	−0.471	0.065
89	−0.094	−0.024	1.329	−0.186	0.053	−0.058	NA	NA	0.767	0.008
90	−0.066	0.011	0.177	−0.046	−0.413	0.087	NA	NA	−0.283	0.064
91	0.663	−0.013	−1.463	0.221	0.981	−0.068	NA	NA	0.327	−0.113
92	−1.319	0.124	−2.088	0.195	−5.011	0.715	NA	NA	−0.134	0.035
93	0.04	0.009	−1.165	0.167	0.182	0.004	NA	NA	−0.682	0.085
94	0.011	−0.02	−1.298	0.144	0.422	−0.098	NA	NA	0.155	0.054
95	−0.162	0.027	3.822	−0.414	1.265	−0.255	NA	NA	0.587	−0.034

Table S19.5 (cont.)

Food web	All species		Tg 1		Tg 2		Tg 3		Tg 4	
	β	ρ	β	ρ	β	ρ	β	ρ	β	ρ
96	0.223	0.009	−1.05	0.174	1.304	−0.185	NA	NA	0.618	0.03
97	0.211	0.007	−0.482	0.086	−0.237	0.156	NA	NA	−0.055	0.02
98	1.017	−0.02	3.397	−0.311	0.115	0.159	NA	NA	0.365	−0.059
99	−0.263	0.051	−0.375	0.062	−0.387	0.074	NA	NA	NA	NA
100	0.485	−0.029	−0.851	0.099	0.292	0.017	NA	NA	NA	NA

Table S19.6 Random effects on intercept (β) for each food web of the relationship between equilibrium abundances and predation pressure (Figure 19.3) for each trophic group.

Food web	Tg 2 β	Tg 3 β	Tg 4 β	Food web	Tg 2 β	Tg 3 β	Tg 4 β	Food web	Tg 2 β	Tg 3 β	Tg 4 β
1	0.111	−0.856	−0.635	36	−0.437	−5.672	0.161	71	0.141	−3.508	−0.799
2	0.153	0.754	−0.202	37	−0.62	2.601	0.629	72	0.617	−6.395	−0.007
3	−1.027	−3.887	0.064	38	0.405	6.361	0.771	73	−0.337	1.132	1.158
4	0.685	1.384	−0.407	39	0.455	0.668	−1.077	74	−0.107	−6.929	0.258
5	−0.437	−0.182	0.14	40	1.469	1.351	1.321	75	0.535	−7.69	0.237
6	1.389	−0.841	−0.966	41	−0.019	4.022	1.381	76	−0.407	1.946	0.477
7	0.601	−2.494	−1.701	42	0.348	0.494	0.492	77	−0.125	13.13	1.794
8	0.241	1.775	−1.502	43	0.382	−3.227	−0.167	78	−0.116	8.934	2.014
9	−0.595	−0.842	0.279	44	−0.219	−33.667	2.914	79	0.48	−1.19	2.037
10	−0.065	−0.39	0.598	45	−0.293	3.417	−0.7	80	−0.382	−3.686	−1.935
11	0.683	0.83	0.309	46	−0.049	1.256	−1.272	81	0.476	−3.952	2.189
12	−0.269	−1.995	−0.141	47	0.062	0.504	2.66	82	−0.285	−1.673	1.069
13	−0.441	0.496	0.065	48	−0.284	−3.21	−3.249	83	1.056	−2.732	−6.689
14	−0.16	1.476	−0.014	49	0.352	−1.223	0.872	84	−0.379	−4.891	−1.295
15	−0.201	−0.857	−1.608	50	0.368	−1.003	0.892	85	0.077	3.498	−0.114
16	−0.097	1.975	−14.928	51	1.091	−3.004	0.112	86	0.111	−5.152	1.13
17	−0.128	36.78	1.56	52	0.532	−3.952	−2.73	87	0.153	NA	1.699
18	1.678	27.346	−0.819	53	−0.783	0.914	0.935	88	−1.027	NA	1.418
19	0.26	11.758	−0.647	54	−0.725	−0.335	−0.62	89	0.685	NA	−1.433
20	−0.029	−4.538	0.848	55	−0.09	−5.502	−0.888	90	−0.437	NA	−1.673
21	−0.346	1.702	−0.81	56	−0.101	2.935	2.391	91	1.389	NA	0.103
22	−0.12	−4.904	1.037	57	−0.784	−2.197	−0.701	92	0.601	NA	0.53
23	−0.1	1.036	−0.912	58	−0.038	−0.203	0.216	93	0.241	NA	−0.495
24	1.09	−18.488	0.782	59	−0.367	−1.238	−1.256	94	−0.595	NA	−1.205
25	0.563	−1.843	2.421	60	−0.426	−2.208	2.32	95	−0.065	NA	−0.79
26	0.011	2.017	−0.395	61	0.522	1.288	0.619	96	0.683	NA	0.18
27	−0.386	−1.492	−0.307	62	0.422	3.225	0.173	97	−0.269	NA	−0.803
28	−0.414	1.096	13.399	63	−0.384	−2.868	−1.7	98	−0.441	NA	0.674
29	0.51	−3.759	2.573	64	−0.417	−7.837	−0.264	99	−0.16	NA	NA
30	0.606	0.528	1.083	65	−0.722	2.889	0.49	100	−0.201	NA	NA
31	0.275	−3.446	−0.447	66	1.01	0.811	−1.112				

Table S19.6 (cont.)

Food web	Tg 2 β	Tg 3 β	Tg 4 β	Food web	Tg 2 β	Tg 3 β	Tg 4 β	Food web	Tg 2 β	Tg 3 β	Tg 4 β
32	0.947	−3.296	0.087	67	0.863	−3.616	1.057				
33	−0.137	−1.744	0.016	68	0.623	−5.098	0.038				
34	−0.026	−9.212	1.95	69	−0.062	1.165	1.897				
35	0.314	−6.533	0.484	70	0.035	0.225	1.023				

References

Allesina, S. and Tang, S. (2012). Stability criteria for complex ecosystems. *Nature*, **483**, 205–208.

Barnosky, A. D., Matzke, N., Tomiya, S., *et al.* (2011). Has the Earth's sixth mass extinction already arrived? *Nature*, **471**, 51–57.

Beddington, J. R., Free, C. A., and Lawton, J. H. (1978). Characteristics of successful natural enemies in models of biological control of insect pests. *Nature*, **273**, 513–519.

Bender, E. A., Case, T. J., and Gilpin, M. E. (1984). Perturbation experiments in community ecology: theory and practice. *Ecology*, **65**, 1–13.

Berg, S., Christianou, M., Jonsson, T., and Ebenman, B. (2011). Using sensitivity analysis to identify keystone species and keystone links in size-based food webs. *Oikos*, **120**, 510–519.

Berg, S., Pimenov, A., Palmer, C., Emmerson, M. C., and Jonsson, T. (2015). Ecological communities are vulnerable to realistic extinction sequences. *Oikos*, **124**, 486–496.

Berlow, E. L., Dunne, J. A., Martinez, N. D, *et al.* (2009). Simple prediction of interaction strengths in compex food webs. *PNAS*, **6**, 187–191.

Borrvall, C. and Ebenman, B. (2006). Early onset of secondary extinctions in ecological communities following the loss of top predators. *Ecology Letters*, **9**, 435–442.

Borrvall, C., Ebenman, B., and Jonsson, T. (2000). Biodiversity lessens the risk of cascading extinction in model food webs. *Ecology Letters*, **3**, 131–136.

Bracken, M. E. S. and Low, N. H. N. (2012). Realistic losses of rare species disproportionately impact higher trophic levels. *Ecology Letters*, **15**, 461–467.

Brown, J. H., Gillooly, J. F., Allen, A. P., and Savage, V. M. (2004). Toward a metabolic theory of ecology. *Ecology*, **85**, 1771–1789.

Case, T. J. (2000). *An Illustrated Guide to Theoretical Ecology*. Oxford, UK: Oxford University Press.

Chesson, P. and Kuang, J. J. (2008). The interaction between predation and competition. *Nature*, **456**, 235–238.

Cottee-Jones, H. E. W. and Whittaker, R. J. (2012). The keystone species concept: a critical appraisal. *Frontiers of Biogeography*, **4**, 117–127.

Curtsdotter, A., Binzer, A., Brose, U., *et al.* (2011). Robustness to secondary extinctions: comparing trait-based sequential deletions in static and dynamic food webs. *Basic and Applied Ecology*, **12**, 571–580.

Ebenman, B. and Jonsson, T. (2005). Using community viability analysis to identify fragile systems and keystone species. *Trends in Ecology and Evolution*, **20**, 568–575.

Eklöf, A. and Ebenman, B. (2006). Species loss and secondary extinctions in simple and complex model communities. *Journal of Animal Ecology*, **75**, 239–246.

Ellison, A. M., Bank, M. S., Clinton, B. D., *et al.* (2005). Loss of foundation species: consequences for the structure and dynamics of forested ecosystems. *Frontiers in Ecology and the Environment*, **3**, 479–486.

Estes, J. A., Terborgh, J., Brashares, J. S., *et al.* (2011). Trophic downgrading of planet Earth. *Science*, **333**, 301–306.

Fowler, M. (2010). Extinction cascades and the distribution of species interactions. *Oikos*, **119**, 864–873.

Gaston, K. J. (2010). Valuing common species. *Science*, **327**, 154–155.

Gaston, K. J. (2011). Common ecology. *BioScience*, **61**, 354–362.

Gaston, K. J. and Fuller, R. A. (2008). Commonness, population depletion and conservation biology. *Trends in Ecology and Evolution*, **23**, 14–19.

Hol, W. H. G., de Boer, W., Termorshuizen, A. J., *et al.* (2010). Reduction of rare soil microbes modifies plant-herbivore interactions. *Ecology Letters*, **13**, 292–301.

Hooper, D. U., Adair, E. C., Cardinale, B. J., *et al.* (2012). A global synthesis reveals biodiversity loss as a major driver of ecosystem change. *Nature*, **486**, 105–108.

Hubbell, S. P. (2001). *The Unified Neutral Theory of Biodiversity and Biogeography.* Princeton, NJ: Princeton University Press.

Isbell, F., Calcagno, V., Hector, A., *et al.* (2011). High plant diversity is needed to maintain ecosystem services. *Nature*, **477**, 199–202.

Ives, A. R. and Cardinale, B. J. (2004). Food-web interactions govern the resistance of communities after non-random extinctions. *Nature*, **429**, 174–177.

Jonsson, T., Cohen, J. E., and Carpenter, S. R. (2005). Food webs, body size, and species abundance in ecological community description. *Advances in Ecological Research*, **36**, 1–84.

Jonsson, T., Berg, S., Pimenov, A., and Emmerson, M. C. (2015). The context dependency of species keystone status during food web disassembly. *Food Webs*, **5**, 1–10.

Lande, R. (1993). Risks of population extinction from demographic and environmental stochasticity and random catastrophes. *American Naturalist*, **142**, 911–927.

Lyons, K. G. and Schwartz, M. W. (2001). Rare species loss alters ecosystem function: invasion resistance. *Ecology Letters*, **4**, 358–365.

Lyons, K. G., Brigham, C. A., Traut, B. H., and Schwartz, M. W. (2005). Rare species and ecosystem functioning. *Conservation Biology*, **19**, 1019–1024.

MacArthur, R. H. and Wilson, E. O. (1967). *The Theory of Island Biogeography.* Princeton, NJ: Princeton University Press.

May, R. M. (1975). Patterns of species abundance and diversity. In *Ecology and Evolution of Communities*, ed. M. L. Cody and J. M. Diamond, Cambridge, UK: Belknap Press, pp. 81–120.

McCoy, M. W. and Gillooly, J. F. (2008). Predicting natural mortality rates of plants and animals. *Ecology Letters*, **11**, 710–716.

Montoya, J. M., Emmerson, M. C., Solé, R. V., and Woodward, G. (2005). Perturbations and indirect effects in complex food webs. In *Dynamic Food Webs: Multispecies*

Assemblages, Ecosystem Development and Environmental Change, ed. P. C. de Ruiter, V. Wolters, and J. C. Moore, Burlington, MA: Academic Press, pp. 369–380.

Montoya, J. M., Pimm, S. L., and Sole, R. V. (2006). Ecological networks and their fragility. *Nature*, **442**, 259–264.

Montoya, J. M., Woodward, G., Emmerson, M. C., and Solé, R. V. (2009). Press perturbations and indirect effects in real food webs. *Ecology*, **90**, 2426–2433.

Morin, P. (2011). *Community Ecology*. Oxford, UK: Wiley-Blackwell.

Mougi, A. and Kondoh, M. (2012). Diversity of interaction types and ecological community stability. *Science*, **337**, 349–351.

Novak, M. (2010). Estimating interaction strengths in nature: experimental support for an observational approach. *Ecology*, **91**, 2394–2405.

O'Gorman, E. J., Jacob, U., Jonsson, T., and Emmerson, M. C. (2010). Interaction strength, food web topology and the relative importance of species in food webs. *Journal of Animal Ecology*, **79**, 682–692.

Paine, R. T. (1966). Food web complexity and species diversity. *American Naturalist*, **100**, 65–75.

Paine, R. T. (1969). A note on trophic complexity and community stability. *American Naturalist*, **355**, 73–75.

Pereira, H. M., Leadley, P. W., Proenca, V., *et al.* (2010). Scenarios for global biodiversity in the 21st century. *Science*, **330**, 1496–1501.

Petchey, O., Eklöf, A., Borrvall, C., and Ebenman, B. (2008). Trophically unique species are vulnerable to cascading extinction. *American Naturalist*, **171**, 568–579.

Peters, R. H. (1983). *The Ecological Implications of Body Size*. New York, NY: Cambridge University Press.

Pinheiro, J. C. and Bates, D. M. (2004). *Mixed-Effects Models in S and S-PLUS*. New York, NY: Springer-Verlag.

Pinheiro, J., Bates, D., DebRoy, S., Sarkar, D., *et al.* (2012). nlme: Linear and Nonlinear Mixed Effects Models. R package version 3.1–104.

R Development Core Team (2012). *R: A Language and Environment for Statistical Computing*. Vienna, Austria: R Foundation for Statistical Computing.

Reuman, D. C., Mulder, C., Raffaelli, D., and Cohen, J. E. (2008). Three allometric relations of population density to body mass: theoretical integration and empirical tests in 149 food webs. *Ecology Letters*, **11**, 1216–1228.

Saavedra, S., Stouffer, D. B., Uzzi, B., and Bascompte, J. (2011). Strong contributers to network persistence are the most vulnerable to extinction. *Nature*, **478**, 233–236.

Sahasrabudhe, S. and Motter, A. E. (2011). Rescuing ecosystems from extinction cascades through compensatory perturbations. *Nature Communications*, **2**, 1–8.

Säterberg, T., Sellman, S., and Ebenman, B. (2013). High frequency of functional extinctions in ecological networks. *Nature*, **499**, 468–470.

Thompson, K. (2010). *Do We Need Pandas?* Foxhole, Dartington: Green Books.

Wood, S. A., Lilley, S. A., Schiel, D. R., and Shurin, J. B. (2010). Organismal traits are more important than environment for species interactions in the intertidal zone. *Ecology Letters*, **13**, 1160–1171.

Wootton, T. and Emmerson, M. (2005). Measurement of interaction strength in nature. *Annual Review of Ecology, Evolution and Systematics*, **36**, 419–444.

Yodzis, P. and Innes, S. (1992). Body size and consumer-resource dynamics. *American Naturalist*, **139**, 1151–1175.

20 Food-Web Simulations: Stochastic Variability and Systems-Based Conservation

Ferenc Jordán, Marco Scotti, and Catherine M. Yule

20.1 Individual-Based Modeling in Conservation Biology

A major challenge for food-web research is studying diversity and variability more explicitly. This means a focus on individual-level variability in populations (Bolnick *et al.*, 2011) that hopefully might help to better understand how structural properties predict dynamical behavior (Dunne, 2006). One reason why linking structure to dynamics is still a hard challenge can be that intrapopulation variability is relatively poorly considered in most models. Yet defining developmental stages as graph nodes is a step toward managing this challenge: for example, in many food-web models certain species are represented by separate graph nodes that include juveniles and adults. Trophic status and network dynamics can be quite sensitive to this kind of demographic aggregation (or resolution) of the web and are expected to be influenced by individual-level differences in terms of behavior and feeding habits.

Individual-level variability includes genetic, demographic, and stochastic factors and to date it is not easy to incorporate all these in most modeling frameworks. Yet developing the methodological background of individual-based modeling seems to be very useful for future research and applications. Since individual-level differences are more important in smaller populations (Lande, 1988), studying their effects explicitly is relevant for conservation efforts.

Using network metrics as proxies or predictors of food-web dynamics is an old but still open issue. For an example, network hubs are supposed to be species of key importance and hubbish networks are thought to be safe against errors but vulnerable against attacks (Montoya and Solé, 2002). We make structural predictions routinely but we are very poor in testing these on either real-time series or simulation models. One way to make our knowledge more robust here is by adopting a comparative approach: studying spatio-temporal food-web gradients can inform about possible relationships between the position occupied by species in trophic networks and their dynamics. Based on spatio-temporal data reporting on various ecological gradients (e.g., plankton biomass: Siokou-Frangou *et al.*, 2002), different versions of food webs can be constructed to represent changes in space and time (Warren, 1989; Ulanowicz, 1996; Winemiller, 1996; Bondavalli *et al.*, 2006), and ecosystem management and restoration can be based on this kind of comparative knowledge (Tallberg *et al.*, 1999). Food-web studies are being improved by the recent development of constructing series of food webs describing a community along an environmental gradient (Lafferty and Dunne, 2010). These data and models make it possible to focus on comparative studies: some notorious methodological problems of food-web research (e.g., aggregation, weighting links) can be slightly reduced and partly managed if network variants of the same ecosystem are described in the same way and the differences among them are emphasized. For example, lumping two species A and B into a single functional group F may be a question with no perfect solution (because of context dependency, for example). The two possibilities will result in two networks of different properties. It is not easy to decide which one is more appropriate. But if we consistently treat these two species in the same way (e.g., always lumped) along a food-web gradient, the differences between the network neighborhood of F can still be informative about the community.

In searching for laws and rules, we tend to simplify and overemphasize the mean behavior of the system under investigation. Explicit studies on variability can provide the scientific basis for functional diversity thus helping to better understand adaptability and evolvability (both stemming from variability). If a certain population gives very consistent responses to the same disturbance, its dynamical behavior can be considered more constrained and predictable (less flexible). The reasons may be internal (type of functional response) or external (the topology of the ecosystem). On the contrary, populations being able to give various responses are less constrained and can be expected to react to external changes in a more flexible way. Food-web simulations performed using an individual-based approach can give one kind of assessment for this dynamical variability, within the limits of the model.

From a conservation biology viewpoint, there is an increasing need to shed light on the quality of biodiversity (Eggers *et al.*, 2014), also in terms of the nature and function of variability (Feest *et al.*, 2010). It is increasingly recognized that natural populations are not homogeneous boxes and, especially in the case of large-bodied animals, population structure and individual-level variability (genetic, behavioral, and trophic) do matter. Effective population size and social network structure may result in situations where only a few individuals may be able to determine the success of the population. In this chapter, we present an individual-based, stochastic food-web simulation model and we evaluate two response variables: the mean and variability of population sizes. Assessing the variability of population dynamics can contribute to better understanding the dynamical background of adaptability in a changing environment. We illustrate possible ways to tackle some of the above challenges (aggregation, variability) and suggest future directions for extending the gradient-based, comparative approach to food webs.

20.2 The Kelian River Ecosystem

The Kelian River (Borneo) is one of the few ecosystems where food webs have been described at several locations. The river was sampled and studied intensely at six sites,

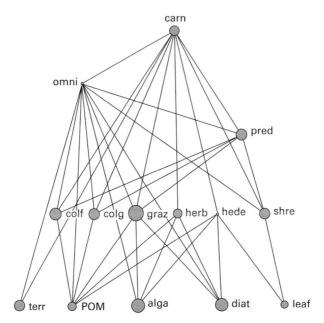

Figure 20.1 The food web at site 1. Node size is proportional to the value of the I_H(V) index. Drawn by CoSBiLab Graph (Valentini and Jordán, 2010).

The code for organisms – leaf: leaf litter; POM: settled and suspended, coarse and fine particulate organic matter; diat: diatoms; alga: algae; terr: terrestrial insects; fila: filamentous bacteria; colf: collector–filterers; colg: collector–gatherers; shre: shredders; hede: herbivore–detritivore fish; herb: herbivorous fish; graz: grazers; omni: omnivorous fish; pred: predators; carn: carnivorous fish.

starting from the pristine tropical dipterocarp rainforest and ending at a human settlement. Long-term sampling was conducted at regular intervals between 1990 and 1995. This makes such a river highly useful for investigating food-web gradients. Apart from the village, during the period of sampling, there was gold-mining activity (alluvial and open cut) along the river, so the system is even useful for studying human impacts.

Following long-term sampling (Yule, 1995; Yule *et al.*, 2010), different food webs were constructed for six sites along the river (Jordán *et al.*, 2012). These were highly aggregated networks, mainly composed of nodes that lump together functionally equivalent components (the number of nodes, *N*, range from 12 to 15). The food webs are composed of two groups of primary producers (diatoms, and green and blue-green algae), three non-living compartments (leaf litter; settled and suspended, coarse and fine particulate organic matter; human waste), terrestrial insects, filamentous bacteria, six herbivorous groups (collector–filterers, collector–gatherers, shredders, herbivore–detritivore fish, herbivorous fish, grazers), and three higher level consumers (omnivorous fish, predators, carnivorous fish; see abbreviated names in Figure 20.1).

Although the data used to construct the models refer to different years and the food-web components (i.e., species and non-living matter) were lumped in highly aggregated functional groups, the six food webs preserved spatial variability.

Such variability is due to longitudinal changes in the river resulting from the anthropogenic impacts. For this reason, we decided not to aggregate the data from the different locations in order to exploit the wealth of information stored in the dataset. Some groups appeared only in a single site, while others were present all along the river (e.g., omnivorous fish).

20.3   An Individual-Based Approach to Simulate Kelian River Dynamics

We built individual-based models and simulated their behavior using a stochastic simulator. We used a stochastic simulation model written in BlenX, a process algebra-based language (Dematté *et al.*, 2007, 2008). This makes it possible to simulate parameter-rich models of parallel ecological processes in a truly stochastic way, using the Gillespie algorithm (Gillespie, 1977; Priami, 2009; Jordán *et al.*, 2011; Livi *et al.*, 2011). Based on the assessed number of individuals in functional groups and the interaction strengths among functional groups approximated by gut content analysis, we modeled their dynamics similarly to a well-stirred chemically reacting system. Such an approach takes account of the system's discreteness and stochasticity. Prey–predator interactions happened when two individuals x and y met and were able to interact according to the defined sets of potential partners. The probability of interaction was a function of their "concentration" (number of individuals) and "affinity" (interaction strength). The dynamic behavior was kept simple in order to be able to focus on community responses. When prey A and predator B interact, there are two possibilities: (1) the predator individual is unchanged and the prey individual disappears with k_1 or (2) the predator individual is duplicated (i.e., this mimics reproduction) and the prey individual disappears with k_2. The k_1/k_2 ratio determines how much prey to consume before being able to reproduce. This approach considers only linear responses (i.e., no functional responses of Holling's type II or III; see Jordán *et al.*, 2012). Beyond the number of individuals and interaction strength, we also needed birth and death rates for each functional group. These parameters are very hard to get but field experience was used to make educated guesses. Plants have a positive reproduction rate and top predators have a positive death rate that are independent of prey–predator interactions and serve to maintain the system at equilibrium.

A genetic algorithm (based on particle swarm optimization; see Kennedy and Eberhart, 1995; Forlin, 2010) was used to search for an initial parameter set that keeps the behavior of the whole community quasi-stable, i.e., no trophic group increases exponentially or becomes extinct (reference simulations). Reference simulation runs were sampled at time t (corresponding to roughly 30 years). At time t, the mean and the standard deviation of all components were registered, based on s simulations. We note here that the mean of many stochastic simulations approaches the results obtained from deterministic simulations, but the variability of dynamical behavior is relatively more informative and important to consider. In fact, by capitalizing on the stochastic nature of the simulations, here we focus on variability.

The reference model was then subject to sensitivity analysis. For estimating the effect of species i on the population size variability of species j, we first calculate the reference value of the standard deviation of the population size for species j in absence of any disturbance, at time t in the simulation run (A_j). Then the initial number of individuals for each component is halved, one by one, and the standard deviation of the population size of all other components recorded after the same time t, for the same number of simulations for each other disturbed species. The standard deviation of the population size for species j, after disturbing species i is $A_j(i)$ and the relative response of species j to disturbing species i is

$$RR_j(i) = \frac{|A_j - A_j(i)|}{A_j}.$$ (20.1)

The relative response is normalized over all the trophic groups (n):

$$NRR_j(i) = \frac{RR_j(i)}{\sum_{i=1}^{n} RR_j(i)}.$$ (20.2)

The community importance of species i equals

$$I_H(V)_i = \sum_{j=1}^{n} NRR_j(i).$$ (20.3)

Instead of considering the standard deviation of population size values, their mean can also be calculated and studied. Moreover, several disturbance regimes can be studied (see Scotti et al., 2012, where the initial population size of the disturbed species was multiplied by $m = 0.25, 0.5, 2,$ and 4). These normalized relative response metrics (in mean and standard deviation) measure the sensitivity of the system to disturbing component i (I_H, where H stands for the Hurlbert response function; Hurlbert, 1997). These simulation-based values are dynamical measurements of community importance.

20.4   Discussion: Top–Down Domination Increases Along the River Gradient

In the case of rare species (i.e., species characterized by small population size), individual-level differences, demographic noise, and random drift may have a large influence on dynamics (Lande, 1988). Both individual-based models and stochastic approaches are more reasonable in this case than classical methods considering homogeneous populations (Judson, 1994; Grimm and Railsback, 2005). In this framework, variability can be explicitly studied and quantified. As variability (also in terms of population behavior) is considered as the source of adaptability in evolutionary ecology, we suggest that this is important information for conservation science. Community responses that are measured in terms of changes in mean population size can be more informative about certain genetic processes (e.g., inbreeding) but when the response is quantified by the variability of population size it can provide more information about some evolutionary ecological processes (e.g., adaptability). We suggest that a particular type of key species in the model is the one that causes the largest changes in others' dynamical variability during the simulations. This approach may help link structure to function in studying complex food webs (de Ruiter *et al.*, 2005; Dunne, 2006), by identifying key elements of ecosystem dynamics: if trophic groups of highest dynamical effects in the simulations are quantified, one can see the functional link between critical positions and key roles in ecological interaction networks.

A key interest in community ecology is to quantify weak and strong interactions (Berlow, 1999). Beyond measuring interaction strength, it is also of great importance to understand the consequences and effects of particular interspecific interactions. Earlier research has shown that strong trophic links (in the sense of carbon transfers) cause large dynamical effects in a predictable way. On the contrary, the consequences of changing weak links is unpredictable: these can result in either larger or smaller dynamical effects (Scotti *et al.*, 2012). Based on our dynamical food-web model, we determined the strength of simulated effects of a disturbed group i on other j groups. Then by summing each effect of i on others, we quantified its community effect. Figure 20.1 shows the relative simulated importance of trophic groups at site 1. If a trophic group represented by a larger circle is disturbed, then the population size values of the other groups will be less consistent (more variable) during the simulations (at time t). Disturbing producers generates larger community responses in this network model. However, in other sampling sites along the river this pattern is changed (Figure 20.2; Table 20.1). In the most downstream sites (i.e., where human impact is much higher) omnivores and carnivores tend to be more important. Leaf particles and algae are more important in the upstream, pristine sites.

This change in the community importance of different groups might be explained by the fact that leaf litter was lacking at the downstream sites due to deforestation and it was also smothered by sediment at the downstream sites, while algal growth was impeded by high turbidity and sedimentation. Also, tropical shredding invertebrates (such as crabs, stoneflies, and caddisflies), which are major processors of leaf litter, are very sensitive to pollution. The lack of shredders to process leaves meant that collector–filterers and collector–gatherers had nothing to filter (except sediment). So, bottom–up domination is mostly related to leaf litter processing and algal production. Some fish groups (e.g., the herbivore group) are tolerant to pollution and can eat the algal-sediment mat present at the polluted downstream sites, which the invertebrates do not consume in large quantities. Omnivores gain increasing importance along the river. At site 6, this node is the key group when importance is measured in terms of variability induced on other systems' compartments. This means that disturbing omnivorous fish will have a non-consistent, non-predictable response, according to the model. One conservation biology implication of this finding may be that protecting this group may have a relatively large stabilizing effect on the whole ecosystem.

Table 20.1 The importance rank of trophic groups according to the $I_H(V)$ index in each of the six sites (see Figure 20.1 for abbreviations).

site 1	$I_H(V)$	site 2	$I_H(V)$	site 3	$I_H(V)$	site 4	$I_H(V)$	site 5	$I_H(V)$	site 6	$I_H(V)$
leaf	0.0915	**terr**	0.0940	**carn**	0.0863	**hede**	0.1283	**carn**	0.1294	**omni**	0.1463
diat	0.0903	colg	0.0801	fila	0.0813	diat	0.1271	fila	0.0769	fila	0.1126
alga	0.0860	pred	0.0800	alga	0.0789	colg	0.0879	diat	0.0768	diat	0.1110
POM	0.0821	shre	0.0784	leaf	0.0746	POM	0.0813	colg	0.0763	carn	0.1010
hede	0.0813	graz	0.0777	colf	0.0726	colg	0.0797	terr	0.0744	herb	0.0807
herb	0.0755	herb	0.0772	pred	0.0724	terr	0.0759	omni	0.0666	pred	0.0761
terr	0.0748	POM	0.0763	POM	0.0702	colf	0.0759	alga	0.0664	POM	0.0690
colg	0.0723	leaf	0.0741	shre	0.0654	herb	0.0755	colf	0.0632	hede	0.0672
pred	0.0666	carn	0.0713	terr	0.0654	leaf	0.0754	herb	0.0616	colg	0.0650
shre	0.0663	alga	0.0653	omni	0.0642	hede	0.0666	leaf	0.0586	alga	0.0611
graz	0.0661	omni	0.0595	graz	0.0642	omni	0.0639	POM	0.0512	terr	0.0556
colf	0.0593	hede	0.0575	hede	0.0579	fila	0.0625	fila	0.0511		
omni	0.0524	colf	0.0570	herb	0.0521			pred	0.0509		
carn	0.0355	diat	0.0516	diat	0.0473			shre	0.0496		
				colg	0.0472			graz	0.0472		

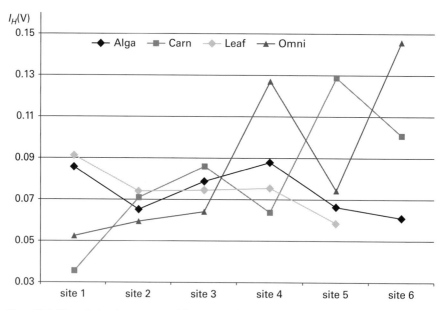

Figure 20.2 The relative importance of four particular trophic groups during the simulations, along the food-web gradient. In site 1, algae and leaf particles have larger effects on community dynamics than all other nodes. In site 6, omnivores and carnivores are the most influential group. This suggests a transition from bottom–up to top–down control. The *x* axis shows space (sampling sites along the river) and the *y* axis shows the $I_H(V)$ simulated importance values. (A black and white version of this figure will appear in some formats. For the color version, please refer to the plate section.)

20.5   Conclusions: Variability in a System Context

We suggest that this kind of system-level simulation approach provides holistic and quantitative indicators for identifying key components in the studied system. Systems-based conservation reveals information otherwise hidden for single-species or sub-community-level analyses.

Especially in the case of higher organisms with large body size and small population size, individual-level differences may dominate population-level behavior. Conservation biology is typically focusing on rare species, and for these, the disproportionate effect of a few individuals is not easy to study by classical population models. Moreover, our modeling framework makes it relatively simple to consider additional sources of information generally under-represented in food-web research (e.g., spatial and social processes, see Scotti *et al.*, 2013).

The limitations of this approach are (1) the need for a large number of parameters typically not easy to measure, so generally approximated; (2) a relatively long process of building a reference model (however, additional versions are very simple to produce because of the composability of the programming language); and (3) the hard work to balance the simulation model (however, sensitivity analysis and the evaluation of the results are faster). The most important future perspective is to implement genetic or demographic variability and make the model really reflect individual-level variability (also in a biological, not only in a technical, sense).

Acknowledgments

We thank Nerta Gjata and Shu Mei for their collaboration.

References

Berlow, E. L. (1999). Strong effects of weak interactions in ecological communities. *Nature*, **398**, 330–334.

Bolnick, D. I., Amarasekare, P., Araújo, M. S., *et al.* (2011). Why intraspecific trait variation matters in community ecology. *Trends in Ecology and Evolution*, **26**, 183–192.

Bondavalli, C., Bodini, A., Rossetti, G., and Allesina, S. (2006). Detecting stress at the whole-ecosystem level: the case of a mountain lake (Lake Santo, Italy). *Ecosystems*, **9**, 768–787.

Dematté, L., Priami, C., Romanel, A., *et al.* (2007). BetaWB: modelling and simulating biological processes. In *Proceedings of Summer Computer Simulation Conference* (SCSC 2007), ed. G. A. Wainer and H. Vakilzadian, pp. 777–784.

Dematté, L., Priami, C., Romanel, A., *et al.* (2008). The Beta Workbench: a computational tool to study the dynamics of biological systems. *Briefings in Bioinformatics*, **9**, 437–449.

de Ruiter, P. C., Wolters, V., and Moore, J. C. (2005). *Dynamic Food Webs: Multispecies Assemblages, Ecosystem Development and Environmental Change*. Amsterdam: Academic Press.

Dunne, J. A. (2006). The network structure of food webs. In *Ecological Networks: Linking Structure to Dynamics in Food Webs*, ed. M. Pascual and J. A. Dunne, Oxford: Oxford University Press, pp. 27–86.

Eggers, S. L., Lewandowska, A. M., Barcelos e Ramos, J., *et al.* (2014). Community composition has greater impact on the functioning of marine phytoplankton communities than ocean acidification. *Global Change Biology*, **20**, 713–723.

Feest, A., Aldred, T. D., and Jedamzik, K. (2010). Biodiversity quality: a paradigm for biodiversity. *Ecological Indicators*, **10**, 1077–1082.

Forlin, M. (2010). Knowledge discovery for stochastic models of biological systems. University of Trento, Ph.D. Thesis.

Gillespie, D. T. (1977). Exact stochastic simulation of coupled chemical reactions. *Journal of Physical Chemistry*, **81**, 2340–2361.

Grimm, V. and Railsback, S. F. (2005). *Individual-Based Modeling and Ecology*. Princeton, NJ: Princeton University Press.

Hurlbert, S. H. (1997). Functional importance vs keystoneness: reformulating some questions in theoretical biocenology. *Australian Journal of Ecology*, **22**, 369–382.

Jordán, F., Scotti, M., and Priami, C. (2011). Process algebra-based models in systems ecology. *Ecological Complexity*, **8**, 357–363.

Jordán, F., Gjata, N., Mei, S., and Yule, C. M. (2012). Simulating food web dynamics along a gradient: quantifying human influence. *PLoS ONE*, **7**(7), e40280.

Judson, O. P. (1994). The rise of the individual-based model in ecology. *Trends in Ecology & Evolution*, **9**, 9–14.

Kennedy, J. and Eberhart, R. (1995). Particle swarm optimization. *Proceedings of IEEE International Conference on Neural Networks IV*, pp. 1942–1948.

Lafferty, K. D. and Dunne, J. A. (2010). Stochastic ecological network occupancy (SENO) models: a new tool for modelling ecological networks across spatial scales. *Theoretical Ecology*, **3**, 123–135.

Lande, R. (1988). Genetics and demography in biological conservation. *Science*, **241**, 1455–1460.

Livi, C. M., Jordán, F., Lecca, P., and Okey, T. A. (2011). Identifying key species in ecosystems with stochastic sensitivity analysis. *Ecological Modelling*, **222**, 2542–2551.

Montoya, J. M. and Solé, R. V. (2002). Small world patterns in food webs. *Journal of Theoretical Biology*, **214**, 405–412.

Priami, C. (2009). Algorithmic systems biology. *Communications of ACM*, **52**, 80–89.

Scotti, M., Gjata, N., Livi, C. M., and Jordán, F. (2012). Dynamical effects of weak trophic interactions in a stochastic food web simulation. *Community Ecology*, **13**, 230–237.

Scotti, M., Ciocchetta, F., and Jordán, F. (2013). Social and landscape effects on food webs: a multi-level network simulation model. *Journal of Complex Networks*, **1**, 160–182.

Siokou-Frangou, I., Bianchi, M., Christaki, U., *et al.* (2002). Carbon flow in the planktonic food web along a gradient of oligotrophy in the Aegean Sea (Mediterranean Sea). *Journal of Marine Systems*, **33–34**, 335–353.

Tallberg, P., Horppila, J., Väisänen, A., and Nurminen, L. (1999). Seasonal succession of phytoplankton and zooplankton along a trophic gradient in a eutrophic lake: implications for food web management. *Hydrobiologia*, **412**, 81–94.

Ulanowicz, R. E. (1996). Trophic flow networks as indicators of ecosystem stress. In *Food Webs: Integration of Patterns and Dynamics*, ed. G. A. Polis and K. O. Winemiller, London: Chapman and Hall, pp. 358–368.

Valentini, R. and Jordán, F. (2010). CoSBiLab Graph: the network analysis module of CoSBiLab. *Environmental Modeling Software*, **25**, 886–888.

Warren, P. H. (1989). Spatial and temporal variation in the structure of a freshwater food web. *Oikos*, **55**, 299–311.

Winemiller, K. O. (1996). Factors driving temporal and spatial variation in aquatic floodplain food webs. In *Food Webs: Integration of Patterns and Dynamics*, ed. G. A. Polis and K. O. Winemiller, London: Chapman and Hall, pp. 298–312.

Yule, C. M. (1995). The impact of sediment pollution on the benthic invertebrate fauna of the Kelian River, East Kalimantan, Indonesia. In *Tropical Limnology, Vol. III*, ed. K. H. Timotius and F. Göltenboth, Salatiga, Indonesia: SatyaWacana University Press, pp. 61–75.

Yule, C. M., Boyero, L., and Marchant, R. (2010). Effects of sediment pollution on food webs in a tropical river (Borneo, Indonesia). *Marine and Freshwater Research*, **61**, 204–213.

21 An Individual-Based Simulation Model to Link Population, Community, and Metacommunity Dynamics

Marco Scotti and Ferenc Jordán

21.1 Introduction

The dynamical behavior of individuals in ecosystems involves a multifaceted set of interaction types and processes that take place at different hierarchical levels. We present an individual-based, stochastic model that considers species dynamics at three hierarchical levels: population, community, and metacommunity. We use an individual-based model to show how the consequences of mechanisms that are specific to each hierarchical level may interact with processes that belong to other hierarchical levels. The strength of these effects is quantified in terms of impacts on metapopulation sizes and spatial distribution of populations. Results indicate the following: (1) the cohesion of the social network structure among conspecific individuals heavily affects their feeding efficiency at food-web level; (2) more generalist feeding habits trigger homogeneous spatial distribution of species at the landscape scale; and (3) high frequency of migration movements limits the local success of a generalist species thus leading to small metapopulation sizes. We illustrate how such a hierarchical framework may contribute to understanding the emergence of macroscopic patterns (i.e., metapopulation size and spatial heterogeneity) starting from elementary, bottom–up rules defined at the individual level.

21.2 Hierarchical Organization and Individual-Based Modeling in Ecology

Concurrent processes and interactions occur at different hierarchical levels in ecosystems (i.e., individual, population, community, and metapopulation/metacommunity) and do often spread their effects beyond the levels in which they actually originate. Some studies describe how ecological dynamics involving two hierarchical layers may interplay with each other. Social interactions among conspecific individuals may be regulated by metapopulation and community dynamics, community composition may be molded by landscape fragmentation, and species coexistence in metacommunities may result from the trade-off between spatial dispersal and multiple interaction types in food webs. Association rates in a population of wild Asian elephants depend on environmental conditions and seasonality (de Silva *et al.*, 2011). The rates at which social ties are formed peak in dry periods and resident elephants tend to maintain over time a stable pool of interactions with the same individuals. The cohesion of social groups of baboons may vary in response to predation pressure or spatial food distribution (Barton *et al.*, 1996). When predation pressure is high, the distances between conspecific individuals are smaller and social groups are more cohesive; this raises the chances of contest competition. Also spatial food distribution affects the quality of social interactions as too densely populated landscape patches, with unevenly distributed food sources, may trigger mechanisms of contest in addition to scramble competition. In the fragmented landscape of coastal southern California, coyotes are particularly vulnerable to extinction from local patches. Local extinctions of the coyote release the pressure on meso-predators and lead to a cascade effect that ends with the disappearance of avian prey from landscape fragments (Crooks and Soulé, 1999). In the butterfly-associated meta-community of Glanville fritillary in Åland Islands (southwestern Finland), the trade-off between spatial dispersal and competition (as well as the apparent competition and the tritrophic interactions between host plants, butterflies, and parasitoids) links the community and the landscape levels (van Nouhuys and Hanski, 2005).

The presence of vertical links that connect the dynamics of horizontal levels (of the ecological hierarchy) calls for a better understanding of their mechanisms of action. Population size and density together with individual-level variability play a crucial role. For example, the decline of the dominant apical predator and its extinction from local patches (i.e., its rarity) in fragmented landscapes are responsible for the release of meso-predators (Crooks and Soulé, 1999). Also, an excessive rise in the frequency of social interactions may be associated with a lower ability to adapt to environmental conditions and this risk is particularly sharp for large populations (Pacala *et al.*, 1996). This is because conspecific individuals that invest a disproportionate amount of time in the exchange of social information may be less effective in predator avoidance or foraging, and such an issue is attenuated in smaller groups for which the chance of an encounter with other individuals is lower. For the same probabilistic reasons, the consequences of stochastic, individual-level variability are emphasized in small population size (Bolnick *et al.*, 2011; Jordán *et al.*, 2011). Given the massive variability that characterizes the frequency of different types of interactions (both within and between populations; see Bolnick *et al.*, 2003) and the relevance of low population densities, individual-based modeling seems the most suitable option for studying the dynamics of hierarchical ecological systems.

Individual-based models (IBMs) are used to simulate the dynamics of complex systems starting from the elementary properties of individual entities (Grimm *et al.*, 2006). IBMs are often defined in terms of computer algorithms; they are complementary to deterministic methods based on ordinary differential equations and allow implementing stochastic dynamics. In IBMs each entity locally interacts with the environment and other individuals; its internal behavior and the relationships with other entities may change in response to external stimuli, interactions, or on the basis of growth mechanisms (e.g., during simulation individuals may be assigned to different age classes). The intriguing perspective underlying IBMs is that global properties at system level may emerge from the dynamics of individual entities. These individual entities behave

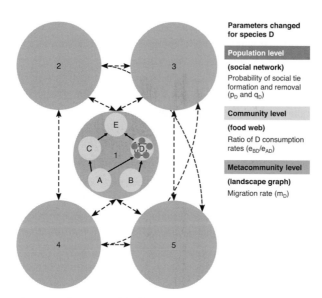

Figure 21.1 Structure of the hierarchical model. Five metapopulations occupy and migrate through five patches in a complete landscape graph. In each patch, individuals of the five species interact in a food web and conspecific individuals form and remove social interactions. The food web and social network (of species D only) are illustrated here for the central patch but they apply to the whole landscape graph. (A black and white version of this figure will appear in some formats. For the color version, please refer to the plate section.)

according to simple elementary rules assigned during model construction (e.g., fish schooling depends on local-scale perception of individual fish; Kunz and Hemelrijk, 2003). The main applications of IBMs in ecology refer to population dynamics (DeAngelis *et al.*, 1980, 1997). Although the use of IBMs in ecology dates back to the 1970s (e.g., Botkin *et al.*, 1972), their spreading has been partially impaired by: (1) structural complexity and lack of clarity (although the adoption of a standard protocol may address this issue; see Grimm *et al.*, 2006, 2010); (2) uncertainties in model design (i.e., the need to identify which key traits are the most relevant for individual-level dynamics, a task which is far from being trivial); (3) data availability (IBMs often rely on highly parameterized descriptions of individual dynamics); (4) lack of generality (IBMs are well suited for specific ecosystems, on the other hand extending them to other contexts may not always be straightforward or their application to answer general questions of theoretical ecology may be challenging); and (5) difficulties in validating the simulation's outcomes (model outcomes are complex and independent empirical data for their testing are often missing). Scotti *et al.* (2017) reviewed the use of IBMs to simulate ecological network dynamics; they discussed limits, challenges, and possible future developments of the application of IBMs in community ecology.

In this chapter we illustrate an IBM that simulates the dynamics of a hierarchical (ecological) system composed of three levels: population, community, and metapopulation/metacommunity (Figure 21.1). The model is constructed with parameters that refer to different species and ecosystems (e.g., social interactions between dolphins, feeding rates in aquatic food webs, dispersal rates of zooplankton). It outlines a conceptual framework for vertically linking three otherwise well-studied horizontal levels, without being concerned about the correspondence with the reality. The model represents a metacommunity of five species distributed in a complete graph (each patch can be directly reached from each other) of five landscape patches (i.e., each patch includes a food web of five species). Individuals may take part in processes at three hierarchical levels. First, demographical processes (birth and death) as well as social-network relationships (formation and removal of social links) describe the dynamics at the level of populations. Second, trophic interactions between populations representing different species define ecosystem dynamics at the level of communities in particular habitat patches. Third, dispersal of individuals between habitat patches represents the spatial movements at the level of the landscape (metacommunity dynamics). By vertically linking these three levels in an IBM, our main objectives are: (1) to quantify at which extent the dynamics that originate in each hierarchical level spread and affect the other levels (e.g., the model can be applied to investigate how traits of individuals from different taxa influence dynamics across the hierarchy) and (2) to illustrate possible applications of such a multilayer modeling framework in the context of ecosystem management and environmental planning.

21.3   The Hierarchical Model

We provide a standard description of the IBM following the ODD (overview, design concepts, and details) protocol (Grimm *et al.*, 2006, 2010). The "overview" consists of three elements: (1) purpose; (2) entities, state variables, and scales; and (3) process overview and scheduling. In Section 21.3.2 (Design Concepts) we illustrate general concepts that represent the basis of model design (i.e., emergence, adaptation, interaction, stochasticity, collectives, and observation). Technical details concerning the initial values of the state variables and the mathematical backbone of the model (i.e., model equations and rules) are summarized in the third building block (i.e., Section 21.3.3: Details); in the hierarchical model there is no environmental input (i.e., no external conditions influence state variables). Section 21.3.4 (Simulation Experiments) illustrates the scheme of the sensitivity analysis performed. Sensitivity analysis serves to quantify changes in metapopulation size and spatial heterogeneity that are triggered by changes in parameters that regulate individual-level behavior at three levels of the ecological hierarchy.

21.3.1   Overview

21.3.1.1   Purpose of the Model
The purpose of the model is to understand how processes and interactions that occur at population, community, and metacommunity/metapopulation levels may interact and have an impact on metapopulation size and spatial heterogeneity of five species in a hierarchical (ecological) system.

21.3.1.2   Entities, State Variables, and Scales

Three hierarchical levels compose the model (Figure 21.1): individual (population), food web (community), and landscape graph (metapopulation/metacommunity). An identity code characterizes which species each individual belongs to; an additional identifier informs about the landscape patch where individuals are located (i.e., the model is not spatially explicit), while the propensity to form or remove social interactions depends on a constant rate (Table 21.1). The demography of individuals is regulated by constant birth and death rates, while constant feeding and migration rates control food-web dynamics and spatial movements, respectively; all these constant rates are species specific, spatially homogeneous, and defined in terms of probabilities (Table 21.1). Five food webs (one for each landscape patch) include individuals of five species (two primary producers = A, B; two herbivores = C, D; and an apical predator = E).

The food-web topology is definite (i.e., the quality of feeding interactions is set; e.g., C and D are specialist and generalist species, respectively) and has been characterized in previous studies (Jordán and Molnár, 1999); food-web connectance (i.e., the ratio between existing trophic links and all possible trophic interactions) equals 0.2 (5 trophic links/5^2 possible trophic interactions, including self-loops = 0.2). A complete network of five patches constitutes the landscape graph (i.e., each patch is connected to all other patches).

Local populations consist of all individuals of a species in a landscape patch. Local communities include all individuals of the five species in a landscape patch. A metapopulation corresponds to all the individuals that belong to a species and are distributed in the five patches of the landscape graph. All individuals of the five species in the five patches form the metacommunity.

21.3.1.3   Process Overview and Scheduling

Simulations were carried out on a time scale of two months, which corresponds to 120 discrete time steps (i.e., two steps per day). We used a time scale of two months because this allowed all of the species to reach a new equilibrium and stabilize around the average value when sensitivity analysis experiments were performed (i.e., when the value of a state variable was changed). In two months the system always leaves the transient phase (that never exceeded five weeks). Fully stochastic simulations were performed, without any order or priority for different interactions or processes. During simulations each individual is involved in interactions and processes that occur at three hierarchical levels. At population level, individuals form and remove social ties to create a social network with conspecific individuals; demography is regulated by reproduction and mortality (this latter may be caused by "spontaneous" events or feeding activities of the consumers). At community level, feeding interactions contribute to shape the food-web composition (i.e., together with migration and demography, feeding interactions determine the numbers of individuals per species, in each patch). Spatial dynamics is regulated by migration rates and modeled adopting a mass effect perspective (i.e., the focus is only on immigration/emigration of local populations and there is no topology

Table 21.1 Low-level state variables. Empirical values gathered from the literature to construct the "base model" and evolved parameter set used in the "reference model." Initial population size (*init*) consists of raw numbers of individuals per patch. Interaction rates and demography are expressed as probabilities. Population dynamics is modeled in terms of birth (*b*) and death (*d*) rates, and also refers to (equivalent) rates of social tie formation (*p*) and removal (*q*). Community dynamics is simulated starting from the probabilities that column-consumer j feeds (e_{ij}) or feeds and reproduces (er_{ij}) on row-resource i. Migration rates (m) drive metapopulation dynamics.

Parameter	"Base model" Value and reference	"Reference model" Species A	Species B	Species C	Species D	Species E
Demography						
init	14–127 Faust and Skvoretz, 2002; Croft *et al.*, 2005					
Patch I		250	202	150	170	60
Patch II		249	203	145	180	56
Patch III		253	195	153	190	53
Patch IV		245	196	148	170	54
Patch V		252	198	154	170	56
b		0.2722	0.1199	0.449	0.2369	0.0810
d		0.0003	0.0001			
Social interactions						
p (and *q*)	0.77 Lusseau *et al.*, 2006	0.1	0.1	0.1	0.1	0.1
Trophic relationships						
e_{ij} (and er_{ij})	0.01–0.52 Kokkoris *et al.*, 1999; Bascompte *et al.*, 2005					
Species A				$e_{AC} = 0.0003$ $er_{AC} = 0.0008$	$e_{AD} = 5.602E{-}5$ $er_{AD} = 0.0003$	
Species B					$e_{BD} = 2.899E{-}6$ $er_{BD} = 0.0006$	
Species C						$e_{CE} = 0.0003$ $er_{CE} = 6.992E{-}5$
Species D						$e_{DE} = 1.734E{-}5$ $er_{DE} = 0.0004$
Species E						
Landscape dynamics						
m	0.01–0.5 Gaines *et al.*, 1979; Michels *et al.*, 2001	0.0001	0.001	0.03	0.01	0.08

effect as the landscape consists of a complete graph; Leibold *et al.*, 2004): individuals of all species freely move in the five landscape patches.

21.3.2   Design Concepts

21.3.2.1   Emergence

Metapopulation (and metacommunity) size and spatial heterogeneity of populations emerge from the elementary rules that define the behavior of individuals at three hierarchical levels. The elementary rules are based on probabilities (i.e., rates) that implicitly take into account adaptation and fitness. For example, the frequency of social tie formation/removal has an effect on the structure of the social network, in each landscape patch. Since the density of the social network influences feeding interactions (i.e., denser social networks raise the chance of predator avoidance, but also increase the foraging efficiency of consumers), it also indirectly supports/impairs population size in each patch (through its effect on food-web dynamics). Then, the elementary rule on social network dynamics may have an impact on the emergence of a macroscopic property as population size in a landscape patch.

21.3.2.2   Sensing

Individuals are assumed to distinguish between conspecific individuals (to establish social interactions) and individuals belonging to other species (that represent potential prey or predators). Individuals may sense the density of the social network (at population level, in each patch) as this changes the propensity to form/remove social ties. Also the population density in each patch may be perceived by individuals and influence migration probabilities.

21.3.2.3   Interaction

Two types of interactions are modeled implicitly: (1) the probability of social tie formation/removal changes according to the density of the social network in a given patch (i.e., low densities promote social tie formation while removal is enhanced by high densities); and (2) chances of trophic interactions are characterized by a positive and linear relationship with population densities of prey and predators in the same patch (i.e., they increase with higher densities of prey and predators), but also regulated by social network densities of predators (positive and linear dependence) and prey (negative and linear relationship).

21.3.2.4   Stochasticity

All interactions and processes (i.e., demography and migration) are interpreted as probabilities.

21.3.2.5   Collectives

Individuals are classified according to their species and may occupy five landscape patches. In each patch, conspecific individuals are grouped in social networks.

21.3.2.6   Observation

For model testing we checked which parameter set allowed all populations in the five patches to persist and be stable around the average (without growing exponentially or becoming extinct during simulations). We applied sensitivity analysis to determine system-level responses to changes in the dynamics at three hierarchical levels (i.e., model analysis). Metapopulation size and spatial heterogeneity are the system-level responses measured throughout simulations (i.e., they are recorded at each step during simulation); they represent whole-system properties that emerge from simple individual-level rules (e.g., propensity to form/remove social interactions, feeding preferences, and migration probabilities). Metapopulation size is computed as the sum of all individuals (of each species) in the five landscape patches. Standard deviation of population size in the five patches accounts for the spatial heterogeneity (i.e., of species distribution).

21.3.3   Details

21.3.3.1   Initialization

At the beginning of the simulations the population size of all five species was homogeneously distributed in the five landscape patches (Table 21.1). The evaluation of each simulation run started in the second half of the time series (i.e., during the second month) in order to avoid risks related to transient dynamics, especially when performing sensitivity analysis (i.e., the system was expected to reach a new equilibrium after changing parameters that describe individual behavior in a specific hierarchical level). State variables (e.g., the probability of social tie formation/removal and the migration rates) were deduced from literature and served to construct the "base model" (Table 21.1). The parameter set of the "base model" was evolved through particle swarm optimization (PSO; Kennedy and Eberhart, 1995), in order to define a "reference model" where stable populations persist in the five patches during simulations. The preservation of stable populations in the five patches was set as the fitness goal and the PSO algorithm has iteratively selected regions of the parameter space until convergence criteria were met. The application of the PSO algorithm was essential, as it allowed for a sensitivity analysis to study dynamics near equilibrium.

21.3.3.2   Input

Environmental conditions were not explicitly modeled and should be considered as homogeneous through time and space, for all simulations.

21.3.3.3   Submodels

Global properties of the system result from simple rules that describe the behavior of individuals at three hierarchical levels: population (demography and social interactions), community (trophic relationships), and metapopulation/metacommunity (migrations in the landscape graph). A complete description of submodels together with the code for simulating the "reference model" is provided by Scotti et al. (2013). Values of the parameters used in equations and reference literature sources are shown in Table 21.1. Equations that define all dynamical processes were formulated in accordance with the simulation framework adopted (i.e., the BlenX programming language; Dematté et al., 2008) and are based on simple kinetic laws. The modeling approach used to describe the dynamics related to the encounter of two individuals (i.e., social or trophic interactions) is individual based (although the probability of trophic interactions is scaled according to the social-network density of predators and prey, a global property that reflects the average number of social interactions per individual), while a hybrid approach was implemented for processes that involve isolated individuals (e.g., migration). The choice of a hybrid approach depends on how BlenX prompts for rules.

Demography. The reproduction rate of a generic primary producer i at time $t + 1$ ($rrPP_{i,t+1}$) depends on constant birth rate (b_i) and population size at time t ($popPP_{i,t}$) (Eq. 21.1).

$$rrPP_{i,t+1} = popPP_{i,t} \cdot b_i \tag{21.1}$$

Consumers may reproduce only after a feeding event; the probability that consumer i feeds on resource j and reproduces at time $t + 1$ ($rrC_{i,t+1}$) is obtained by multiplying a constant "eat and reproduce" rate (er_{ji}) with the population size of consumer i ($popC_{i,t}$) and resource j ($popR_{j,t}$) at time t, further scaling this value with the ratio between social network densities of consumer ($D_{i,t}$) and resource ($D_{j,t}$) at time t (Eq. 21.2). The positive and linear relationship that links the probability of a feeding event with the density of the consumer's social network mimics more efficient foraging for more cohesive populations (e.g., cooperative hunting; Stander, 1994; Pacala *et al.*, 1996), while the negative and linear dependence between the probability of a feeding event and the density of the resource's social network may be explained with vigilance strategies (e.g., alarm calls and predator inspection; Sullivan, 1984).

$$rrC_{i,t+1} = popC_{i,t} \cdot popR_{j,t} \cdot er_{ji} \cdot \frac{D_{i,t}}{D_{j,t}} \tag{21.2}$$

Mortality depends on either predation or internal mechanisms (e.g., aging; we call this "natural mortality"). At time $t + 1$, the rate of natural mortality for primary producers ($mortPP_{i,t+1}$) and consumers ($mortC_{i,t+1}$) is computed by multiplying the constant death rate (d_i) with either squared population size of primary producers ($popPP_{i,t}$; for using non-linear functions for phytoplankton mortality see Murray and Parslow, 1999) (Eq. 21.3) or population size of consumers ($popC_{i,t}$) (Eq. 21.4) at time t. Feeding interactions always lead to the death of resource j (i.e., we did not consider the activity of small herbivores that may damage large and modular plants by reducing their mass, without leading to death). At time $t + 1$, the rate of mortality of individuals of species j ($mort_{j,t+1}$) is regulated by the probability that consumer i eats (e_{ji}) (Eq. 21.5) or "eats and reproduces" (er_{ji}) (Eq. 21.6); such probability is further multiplied with population size of consumer ($popC_{i,t}$) and resource ($popR_{j,t}$) at time t, and scaled with the ratio between social network densities of consumer ($D_{i,t}$) and resource ($D_{j,t}$) at time t. When the density of the predator's social network increases the chances of successful predation are raised (e.g., cooperative hunting), while more cohesive social networks among prey increase the chances of predator avoidance (e.g., vigilance strategies).

$$mortPP_{i,t+1} = popPP_{i,t}^2 \cdot d_i \qquad (21.3)$$

$$mortC_{i,t+1} = popC_{i,t} \cdot d_i \qquad (21.4)$$

$$mort_{j,t+1} = popC_{i,t} \cdot popR_{j,t} \cdot e_{ji} \cdot \frac{D_{i,t}}{D_{j,t}} \qquad (21.5)$$

$$mort_{j,t+1} = popC_{i,t} \cdot popR_{j,t} \cdot er_{ji} \cdot \frac{D_{i,t}}{D_{j,t}} \qquad (21.6)$$

Social interactions. Conspecific individuals that occupy the same landscape patch may form or remove social ties, and this even applies to primary producers (e.g., consider the production of toxic and inhibitory compounds that, in case of phytoplankton, reduces the grazing pressure of zooplankton and compensates for the competitive disadvantages with respect to other non-toxic phytoplankton species; see Roy and Chattopadhyay, 2007). If an individual moves between patches it loses all of its social ties. At time $t + 1$, an individual belonging to species i has the probability to form ($linkF_{i,t+1}$) (Eq. 21.7) or remove ($linkR_{i,t+1}$) (Eq. 21.8) a social link; this probability depends on constant rates for social tie formation (p_i) and removal (q_i), and on the number of potential (i.e., missing; $non\_links_{i,t}$) or existing ($links_{i,t}$) links at time t, respectively.

$$linkF_{i,t+1} = p_i \cdot non\_links_{i,t} \qquad (21.7)$$

$$linkR_{i,t+1} = q_i \cdot links_{i,t} \qquad (21.8)$$

In each patch, social network density is constantly recorded during simulations (all social links are equivalent and do not change their role or value with time). At time t, the social network density of species i ($D_{i,t}$) is computed by doubling the existing links (i.e., in undirected graphs each social interaction involves two individuals; $2\ links_{i,t}$) and dividing this value with the number of all possible connections (self-loops are excluded; $pop_{i,t}\ (pop_{i,t} - 1)$) (Eq. 21.9). Social network density is computed when, at least, two individuals of the same species are present in the same patch.

$$D_{i,t} = \frac{2 \cdot links_{i,t}}{pop_{i,t} \cdot (pop_{i,t} - 1)} \qquad (21.9)$$

Since the constant rates of social tie formation (p_i) and removal (q_i) are equal, the social network density will change during simulations in a stochastic way around the initial value. Bigger rates can produce larger variability and more frequent extreme values.

Trophic relationships. Food-web topology defines possible trophic relationships between individuals (i.e., it characterizes which resources may be sensed by different consumers). All of the trophic relationships lead to prey death. They are regulated by consumer ($popC_{i,t}$) and resource ($popR_{j,t}$) densities, and by the ratio between social network densities of consumer and resource ($D_{i,t}/D_{j,t}$)

(Eq. 21.10), at time t. At time $t + 1$, trophic relationships may lead to consumer reproduction ($eat\_repr_{i,t+1}$) (Eq. 21.10) or simply satisfy its energy demand ($eat_{i,t+1}$) (Eq. 21.11); these probabilities depend on "eat and reproduce" (er_{ji}) and "eat" (e_{ji}) rates, respectively. These two equations also regulate demography; they have an effect on reproduction rates of consumer i (Eq. 21.2; Eq. 21.10) and control the mortality of resource j (Eqs. 21.5 and 21.6, Eqs. 21.10 and 21.11).

$$eat\_repr_{i,t+1} = popC_{i,t} \cdot popR_{j,t} \cdot er_{ji} \cdot \frac{D_{i,t}}{D_{j,t}} \tag{21.10}$$

$$eat_{i,t+1} = popC_{i,t} \cdot popR_{j,t} \cdot e_{ji} \cdot \frac{D_{i,t}}{D_{j,t}} \tag{21.11}$$

Migrations in the landscape graph. The landscape graph consists of a complete network. For a given species i, spatial homogeneous migration rates (m_i) regulate the effective probabilities of migration at time $t + 1$ ($mr_{i,t+1}$); effective migration probability is also proportional to the population density of species i ($pop_{i,t}$) in the patch from which the individual migrates, at time t (Eq. 21.12).

$$mr_{i,t+1} = pop_{i,t} \cdot m_i \tag{21.12}$$

21.3.4   Simulation Experiments

Simulations of the IBM were performed with BlenX, a process algebra-based programming language (Dematté *et al.*, 2008). The BlenX language is stochastic and is executed with the Gillespie SSA algorithm (Gillespie, 1977). In BlenX an individual is represented by a box with labels that inform about its species and spatial location (i.e., presence in one of the five landscape patches). The behavior of individuals may be driven by internal programs that determine demography (i.e., birth of primary producers and "natural mortality") and effective migration rates (these also depend on sensing population density of conspecific individuals in a patch). Moreover, each individual perceives the species of other individuals through external interfaces (i.e., binders). Interfaces allow establishing (or removing) social interactions with conspecific individuals or feeding on prey that are included in the dietary spectrum of the individual.

After having assembled the "reference model" we carried out sensitivity analysis by focusing on D as a target species. We quantified the consequences of changing the values of parameters that describe species D dynamics at three hierarchical levels: (1) at population level we modified the probability of social tie formation/removal (p_D and q_D); (2) at community level we investigated different feeding habits (i.e., generalist vs. specialist trophic activity of D measured as the ratio of consumption rates: e_{BD}/e_{AD}); and (3) at landscape level we modulated migration rates (m_D). For each parameter we scanned 15 values (surrounding of the parameter value of the "reference model") and performed 100 simulations per value (altogether, for each parameter we carried out 1500 simulations = 1 parameter × 15 values per parameter × 100 simulations per value). We investigated the response variables for all species (even if they were always triggered by changes specific to species D). Response variables consist of metapopulation size and spatial heterogeneity, in the second half of simulations. For each value checked, average numbers of individuals per species are computed, in each landscape patch, over the second half of 100 simulations. Metapopulation size is the sum of average numbers of individuals in the five patches. Spatial heterogeneity is the coefficient of variation of average numbers of individuals in the five patches.

21.3.5 Macroscopic Patterns That Emerge from Vertical Interactions

Changes in social tie formation/removal and migration rates have commensurable effects on metapopulation size of all five species (see Figure 21.2 for species D). Less pervasive consequences are observed when modulating the feeding activity of species D (i.e., specialist vs. generalist feeding behaviors are checked). Both the probability of social tie formation/removal and the feeding rates influence the feeding interactions between species (see Eqs. 21.10 and 21.11). Thus, when the relevance of these two parameters was investigated with sensitivity analysis we expected to observe comparable consequences at the level of both response variables (metapopulation size and spatial heterogeneity). However, the model shows that changing the rates of social tie formation/removal outscores the effects of varying feeding preferences when the response is measured in terms of metapopulation size (Figure 21.2). Such a difference vanishes when spatial heterogeneity is used as response variable (Figure 21.3).

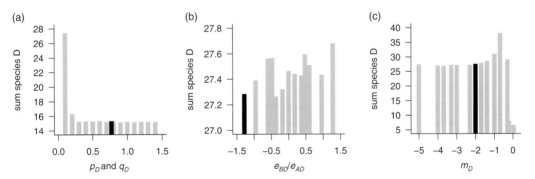

Figure 21.2 Metapopulation size of species D after sensitivity analysis. Metapopulation size is the sum of the average number of conspecific individuals in the five patches. Sensitivity analysis has been applied to check three parameters that describe species D's behavior: (a) probabilities of social tie formation/removal (p_D and q_D); (b) ratio of consumption rates (e_{BD}/e_{AD}; species D may feed on both species A and B: $e_{BD}/e_{AD} = -1.284$ when species D specializes on species A, $e_{BD}/e_{AD} = 1.279$ when species D specializes on species B, and intermediate values explore more generalist feeding strategies); and (c) migration rates (m_D). Histograms in black refer to the "reference model," and the results are on a log10-scale x axis in (b) and (c).

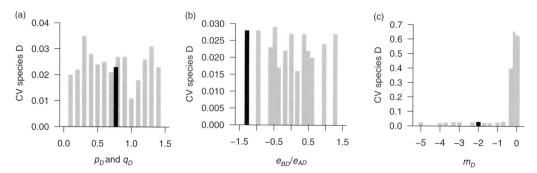

Figure 21.3 Spatial heterogeneity of species D after sensitivity analysis. Spatial heterogeneity is the coefficient of variation of the average population size in the five patches. Parameters checked with sensitivity analysis: (a) probability of social tie formation/removal (p_D and q_D); (b) ratio of consumption rates (e_{BD}/e_{AD}; $e_{BD}/e_{AD} = -1.284$ means that species D specializes on species A, $e_{BD}/e_{AD} = 1.279$ stands for species D being specialized on species B, and intermediate values refer to a gradient of generalist feeding strategies); and (c) migration rates (m_D). Black histograms identify the "reference model," and the results are on a log10-scale x axis in (b) and (c).

At population level, we found that an increase in the frequency of social tie formation and removal leads to less cohesive social networks. This means that the food-web persistence of populations that invest too much energy (and time) in social relationships can be impaired due to: (1) the reduction of foraging efficiency (e.g., less cohesive social networks can indicate less effective cooperative hunting) and (2) the exposure to higher risks of predation (e.g., less cohesive social networks can correspond to inefficient predator avoidance). Our model showed that less cohesive social networks impair predator feeding efficiency more than predator avoidance, resulting in a smaller metapopulation size of species D. Pacala *et al.* (1996) observed how increasing the frequency of social interactions may be particularly detrimental in the case of larger groups. This is because individuals investing a disproportionate amount of time in social interactions and information exchange reduce their adaptability to environmental changes. Our results confirm this trend with metapopulation size that stabilizes around lower average values when the frequency of social tie formation/removal is increased. Although community dynamics trigger less marked changes in metapopulation size (in comparison to population and metapopulation level mechanisms), a clear pattern can be detected. Trophic specialization of species D maximizes its metapopulation size when it contributes to escape competition with species C (i.e., when species D is specialized on consuming species B) being harmful otherwise (i.e., when species D feeds only on species A and the competition with species C reaches the maximum). Our results are consistent with the fact that coexistence is fostered by feeding specialization when this avoids competition. Feeding specialization on slightly different food sources has been shown to be an effective strategy to pursue coexistence in a deep-sea benthic community (Iken *et al.*, 2001); and consists of an important mechanism of resource partitioning between two species of bats with similar morphology and feeding habits (Marinho-Filho, 1991). The largest metapopulation size was found at low/intermediate migration rates and a sharp decline was observed when raising the migration frequencies through simulations. In the presence of highest migration rates, the generalist species D has a more inhomogeneous distribution in the landscape and is less abundant in some habitat patches. Such condition results in the competitive advantage for species C, leading to the exclusion of species D from some habitat patches. Moreover, when migration rates are high the individuals belonging to species D do not have enough time to form a cohesive social network in all habitat patches. This translates into lower efficiency for what concerns predator avoidance strategies, thus representing a limit to the establishment of a large metapopulation. Metapopulation persistence is expected to benefit from intermediate migration rates, and this has been found to be particularly true in the case of fragmented habitats (Hanski and Zhang, 1993). In our hierarchical model the landscape graph is complete (i.e., no topological fragmentation). By applying sensitivity analysis we checked a set of migration rates, thus mimicking a gradient of landscape fragmentation (i.e., changes in the probabilities of migration implicitly reflect a modulation of habitat heterogeneity). Absence of a clear trend characterizes the outcomes of sensitivity analysis for social tie formation/removal with respect to spatial distribution of species (see Figure 21.3 for species D). At community level, feeding specialization goes with spatial specialization but the greatest impact on spatial heterogeneity was observed for migration rates. Specialist species occupy well defined habitat patches with inhomogeneous spatial distribution (i.e., their spatial distribution displays a high coefficient of variation), while generalist species use a greater diversity of habitat patches and are more homogeneously distributed (Kneitel and Chase, 2004). Also, generalist species prevail over specialist species in heterogeneous landscapes, especially when supported by high migration rates (Brown and Pavlovic, 1992). However, we found that at highest migration rates the specialist competitor of species D (i.e., species C) over-competes on it (i.e., on species D that has a generalist feeding strategy) in some patches, leading to a high spatial heterogeneity of species D. The advantage of the specialist consumer C is primarily due to lower population densities of species D in some patches (i.e., this happens when migration frequencies of species D are high). Moreover, we found that population density and migration frequency may affect the cohesiveness of social networks: more cohesive social networks of species C raise the efficacy of predator-avoidance strategies and sharpen top-predator (i.e., species E) pressure on species D. This pattern develops from the trade-off between low and high migration rates of the specialist and generalist species, respectively. In our case a homogeneous landscape was considered while other studies focused on heterogeneous environments (Levins, 1968; Brown and Pavlovic, 1992). In two different habitats with a number of competing species, Brown and Pavlovic (1992) found that two specialist species were the most successful in the presence of low migration rates but high migration rates encouraged the evolution of a single generalist species.

21.4   Current Drawbacks and Future Developments

We implemented the theoretical model of a hierarchical ecological system to quantify the emergence of macroscopic patterns from simple rules that govern individuals' behavior. Once the structure of the equations driving individuals' behavior has been defined, empirical data from the literature have been collected to assemble a realistic model (i.e., a "base model" has been constructed). The parameter set of the "base model" has been evolved to obtain a "reference model." In the "reference model" all species persist and display stochastic dynamics around equilibrium values. The "reference model" was perturbed during sensitivity analysis in order to quantify system-level impacts of changes at population, community, and metapopulation levels. The "reference model" is composed of an evolved parameter set (i.e., that does not necessarily include empirical parameters even if it has been developed from them) and is analogous to a mathematical pendulum for which principles of dynamics can be found without worrying about idiosyncrasies of the experimental set-up. In this section we discuss current issues to the development of such an IBM of a hierarchical system, and emphasize possible applications. The main concerns are related to model construction and validation, while the use of this framework in the context of environmental management depends on the way conservation practices and actions may be incorporated in the model.

First, model construction requires identifying the most relevant agents and processes underlying system dynamics (i.e., characterizing which species/agents, equations, and parameters should be included in the model to capture the emergence of global properties). Algorithms for data mining may contribute to achieve this goal (Tirelli and Pessani, 2011) but their outcomes should be complemented with interpretations of biologists and ecologists that have prior knowledge about the system under investigation. The idea of selecting key features is not only feasible from a technical point of view but also supported by empirical data on ecological networks. Eklöf *et al.* (2013) have suggested that a few dimensions (where each dimension might represent a combination of functional traits) are needed to completely explain all interactions in different types of ecological networks (i.e., food webs, antagonistic and mutualistic networks). In our hierarchical model we adopted a simple but realistic description of processes that occur at three layers of the ecological hierarchy. Model construction was mainly literature-driven (e.g., the food-web topology that represents the backbone for trophic interactions was previously characterized by Jordán and Molnár, 1999) and the following aspects shall need a more accurate description: (1) rates of social tie formation and removal might differ and be modulated by predator density; (2) trophic interactions might be regulated by Holling's type II or III functional responses instead of being linearly controlled by constant feeding rates and population densities of prey and predators (by keeping the role of social network densities for both prey and predators); (3) the consequences of social interactions on feeding relationships might be modeled as a function of social link longevity and social network plasticity (e.g., see Bhadra *et al.*, 2009); and (4) the consequences of a fragmented landscape topology might be considered in addition to changes in migration rates.

Another important matter to consider during model construction is represented by the possible lack of data. Ecological systems are studied with great details at different levels of their hierarchy. However, there are no examples for which the behavior of individuals that belong to different species is characterized in a comprehensive way at three hierarchical levels. There are studies that link two levels of the ecological hierarchy, for example by investigating the trade-off between food-web interactions and dispersal frequencies (van Nouhuys and Hanski, 2005) or exploring the causal relationships connecting habitat fragmentation (and local extinction) to food-web dynamics (Crooks and Soulé, 1999). We would recommend collecting ecological data with public and standard web-repositories (e.g., equivalent to the ones of the *omics* fields in systems biology) as this might alleviate the issue of missing data. Besides summarizing average population properties (e.g., average body length and body mass of a fish species), such databases should also include details on individual-level variability of functional traits. Individual-level variability represents an essential requirement for implementing (and exploiting the potential of) IBMs. For example, individual-level variability of feeding habits has to be known to describe interactions at community level properly and cannot be replaced by average information concerning species' feeding preferences. Variability has been shown to be particularly relevant as many species are composed of individuals that are specialized among their own population (Bolnick *et al.*, 2003, 2011). Possible ways to catch the individual-level variability of trophic habits include the use of fatty-acid biomarkers and stable isotopes (Alfaro *et al.*, 2006; Mittermayr *et al.*, 2014a, b). However, we emphasize that issues on model construction and data availability cannot be simply addressed by public repositories and ad hoc technical solutions. In the case of IBMs the knowledge of experts is fundamental for successful model constructions and outcome interpretations. Our present model does not include data on individual-level variability but it is very straightforward to implement this aspect (e.g., age classes) in potential future extensions, given that appropriate databases will exist. This is because the model is written in BlenX, a process algebra-based language that enables us to begin with a simple, well-characterized set of rules and increase it in complexity gradually by adding additional elements (i.e., this feature is called "composability"; see Jordán *et al.*, 2011). Moreover, for small population size (i.e., rare species) the IBM approach makes it possible to study the effects of stochasticity even in the absence of individual-level variability.

IBMs are often implemented for specific systems. This impairs their wide application to answer general questions of theoretical ecology and poses serious concerns on how to test and validate their predictions. Mathematical analysis of the master equations that regulate IBMs has been suggested as a strategy to mitigate the lack of generality (Black and McKane, 2012). The master equation describes the temporal evolution of a system and may be used to study the properties and the dynamics of the stochastic model. Compared to the statistical analysis of the results obtained through simulations, the mathematical analysis of the master equation avoids issues related to computational costs and is not biased by difficulties in investigating rare events. In most of the cases the master equation cannot be solved analytically and approximation methods have been formulated (Newman *et al.*, 2004).

Alternative approaches rely on sensitivity analysis of IBMs and serve to quantify the whole-system response to changes in parameters that govern individual-level dynamics. This is analogous to the approach we have implemented to analyze the hierarchical model. In our case the model was purely theoretical and was constructed using heterogeneous data sources (i.e., frequencies of social interactions were deduced from dolphins, feeding preferences were mainly inferred from aquatic food webs, and migration rates refer to voles and zooplankton). For model validation and the interpretation of the results we compared our findings with the literature (i.e., our main objective was to verify which emergent properties extracted through simulations matched the patterns described by previous studies). Nevertheless, when IBMs are developed for a specific ecological system and their purpose is predicting its dynamics (e.g., modeling *Mnemiopsis leidyi* dynamics in the North Sea in order to predict which environmental conditions and community-level interactions are likely to foster its invasion in the Baltic Sea), model validation should be carried out using ad hoc experimental data and sampling campaigns (with the support of an intense collaboration between field and theoretical ecologists).

Here we performed individual-based simulations to investigate theoretical aspects, with the goal of showing interdependencies among different levels of the ecological hierarchy. Nevertheless, the same modeling framework might be applied in the context of environmental management and planning (e.g., it might measure consequences of habitat fragmentation, predict changes due to climate change, and quantify the impacts triggered by the reintroduction of a given species). To support concrete strategies of management: (1) model development and validation should be carried out with reference to a specific system (i.e., by using empirical data that describe the system under investigation and not relying on heterogeneous information from literature); and (2) a standard protocol should enable us to link management strategies and disturbance regimes to changes in model structure and parameters (e.g., the construction of a new highway might be represented by the removal of one or more connections from the landscape graph, and the consequences of global warming might be simulated by heterogeneous migration rates that mimic different ranges of seed dispersal). The use of a stochastic, individual-based approach represents a further asset especially for understanding the dynamics of rare species in relation to the different types and severity of disturbances.

References

Alfaro, A. C., Thomas, F., Sergent, L., and Duxbury, M. (2006). Identification of trophic interactions within an estuarine food web (northern New Zealand) using fatty acid biomarkers and stable isotopes. *Estuarine, Coastal and Shelf Science*, **70**(1), 271–286.

Barton, R. A., Byrne, R. W., and Whiten, A. (1996). Ecology, feeding competition and social structure in baboons. *Behavioral Ecology and Sociobiology*, **38**(5), 321–329.

Bascompte, J., Melián, C. J., and Sala, E. (2005). Interaction strength combinations and the overfishing of a marine food web. *Proceedings of the National Academy of Sciences of the United States of America*, **102**(15), 5443–5447.

Bhadra, A., Jordán, F., Sumana, A., Deshpande, S. A., and Gadagkar, R. (2009). A comparative social network analysis of wasp colonies and classrooms: linking network structure to functioning. *Ecological Complexity*, **6**(1), 48–55.

Black, A. J. and McKane, A. J. (2012). Stochastic formulation of ecological models and their applications. *Trends in Ecology and Evolution*, **27**(6), 337–345.

Bolnick, D. I., Svanbäck, R., Fordyce, J. A., *et al.* (2003). The ecology of individuals: incidence and implications of individual specialization. *American Naturalist*, **161**(1), 1–28.

Bolnick, D. I., Amarasekare, P., Araújo, M. S., *et al.* (2011). Why intraspecific trait variation matters in community ecology. *Trends in Ecology and Evolution*, **26**(4), 183–192.

Botkin, D. B., Janak, J. F., and Wallis, J. R. (1972). Some ecological consequences of a computer model of forest growth. *Journal of Ecology*, **60**(3), 849–872.

Brown, J. S. and Pavlovic, N. B. (1992). Evolution in heterogeneous environments: effects of migration on habitat specialization. *Evolutionary Ecology*, **6**(5), 360–382.

Croft, D. P., James, R., Ward, A. J. W., *et al.* (2005). Assortative interactions and social networks in fish. *Oecologia*, **143**(2), 211–219.

Crooks, K. R. and Soulé, M. E. (1999). Mesopredator release and avifaunal extinctions in a fragmented system. *Nature*, **400**(6744), 563–566.

DeAngelis, D. L., Cox, D. K., and Coutant, C. C. (1980). Cannibalism and size dispersal in young-of-the-year largemouth bass: experiment and model. *Ecological Modelling*, **8**, 133–148.

DeAngelis, D. L., Loftus, W. F., Trexler, J. C., and Ulanowicz, R. E. (1997). Modeling fish dynamics and effects of stress in a hydrologically pulsed ecosystem. *Journal of Aquatic Ecosystem Stress and Recovery*, **6**(1), 1–13.

Dematté, L., Priami, C., and Romanel, A. (2008). The BlenX language: a tutorial. In *Formal Methods for Computational Systems Biology*, ed. M. Bernardo, P. Degano, and G. Zavattaro, Berlin Heidelberg: Springer, pp. 313–365.

de Silva, S., Ranjeewa, A. D., and Kryazhimskiy, S. (2011). The dynamics of social networks among female Asian elephants. *BMC Ecology*, **11**(1), 17.

Eklöf, A., Jacob, U., Kopp, J., *et al.* (2013). The dimensionality of ecological networks. *Ecology Letters*, **16**(5), 577–583.

Faust, K. and Skvoretz, J. (2002). Comparing networks across space and time, size and species. *Sociological Methodology*, **32**(1), 267–299.

Gaines, M. S., Vivas, A. M., and Baker, C. L. (1979). An experimental analysis of dispersal in fluctuating vole populations: demographic parameters. *Ecology*, **60**(4), 814–828.

Gillespie, D. T. (1977). Exact stochastic simulation of coupled chemical reactions. *Journal of Physical Chemistry*, **81**(25), 2340–2361.

Grimm, V., Berger, U., Bastiansen, F., *et al.* (2006). A standard protocol for describing individual-based and agent-based models. *Ecological Modelling*, **198**(1–2), 115–126.

Grimm, V., Berger, U., DeAngelis, D. L., *et al.* (2010). The ODD protocol: a review and first update. *Ecological Modelling*, **221**(23), 2760–2768.

Hanski, I. and Zhang, D. Y. (1993). Migration, metapopulation dynamics and fugitive co-existence. *Journal of Theoretical Biology*, **163**(4), 491–504.

Iken, K., Brey, T., Wand, U., Voigt, J., and Junghans, P. (2001). Food web structure of the benthic community at the Porcupine Abyssal Plain (NE Atlantic): a stable isotope analysis. *Progress in Oceanography*, **50**(1), 383–405.

Jordán, F. and Molnár, I. (1999). Reliable flows and preferred patterns in food webs. *Evolutionary Ecology Research*, **1**(5), 591–609.

Jordán, F., Scotti, M., and Priami, C. (2011). Process algebra-based computational tools in ecological modelling. *Ecological Complexity*, **8**(4), 357–363.

Kennedy, J. and Eberhart, R. (1995) Particle swarm optimization. *Proceedings of IEEE International Conference on Neural Networks*, **4**, 1942–1948.

Kneitel, J. M. and Chase, J. M. (2004). Trade-offs in community ecology: linking spatial scales and species coexistence. *Ecology Letters*, **7**(1), 69–80.

Kokkoris, G. D., Troumbis, A. Y., and Lawton, J. H. (1999). Patterns of species interaction strength in assembled theoretical competition communities. *Ecology Letters*, **2**(2), 70–74.

Kunz, H. and Hemelrijk, C. K. (2003). Artificial fish schools: collective effects of school size, body size, and body form. *Artificial Life*, **9**(3), 237–253.

Leibold, M. A., Holyoak, M., Mouquet, N., *et al.* (2004). The metacommunity concept: a framework for multi-scale community ecology. *Ecology Letters*, **7**(7), 601–613.

Levins, R. (1968). *Evolution in Changing Environments: Some Theoretical Explorations*. Princeton, NJ: Princeton University Press.

Lusseau, D., Wilson, B. E. N., Hammond, P. S., *et al.* (2006). Quantifying the influence of sociality on population structure in bottlenose dolphins. *Journal of Animal Ecology*, **75**(1), 14–24.

Marinho-Filho, J. S. (1991). The coexistence of two frugivorous bat species and the phenology of their food plants in Brazil. *Journal of Tropical Ecology*, **7**(1), 59–67.

Michels, E., Cottenie, K., Neys, L., and De Meester, L. (2001). Zooplankton on the move: first results on the quantification of dispersal of zooplankton in a set of interconnected ponds. *Hydrobiologia*, **442**(1), 117–126.

Mittermayr, A., Hansen, T., and Sommer, U. (2014a). Simultaneous analysis of $\delta 13C$, $\delta 15N$ and $\delta 34S$ ratios uncovers food web relationships and the trophic importance of epiphytes in an eelgrass, *Zostera marina* community. *Marine Ecology Progress Series*, **497**, 93–103.

Mittermayr, A., Fox, S. E., and Sommer, U. (2014b). Temporal variation in stable isotope composition ($\delta 13C$, $\delta 15N$ and $\delta 34S$) of a temperate *Zostera marina* food web. *Marine Ecology Progress Series*, **505**, 95–105.

Murray, A. G. and Parslow, J. S. (1999). The analysis of alternative formulations in a simple model of a coastal ecosystem. *Ecological Modelling*, **119**(2), 149–166.

Newman, T. J., Ferdy, J. B., and Quince, C. (2004). Extinction times and moment closure in the stochastic logistic process. *Theoretical Population Biology*, **65**(2), 115–126.

Pacala, S. W., Gordon, D. M., and Godfray, H. C. J. (1996). Effects of social group size on information transfer and task allocation. *Evolutionary Ecology*, **10**(2), 127–165.

Roy, S. and Chattopadhyay, J. (2007). Towards a resolution of "the paradox of the plankton": a brief overview of the proposed mechanisms. *Ecological Complexity*, **4**(1), 26–33.

Scotti, M., Ciocchetta, F., and Jordán, F. (2013). Social and landscape effects on food webs: a multi-level network simulation model. *Journal of Complex Networks*, **1**(2), 160–182.

Scotti, M., Hartvig, M., Winemiller, K. O., *et al.* (2017). Trait-based and process-oriented modeling in ecological network dynamics. In *Adaptive Food Webs: Stability and Transitions of Real and Model Ecosystems*, ed. J. C. Moore, P. C. de Ruiter, K. S. McCann, and V. Wolters, Cambridge, UK: Cambridge University Press, pp. 228–256.

Stander, P. E. (1994). Cooperative hunting in lions: the role of the individual. *Behavioral Ecology and Sociobiology*, **29**(6), 445–454.

Sullivan, K. A. (1984). The advantages of social foraging in downy woodpeckers. *Animal Behaviour*, **32**(1), 16–22.

Tirelli, T. and Pessani, D. (2011). Importance of feature selection in decision-tree and artificial-neural-network ecological applications. *Alburnus alburnus alborella*: a practical example. *Ecological Informatics*, **6**(5), 309–315.

van Nouhuys, S. and Hanski, I. (2005). Metacommunities of butterflies, their host plants, and their parasitoids. In *Spatial Dynamics and Ecological Communities*, ed. M. Holyoak, M. A. Leibold, and R. D. Holt, Chicago: University of Chicago Press, pp. 99–121.

22 Structural Instability of Food Webs and Food-Web Models and Their Implications for Management

Axel G. Rossberg, Amanda L. Caskenette, and Louis-Félix Bersier

22.1 Introduction

By *structural instability* we mean a sensitivity of the structure of ecological communities to changes in biological or ecological parameters – including external pressures – that is so high that small pressures or changes in the environment can have such large effects on the community that they lead to species extinctions (Rossberg, 2013; Rohr *et al.*, 2014). Since sensitivity increases with species richness in complex communities (Novak *et al.*, 2011), structural instability imposes, at least in principle, a limit to local species richness. There is good evidence, discussed below, that structural instability does indeed determine local species richness: the number of species in a community increases through invasions to the point where it becomes structurally unstable, and then cannot increase much more because the invasion of any further species leads to the extinction of another species on average. Structural instability unifies, among others, observations of high sensitivity of complex food-web models, made prominent by Yodzis (1988), with observations of difficulties in getting species to coexist with reasonable abundances in empirically parameterized food webs, perhaps first reported by Andersen and Ursin (1977). We shall briefly review these lines of thought below.

The ecological concept of structural instability describes a phenomenon that is, in principle, just an instance of particularly high sensitivity. Indeed, Flora *et al.* (2011) coined the term structural sensitivity for a closely related but different concept (small changes in the functional representation of processes in a model lead to large changes in attractors). Ecological structural instability is closely related to but different from the mathematical concept known as "structural instability" in dynamical systems theory (infinitesimal changes in flow fields lead to large changes in flow trajectories). Indeed, this mathematical concept (Pugh and Peixoto, 2008), which is narrower than the ecological understanding, has also been employed in the context of community ecology (Meszéna *et al.*, 2006).

Despite this ambiguity, the established laxer understanding of "structural instability" in ecology (Rossberg, 2013; Rohr *et al.*, 2014) is adequate, because it captures the important role the phenomenon plays among many other forms of instability in the diversity–stability debate (Ives and Carpenter, 2007). While a good mathematical understanding of structural instability and of the reasons why it is common in ecology is now emerging (Rossberg, 2013; Rohr *et al.*, 2014), we consider it too early to decide which specific, rigorous mathematical definition is the most useful for ecology.

This chapter will first discuss two simple models that illustrate the concept of structural stability in ecology and its implications, especially for management. Next, the relationship of structural stability to other notions of stability commonly used in ecology is discussed. It is argued that some of the other kinds of "instability" considered in community ecology (Ives and Carpenter, 2007) naturally co-occur with structural instability, that some tend to be stronger when structural instability is weak, and that some others are causally unrelated to structural instability. On this basis, evidence from the empirical and modeling literature is then presented, which suggests that structural instability is ubiquitous in ecology, especially in food webs.

22.2   Structural Instability in Two-Species Competition

A minimal model for structural instability, useful for building intuition, is a community consisting of exactly two species that are ecologically nearly identical and compete, but coexist in a stable equilibrium. If one modifies the system by relatively small changes in population-dynamical parameters or by applying a weak pressure on one of the species, this is likely to tip the delicate balance between the two, leading to competitive exclusion of one species by the other. While the two species could originally coexist over indefinite time, the modification leads to the inevitable extinction of one of the species. One also says that the two-species community, which was feasible before, has become unfeasible.

In this transition, gradual changes in pressures or parameters lead to a gradual decline of the equilibrium abundance of one of the species to zero, hence to its extinction. Once the abundance has reached zero, further-going changes in pressures or parameters do not change its abundance further (as ecologically obvious, the abundance cannot become negative). In dynamical systems theory, this kind of transition is called a transcritical bifurcation. Occurrence of such transcritical bifurcations differentiates structural instability from other kinds of instability of ecological communities. Among the few alternative types of bifurcations common in ecology are the saddle-node (or fold) bifurcation, where an equilibrium disappears as parameters are changed, resulting in sudden, discontinuous transitions (and often bistability) as parameters or pressures change, and Hopf bifurcations, where the system state begins to oscillate around a then unstable equilibrium configuration (as typical for predator–prey oscillations), either gradually as parameters or pressures change, or abruptly. The classification of bifurcations is a useful tool for discussing ecological stability, because each bifurcation type is associated with its own characteristic ecological phenomenology (Rossberg, 2013) and dynamic systems theory reveals that the number of generic types of bifurcations is very limited (Guckenheimer, 2007).

22.3   A Toy Community Model Exhibiting Structural Instability

We shall now discuss another structurally unstable community model, which is slightly more complex than the two-species model, to illustrate how structural instability can naturally arise in ecological communities and what its implications are for management.

The model discussed here is a member of a family of community assembly models with varying degrees of competition symmetry introduced by Rossberg (2013). Population dynamics in the model are given by a simple Lotka–Volterra competition model with S species. Starting with an empty community ($S = 0$), model communities are constructed through gradual assembly with only one species invading at a time. Species with biomasses decaying to zero as a result of another species' invasion are removed as extinct.

The model makes the simplifying assumption that in isolation all species would have identical, positive population growth rates at low abundance and reach identical population biomass at their carrying capacity. Population biomasses are measured as proportions of this carrying capacity, and time in units of the inverse population growth rate. Denoting the biomass of species i by B_i and the elements of the competition matrix by G_{ij}, this leads to the Lotka–Volterra system

$$\frac{dB_i}{dt} = \left(1 - \sum_{j=1}^{S} G_{ij}B_j\right)B_i.$$

For each newly invading species j, the entries of the jth row and column of the competition matrix are chosen at random by the following rule: when there are $S - 1$ species in the community before the invasion, sample independently two S-component vectors ψ and ψ', such that each entry of the vectors is set to 0.2 with probability 0.2 and to zero otherwise. Then set $G_{jj} = 1$ and, for all $i \neq j$, $G_{ij} = 0.75\psi_i + 0.25\psi'_i$ and $G_{ji} = 0.75\psi'_i + 0.25\psi_i$. This leads to competition matrices with a moderate degree of symmetry.

After about 1000 successful invasions of species so sampled, the model community satiates. Species richness S begins to fluctuate around a value of about 119 (S.D. 5.6) as species turnover continues. As illustrated in Figure 22.1, the eigenvalues λ_i of the competition matrix $\mathbf{G}$ in these satiated communities cover an ellipsis in the complex

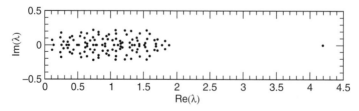

Figure 22.1 Locations in the complex plane of the eigenvalues of the competition matrix **G** for a community sampled from the satiated steady state of the toy community model described here.

plane (Sommers *et al.*, 1988), centered near $\lambda = 1$ and extending such that it closely approaches or, in rare cases, slightly straddles the complex plane's origin, corresponding to $\lambda = 0$ (Rossberg, 2013). In addition, the competition matrix has a single, real eigenvalue of around 4, corresponding to the average row sum of off-diagonal entries of $\mathbf{G}$.

Because the matrix $\mathbf{G}$ has both eigenvalues that are close to zero and eigenvalues further away, it is, by definition, *ill conditioned* (Golub and Van Loan, 1996). The matrix inverse of ill-conditioned matrices is known to be dominated by large contributions corresponding to the eigenvalues near zero and to depend sensitively on the precise values of the matrix entries (Golub and Van Loan, 1996). Since in our model population biomasses are given through the inverse $\mathbf{G}^{-1}$ of $\mathbf{G}$ as $B_i = \sum_{j=1}^{S} G_{ij}^{-1}$ (recall that the linear growth rate is one for all species), population biomasses depend sensitively on the entries of $\mathbf{G}$.

Responses of the model community to pressures are also given by $\mathbf{G}^{-1}$. Specifically, if $\mathbf{F}$ denotes a vector of the exploitation rates (dimension 1/time) by which the biomasses of the S species in the model are hypothetically removed, then the corresponding changes in the equilibrium abundances of species are given as $-\mathbf{G}^{-1}\mathbf{F}$ (assuming pressures are small enough not to lead to extinctions). When $\mathbf{G}$ has eigenvalues near zero, the community's response to pressures is therefore sensitive in two ways: it depends sensitively on the entries of $\mathbf{G}$ because the matrix is ill conditioned, and the large contributions to $\mathbf{G}^{-1}$ related to small eigenvalues make $-\mathbf{G}^{-1}\mathbf{F}$ sensitive to small changes in $\mathbf{F}$. Because each invasion leads to new pressures on the residents, the invasion of one species into such a community often leads to the extinction of other species (Rossberg, 2013). That is, the model community has self-organized to be structurally unstable.

How has this self-organized state been reached? The theory of Sommers *et al.* (1988) allows us to approximately compute the area covered by the distribution of the eigenvalues of $\mathbf{G}$ in the complex plane. Fundamentally, it depends on mean size of the off-diagonal entries of $\mathbf{G}$, their variance, entry-wise correlations between $\mathbf{G}$ and its transpose $\mathbf{G}^{\mathrm{T}}$, and species richness. Independent of how off-diagonal elements are sampled, the area always increases with increasing species richness S in such a way that it reaches and then covers the origin of the complex plane for values of S that exceed some threshold (Rossberg, 2013). For smaller S, all eigenvalues remain separated from zero, implying the community is less sensitive to perturbation. When simulating the process of gradual assembly in this model, S gradually increases until structural instability sets in (i.e., some eigenvalues get close to zero) at which point extinctions resulting from structural instability prevent further increases in species richness S. Community satiation therefore always coincides with the onset of structural instability.

In the absence of an effective mechanism to control abundances, the abundance distribution of resident species spreads out widely as structural instability sets in (Rossberg, 2013). As a result, there will always be some species with abundances much lower than the average, and these are the ones most likely to become extinct when the next species invades (Rossberg, 2013).

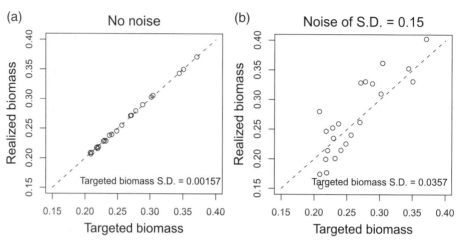

Figure 22.2 Simple illustration of the effect of parameter uncertainty on controllability in a structurally unstable model community. The management problem studied is that of exploiting the subset of most abundant species in the community at such rates that the biomasses of these species decline to 50% of their unexploited equilibrium biomass. For each abundant species, we represent the realized biomass as a function of the targeted biomass. In (a) we assumed perfect knowledge of the competition matrix (i.e., no noise). In that case, the targets are met to high accuracy. In (b) we assumed that the competition matrix is known up to an observation error given by a log-normally distributed factor with median one and spread (S.D. of natural logarithm) equal to 0.15. As a result, the accuracy of meeting the targets is reduced. The absolute S.D. between targeted and realized biomasses is quoted in the insets.

To illustrate the model's sensitivity to pressures, we shall now use this toy model to play fisheries managers. The studied fish communities are sampled from the steady state of the satiated toy model as species composition turns over. To avoid complications related to the extinction of rare species, we assume that fisheries target the most abundant species only, here taken to be all species with $B_i > 0.4$ in the unfished equilibrium. In the case illustrated in Figure 22.2, these are 23 out of a total of 107 species. Management attempts to control the abundances of the targeted species by managing fishing effort, which is here expressed in terms of the corresponding exploitation rates F_i of the targeted species. The management objective is to fish such that all targeted species reach exactly half of their natural abundances. The correct choice of the exploitation rates F_i depends on the entire interaction matrix. In Figure 22.2 we illustrate how misspecifications of the off-diagonal entries of **G** (while keeping ones on the diagonal) lead to deviations of the targeted species abundances from the management goal.

Absolute deviations are of similar magnitude for all species, so that relative deviations are larger for the less abundant species. As we demonstrate in Figure 22.3, these errors are a consequence of structural instability. The ability of management to achieve its goal despite inaccurate knowledge of interactions changes dramatically depending on whether, for the particular model community sampled from the steady state, the community is more or less structurally unstable, i.e., the eigenvalue with the smallest real part is closer or further away from zero.

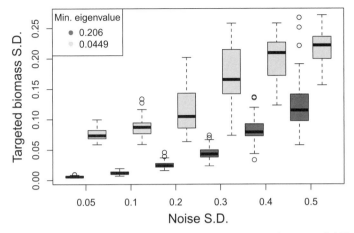

Figure 22.3 The effect of structural instability on community controllability. For two communities sampled from the steady state of the toy model (represented by dark gray boxes to the left and pale gray boxes to the right, respectively), the accuracy of meeting the management goal (halving the populations of the most abundant species) was evaluated for varying degrees of imperfection of knowledge of the competition matrix. Imperfect knowledge was simulated by computing the exploitation rates meant to achieve the management target after multiplying the off-diagonal elements of the competition matrix with log-normally distributed random numbers with median one and a spread (S.D. of natural logarithm) given by noise S.D. (horizontal axis). Controllability was measured by the S.D. of the differences between targeted and realized biomasses, after simulating the relaxation of the community to its equilibrium under the calculated exploitation pressures. For each community and each level of noise S.D., 50 independent replicates were evaluated. The competition matrices of the sample communities differ in the magnitude of the eigenvalue with the smallest real part. When this eigenvalue is close to zero, controllability is much worse than when it is further away. We find this to be a general pattern.

In Lotka–Volterra competition models, such as the one studied here, feasible equilibria are always linearly stable. However, the dynamical relaxation to these equilibria can be very slow, as we demonstrate in Figure 22.4. Even though the linear growth rate of each species in isolation equals one, it lasts over 100 time units for the system to reach its new equilibrium state after fishing pressure has been applied to the targeted species. In principle, there are two reasons for this. The first is that eigenvalues of the competition matrix near zero always imply that, similar to the case of the two-species model, there are certain mixtures of one group of species that can gradually replace an adequate mixture of another group of species without bringing the community much out of equilibrium. Recovery from states where such a replacement has taken place to some degree therefore tends to be slow. The second reason is that in satiated, structurally unstable communities some species – those likely to become extinct soon – have population biomasses much smaller than the average. Because of these species, the community as a whole is always close to a transcritical bifurcation. Critical slowing down in the vicinity of this bifurcation contributes further to delaying community relaxation to equilibrium. The relative importance of these two mechanisms and conceivable interaction between them, however, do not appear to be fully understood yet.

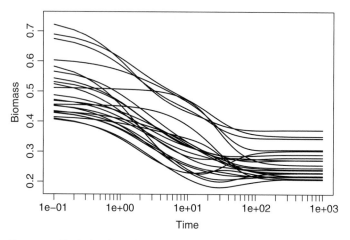

Figure 22.4 Typical trajectories of population biomasses in response to exploitation. Only the trajectories of the most abundant species, subject to exploitation, are shown. A logarithmic time axis is used to visualize the broad range of relevant time scales.

22.4   Relation of Structural Stability to Other Concepts for Community Stability

Ives and Carpenter (2007) discuss a useful selection of (in)stability measures for ecological communities, covering a range of different stability concepts. Here we discuss how these are related to structural (in)stability.

Going through the list of instability measures by Ives and Carpenter (2007), structural instability is only weakly related to the *number of alternative stable states* and the *rate at which intermediate, unstable equilibrium states destabilize* (assuming that, as conventional, extinctions are not counted among the "alternatives"); structural instability is also unrelated to the *probability of occurrence of periodic or chaotic oscillations* or the *amplitude of these oscillations*. The reason is that the underlying bifurcations are different, as explained above: saddle-node bifurcations in the case of alternative stable states and Hopf bifurcations in the case of transitions to oscillations. The phenomenology of these kinds of destabilization is different, and so also the appropriate quantitative characterization.

On the other hand, structural instability tends to go along with a *slow rate of return of model communities to equilibrium* (Figure 22.4) as discussed above, and so with *higher variability of population abundances* in response to environmental stochasticity (May, 1973), even though neither are the defining characteristics. Potentially indicative of structural instability is a low value of what Ives and Carpenter (2007) call Δr_{crit}, i.e., *the minimal rate of indiscriminate removal of individuals that destabilizes a community equilibrium and/or leads to an extinction*, because this represents sensitivity to a particular kind of pressure. Note, however, that in the special case of our toy model pressure applied evenly to all species, which Ives and Carpenter (2007) choose for their definition, is exactly the perturbation that leaves all relative abundances unchanged;

sensitivity to other pressures is more relevant. In satiated communities *the average number of secondary extinctions following an invasion* will always be one, provided the community can be invaded at all. Characteristic for structural instability is that, because each invasion acts essentially as a random perturbation and has an effect of similar magnitude as forced removal of a random species, the *number of secondary extinctions following removal of a random species* will be close to one as well. As exemplified by the minimal two-species example we gave, a high degree of *compensation after species loss* is typical for structurally unstable communities. Along with structural instability comes also enhanced *invasion resistance*. Typically, only around 50% of randomly sampled species will succeed at invading a community that is satiated via structural instability (Rossberg, 2013).

Summarizing, structural instability naturally goes along with a range of other phenomena, some of which can be interpreted as declines in stability (e.g., fluctuating population sizes), others as manifestations of increased stability (e.g., resistance to invasion).

22.5   Evidence for Structural Instability from Models

Yodzis (1988) was probably the first to systematically describe the high sensitivity to misspecifications of food-web models constructed from detailed observation data:

"When the sizes of direct interactions [in an ecosystem] are determined to within an order of magnitude, the long-term outcomes of press perturbations are highly indeterminate, in terms both of whether species density increases or decreases, and of which interactions have the largest effects."

This line of thinking was later followed up, among others, again by Yodzis (1998), who highlighted the consequences of this indeterminacy for ecosystem management, and Novak *et al.* (2011), who noted that this indeterminacy increases with increasing species richness and connectance (proportion of species pairs that directly interact). We reproduced a similar phenomenon above using our toy model.

Constraints imposed to the feasibility of model communities, implied by this high sensitivity, have been noted since the early days of fish-community modeling (Andersen and Ursin, 1977). A common response by modelers to this problem is to increase the strength of intraspecific competition in some form (Andersen and Ursin, 1977; Andersen and Pedersen, 2010; Speirs *et al.*, 2010), e.g., through built-in limits to the number of recruits to a stock per year (Blanchard *et al.*, 2014). In the picture of the toy model discussed above, this corresponds to increasing the values of the entries on the diagonal of the competition matrix **G**, thus shifting its eigenvalues to the right in the complex plane (see Figure 22.1) away from zero, and so reducing structural instability. This operation can be justified in some cases (Farcas and Rossberg, 2016), but generally carries a risk of systematic model misspecification.

Coexistence of species in model communities can often be enforced by appropriate choices of free model parameters. Structural instability then manifests itself through tight limits to the region in parameters space where communities are feasible (Meszéna *et al.*, 2006). Rohr *et al.* (2014) analyzed this phenomenon in detail for empirical mutualistic systems, and could show that these were structured so as to mitigate structural instability.

22.6   Evidence for Structural Instability from Natural Food Webs

In the empirical literature, evidence for emergent high sensitivity of natural communities to perturbations is frequently found under the keyword "indirect effects," as reviewed by Wootton (1994). A prominent example is the study of manipulations of rocky shore communities by Menge (1995), which revealed that indirect effects consistently accounted for about 40% of change in community structure after perturbations, independent of species richness. Analytic theory is consistent with this observation, provided that the interactions among species in these communities are largely asymmetric (Rossberg, 2013). Strong indirect effects manifest themselves, for example, through the idiosyncratic forms in which populations respond to climate change (Cahill *et al.*, 2013).

Consistent with structural instability are observations that large changes in communities of competing species can result from introduction or exclusion of a single predator species (Paine, 1980). Supportive of structural instability, in view of the associated long relaxation times exemplified in Figure 22.4, is also the often strongly felt empirical intuition that natural ecological communities generally respond too slowly to ever reach the population-dynamical equilibria so central to ecological theory (Sousa, 1984). Further indirect support for structural instability comes from the observation that the establishment probability for species that invaded Europe from North America or vice versa, averaged over a wide range of species, is close to 50% (Jeschke and Strayer, 2005), in agreement with theoretical expectations mentioned above.

Less obvious but probably even stronger evidence comes from observations of approximately constant ratios between the species richness at one trophic level in a food web and the richness at the next higher level. For community food webs with several trophic levels, independent studies report that species richness at a given trophic level tends to be about three times as high as richness at the next higher level (Jeffries and Lawton, 1985; Jennings *et al.*, 2002), or, correspondingly, twice as high as the combined species richness at all higher trophic levels (Warren and Gaston, 1992). To relate these observations to structural instability, we recall first that, according to classical theory, the richness of a community of consumers competing for resources cannot be larger than the number of kinds of resources they compete for (MacArthur and Levins, 1964; Rescigno and Richardson, 1965). If the richness of consumers was larger than this limit, their (appropriately defined) competition matrix would have eigenvalues equal to zero, so rendering the consumer community structurally unstable (Meszéna *et al.*, 2006). In food webs, this limit is modified because for most conceivable consumer–resource pairings trophic interaction strength is so weak that it is effectively zero. The small number of relevant trophic interaction pairs leads to a competition matrix (precisely, a competitive overlap matrix; Rossberg, 2013) in which just a few off-diagonal entries have a magnitude comparable to that of the entries on the diagonal. Most off-diagonal entries are effectively zero. Such so-called "sparse" competition matrices can have eigenvalues very close to zero even when the number of resources clearly exceeds the number of competing consumers. For multilevel food webs, where complications due to competition among resources need to be taken into account, one finds numerically that accumulation of eigenvalues near zero and the resulting transition to structural instability set in when species richness declines by a factor of three with each higher trophic level. This is the empirically observed value.

This explanation of the observed fixed richness ratios as representing a limit to species richness due to structural instability is consistent with their interpretation by Arnold (1972) as manifestations of competitive exclusion among consumers sharing common resources. The consistent observations of richness ratios close to the numerical values where structural instability is predicted to set in (Warren and Gaston, 1992; Wright and Samways, 1998; Jennings *et al.*, 2002; Santos de Araújo, 2011) leaves little room for other explanations than the one that the observed communities are indeed structurally unstable.

22.7   Conclusions: Especially for Management

We have demonstrated the convergence of a number of lines of reasoning leading to the conclusion that structural instability is a ubiquitous phenomenon in natural ecological communities. There is good evidence that structural instability is the mechanism limiting and controlling species richness in community food webs, at least for non-basal species, which implies that most natural communities are self-organized to be structurally unstable.

Modelers and managers of ecological communities should be conscious of structural instability in their work, as it imposes constraints on what can and what can not be achieved. Strategies to work around structural instability in management have recently been proposed (Rossberg, 2013) and successfully tested (Farcas and Rossberg, 2016).

Acknowledgments

The authors are in great debt to Rudolf Rohr for assistance with simulations and figures, and thank John Moore and two anonymous referees for valuable comments on earlier versions of this chapter. This work was partially supported by NERC and Defra through the UK's Marine Ecosystems Research Program (www.marine-ecosystems.org.uk).

References

Andersen, K. H. and Pedersen, M. (2010). Damped trophic cascades driven by fishing in model marine ecosystems. *Proceeding of the Royal Society B: Biological Sciences*, **277**(1682), 795–802.

Andersen, K. and Ursin, E. (1977). A multispecies extension to the Beverton and Holt theory of fishing, with accounts of phosphorus circulation and primary production. *Meddelelser fra Danmarks Fiskeri- og Havundersøgelser*, **7**, 319–435.

Arnold, S. J. (1972). Species densities of predators and their prey. *American Naturalist*, **106**(948), 220–236.

Blanchard, J. L., Andersen, K. H., Scott, F., *et al.* (2014). Evaluating targets and trade-offs among fisheries and conservation objectives using a multispecies size spectrum model. *Journal of Applied Ecology*, **51**(3), 612–622.

Cahill, A. E., Aiello-Lammens, M. E., Fisher-Reid, M. C., *et al.* (2013). How does climate change cause extinction? *Proceeding of the Royal Society B: Biological Sciences*, **280**(1750), 20121890.

Farcas, A. and Rossberg, A. G. (2016). Maximum sustainable yield from interacting fish stocks in an uncertain world: two policy choices and underlying trade-offs. *ICES Journal of Marine Science*, **73**(10), 2499–2508.

Flora, C., David, N., Mathias, G., and Jean-Christophe, P. (2011). Structural sensitivity of biological models revisited. *Journal of Theoretical Biology*, **283**(1), 82–91.

Golub, G. H. and Van Loan, C. F. (1996). *Matrix Computations, third edition*. Baltimore: Johns Hopkins University Press.

Guckenheimer, J. (2007). Bifurcation. *Scholarpedia*, **2**(6), 1517.

Ives, A. R. and Carpenter, S. R. (2007). Stability and diversity of ecosystems. *Science*, **317**, 58–62.

Jeffries, M. J. and Lawton, J. H. (1985). Predator–prey ratios in communities of freshwater invertebrates: the role of enemy free space. *Freshwater Biology*, **15**(1), 105–112.

Jennings, S., Warr, K. J., and Mackinson, S. (2002). Use of size-based production and stable isotope analyses to predict trophic transfer efficiencies and predator–prey body mass ratios in food webs. *Marine Ecology Progress Series*, **240**, 11–20.

Jeschke, J. M. and Strayer, D. L. (2005). Invasion success of vertebrates in Europe and North America. *Proceedings of the National Academy of Sciences of the United States of America*, **102**(28), 7198–7202.

MacArthur, R. and Levins, R. (1964). Competition, habitat selection, and character displacement in a patchy environment. *Proceedings of the National Academy of Sciences of the United States of America*, **51**(6), 1207.

May, R. M. (1973). *Stability and Complexity in Model Ecosystems*. Princeton, NJ: Princeton University Press.

Menge, B. A. (1995). Indirect effects in marine rocky intertidal interaction webs: patterns and importance. *Ecological Monographs*, **65**(1), 21–74.

Meszéna, G., Gyllenberg, M., Pásztor, L., and Metz, J. A. J. (2006). Competitive exclusion and limiting similarity: a unified theory. *Theoretical Population Biology*, **69**, 68–87.

Novak, M., Wootton, J. T., Doak, D. F., *et al.* (2011). Predicting community responses to perturbations in the face of imperfect knowledge and network complexity. *Ecology*, **92**(4), 836–846.

Paine, R. T. (1980). Food webs: linkage, interaction strength and community infrastructure. *Journal of Animal Ecology*, **49**, 667–685.

Pugh, C. and Peixoto, M. M. (2008). Structural stability. *Scholarpedia*, **3**(9), 4008.

Rescigno, A. and Richardson, I. W. (1965). On the competitive exclusion principle. *Bulletin of Mathematical Biology*, **27**, 85–89.

Rohr, R. P., Saavedra, S., and Bascompte, J. (2014). On the structural stability of mutualistic systems. *Science*, **345**(6195), 1253497.

Rossberg, A. G. (2013). *Food Webs and Biodiversity: Foundations, Models, Data.* Oxford, UK: Wiley.

Santos de Araújo, W. (2011). Can host plant richness be used as a surrogate for galling insect diversity? *Tropical Conservation Science*, **4**(4), 420–427.

Sommers, H. J., Crisanti, A., Sompolinsky, H., and Stein, Y. (1988). Spectrum of large random asymmetric matrices. *Physical Review Letters*, **60**(19), 1895–1898.

Sousa, W. P. (1984). Intertidal mosaics: patch size, propagule availability, and spatially variable patterns of succession. *Ecology*, **65**, 1918–1935.

Speirs, D. C., Guirey, E. J., Gurney, W. S. C., and Heath, M. R. (2010). A length-structured partial ecosystem model for cod in the North Sea. *Fisheries Research*, **106**(3), 474–494.

Warren, P. H. and Gaston, K. J. (1992). Predator–prey ratios: a special case of a general pattern? *Philosophical Transactions of the Royal Society B: Biological Sciences*, **338** (1284), 113–130.

Wootton, J. T. (1994). The nature and consequences of indirect effects in ecological communities. *Annual Review of Ecology and Systematics*, **25**, 443–466.

Wright, M. G. and Samways, M. J. (1998). Insect species richness tracking plant species richness in a diverse flora: gall-insects in the Cape Floristic Region, South Africa. *Oecologia*, **115**(3), 427–433.

Yodzis, P. (1988). The indeterminacy of ecological interactions as perceived through perturbation experiments. *Ecology*, **69**(2), 508–515.

Yodzis, P. (1998). Local trophodynamics and the interaction of marine mammals and fisheries in the Benguela ecosystem. *Journal of Animal Ecology*, **67**(4), 635–658.

23 Linking Ecology and Epidemiology: The Case of Infected Resource

Sanja Selaković, Peter C. de Ruiter, and Hans Heesterbeek

23.1 Introduction

Interspecific interactions in ecological communities are the main mechanisms that determine structure, functioning, and stability of ecosystems (May, 1972, 1973; Neutel *et al.*, 2002; Alessina and Tang, 2012; Mougi and Kondoh, 2012, 2014). These interactions can be qualitatively positive, negative, or neutral, and pairs of these interactions between two species may be of opposite sign (e.g., trophic, parasitic) or of equivalent sign (e.g., mutualistic, competitive). Most of the research on ecological interactions has focused on feeding relations (Odum, 1971; Pimm, 1982; Levin *et al.*, 2009; McCann, 2011; Moore and de Ruiter, 2012), but in recent studies of ecological communities this was extended to parasitic (Huxham *et al.*, 1995; Thompson *et al.*, 2004; Lafferty *et al.*, 2006; Kuris *et al.*, 2008) and non-parasitic non-trophic relations (Thebault and Fountaine, 2010; Fontaine *et al.*, 2011; Kéfi *et al.*, 2012; Mougi and Kondoh, 2012; Sauve *et al.*, 2014).

In this chapter, we focus on parasitic relations and notably on the question of how trophic interactions and infectious agents mutually influence each other. Here we will refer to the combined classes of infectious species as parasites (see next section for details). The impact of parasites in an ecological community can be quantified through their direct influence on the food-web structure, as well as more indirectly through the way they influence physiological traits of host species and trophic relations of the host and non-host species (Kéfi *et al.*, 2012; Selakovic *et al.*, 2014). In this chapter we first briefly discuss the diversity of parasitic interactions, their relationships with host and non-host species, as well as their effects on a simple consumer–resource relationship consisting of one host and one non-host species. The largest part of the chapter is devoted to exploring a basic model, to show how intricately ecological and epidemiological effects are interwoven, even in the simplest possible ecosystem consisting of two species. Even though this model is basic in the sense that it is low dimensional and not meant to realistically represent any particular system, the analysis does hint at broader ecological insight, for example into possible differences between terrestrial and aquatic ecosystems based on parasitic interaction. The simple analysis highlights the need to study the link between ecology and infectious disease epidemiology in more realistic models.

23.2   Parasitic Interactions, Diversity, Types, Functional Roles, and Modeling

The relationship between parasites and their hosts can have aspects of both trophic and non-trophic interactions. Parasites are in essence consumers of resources, but they are different from typical consumers in several ways. For example, while a typical consumer has more than one victim during its life, parasites as a rule have only one victim per life stage (Lafferty and Kuris, 2002). Also, parasites do not necessarily kill or fully consume their victims. Parasites may also act as prey in a food web, and can be seen as part of trophic interaction in this way as well (Faust, 1975; Marcogliese and Cone, 1997; Johnson et al., 2010; Thieltgets et al., 2013). Inclusion of 47 parasites in an aquatic food web, for example, gave rise to 1093 new interactions of parasites that were prey for other species (Lafferty, 2013). Parasitic interaction can directly or indirectly influence attributes of species in ecological networks, comparable to other non-trophic interactions. The non-trophic interaction that affects attributes of nodes (hosts), and in that way influences the consumer–resource relation, is called "interaction modification" (Kéfi et al., 2012). The attributes could have a direct or indirect influence on the behavior of hosts and non-hosts, handling time of prey, prey preference, assimilation efficiency, conversion efficiency, mortality, reproduction, and growth, and they are common for the different classes of parasites.

Parasites are diverse but the magnitude of this diversity is unknown and it is impossible to estimate the number of species (Dobson et al., 2008). The main characteristic is that they use a host individual's energy for growth, reproduction, and survival. They have, however, very different life histories and sizes. We distinguish microparasites (viruses, bacteria, fungi, protozoa), macroparasites (endoparasites such as helminthes), ectoparasites (fleas and ticks), parasitic castrators, and parasitoids (Kuris and Lafferty, 2000; Lafferty and Kuris, 2002). At one end of the size spectrum, viruses vary in length from 30 to 200 nm. For example, rabies virus has a length of 180 nm (Baer, 1991). At the other side of the spectrum, tapeworms, such as *Diphyllobothrium*, vary from 1 mm to several meters (Faust et al., 1968). Furthermore, the sizes and masses of parasites compared to their hosts are very diverse and depend on the type of parasite. While most microparasites have ratios between $1:10^8$ and $1:10^2$, parasitoids and parasitic castrators are sometimes of mass and size comparable to their host (Lafferty and Kuris, 2002).

Ectoparasites affect their hosts through energy drain by sucking their blood and by activation of a host's immune response with their saliva. This drain of energy can produce subtle subclinical responses, even when these parasites do not by themselves cause disease in their hosts. However, ticks and fleas can also transmit other parasite species, notably microparasites, initiating infection inside of the host that can lead to strong clinical effects, including substantial morbidity, impairing normal ecological functioning, and mortality. In Ngorongoro Conservation Park in 2000 and 2001 there occurred significant mortality among buffaloes, wildebeests, lions, and rhinoceros that had showed infection with *Babesia* species transmitted by ticks (Nijhof et al., 2003;

Munson *et al.*, 2008). But parasites carried by ectoparasites can also cause only subtle subclinical effects in host species to which they have strongly adapted.

Subclinical and clinical effects of parasites impact on the overall fitness of the host. Microparasites and macroparasites often negatively influence the fitness of the host, while parasitic castrators and parasitoids reduce fitness of the host to zero (Kuris and Lafferty, 2000; Lafferty and Kuris, 2002). Fitness reduction (e.g., reduced growth and reproduction) of the host originates from the effect of the parasites on the host's ability to feed and on the efficiency of using ingested food for maintenance and production (Anderson and May, 1979; Lafferty *et al.*, 2006). Workers of the bumblebee, *Bombus terrestris*, challenged with lipopolysaccharides in order to induce their immune system under starvation, reduced survival by 50 to 70% (Moret and Schmid-Hempel, 2000). Further, multiple parasite infections in North American red squirrels, *Tamiasciurus hudsonicus*, negatively impact reproductive success due to allocation of the energy toward immune response (Gooderham and Schulte-Hostedde, 2011). Similarly, experimental removal of ectoparasites (mainly fleas) in Columbian ground squirrels, *Spermophilus columbianus*, led to an increase of female body condition (Neuhaus, 2003). Parasitic castrators and parasitoids extend this effect even more utilizing almost completely the host's energy that is assimilated through trophic interaction directly for its own reproduction and growth, leading to zero reproduction or death for the host itself (Kuris and Lafferty, 2000; Lafferty and Kuris, 2002, 2009; Hechinger *et al.*, 2008).

Parasites can have many different functional roles (Poulin, 1999; Selaković *et al.*, 2014). The difference in the susceptibility of possible hosts, gives to the parasite a role in shaping the population abundance of the host species, thereby affecting the other types of non-trophic interactions. The difference in susceptibility to the malarial parasite, *Plasmodium azurophilum*, of two species of lizards in the Caribbean plays an important role in their coexistence (Schall, 1992). Some species of parasite affect their host by changing the host's behavioral and physical characteristics and by altering the feeding relationship of the host with its consumers and predators (Moore, 2002). An experiment of three-spined sticklebacks, *Gasterosteus aculeatus*, that received the same amount of uninfected prey and prey infected with *Pomphorhynchus laevis* showed significant difference in the predation rate on infected individuals due to the parasite's impact on color and behavior (Bakker *et al.*, 1997). Sometimes parasites lead host species to functional extinction, an example of which will be given below. Many additional interesting examples from the literature that illustrate other functional roles of parasites have been reviewed elsewhere (Selaković *et al.*, 2014).

The above motivates a closer look at how, in ecological theory, parasites of all types can be integrated, and what can then be learned from studying the combined ecological and epidemiological dynamics. There is a large and rich literature on infectious disease dynamics and its mathematical and computational tools and models (see Diekmann *et al.*, 2013 and Heesterbeek *et al.*, 2015, for recent overviews). Mostly, this literature has developed around combinations of one parasite species and one host species. Broadly speaking, there are two modeling approaches depending on the nature of the parasite. If the extent of infection in a host individual and its effects on the host's life history can be quantified at the level of the parasite and is influenced by, or even depends on, re-infection, models are in terms of the "degree" of infection (for example, number of hosts carrying n parasites, or the mean parasite load of infected individuals or the environment). These models typically relate to macroparasites. Typically, such parasites and models involve distinct stages in the life cycle, related to different host species or free living in the environment. If the course of infection and its effects on the host are a more or less autonomous process from first successful exposure, models are in terms of generalized and uniform epidemiological states for host individuals (for example, susceptible, latently infected, infectious, immune). Such models typically relate to microparasites.

Work relating to multiple host species interacting with a single parasite species has emerged, but initially ignoring relations within and between host species that were not linked to infection. Only in recent years has there been more substantial effort to regard parasites in systems of multiple host species that also interact *ecologically*. There is a growing literature, with studies ranging from specific models to more general theory (see Roberts and Heesterbeek, 2012 and the references therein).

In addition to the distinction in approaches in epidemiological models for macro- and microparasites, the added ecological dimension introduces another choice to be made: infectious disease agents can be studied directly or indirectly (Selaković *et al.*, 2014). In the direct approach, parasites are studied as species in food webs, represented by nodes in the web with links to species that are their hosts. In an indirect approach, parasites are studied through their effects on hosts, for example by recognizing different epidemiological states for host individuals of the host species involved, or by recognizing individual hosts with different dynamic infection levels. The indirect approach would combine well with existing epidemiological modeling frameworks for both macro- and microparasites. The direct approach could combine with macroparasitic epidemiological models, especially for systems with a free-living stage of the parasite.

Here, we aim to give an idea of the intricate way in which ecological and epidemiological processes interact in determining dynamic behavior, using an indirect approach. We do so by studying a model for a simple situation for which the ecological dynamics, in the absence of a parasite, are well known. To motivate our model and analysis, we give a few examples of infection in the resource species and its effect on the consumer–resource relationship as an introduction to our next section where we discuss the influence of infection in the resource to the interaction with the consumer. Gerbils, *Gerbillus andersoni*, affected by the higher abundance of the fleas *Synosternus cleopatrae* than in nature had higher rates of body mass lost than non-parasitized control individuals (Hawlena *et al.*, 2006). This was probably due to their reduced attention to forage (Raveh *et al.*, 2011). The loss of body mass in gerbils influences their consumers by their need to use more energy to catch additional prey, but on the other side it makes prey more available because of the lack of attention to detect a predator. Further, a nematomorph parasite infects crickets and changes their behavior leading crickets to enter streams and become a new prey connection for predators changing the strength of their neighboring interactions in the trophic network (Sato *et al.*, 2012), where we see the indirect impact of the rest of the trophic network community. Infection in resource can lead to functional extinction of consumer species: the Asian chestnut fungus effectively extirpated the American chestnut from eastern US forests, causing the apparent extinction of several phytofagus insects (Anagnostakis, 1987).

We do not claim realism in our model that allows insight into specific disease systems, but at a general level we can discuss the bidirectional influence between a parasite and a consumer–resource relation in different types of ecosystems in terms of energy flow, mirroring Rip and McCann (2011) who looked at non-parasitic systems. We explore the influence of non-trophic parasitic behavior going from aquatic to terrestrial ecosystems. Although terrestrial and aquatic parasites influence their hosts and non-hosts in similar ways, there are distinctions between terrestrial and aquatic environments that influence, for example, parasite biodiversity. Only nine animal phyla are found in terrestrial ecosystems compared to 34 in aquatic ecosystems. This indicates that biodiversity of hosts and parasites may be higher in aquatic ecosystems (McCallum *et al.*, 2004). Further, there are differences in the types of parasites that appear in the two ecosystems. Parasitoids are relatively common for terrestrial ecosystems (Godfray, 1994), while the opposite applies for parasitic castrators (Kuris, 1974). The differences between the two environments and their parasites extends to ways of transmission. Rates of spread of infection in marine ecosystems are higher than those observed in terrestrial ecosystems (McCallum *et al.*, 2003). Also, vertical transmission is very rare in aquatic ecosystems, as well as vector transmission of the diseases (although there are some examples: fireworms spreading *Vibrio* sp. among corals).

23.3 Including a Simple Microparasite Affecting Feeding Behavior in a Simple Consumer–Resource Relationship

Taking into consideration the above examples illustrating how different types of parasites affect their hosts, we analyzed a simple Lotka–Volterra consumer–resource model. Mathematically more sophisticated, and ecologically more realistic, models have been studied, but not in a detailed way exploring the interplay between parameters typically involved in ecology and in epidemiology. Our aim is not to provide maximal realism, but to explore the interplay in the system satisfying the minimum requirements to make it non-trivial.

A simple Lotka–Volterra system, for a consumer interacting with a resource, is used for a broader discussion on stability and energy flux. We use the notation in Rip and McCann (2011), who analyze the stability of the simple system without parasites, and concentrate on "relative energy ratio" (defined below). They regard the largest real part of the eigenvalues of the Jacobian evaluated at the steady state where resource (R) and consumer (C) coexist. The system is given by

$$\frac{dR}{dt} = rR\left(1 - \frac{R}{k}\right) - aCR$$
$$\frac{dC}{dt} = eaCR - mC$$

(23.1)

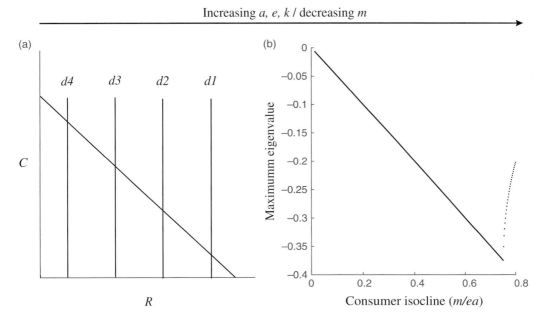

Increasing *a, e, k* / decreasing *m*

Figure 23.1 Simple Lotka–Volterra model analyzed as in Rip and McCann (2011). (a) Increasing *d*, i.e., decreasing the relative energy ratio as defined in Eq. (23.1) in the text, shifts the consumer isocline (vertical line) relative to the resource isocline from right to left (*d1* → *d4*). (b) Shifting the consumer isocline to the left, increases the maximum eigenvalue (decreases stability).

with resource growth rate *r* (biomass time^{-1}), resource carrying capacity *k* (biomass), consumption coefficient *a* (biomass^{-1} time^{-1}), conversion efficiency *e* (dimensionless), and consumer mortality *m* (time^{-1}). The consumer is assumed not to have alternative sources of food. Rip and McCann (2011) take *r* = *k* = 1 for convenience and we will do the same. The dynamics around the non-trivial steady state (*R**, *C**) are governed by the value of a combined parameter that we shall denote by *d* : = *m*/(*ea*), i.e., *d* (unit: biomass) denotes the *C*-isocline (see Figure 23.1, left graph). We mirror Rip and McCann (2011) and call the "predation rate" *ea* between the consumer and the resource relative to the consumer loss rate the *relative energy ratio (ea/m = 1/d)* (note that Rip and McCann, 2011, use the word "flux," but we prefer to avoid that because it suggests units time^{-1}). They argue that aquatic ecosystems have a higher relative energy ratio, a high herbivore/ plant ratio and more variable population dynamics, compared to terrestrial ecosystems. Therefore low values of *d* would relate more to the behavior of an aquatic ecosystem and unstable dynamics, while high values of *d* would relate more to the behavior of a terrestrial system with a low herbivore/plant ratio and stable population dynamics.

The steady state of the simple model (*R**, *C**) = (*d*, 1/*a* − *d*/*a*) is stable when it exists, i.e., for 0 < *d* < 1. Because for this simple system the eigenvalues of the Jacobian evaluated at the steady state can be given analytically, one can easily show that the largest real part (and hence the stability) depends on *d* in a way described in Figure 23.1, right graph. The Jacobian is given by

$$J = \begin{pmatrix} -d & -ad \\ e(1-d) & 0 \end{pmatrix} \qquad (23.2)$$

If $0 < d < 4ae/(1 + 4ae)$, the eigenvalues are complex and the largest real part is linearly decreasing in d; for $4ae/(1 + 4ae) < d < 1$, the eigenvalues are real and a non-linear increasing function of d. So although the non-trivial steady state is stable where it exists, the return time to equilibrium, as provisionally measured by the absolute value of the largest negative real part of the eigenvalues, is a non-linear function of the combined parameter d, describing the ecological balance for the consumer of death and recruitment via resource consumption.

This simple system, with clear behavior, is an interesting null model to explore the influence of infectious agents on consumer–resource interaction. We now regard a non-lethal parasite for which only the resource species is a host. We study the dynamics and stability of the parasite–resource–consumer system. In this specific case we model consumer–resource–microparasite interaction. One could model the epidemiology in many different ways but we choose to keep things simple as an initial exploration and allow some analytic tractability. The consumer–resource–microparasite system we study is as follows:

$$\frac{dR_s}{dt} = rR_s\left(1 - \frac{R_s}{k}\right) - aCR_s - \beta R_s R_i \qquad (23.3a)$$

$$\frac{dC}{dt} = eaR_sC + qepaR_iC - mC \qquad (23.3b)$$

$$\frac{dR_i}{dt} = \beta R_s R_i - paR_iC \qquad (23.3c)$$

Here, R_s denotes the susceptible resource population and R_i the infected ($=$ infectious) resource population. The transmission rate is denoted by β (time^{-1}). It is the probability per unit of time for one susceptible individual to become infected, i.e., the infection pressure that one infectious individual exerts on susceptible individuals. The dimension-less factors $p > 0$ and $q > 0$ describe the influence that infection has on the consumption coefficient and conversion efficiency, respectively.

It is important to first note that now we have two different biological points of view for stability. There is *ecological stability* and *epidemiological stability*. Ecological stability refers to the balance in the system in terms of intra- and interspecies interaction in the absence of infection; epidemiological stability refers to the balance in the system in terms of the parasite and its hosts. In the absence of the parasite ($R_i = 0$), the system (Eqs. 23.3a–c) is equal to the Lotka–Volterra system (1) in the (R_s, C)-plane. The non-trivial steady state (R_s^*, C^*, 0) of system (3) is ecologically stable in that plane. The first question is when this steady state is also epidemiologically stable, i.e., able to withstand invasion by the infectious agent of the resource. For situations where the agent is able to invade, i.e., where the steady state (R_s^*, C^*, 0) is epidemiologically unstable, one can then ask the next question under which conditions an endemic steady state is stable, where susceptible and infected resources and consumers all coexist, and how does this stability depend on the values of the ecological and epidemiological parameters.

In Roberts and Heesterbeek (2012), the invasion problem is studied for systems where any number of host and non-host species can ecologically interact in a food web, and where a microparasite interacts epidemiologically with its host species. If the uninfected states for all species are listed first (characterizing individuals of non-host species as being always in the uninfected state), followed by the infected states in the same order of species, a general system has a Jacobian matrix of the following form

$$J = \begin{pmatrix} A & B \\ D & T \end{pmatrix} \tag{23.4}$$

where matrix A is the ecological community matrix, i.e., the Jacobian of the reduced system when the parasite is absent, and where matrix T is the epidemiological matrix describing transmission among the host species. When J is evaluated at a given steady state, the eigenvalues of J as usual determine the stability of that steady state. In the case of invasion of an infectious agent (i.e., when looking at the infection-free steady state), we have that matrix D is the zero matrix, and the eigenvalue problem decouples in the eigenvalues of the community matrix A, now fully governing ecological stability of the infection-free steady state, and the eigenvalues of the epidemiological matrix T, governing the epidemiological stability of the infection-free steady state (Roberts and Heesterbeek, 2012). The characteristic equation of J is then the product of the characteristic equations of A and T.

In the system described by Eqs. (23.3a–c), where we assume for convenience that $r = 1$ and $k = 1$, the Jacobian at the infection-free steady state $(R_s{}^*, C^*, 0) = (d, (1 - d)/a, 0)$ is given by

$$J = \begin{pmatrix} -d & -ad & -\beta d \\ e(1 - d) & 0 & pqe(1 - d) \\ 0 & 0 & \beta d - p(1 - d) \end{pmatrix} \tag{23.5}$$

and we see that the ecological stability is governed by matrix (23.2), as expected, and that the epidemiological stability is governed by the one-dimensional matrix $T = \beta d - p(1 - d)$. Hence the infection-free steady state is epidemiologically stable as long as $\beta d - p(1 - d) < 0$, or when $R_0 < 1$, where

$$R_0 := \frac{\beta d}{p(1 - d)} \tag{23.6}$$

is the basic reproduction number of the infection system. The basic reproduction number is the average number of new cases of an infection caused by a typical infected individual in a fully susceptible population of hosts in steady state (see Diekmann et al., 2013).

Note that R_0 in Eq. (23.6) is a combination of ecological and epidemiological parameters because a non-host species (the consumer) influences through ecological

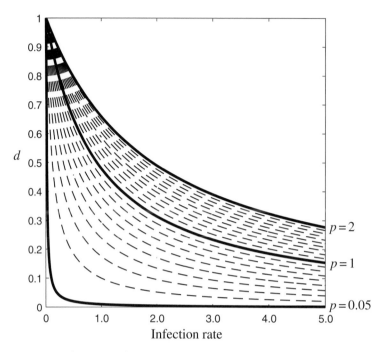

Figure 23.2 The ability of the parasite to invade the consumer–resource system, as a function of infection rate β, the steady-state population size of the resource in the absence of infection, and for a range of values for the influence of the parasite on the feeding of the host (p; three specific values indicated). For a given value of p, the parasite is able to invade the system ($R_0 > 1$) for (β, d)-combinations above the corresponding line.

interaction, epidemiologically relevant aspects of resource individuals (in this case their life expectancy). The biological interpretation is that an infected resource individual is expected to produce βd new cases per unit of time during its infectious period with expected length $1/p(1 - d)$. The latter is 1 divided by the probability per unit of time of dying (i.e., by being eaten by a consumer in our model), in the steady state at invasion of the parasite. When $R_0 > 1$, the steady state is epidemiologically unstable and the agent can invade. In Figure 23.2, we show curves in a feasible part of the (β, d)-plane where $R_0 = 1$, for various values of p. We see that, for parasites of limited infectiousness, successful invasion needs more severe ecological effects (smaller values for p) and higher values for the resource steady state, compared to parasites that induce high infectiousness. Increasing the severity of the ecological effect of the parasite on the consumer–resource interaction (i.e., decreasing the value of p) increases the area of the (β, d)-parameter space where the parasite can invade. In ecological terms, if p is small, consumers eat a relatively small proportion of the infected resource population that can hence contact a relatively larger part of the susceptible resource population, leading to more transmission.

Upon successful invasion, the system moves away from the state (R_s^*, C^*, 0), which is ecologically stable in the (R, C)-plane, and moves into the three-dimensional space with variables: susceptible resource, consumer, and infectious resource. The system can then converge to the steady state (R_s^*, C^*, R_i^*) given by

$$R_s^* = \frac{pq - \beta d}{pq + (q - 1)\beta} \tag{23.7a}$$

$$C^* = \frac{\beta R_s^*}{pa} \tag{23.7b}$$

$$R_i^* = \frac{d}{pq} - \frac{R_s^*}{pq} \tag{23.7c}$$

where resource and consumer coexist and the parasite is endemic (endemic steady state).

Some algebra shows that the endemic steady state is feasible, i.e., exists in the sense that all three variables are non-negative, if β takes a value in the interval

$$\frac{p(1-d)}{d} =: \beta_1 < \beta < \beta_2 := \frac{pq}{d} \tag{23.8}$$

In this range we will, in the next section, numerically explore the interaction between ecology and epidemiology for the stability of the coexistence of the susceptible resource, the infectious resource, and the consumer population.

23.4   Numerical Exploration of the Stability of Coexistence

The analysis of the system described by Eqs. 23.7a–c shows how a microparasite that does not produce mortality but only influences the behavior of its host, affects simple consumer–resource systems in different types of environment. In a series of figures, we explore how the stability of the endemic steady state with coexisting consumer, resource, and microparasite changes if we vary epidemiological aspects (the infection transmission rate β and the effect of the parasite on the consumption coefficient and the conversion efficiency) and ecological aspects (notably the resource population size d in steady state in the absence of infection, varying between 0.4 and 1 as a result of a mortality variation between 0.3 and 0.75 in steps of 0.05). In Table 23.1, we give 10 different regimes of values for the combined parameter d that are explored. By varying d we simulate 10 different "types of environment," and for each of these we vary the *infection transmission rate* over a continuous range. Within those combinations for every fixed *type of environment* (d) and *infection transmission rate* (β), we additionally

Table 23.1 Parameter regimes used for the numerical exploration of model (3)

Regime	I	II	III	IV	V	VI	VII	VIII	IX	X
d	0.400	0.467	0.533	0.600	0.667	0.733	0.800	0.867	0.933	1.000

vary two other parameters, the *conversion efficiency* (*qe*) and the *consumption coeffi-cient* (*pa*). We first vary the *conversion efficiency* by increasing the value of *q* between 0 and 1 in steps of 0.1. Finally, for every (*d*, *β*, *qe*)-combination we vary the *consumption coefficient* (*pa*) by changing the parameter *p* between 0 and 2 in steps of 0.1 to simulate the effect of parasite on the feeding behavior of its host, allowing for decreased (*p* < 1) or increased (*p* > 1) consumption of the infected resource by the consumer. We produce graphs of the largest real part of the eigenvalues of the Jacobian matrix, evaluated at endemic steady state, with positive values implying an unstable steady state, and negative values implying stability.

In Figure 23.3, we show stability in the (*β*, *d*)-plane, for a range of values for the epidemiological effects *p* and *q*. The curves in the left panel of Figure 23.3a indicate where the endemic steady exists and is stable, with various shades of gray indicating the size of the largest real part. Darker shades denote smaller values of the real part of the dominant eigenvalues, and tell us when the system is more resilient to perturbations (a higher return time to equilibrium). White in that figure indicates that either the steady state does not exist in that range of parameter space, or that the steady state is unstable. The results show that the reason an endemic steady state does not exist, for a given combination of parameters (*d*, *qe*, *pa*), depends on the value for the infection rate: at low values for the infection rate the parasite is not able to invade the system under the given conditions, whereas at high values for the infection rate the susceptible resource and consumer interaction cannot sustain the high infection pressure. In Figure 23.3a, right panel, the curves in shade of gray indicate where the steady state exists and is unstable, so here "white" means: the steady state does not exist or is stable). The perturbation of any of the parameters that produce an unstable steady state will easily lead to extinction in either infectious resource or susceptible resource and consumer. Figure 23.3b shows that the endemic steady state exists only in the *β* interval given by Eq. (23.6). In a series of additional figures, we examined the model as follows: (1) without any impact of *q* and *p* parameters; (2) with an effect of the disease only on *q* and only on *p* separately; and (3) the effect of *p* < 1 and *p* > 1 on the stability of the system (see Figures S23.1–S23.5 in the appendix).

The stability analyses of the model without any effect of the parasite on conversion efficiency and consumption coefficient of consumer (*p* = *q* = 1) shows smaller ranges for the infection rate where the species can coexist, compared to the model where these influences are included. That implies the importance of the non-trophic parasitic influences on their hosts and non-hosts for the stability of the system to allow a wider ecological range of interaction conditions suitable for coexistence. The influence of the conversion efficiency parameter and the consumption coefficient parameter separately, showed different effects: when we keep the system without change in *consumption coefficient* parameter (*p* = 1), lower values of *conversion efficiency* drive the system sometimes to stable and sometimes to unstable behavior, while when we keep the system without change in *conversion efficiency* (*q* = 1) lower values of *consumption coefficient* always lead the system to stability.

If we assume that consumer–resource interaction is affected by the microparasite through both ways of influence (*p* and *q*), but we are interested in the importance of only

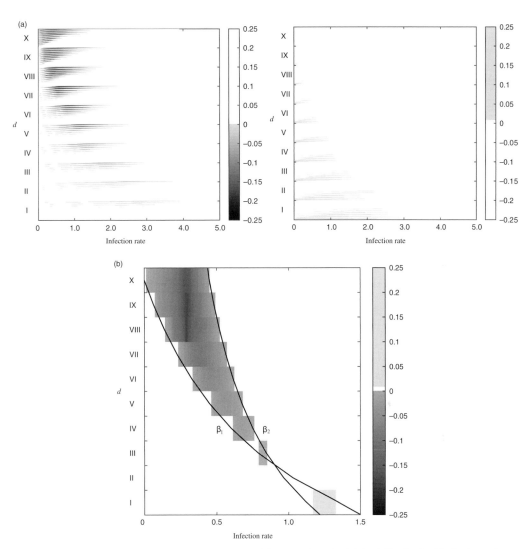

Figure 23.3 Stability analyses of the model (Rs^*, C^*, Ri^*). (a) Stability in the (β, d)-plane, for a range of values for the epidemiological effects p and q. The left panel indicates where the endemic steady state exists and is stable, and the right panel indicates where it is unstable. (b) Endemic steady state of the model exists only between β_1 and β_2 given in Eq. (23.6). In this simulation of the toy model we took the values of rates of change for assimilation efficiency ($p = 0.9$) and conversion efficiency ($q = 0.5$).

the effect of the microparasite on the consumption coefficient, we get other interesting insights. The model shows that greater consumption ($p > 1$) of the infected prey gives greater stability to the system and that coexistence of all species can occur with higher infection rates under this condition.

Further, the effects of microparasite on the consumer–resource interaction in different types of environment are examined using the parameter d as in Rip and McCann (2011).

Table 23.2 Results of percentage of coexistence and stable/unstable distribution ranges of infection rates that support consumer–resource–microparasite coexistence in different types of environment (regimes of predation). Distribution range of infectious rate depends on the consumption coefficient and conversion efficiency parameters.

d	% of coexistence	% of stable	% of unstable
I	12.9 (β[0.01–4.34])	23.9 (β[0.02–4.34])	76.1 (β[0.01–2.86])
II	10.5 (β[0.01–3.72])	35.5 (β[0.13–3.72])	64.5 (β[0.01–2.18])
III	9.0 (β[0.01–3.25])	48.7 (β[0.01–3.25])	51.3 (β[0.01–1.67])
IV	8.0 (β[0.01–2.89])	61.9 (β[0.01–2.89])	38.1 (β[0.01–1.27])
V	7.6 (β[0.01–2.60])	73.9 (β[0.01–2.60])	26.1 (β[0.01–0.95])
VI	7.5 (β[0.01–2.37])	83.7 (β[0.01–2.37])	16.3 (β[0.01–0.69])
VII	7.6 (β[0.01–2.17])	91.2 (β[0.01–2.17])	8.8 (β[0.01–0.47])
VIII	7.9 (β[0.01–2.00])	95.9 (β[0.01–2.00])	4.1 (β[0.01–0.29])
IX	8.3 (β[0.01–1.89])	98.6 (β[0.01–1.89])	1.4 (β[0.01–0.13])
X	8.7 (β[0.01–1.73])	100 (β[0.01–1.73])	0

Following their argumentation, we interpret that going from small to large values of d, means that the ecosystem that is modeled changes from "aquatic" to "terrestrial." The stability of the consumer–resource interaction with the infection in the resource if it exists is more often stable in the terrestrial ecosystems, whereas in the aquatic ecosystems it is more frequently unstable. The non-trophic influence of parasites on their hosts and on non-hosts is different in different types of ecosystem. Lowering the conversion efficiency leads to instability in the aquatic ecosystems while in a more terrestrial ecosystem it does not have a destabilizing effect. A higher consumption coefficient (pa) in the aquatic ecosystem sustained higher infection rates compared to the more terrestrial ones. The coexistence of consumer–resource–microparasite interaction and the stability distributions of this interaction in different types of ecosystem are presented in Table 23.2 and Figure 23.4. In the table, we show the percentage of the coexistence and infection rates that are supported, as well as the stability distributions with infection rates in different regimes of the steady-state susceptible resource population in the absence of the parasite, d (or $1/$(relative energy ratio)). Figure 23.4 shows the stability distributions for each of the ten regimes of d (the left panel) and is comparable to Figure 23.1 where stability decreases if we move the consumer isocline (d) to the left. Further, the right panel of Figure 23.4 singles out one regime for d (regime V) to better observe the influence of the *parasite-related* parameters (p and q) on the stability of the interaction.

Table 23.2 and Figure 23.4 suggest that every type of consumer–resource relationship and thus the type of the ecosystem has a specific range of infection rates that can be sustained, and a specific range of parasite-induced characteristics of resource and consumer that allow coexistence in a stable ecosystem of all three species. Additionally, in Table 23.2 we observe that the ecosystems that support the largest ranges for the infection rate and the feeding influences of parasites are those with lower values for d, i.e., those with higher relative energy ratios, which we interpret as aquatic ecosystems.

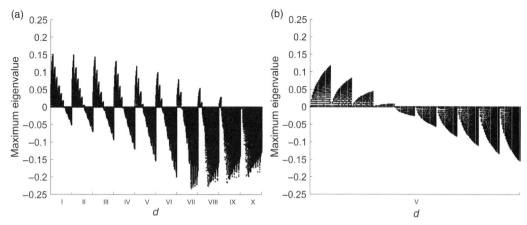

Figure 23.4 Distribution of stable (−)/unstable (+) values in different regimes of d. Every dot represents the real part of the dominant eigenvalue for certain combination of parameters (β, d, qe, pa). Left panel: all regimes for d; right panel: zooming in on regime V (contains steps of the qe parameter, which incorporate steps of the pa parameter that with a certain infectious rate gives the maximum eigenvalue of the consumer–resource–microparasite system).

23.5   Discussion

The research on the inclusion of other than feeding-type interactions in ecological communities, and how these other interactions affect the functioning, structure, and stability of the ecosystems, has developed fast in recent years (Thebault and Fountaine, 2010; Fontaine *et al.*, 2011; Allesina and Tang, 2012; Kéfi *et al.*, 2012; Mougi and Kondoh, 2012, 2014; McQuaid and Britton, 2014; Sauve *et al.*, 2014). Parasitic interaction is one of the first types that was recognized as important in this respect (Lafferty, 1992; Poulin, 1994, 1999; Huxham *et al.*, 1995; Lafferty and Kuris, 2002; Thompson *et al.*, 2004; Lafferty *et al.*, 2006; Kuris *et al.*, 2008; Sukhdeo, 2012; Dunne *et al.*, 2013; Lafferty, 2013; McQuaid and Britton, 2014). One of the proposed ways for studying this particular type of interaction is through the effects parasites have on hosts and on the non-host species their hosts interact with – as non-trophic interaction. This way of incorporating parasites describes their influence via the effect the parasite has on the (physiological/epidemiological) state and behavior of host and non-host species in their ecological network. This is an indirect approach (Kéfi *et al.*, 2012; Selaković *et al.*, 2014), compared to approaches where parasite species are described as biological species, represented directly via their own node in an ecological network where they are linked to nodes representing their host species. Many food-web studies have shown that interaction strengths in food webs are strongly patterned (McCann, 2011), and that both distribution of interaction strengths and the topological structure are important for the stability in ecosystems (Neutel *et al.*, 2002; Allesina and Pascual, 2008). The indirect way of inclusion could be a method to add more accuracy to these analyses by including non-trophic interactions as a real-world simulation.

Similar consumer–resource models have already discussed the effect of parasites on host behavior and the benefits of consumers foraging on a parasitized resource (Lafferty, 1992), as well as the behavioral effect of trophically transmitted parasites on the dynamics of the consumer–resource relationship (Fenton and Rands, 2006). Here we go one step further with a preliminary interpretation of these effects comparing different types of ecosystem and the range of parasitic influence that are supported by these environments. However, our approach is limited to a model that is very general and we use it only for an initial theoretical discussion, and to highlight phenomena that could occur in relation to stability and coexistence. Moreover, we have concentrated on microparasites that affect only feeding-related behavior of their host without a direct disease-induced effect on the mortality of the host. The advantage of a simple model is that the mutual influence between ecology and epidemiology can be more easily explored. These effects may also occur in more realistic settings, and notably food webs should be explored, rather than the simple two-species food chain in this initial analysis, and for a broader range of parasite types and their influences on hosts.

We included knowledge about the influence of infectious agents on their hosts and non-hosts interactions to the simple consumer–resource interaction using the idea of relative energy ratio (sensu Rip and McCann, 2011, who call it relative energy flux). The idea is that any biological trait that increases the relative energy ratio (predation rate of the consumer relative to its loss term) makes the consumer–resource biomass ratio top heavy and the system less stable. Rip and McCann (2011) examine terrestrial versus aquatic ecosystems and predict that aquatic ecosystems tend to have a higher relative energy ratio and decreased stability relative to terrestrial ecosystems. Our analysis shows how the same idea relates to microparasites in such a setting. One can imagine that every combination of parameters we explored is one type of parasite that influences the consumer–resource relationship in its own specific way. The cross-ecosystem analysis shows that consumer–resource interactions with parasites in different environments are stable within certain ranges of parasitic influence on its host, and therefore for smaller or larger sets of potential parasites. Our analysis suggests that aquatic-like systems, in the above sense, support broader ranges of parasites compared to aquatic ones. This agrees with the observation of a higher biodiversity of parasites and their hosts in aquatic ecosystems (McCallum, 2004). For example, oceans contain an estimated 10^{30} virus particles, with 10^{23} infections occurring each second (Suttle, 2007). Our analysis also shows that aquatic ecosystems with parasites are more unstable for coexistence of susceptible and infectious resource sub-populations and consumers, compared to terrestrial systems. This conclusion agrees with the discussion on cross-ecosystem stability from Rip and McCann (2011), for the pure consumer–resource case without parasites.

Indirect inclusion of parasites confirms several general insights. Consumer–resource interaction with infection can be stable over broad ranges of values for epidemiological parameters and of influence on ecological processes. These ranges depend on the ecological characteristics of the consumer–resource system, and parasites can extend the ecological range of coexistence. The analysis gives a clear idea of the importance of non-trophic interaction via parasites in ecosystems. Although our model is basic, it does capture the essentials. In the introduction section we discussed many examples of how different parasites affect their hosts, from direct energy drain to indirect change in feeding interaction between resource and consumer. For example, infected resource individuals may be caught less easily by a consumer or more easily. If part of the resource population is infected, consumers spend either more time in search of a suitable prey, or find infected prey using less energy than in the absence of the parasite. Once caught, infected resource individuals may also affect consumers by reduced feeding value. These trait-mediated effects of parasites can be described with conversion efficiency or consumption coefficient parameters in Lotka–Volterra models. Our results show that including non-trophic influences of parasites increases the stability range and coexistence of the consumer–resource–parasite system compared to the system without non-trophic influence of parasite on its host and non-host. For instance, greater consumption of the infected resource increases the stability of the system and supports higher infection rates. With a higher consumption of infected resource consumers control the infection spread in their resource in our basic setting (in line with the healthy herds hypothesis; Packer *et al.*, 2003).

The next step is to expand this theoretical model to a food web that includes many connected consumer–resource relationships, and to explore trait-mediated parasitic impacts on energy flow, strength of interactions, and stability in different types of ecosystem. Parasites included in food webs in an indirect way can increase and decrease the strength of interaction between neighboring species. It would be interesting to see the consequences of such influences even when they are very weak, as such weak links have been shown to play a role in ecosystem stability (Neutel *et al.*, 2002). Because one can hardly observe or measure ecosystems without parasites playing a role (as every living species is a host to probably several types of parasite; Rossiter, 2013), it may be that interaction strengths are importantly moderated by the omnipresence of parasites and that these parasites, even though having very weak effects on individuals, do play a major role in shaping stability and structure in real ecosystems.

23.6   Summary

In nature, ecological communities exist as a result of different interactions between species determining structure, functioning, and stability. Empirical as well as theoretical studies are mostly focused on trophic (consumer–resource) interactions and non-trophic interactions separately. Recently, in theoretical and field work, studies started to explore combinations of these interaction types, notably looking at the way infectious disease agents affect consumer–resource relationships in food webs. Here we illustrate such influence by looking at a simple model of a microparasite in a very basic Lotka–Volterra consumer–resource system. We show that even in this simplest of settings one can see a diverse range of subtle changes in system behavior if one lets the main trophic parameters for both host and non-host species be influenced by non-trophic interaction.

Appendix

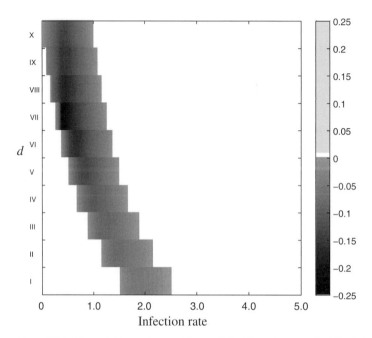

Figure S23.1 The stability analyses of the model without impact of epidemiological parameter p on consumption coefficient (a) of the consumer and epidemiological parameter q on conversion efficiency (e) of the consumer.

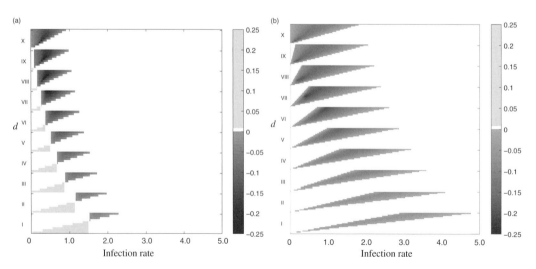

Figure S23.2 Stability analyses of infection impact on (a) only conversion efficiency q (when the p parameter is fixed) and (b) only consumption coefficient p (q parameter is fixed).

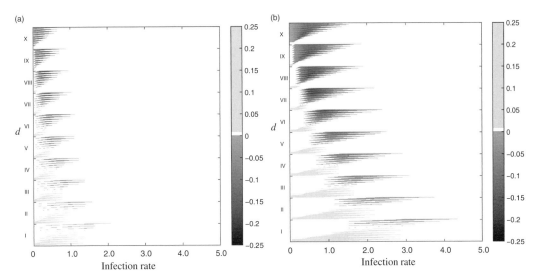

Figure S23.3 Stability analyses for infection impact on consumption coefficient parameter. (a) Left panel shows the stability analyses with combination of parameters d, qe, and $a^*(p < 1)$, while (b) right panel, shows the stability analyses with combination of parameters d, qe, and a^* ($p > 1$).

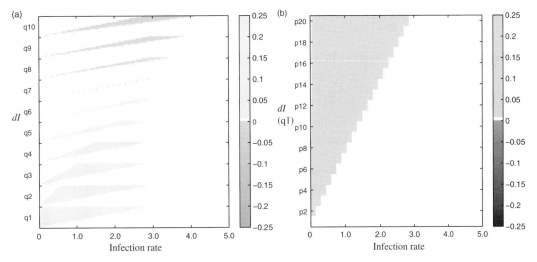

Figure S23.4 Stability analyses of the toy model fixed only on dI where we show the influence of parameters (a) q and (b) p in this situation of energy flux that is compared to energy flux with higher turn-over rates as in aquatic ecosystems.

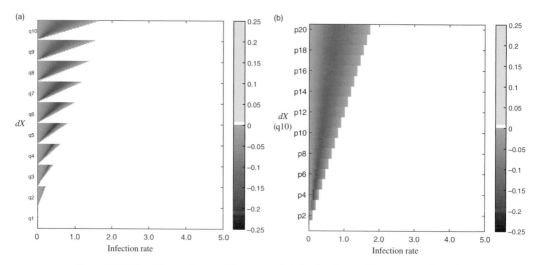

Figure S23.5 Stability analyses of the toy model fixed only on dX where we show the influence of parameters (a) q and (b) p in this situation of energy flux that is compared to energy flux of terrestrial ecosystems.

References

Allesina, S. and Pascual, M. (2008). Network structure, predator–prey modules, and stability in large food webs. *Theoretical Ecology*, **1**, 55–64.

Allesina, S. and Tang, S. (2012). Stability criteria for complex ecosystems. *Nature*, **483**, 205–208.

Anagnostakis, S. L. (1987). Chestnut blight: the classical problem of an introduced pathogen. *Mycologia*, **79**(1), 23–37.

Anderson, R. M. and May, R. M. (1979). Population biology of infectious diseases: Part I. *Nature*, **280**, 361.

Baer, G. M. (1991*). The Natural History of Rabies*. Boca Raton, FL: CRC Press.

Bakker, T. C., Mazzi, D., and Zala, S. (1997). Parasite-induced changes in behavior and color make *Gammarus pulex* more prone to fish predation. *Ecology*, **78**, 1098–1104.

Diekmann, O., Heesterbeek, H., and Britton, T. (2013). *Mathematical Tools for Understanding Infectious Disease Dynamics*, Princeton, NJ: Princeton University Press.

Dobson, A., Lafferty, K. D., Kuris, A. M., Hechinger, R. F., and Jetz, W. (2008). Homage to Linnaeus: How many parasites? How many hosts? *Proceedings of the National Academy of Sciences of the United States of America*, **105**, 11482–11489.

Dunne, J. A., Lafferty, K. D., Dobson, A. P., *et al.* (2013). Parasites affect food web structure primarily through increased diversity and complexity. *PLoS Biology*, **11**, e1001579.

Faust, E. C. (1975). The *Leishmania* parasite of man. In *Animal Agent and Vector of Human Disease*, ed. E. C. Faust, P. C. Beaver, and R. C. Jung, Philadelphia: Lea & Febiger, pp. 33–45.

Fenton, A. and Rands, S. (2006). The impact of parasite manipulation and predator foraging behavior on predator–prey communities. *Ecology*, **87**, 2832–2841.

Fontaine, C., Guimarães, P. R., Kéfi, S., *et al.* (2011). The ecological and evolutionary implications of merging different types of networks. *Ecology Letters*, **14**, 1170–1181.

Godfray, H. C. J. (1994). *Parasitoids: Behavioral and Evolutionary Ecology*, Princeton, NJ: Princeton University Press.

Gooderham, K. and Schulte-Hostedde, A. (2011). Macroparasitism influences reproductive success in red squirrels (*Tamiasciurus hudsonicus*). *Behavioral Ecology*, **22**, 1195–1200.

Hawlena, H., Khokhlova, I., Abramsky, Z., and Krasnov, B. (2006). Age, intensity of infestation by flea parasites and body mass loss in a rodent host. *Parasitology*, **133**, 187–193.

Hechinger, R. F., Lafferty, K. D., and Kuris, A. M. (2008). Diversity increases biomass production for trematode parasites in snails. *Proceedings of the Royal Society B: Biological Sciences*, **275**, 2707–2714. DOI 10.1098/rspb.2008.0875 [doi].

Heesterbeek, H., Anderson, R. M., Andreasen, V., *et al.* (2015). Modeling infectious disease dynamics in the complex landscape of global health. *Science*, **347**(6227), aaa4339. DOI 10.1126/science.aaa4339 [doi].

Huxham, M., Raffaelli, D., and Pike, A. (1995). Parasites and food web patterns. *Journal of Animal Ecology*, **64**(2), 168–176. DOI 10.2307/5752.

Johnson, P. T., Dobson, A., Lafferty, K. D., *et al.* (2010). When parasites become prey: ecological and epidemiological significance of eating parasites. *Trends in Ecology and Evolution*, **25**, 362–371. DOI 10.1016/j.tree.2010.01.005.

Kéfi, S., Berlow, E. L., Wieters, E. A., *et al.* (2012). More than a meal. . . integrating non-feeding interactions into food webs. *Ecology Letters*, **15**, 291–300.

Kuris, A. M. (1974). Trophic interactions: similarity of parasitic castrators to parasitoids. *Quarterly Review of Biology*, **49**(2), 129–148.

Kuris, A. and Lafferty, K. (2000). Parasite–host modeling meets reality: adaptive peaks and their ecological attributes. In *Evolutionary Biology of Host–Parasite Relationships: Theory Meets Reality*, ed. R. Poulin and S. Skorping Morand, New York: Elsevier Science, pp. 9–26.

Kuris, A. M., Hechinger, R. F., Shaw, J. C., *et al.* (2008). Ecosystem energetic implications of parasite and free-living biomass in three estuaries. *Nature*, **454**, 515–518. DOI 10.1038/nature06970.

Lafferty, K. D. (1992) Foraging on prey that are modified by parasites. *American Naturalist*, **140**, 854–867.

Lafferty, K. D. (2013). Parasites in marine food webs. *Bulletin of Marine Science*, **89**, 123–134.

Lafferty, K. D. and Kuris, A. M. (2002). Trophic strategies, animal diversity and body size. *Trends in Ecology and Evolution*, **17**, 507–513.

Lafferty, K. D. and Kuris, A. M. (2009). Parasitic castration: the evolution and ecology of body snatchers. *Trends in Parasitology*, **25**, 564–572.

Lafferty, K. D., Dobson, A. P., and Kuris, A. M. (2006). Parasites dominate food web links. *Proceedings of the National Academy of Sciences of the United States of America*, **103**, 11211–11216. DOI 10.1073/pnas.0604755103.

Levin, S. A., Carpenter, S. R., Godfray, H. C. J., *et al.* (2009). *The Princeton Guide to Ecology*. Princeton, NJ: Princeton University Press.

Marcogliese, D. J. and Cone, D. K. (1997). Food webs: a plea for parasites. *Trends in Ecology and Evolution*, **12**, 320–325.

May, R. M. (1972). Will a large complex system be stable? *Nature*, **238**, 413–414.

May, R. M. (1973). *Complexity and Stability in Model Ecosystems*. Princeton, NJ: Princeton University Press.

McCallum, H., Harvell, D., and Dobson, A. (2003). Rates of spread of marine pathogens. *Ecology Letters*, **6**, 1062–1067.

McCallum, H. I., Kuris, A., Harvell, C. D., *et al.* (2004). Does terrestrial epidemiology apply to marine systems? *Trends in Ecology and Evolution*, **19**, 585–591.

McCann, K. S. (2011). *Food Webs*. Princeton, NJ: Princeton University Press.

McQuaid, C. F. and Britton, N. F. (2014). Parasite species richness and its effect on persistence in food webs. *Journal of Theoretical Biology*, **364**, 377–382.

Moore, J. (2002). *Parasites and the Behavior of Animals*. Oxford, UK: Oxford University Press.

Moore, J. C. and de Ruiter, P. C. (2012). *Energetic Food Webs: An Analysis of Real and Model Ecosystems*. Oxford, UK: Oxford University Press.

Moret, Y. and Schmid-Hempel, P. (2000). Survival for immunity: the price of immune system activation for bumblebee workers. *Science*, **290**, 1166–1168. DOI 8972 [pii].

Mougi, A. and Kondoh, M. (2012). Diversity of interaction types and ecological community stability. *Science*, **337**, 349–351. DOI 10.1126/science.1220529 [doi].

Mougi, A. and Kondoh, M. (2014). Adaptation in a hybrid world with multiple interaction types: a new mechanism for species coexistence. *Ecological Research*, **29**, 113–119.

Munson, L., Terio, K. A., Kock, R., *et al.* (2008). Climate extremes promote fatal co-infections during canine distemper epidemics in African lions. *PLoS One*, **3**, e2545. DOI 10.1371/journal.pone.0002545.

Neuhaus, P. (2003). Parasite removal and its impact on litter size and body condition in Columbian ground squirrels (*Spermophilus columbianus*). *Proceedings of the Royal Society B: Biological Sciences*, **270**(Suppl 2), S213–S215. DOI 10.1098/rsbl.2003.0073 [doi].

Neutel, A. M., Heesterbeek, J. A., and de Ruiter, P. C. (2002). Stability in real food webs: weak links in long loops. *Science*, **296**, 1120–1123. DOI 10.1126/science.1068326.

Nijhof, A., Cludts, S., Fisscher, O., and Laan, A. (2003). Measuring the implementation of codes of conduct: an assessment method based on a process approach of the responsible organisation. *Journal of Business Ethics*, **45**, 65–78.

Odum, E. P. (1971). *Fundamentals of Ecology*, 3rd edn. Philadelphia, PA: Saunders.

Packer, C., Holt, R. D., Hudson, P. J., Lafferty, K. D., and Dobson, A. P. (2003). Keeping the herds healthy and alert: implications of predator control for infectious disease. *Ecology Letters*, **6**, 797–802.

Pimm, S. (1982). *Food Webs*. London, UK: Chapman and Hall.

Poulin, R. (1994). Meta-analysis of parasite-induced behavioural changes. *Animal Behavior*, **48**, 137–146.

Poulin, R. (1999). The functional importance of parasites in animal communities: many roles at many levels? *International Journal of Parasitology*, **29**, 903–914.

Raveh, A., Kotler, B. P., Abramsky, Z., and Krasnov, B. R. (2011). Driven to distraction: detecting the hidden costs of flea parasitism through foraging behaviour in gerbils. *Ecology Letters*, **14**, 47–51.

Rip, J. and McCann, K. (2011). Cross-ecosystem differences in stability and the principle of energy flux. *Ecology Letters*, **14**, 733–740.

Roberts, M. and Heesterbeek, J. A. P. (2012). Characterizing the next-generation matrix and basic reproduction number in ecological epidemiology. *Journal of Mathematical Biology*, **66**(4), 1–20.

Rossiter, W. (2013) Current opinions: zeros in host–parasite food webs: are they real? *International Journal for Parasitology: Parasites and Wildlife*, **2**, 228–234.

Sato, T., Egusa, T., Fukushima, K., *et al.* (2012). Nematomorph parasites indirectly alter the food web and ecosystem function of streams through behavioural manipulation of their cricket hosts. *Ecology Letters*, **15**, 786–793. DOI 10.1111/j.1461–0248.2012.01798.x; 10.1111/j.1461–0248.2012.01798.x.

Sauve, A., Fontaine, C., and Thébault, E. (2014). Structure–stability relationships in networks combining mutualistic and antagonistic interactions. *Oikos*, **123**, 378–384.

Schall, J. J. (1992). Parasite-mediated competition in Anolis lizards. *Oecologia*, **92**, 58–64.

Selaković, S., de Ruiter, P. C., and Heesterbeek, H. (2014). Infectious disease agents mediate interaction in food webs. *Proceedings of the Royal Society B: Biological Sciences*, **281**(1777), 20132709. DOI 10.1098/rspb.2013.2709.

Sukhdeo, M. V. (2012). Where are the parasites in food webs? *Parasites and Vectors*, **5**, 239. DOI 10.1186/1756–3305-5–239 [doi].

Suttle, C. A. (2007). Marine viruses: major players in the global ecosystem. *Nature Reviews Microbiology*, **5**, 801–812.

Thebault, E. and Fontaine, C. (2010). Stability of ecological communities and the architecture of mutualistic and trophic networks. *Science*, **329**, 853–856. DOI 10.1126/science.1188321 [doi].

Thieltges, D. W., Amundsen, P., Hechinger, R. F., *et al.* (2013). Parasites as prey in aquatic food webs: implications for predator infection and parasite transmission. *Oikos*, **122**, 1473–1482.

Thompson, R. M., Mouritsen, K. N., and Poulin, R. (2004). Importance of parasites and their life cycle characteristics in determining the structure of a large marine food web. *Journal of Animal Ecology*, **74**, 77–85.

Index